Technologies for Supporting Reasoning Communities and Collaborative Decision Making:

Cooperative Approaches

John Yearwood
University of Ballarat, Australia

Andrew Stranieri
University of Ballarat, Australia

Information Science REFERENCE

INFORMATION SCIENCE REFERENCE
Hershey · New York

Director of Editorial Content:	Kristin Klinger
Director of Book Publications:	Julia Mosemann
Acquisitions Editor:	Lindsay Johnston
Development Editor:	David DeRicco
Typesetter:	Deanna Jo Zombro
Production Editor:	Jamie Snavely
Cover Design:	Lisa Tosheff

Published in the United States of America by
Information Science Reference (an imprint of IGI Global)
701 E. Chocolate Avenue
Hershey PA 17033
Tel: 717-533-8845
Fax: 717-533-8661
E-mail: cust@igi-global.com
Web site: http://www.igi-global.com

Library of Congress Cataloging-in-Publication Data

Technologies for supporting reasoning communities and collaborative decision making : cooperative approaches / John Yearwood and Andrew Stranieri, editors.
p. cm.
Includes bibliographical references and index.
Summary: "This book covers diverse fields of enquiry including decision science, political science, argumentation, knowledge management, cognitive psychology and business intelligence, giving a perspective on group reasoning that ultimately aims to lead to a greater understanding of reasoning communities and informed technological developments"--Provided by publisher.
ISBN 978-1-60960-091-4 (hbk.) -- ISBN 978-1-60960-093-8 (ebook) 1. Group decision making. 2. Group problem solving. 3. Reasoning. 4. Information technology--Social aspects. 5. Telematics. I. Yearwood, John, 1954- II. Stranieri, Andrew.
HM746.T43 2011
302.30285--dc22
2010041640

British Cataloguing in Publication Data
A Cataloguing in Publication record for this book is available from the British Library.

Simon Niemeyer, *Australian National University, Australia*
Sita Venkatraman, *University of Ballarat, Australia*
Suzanne Zyngier, *LaTrobe University, Australia*
Zhangwei Zhang, *University of Southern Queensland, Australia*
Zhaohao Sun, *University of Ballarat, Australia*

Table of Contents

Section 3
Collaborative Reasoning in the Real World

Section 4
Collaborative Reasoning in the Virtual World

Detailed Table of Contents

Insights into collaborative reasoning in groups have emerged from studies comparing face to face groups with computer mediated groups. Li, Huang and Liu present a comparative study in order to illustrate how the two groups operate quite differently. Media richness theories advocate the use of face to face meetings with problem solving tasks that are complex and highly ambiguous and leaner media such as email when the problem solving tasks are less ambiguous. Li and her colleagues use empirical studies to report on group solutions to a hypothetical complex problem by positing that some parts of the outcome could be objectively rated. The aim was to identify the factors critical for effective group reasoning in face to face and computer mediated groups. They found that problem analysis, generating alternatives, and assessment of positive and negative consequences were critical. Face to face groups performed these tasks better than the virtual groups, thereby confirming the media rich theories. However, contrary to previous studies, elements such as social talk, not directly related to the problem, was found to play an important role in group outcomes.

Another insight from face to face and computer mediated studies involves the sharing of private information. This chapter examines collective information sharing—the tendency to use shared information rather than private information. Evidence from face to face groups has demonstrated that participants face many barriers to sharing all of their information. Some of the barriers seem to be overcome in computer mediated groups; however, a survey of the literature reveals information sharing is less than optimal in either forum. Campbell puts forward a strong case for the discovery of optimal ways to integrate face to face communication with computer mediated communications.

This chapter presents a comprehensive study on communities of practice in Poland. They discover that despite some awareness of the benefits of communities of practice, their use in practice is in fact quite limited. The empirical findings are contrasted against a theoretical perspective drawn from Nonaka's conceptualisation of organisational growth through knowledge evolution.

This chapter derives from the authors' earlier work in computational modeling of reasoning using argumentation and narrative theories. Conceiving of a group as a community that is comprised of participants connected by their need to reason toward a course of action, enables the specification of high level phases the community performs. The phases—engagement, individual reasoning and group coalescing—iterate until the ultimate decision making phase is performed. Adopting this view enables an understanding of group reasoning that facilitates the identification of technological support at each phase.

Section 2
Enhancing Collaborative Reasoning

The second group of chapters focuses more heavily on advancing theoretical and practical approaches that directly enhance group reasoning. Like the chapters in the previous section, they derive from very diverse fields of inquiry.

The influential Analytical Hierarchy Process that Saaty advanced some years ago enables a group to identify criteria that are prominent in decision making. In his chapter, Saaty discusses issues that arise when some criteria are tangible and easily measured and others are not so tangible yet quite important and presents a way to integrate intangible with tangible scales. The approach is elegantly illustrated with a case study involving a hospice decision.

The authors' aim to support group deliberations with the direct use of electronic devices in meetings presents an innovative contrast to other approaches. Their chapter introduces the use of location, sound and video devices that provide feedback to participants about the discussion. Information such as the duration of each participant's contributions and the frequency of exchanges between each pair of participants does help face to face and computer mediated groups identify and correct communicative imbalances.

A key element of any group reasoning exercise involves the integration of diverse and often conflicting views that occur in many group reasoning endeavours. Numerous approaches have been advanced, but most do not derive from the formalism of mathematical theory. In this chapter, Vaníček, Vrana and Aly present an approach for integrating diverse positions that is based on fuzzy set theory. The fuzzy set based approach enables the representation of uncertainty that is required while still retaining some mathematical rigor.

The central assumption that Castellani makes, in drawing on George Kelly's cognitive theory of personal constructs for his chapter is that the full context of reasoning is required to understand a group's reasoning. He includes a case study illustrating how a world view can be elicited from group participants using the repertory grid technique from cognitive psychology. Without this approach, he argues, issues that arise as important for decision making may not be the most pressing for resolution. Solutions are premature or inappropriate, and an opportunity for deeper understanding of the context is lost to participants.

The authors advance a mathematical approach for determining which computer assisted manufacturing technologies are best deployed for a given problem. This task becomes very cumbersome for Analytical Hierarchy Process because there are so many configuration options that the pairwise comparisons of each elements becomes difficult to perform and interpret. Further, many criteria in practice are not independent from each other even though most multi-criteria decision models assume independence. In their chapter, Kahraman and his colleagues use a method known as the Choquet integral that measures the expected utility of an uncertain event. Based on fuzzy set theory, the Choquet Integral can represent interactions between criteria so that they can be taken into account in solving multi criteria decision equations.

Schweighofer focuses on legal reasoning and surveys three decades of research aimed at automating reasoning in law, concluding that most attempts have been far from successful. In his chapter, Schweighofer identifies the role ontologies play in directing information retrieval technologies in law. Ontologies represent agreement by the community on core concepts and their inter-relationships. As such, they provide important components of the infrastructure for individual and group reasoning. The role ontologies play in multilingual legal retrieval and indexing huge legal corpora attest to this.

The authors argue that evidence is mixed that argument visualisation enhances a group's reasoning, deliberations or decisions. However, the case is made that argument visualisation enables a community to re-use the reasoning arrived at by another community. Benefits are achieved over time as the argument visualisation permits greater transparency and re-use of reasoning.

Section 3
Collaborative Reasoning in the Real World

Many initiatives toward enhancing group reasoning have been performed to date in diverse application areas. Insights from application areas can be categorised into two groups; those that seek to enhance or understand group reasoning in real world communities, and those that seek to enhance or understand group reasoning in virtual communities. Significant insight into collaborative reasoning can be gleaned from the experiences in understanding how collaborative reasoning actually occurs in real world groups, and how challenges in developing technologies to support the reasoning have been tackled. This group of chapters presents contributions from a variety of real world applications in diverse industries.

In looking at commerce in France, this chapter explores the role that information sharing has on the collective decisions made in supply chain management. In their chapter, Gonzalez-Feliu and Morana identify that information sharing is the critical factor in reasoning that occurs as stakeholders along the supply chain collaboratively make decisions. However, a shared conceptual model is required for determining what information must be shared. The model they advance identifies six kinds of information related to supply chain enterprises: an enterprise's solutions, their deals, sharing management, organizational features, technological features and information related to the management of shared information. They illustrate the applicability of their model with a case study related to the distribution of newspapers in France.

This chapter makes the case that the kind of reasoning task is fundamentally important. The authors look at a kind of collaborative reasoning often overlooked: reasoning used to navigate through physical space. In one study, they explore the relationship individuals have with their immediate environment in different localities throughout Europe. In a related study they explore the elements that are important when students navigate through a building to find a room. They identify the need for an ontology of concepts related to that task, particularly if intelligent systems are to support or emulate the task.

Mummery and Rodan identify an imperative for appropriate forums for community deliberation. They examine two forums—newspaper Letters to the Editor and political internet blogs—and contend that open airing of views is the critical element in collaborative reasoning, even if the nature of dialogue or representativeness of the participants does not meet theoretical ideals. In a kind of "just do it" stance the authors advocate that accessibility to forums sets community deliberations in place.

This chapter illustrates how technologies can support reasoning when participants are in dispute. The survey of negotiation support systems demonstrate that effective systems can be developed based on heuristics of principled negotiation. They present two systems in family law disputes, FamilyWinner and AssetDivider that implement principled negotiation heuristics in different ways. It is ironic that the development of systems that facilitate the resolution of disputes can provide invaluable insights into the collaborative reasoning process.

Esmahi and Badidi also propose that the assembly of appropriate information is a critical success factor for a collaborative reasoning task in their chapter. Naval operations characterized by combatants reasoning toward an action decision rather than waiting for instructions from a centralized command typically achieve greater success. The chapter presents a multi-agent simulation of naval operations where key reasoning is performed by specialized agents who have roles within the society of agents that define and constrain their activities. The chapter illustrates how insights into the nature of collaborative reasoning can be gleaned from efforts to develop simulated systems.

Section 4
Collaborative Reasoning in the Virtual World

The World Wide Web was once seen as an information repository and resource. However, more recently, it acts as a communication medium and as a forum for the liberal exchange of information and personal expression. As such, it has considerable potential for supporting group reasoning and decision-making. Key considerations needed to move towards implementations of systems to support collaborative reasoning in virtual worlds include a greater understanding of the nature of virtual communities, their differences and similarities with real world communities, the nature of reasoning tasks in both spheres, and the limitations of both. The next group of chapters discusses issues related to collaborative reasoning within or about a virtual world.

Dennen looks closely at group reasoning in blogs and particularly focus on two aspects of behaviour in groups, the propensity individuals have to share knowledge and the extent to which bloggers are prepared to encourage, empathise and provide support to other bloggers. By looking at posts to a single and multiple blogs, she discerned problems that relate uniquely to the individual or are shared by others. Each type of problem is empirically shown to have distinct norms, roles, and individual ownership prescriptions.

This chapter examines the way in which features of the problem informs group reasoning. The authors focus on internet security response, where a decision, taken by an organisation, government or sector about how to respond to a cyber-attack differs from many other decisions in that it must be made quickly, on the basis of very little clear information, and where the consequences of failing to respond appropriately may be devastating. They compare and contrast the use of a number of techniques to facilitate group reasoning including Brainstorming, Reverse brainstorming, Delphi, Wide Delphi and Generic Actual Argument Model for that task and suggest the task environment is particularly well suited to the Delphi approaches.

Miller and Smith align Web 2.0 and emerging Web 3.0 technologies to engagement, individual reasoning, group coalescing and decision making phases of reasoning communities. The comprehensive survey of social networking and collaborative work applications available on the web demonstrate, in a fitting conclusion to the book, that practice is not waiting for theory in the collaborative reasoning realm.

Preface

The information age has enabled unprecedented levels of data to be collected and stored. Consequently, decisions in many facets of human endeavour have become increasingly complex. In a parallel development, decision making is rarely left in the discretionary hands of a single body but is increasingly performed by groups, often not meeting at the same time or in the same place, but always pursuing consultative and often deliberative dialogues toward a decision. These changes result in the emergence of one of the major challenges of the modern era: ensuring that technologies develop to support collaborative reasoning.

It seems unimaginable that information and communication technologies continue to transform the reasoning landscape without some direct contribution to supporting the reasoning process. This book, *Technologies for Supporting Reasoning Communities and Collaborative Decision Making: Cooperative Approaches* includes chapters from diverse fields of enquiry including political science, argumentation, knowledge management, cognitive psychology and business intelligence. Each chapter illustrates a perspective on group reasoning that ultimately leads to a greater understanding of reasoning communities and inform technological developments.

Understanding Collaborative Reasoning

In order to enhance technological support for collaborative reasoning, more must be understood about the underlying processes that occur when groups reason. The first six chapters describe diverse conceptual perspectives on group reasoning.

In the first chapter, *An Argumentation Model of Deliberative Decision-Making*, Doug Walton identifies deliberation as a pivotal type of dialogue for group reasoning. Deliberative dialogues, in contrast to the more combative persuasion dialogues, have not been extensively studied in argumentation theories. He presents a model of deliberative argumentation using a readily accessible example and suggests that computer support systems based on this kind of model are clearly feasible.

As a political scientist, Simon Niemeyer is concerned with understanding collaborative reasoning for the purpose of advancing democracies by enabling greater direct citizen participation in policy generation. The realisation that Habermasian communicative action idealises deliberation in a way that is too prescriptive for practical application led Niemeyer to two conceptual advances: meta-consensus and intersubjective rationality. Meta-consensus describes the process a group undergoes to achieve agreement on the factors that are relevant in the discussion without agreeing on facts or outcomes. Meta-consensus contextualises and places boundaries on the reasoning. Inter-subjective rationality draws on meta-consensus and provides a much needed mechanism to measure the progress a group is making toward rational decisions.

Insights into collaborative reasoning in groups have emerged from studies comparing face to face groups with computer mediated groups. In *Communication and Group Performance: Comparing CMC with FTF Decision-making Groups in Taiwan*, Shu-Chu Sarrina Li, Lin-Mei Huang, & Yi-Ching Liu present a comparative study in order to illustrate how the two groups operate quite differently. Media richness theories advocate the use of face to face meetings with problem solving tasks that are complex and highly ambiguous and leaner media such as email when the problem solving tasks are less ambiguous. Li and her colleagues use empirical studies to report on group solutions to a hypothetical complex problem by positing that some parts of the outcome could be objectively rated. The aim was to identify the factors critical for effective group reasoning in face to face and computer mediated groups. They found that problem analysis, generating alternatives, and assessment of positive and negative consequences were critical. Face to face groups performed these tasks better than the virtual groups, thereby confirming the media rich theories. However, contrary to previous studies, elements such as social talk, not directly related to the problem, was found to play an important role in group outcomes.

Another insight from face to face and computer mediated studies involves the sharing of private information. Jamonn Campbell's *Information Sharing and Processing in Computer Mediated Interations*, examines collective information sharing—the tendency to use shared information rather than private information. Evidence from face to face groups has demonstrated that participants face many barriers to sharing all of their information. Some of the barriers seem to be overcome in computer mediated groups; however, a survey of the literature reveals information sharing is less than optimal in either forum. Campbell puts forward a strong case for the discovery of optimal ways to integrate face to face communication with computer mediated communications.

Celina Olszak and Ewa Ziemba's chapter, *Communities of Practice in Knowledge Management and Organisational Learning* presents a comprehensive study on communities of practice in Poland. They discover that despite some awareness of the benefits of communities of practice, their use in practice is in fact quite limited. The empirical findings are contrasted against a theoretical perspective drawn from Nonaka's conceptualisation of organisational growth through knowledge evolution.

The perspective we bring to the field of collaborative reasoning in the final chapter of the section, *A Reasoning Community Perspective on Deliberative Democracy*, derives from earlier work in computational modeling of reasoning using argumentation and narrative theories. Conceiving of a group as a community that is comprised of participants connected by their need to reason toward a course of action, enables the specification of high level phases the community performs. The phases—engagement, individual reasoning and group coalescing—iterate until the ultimate decision making phase is performed. Adopting this view enables an understanding of group reasoning that facilitates the identification of technological support at each phase.

Enhancing Collaborative Reasoning

The second group of chapters focuses more heavily on advancing approaches that directly enhance group reasoning. Like the chapters in the previous section, they derive from very diverse fields of inquiry.

The influential Analytical Hierarchy Process that Tom Saaty advanced some years ago enables a group to identify criteria that are prominent in decision making. In his chapter, *Group Decision Making: How to Obtain Representative Group Judgments from Individual Judgments*, Saaty discusses issues that arise when some criteria are tangible and easily measured and others are not so tangible yet quite important and presents a way to integrate intangible with tangible scales. The approach is elegantly illustrated with a case study involving a hospice decision.

Taemie Kim and Alex Pentland's aim to support group deliberations with the direct use of electronic devices in meetings presents an innovative contrast to other approaches. Their chapter, *Distributed Collaborations and the Effect of Sociometric Feedback* introduces the use of location, sound and video devices that provide feedback to participants about the discussion. Information such as the duration of each participant's contributions and the frequency of exchanges between each pair of participants does help face to face and computer mediated groups identify and correct communicative imbalances.

A key element of any group reasoning exercise involves the integration of diverse and often conflicting views that occur in many group reasoning endeavours. Numerous approaches have been advanced, but most do not derive from the formalism of mathematical theory. In the next chapter, *Methods of Aggregation of Experts' Opinions*, Jiří Vaníček, Ivan Vrana and Shady Aly present an approach for integrating diverse positions that is based on fuzzy set theory. The fuzzy set based approach enables the representation of uncertainty that is required while still retaining some mathematical rigor.

The central assumption that Marco Castellani makes, in drawing on George Kelly's cognitive theory of personal constructs for his chapter *Cognitive Tools for Group Decision Making: the Repertory Grid Approach Revisited*, is that the full context of reasoning is required to understand a group's reasoning. He includes a case study illustrating how a world view can be elicited from group participants using the repertory grid technique from cognitive psychology. Without this approach, he argues, issues that arise as important for decision making may not be the most pressing for resolution. Solutions are premature or inappropriate, and an opportunity for deeper understanding of the context is lost to participants.

Cengiz Kahraman, Selcuk Cebi and Ihsan Kaya advance a mathematical approach for determining which computer assisted manufacturing technologies are best deployed for a given problem. This task becomes very cumbersome for Analytical Hierarchy Process because there are so many configuration options that the pairwise comparisons of each elements becomes difficult to perform and interpret. Further, many criteria in practice are not independent from each other even though most multi-criteria decision models assume independence. In their chapter, *Group Decision Making For Advanced Manufacturing Technology Selection Using Choquet Integral*, Kahraman and his colleagues use a method known as the Choquet integral that measures the expected utility of an uncertain event. Based on fuzzy set theory, the Choquet Integral can represent interactions between criteria so that they can be taken into account in solving multi criteria decision equations.

Erich Schweighofer focuses on legal reasoning and surveys three decades of research aimed at automating reasoning in law, concluding that most attempts have been far from successful. In his chapter, *Indexing as an Ontological-based Support for Legal Reasoning*, Schweighofer identifies the role ontologies play in directing information retrieval technologies in law. Ontologies represent agreement by the community on core concepts and their inter-relationships. As such, they provide important components of the infrastructure for individual and group reasoning. The role ontologies play in multilingual legal retrieval and indexing huge legal corpora attest to this.

In our chapter, *A Case for the Re-Use of Community Reasoning*, we argue that evidence is mixed that argument visualisation enhances a group's reasoning, deliberations or decisions. However, the case is made that argument visualisation enables a community to re-use the reasoning arrived at by another community. Benefits are achieved over time as the argument visualisation permits greater transparency and re-use of reasoning.

Many initiatives toward enhancing group reasoning have been performed to date in diverse application areas. Insights from application areas can be categorised into two groups; those that seek to enhance or understand group reasoning in real world communities, and those that seek to enhance or understand group reasoning in virtual communities. The next chapters relate to group reasoning in the real world.

Collaborative Reasoning in the Real World

Significant insight into collaborative reasoning can be gleaned from the experiences in understanding how collaborative reasoning actually occurs in real world groups, and how challenges in developing technologies to support the reasoning have been tackled. This group of chapters presents contributions from a variety of real world applications in diverse industries.

In the realm of commerce, Jesus Gonzalez-Feliu and Joëlle Morana in France explore the role that information sharing has on the collective decisions made in supply chain management. In their chapter, *Collaborative transportation sharing: from theory to practice via a case study from France*, they identify that information sharing is the critical factor in reasoning that occurs as stakeholders along the supply chain collaboratively make decisions. However, a shared conceptual model is required for determining what information must be shared. The model they advance identifies six kinds of information related to supply chain enterprises: an enterprise's solutions, their deals, sharing management, organizational features, technological features and information related to the management of shared information. They illustrate the applicability of their model with a case study related to the distribution of newspapers in France.

Dino Borri and Domenico Camarda in their chapter on *Spatial Ontologies in Multi-Agent Environmental Planning* make the case that the kind of reasoning task is fundamentally important. They look at a kind of collaborative reasoning often overlooked: reasoning used to navigate through physical space. In one study, they explore the relationship individuals have with their immediate environment in different localities throughout Europe. In a related study they explore the elements that are important when students navigate through a building to find a room. They identify the need for an ontology of concepts related to that task, particularly if intelligent systems are to support or emulate the task.

In *Chewing the Communal Cud: Community Deliberation in Broadsheet Letters and Political Blogs,* Jane Mummery and Debbie Rodan identify an imperative for appropriate forums for community deliberation. They examine two forums—newspaper Letters to the Editor and political internet blogs—and contend that open airing of views is the critical element in collaborative reasoning, even if the nature of dialogue or representativeness of the participants does not meet theoretical ideals. In a kind of "just do it" stance the authors advocate that accessibility to forums sets community deliberations in place.

In their chapter, Emilia Bellucci and John Zeleznikow provide an illustration of how technologies can support reasoning when participants are in dispute. The survey of negotiation support systems demonstrate that effective systems can be developed based on heuristics of principled negotiation. They present two systems in family law disputes, FamilyWinner and AssetDivider that implement principled negotiation heuristics in different ways. It is ironic that the development of systems that facilitate the resolution of disputes can provide invaluable insights into the collaborative reasoning process.

Larbi Esmahi and Elarbi Badidi also propose that the assembly of appropriate information is a critical success factor for a collaborative reasoning task in their chapter entitled, *A Multiagents System for Coordinating Naval forces*. Naval operations characterized by combatants reasoning toward an action decision rather than waiting for instructions from a centralized command typically achieve greater success. The chapter presents a multi-agent simulation of naval operations where key reasoning is performed by specialized agents who have roles within the society of agents that define and constrain their activities. The chapter illustrates how insights into the nature of collaborative reasoning can be gleaned from efforts to develop simulated systems.

Collaborative Reasoning in the Virtual World

The World Wide Web was once seen as an information repository and resource. However, more recently, it acts as a communication medium and as a forum for the liberal exchange of information and personal expression. As such, it has considerable potential for supporting group reasoning and decision-making. Key considerations needed to move towards implementations of systems to support collaborative reasoning in virtual worlds include a greater understanding of the nature of virtual communities, their differences and similarities with real world communities, the nature of reasoning tasks in both spheres, and the limitations of both. The next group of chapters discusses issues related to collaborative reasoning within or about a virtual world.

Vanessa Dennen looks closely at group reasoning in blogs and particularly focus on two aspects of behaviour in groups, the propensity individuals have to share knowledge and the extent to which bloggers are prepared to encourage, empathise and provide support to other bloggers. By looking at posts to a single and multiple blogs, she discerned problems that relate uniquely to the individual or are shared by others. Each type of problem is empirically shown to have distinct norms, roles, and individual ownership prescriptions.

Andrei Kelarev, Simon Brown, Paul Watters, Xinwen Wu and Richard Dazeley examine the way in which features of the problem informs group reasoning. In their focus on internet security response, a decision, taken by an organisation, government or sector about how to respond to a cyber-attack differs from many other decisions in that it must be made quickly, on the basis of very little clear information, and where the consequences of failing to respond appropriately may be devastating. In their chapter, *Establishing Reasoning Communities of Security Experts for Internet Commerce Security*, they compare and contrast the use of a number of techniques to facilitate group reasoning including Brainstorming, Reverse brainstorming, Delphi, Wide Delphi and Generic Actual Argument Model for that task and suggest the task environment is particularly well suited to the Delphi approaches.

Charlynn Miller and Philip Smith in *Web Technologies and Reasoning Communities* align Web 2.0 and emerging Web 3.0 technologies to engagement, individual reasoning, group coalescing and decision making phases of reasoning communities. The comprehensive survey of social networking and collaborative work applications available on the web demonstrate, in a fitting conclusion to the book, that practice is not waiting for theory in the collaborative reasoning realm.

John Yearwood
Andrew Stranieri
Editors

Acknowledgment

We are grateful to many people who have contributed to this book. First and foremost, to the contributing authors: their effort and goodwill have been invaluable. Our gratitude also extends to the army of reviewers who took on their tasks with dedication and enthusiasm. Closer to home, we would like to acknowledge the support of the Centre for Informatics and Applied Optimisation at the University of Ballarat. The determination and tolerance shown by Yve Rowe, Priscilla Clark and Nadine Gass in helping to prepare the book for publication has been exemplary.

John Yearwood
Andrew Stranieri
Editors

Section 1
Understanding Collaborative Reasoning

In order to enhance technological support for collaborative reasoning, more must be understood about the underlying processes that occur when groups reason. The first six chapters describe diverse conceptual perspectives on group reasoning.

Chapter 1
An Argumentation Model of Deliberative Decision-Making

Douglas Walton
University of Windsor, Canada

ABSTRACT

This chapter presents deliberation dialogue as a framework for argumentation used in group decision-making. Drawing on and summarizing the previous literature in argumentation and artificial intelligence (AI), the chapter: (1) outlines the characteristics of deliberation as a type of dialogue; (2) distinguishes between deliberation dialogue and other types of dialogue it is closely related to and often confused with; (3) refines the existing models of deliberation to make them more useful for supporting reasoning communities engaged in collaborative decision making; (4) provides a worked example to show what the stages and characteristics of a deliberation dialogue are, and show how methods from AI and argumentation can be applied to analyzing it; and (5) outlines some further areas for research on deliberation that are currently being studied.

INTRODUCTION

Argumentation has been mainly concerned with finding or assessing reasons to prove a statement is true or false, but recently there has been some attention paid to the kind of arguments used in deliberative decision-making. Van Gelder (2009, p. 5) contrasts intuitive decision-making, where decisions are made on the basis of what seems right without any systematic review of options, with technical decision-making, where qualitative argumentation is replaced by calculation. He argues (2009, p. 6) that although many important decisions go wrong, deliberative decision-making cannot be replaced by either intuitive or technical decision-making. The aim of this paper is to apply some tools recently developed in argumentation, like argument mapping, that show promise of being helpful to improve both intuitive and technical deliberative decision-making.

The traditional technical method of decision-making weighs the expected costs of a set of alter-

DOI: 10.4018/978-1-60960-091-4.ch001

native course of action costs against its expected benefits, and decides the outcome by selecting the alternative with the greatest expected utility. In this chapter, an alternative is proposed called the argumentation approach. This new approach uses argumentation schemes, defeasible forms of reasoning representing common kinds of arguments, and applies these schemes within contextual frameworks called dialogues, in which both sides (pro and contra) of an argument are taken into account (Bench-Capon & Dunne, 2007). One type of dialogue, called deliberation, is shown to be especially important in this regard, and one scheme, called practical reasoning (Searle, 2001; Bench-Capon, 2003) is chosen to show how the approach works.

This chapter presents an argumentation method of deliberative decision-making that compares the arguments on both sides, using argumentation schemes to identify their premises and conclusions and to find missing assumptions. As critical questions are raised, as new information comes in, and as counter-arguments are put forward, reasons for a proposal can be supported or undercut. The argumentation approach is to construct an argument map that represents the chained argumentation on each side, and shows how arguments support or attack a proposal. A worked example that illustrates this methodology is included. Drawing on and summarizing the previous literature in argumentation and artificial intelligence, this chapter outlines the characteristics of deliberation as a type of dialogue, and refines existing models of deliberation.

Section 1 provides an example of deliberation that is extensive enough that every reader can see how it represents a common kind of decision-making that we are all familiar with on a daily basis. The argumentation in the example is also extensive enough that there is not enough space to analyze all the arguments in it. However, the goal of the chapter is not to provide a complete analysis or evaluation of the argumentation in the example, but only to illustrate some argumentation tools to the reader and give a general idea of how these tools can be applied to realistic cases of deliberative decision-making. The hope is to convince the reader that the argumentation approach has some advantages over intuitive methods and some technical methods when used alone.

BACKGROUND

Alice and Bob have moved from Edmonton to Windsor, and their real estate agent, Carol, has been showing them houses. They are under some pressure to make a decision within two months, as their furniture is then to be shipped from their previous house, and moving this amount of furniture to a storage unit, and then moving again to their new home, is a costly procedure. Their agent has provided them with a lot of information, and they have been looking through housing ads, and collecting as much information as they can about the housing situation in Windsor at the present time. Now they have narrowed the choice down to three homes, each of which would be suitable. The first is a condominium listed at \$200,000. The second is a two-story house listed at \$340,000. The third is a bungalow listed at \$270,000. Another option is to select none of the three and keep searching, but they hope to avoid the potentially costly consequences of this action.

Before they sit down to discuss which one they should choose, they began by discussing general goals and values that are important to them in making this kind of decision. Both agree that having a quiet atmosphere is an important goal, and both agree that not being too far from the downtown area where Bob works is a goal. These goals could, of course, conflict in some instances, since there tends to be more traffic and noise in the downtown area. Costs are always important in such a large investment, and they agree that they want to buy a house that will keep its value, and to avoid buying a house that will lose value, for example because it is over-valued or might have problems that could be costly to fix. Having discussed these matters, they proceed to have the following conversation about how to proceed at this point.

Alice: The two-story house is only four years old. It would not need any renovations. I like it.
Bob: Me too. It's right on the river and has a nice view. If it has a nice a view, that is good for resale value.
Alice: Yes, it would have good resale value, but since it's right next to the river, it could be susceptible to water problems in the basement. Even worse, flooding could be a financial disaster, because it is not covered by insurance.
Bob: There has only been one serious flood in the recorded history of Windsor, a long time ago. Also, this house does have a sump pump that can prevent minor water problems in the basement.
Alice: The condominium also has a nice view, but it is on the seventh floor. There is no danger of flooding at that level.
Bob: The bungalow is 27 years old, and would need some renovating.
Alice: Yes, we want to avoid renovations if possible, but it has been kept in very good condition. The renovations required would not be too costly.
Bob: The bungalow has no view of the river, however. Watching the boats go down the river is so relaxing to the mind.
Alice: The condo is the least expensive of the three homes, and would require little or no renovations.
Bob: Of the three, the two story house is the most expensive. Even if we renovated the bungalow, it would still cost less than the two story house.
Alice: What about getting to work for you? That is a factor.
Bob: The condo and the bungalow are both are in the same area, where I can ride to work on my bike in 40 minutes. The bike path goes right along the river straight to my office.
Alice: The problem with the two story house is that it is twice as far away as the two other homes. It would take at least 40 minutes for you to ride in from there.
Bob: If we lived there, I would have to drive the car to work most days.
Alice: Riding the bike to work is more environmentally friendly. Also, you really like riding your bike to work, and it is good exercise.
Bob: Yes, and there is also the factor of the cost of gas. The cost of driving to work adds up to a significant amount over a year. Also, neither of us likes spending a lot of time in the car. It is wasted time, and spending that time outside getting exercise is really a good way to spend that time. When you spend your whole day working on a computer, it is really important to get some exercise to break up your day.
Alice: That is offset by the lower taxes of the two story house, because it is outside the city, even though it is a larger house.
Bob: I say let's make an offer on the condo.
Alice: I say we make an offer on the bungalow.
Bob: The bungalow does not have the view of the boats!
Alice: There might be noise in the condo. It is right on the main street. Also, your neighbours could be noisy. Not only are you closer to your neighbours in one of those condos, but it is also hard to predict who might be sharing your common wall.
Bob: I agree that either of the houses is more likely to be quiet.
Alice: In the condo there is a central fee, and that can go up.
Bob: The central fee in this condo $180. That is not too high right now.
Alice: Yes, but the central fee can go up in a condo, and setting the fee is out of our control. It could go up at any time.
Bob: But there are also maintenance costs in houses. If you need a new roof, you have to pay for it. Houses continually need a lot of repair and maintenance work.
Alice: Yes, that is true, but in a condo you do not have the same control over these things. Decisions

about what upgrades you can do are voted on by the condo committee.

The example above represents only a short part of the deliberations that Alice and Bob have engaged in before they might finally arrive at a decision to make an offer on some home. It is based on previous discussions and collecting of information at the opening stage. It will presumably be followed by the making of more specific proposals once the stage of making a specific offer has been reached. The part of the conversation shown above, however, represents a realistic enough case to give us some materials to work with. We can see that it is based on some existing information that Alice and Bob share, but that will change as the deliberations proceed. They each presented some arguments, and in some instances they put forward arguments against the previous arguments of the other party. Towards the end, each made a proposal and supported it with arguments. The two proposals are opposed to each other, since they need to buy only one home. The possible option of buying both house and condo is not being considered.

THE FACTORS LIST METHOD OF STRUCTURING DELIBERATION

One method that is commonly used where the deliberation is only between two choices of actions is to summarize the deliberations by writing down two columns, one citing all the reasons in favor of the one action and the other citing all the reasons in favor of the other action. But there can also be reasons for not doing something. Hence a variant would be to cite all pro and contra reasons on both sides. For example, suppose I am trying to decide whether to take my umbrella. On the one side, I might list reasons for taking the umbrella like 'It might rain.', 'If it rains and I don't take my umbrella, I will get my clothes wet.' On the other side, I might list reasons for not taking the umbrella, like, 'There is a good chance it will not rain.', 'If it doesn't rain I won't need the umbrella.', and 'Carrying the umbrella is a nuisance'. Normally we wouldn't bother to build such an itemized list of all arguments for and against a proposed course of action in a case like the umbrella one, because the factors involved are fairly simple.

However, in a case like a decision to buy a house, the factors could be more numerous and complex. What might help is to draw up a list of positive and negative factors under each option that is being considered. Factors have been used in case-based reasoning (CBR) in AI in an intelligent tutoring system called CATO (Ashley, 2006, pp. 39-43). Tracking through the factors cited in the decision to buy a house example above, we could draw up such a list as in Table 1.

The pro and contra list, of the kind shown in Table 1, could be a simple method of arriving at a decision. We could pick the option that has the greatest number of positive factors and the least number of negative factors. Let's see by counting the factors under each option. The two-story house has six positive factors and three negative factors. The condo has five positive factors and three negative factors. Here we do not count the factor 'moderately expensive' as either positive or negative, because it falls between the positive factor of 'expensive' and the negative factor of 'inexpensive'. The bungalow has five positive factors and three negative factors. The winner is the two-story house, since it has the highest ratio of positive or negative factors.

Does this outcome reflect the best choice for Alice and Bob? It would seem not, for as the dialogue proceeds, Bob proposes making an offer on the condo and Alice proposes making an offer on the bungalow. Neither of them puts forward the proposal of making an offer on the two-story house, and the arguments they subsequently advance in support of their proposals suggest that the two-story house might not be the best choice for them. What has gone wrong here?

Table 1. List of positive and negative factors in the house-buying example

Two-Story House	*Condominium*	*Bungalow*
Needs no renovations.	Needs no renovations.	Needs some renovations.
Nice view of boats.	Nice view of boats.	No nice view of boats.
Expensive.	Inexpensive.	Moderately expensive.
Problem of water leaks.	No water leak problem.	No water leak problem.
No bike-riding to work.	Bike-riding to work.	Bike-riding to work.
Favorable for taxes.	Less favorable for taxes.	Less favorable for taxes.
Quiet atmosphere.	Less quiet atmosphere.	Quiet atmosphere.
Control over repair cost.	No control over repair cost.	Control over repair cost.
Outstanding resale value.	Lesser resale value.	Lesser resale value.

By simply counting the number of factors, we might be overlooking the importance of each factor. It might be helpful to try to see that we have taken all of the important reasons into account before making the decision. Summing them up with an itemized list of reasons pro and con could help when we come to review the factors and finally make the decision. It would be better if we could list not only all the arguments on both sides, but also give some account of how these arguments interact with each other. This is the method of argumentation, which analyzes all of the arguments on both sides of an issue under discussion, taking into account which arguments support each other and which arguments attack each other.

BASIC METHODS OF ARGUMENTATION

The emerging methodology in argumentation theory is to study an argument not only as a set of premises and conclusion, but also as statements that can be challenged and that may require support in a context of dialogue in which two or more parties reason together in order to achieve some common goal (Walton, 2007). A dialogue is a structure in which two parties (in the simplest case) take turns making moves that take the form of speech acts, like asking a question, making an assertion, or putting forward an argument. Dialogue rules define what types of moves are allowed (Walton & Krabbe, 1995). As each party makes a move, statements are inserted into or retracted from his/her commitment store. Other dialogue rules determine such insertions and retractions (Walton & Krabbe, 1995). Still other dialogue rules determine which side has won or lost, once the closing stage of the dialogue is reached. Each party has an individual goal and the dialogue itself has a collective goal. Several different types of dialogue are defined, explained and classified in (Walton, 2007, chapter 1) that represent familiar kinds of goal-directed conversations in which argumentation is used to contribute to the goal of the dialogue. Some dialogues are composites that combine features of the basic types.

During the argumentation stage of a dialogue, two parties (in the simplest case) take turns making moves that take the form of speech acts, like asking a question, making an assertion, or putting forward an argument. Dialogue rules define what types of moves are allowed (Walton & Krabbe, 1995). As each party makes a move statements are inserted into or retracted from his/her commitment store. The six basic types of dialogue previously recognized in the argumentation literature are persuasion dialogue, inquiry, negotiation dialogue, information-seeking dialogue, deliberation, and

Table 2. Seven basic types of dialogue

Type of Dialogue	Initial Situation	Participant's Goal	Goal of Dialogue
Persuasion	Conflict of Opinions	Persuade Other Party	Resolve or Clarify Issue
Inquiry	Need to Have Proof	Find and Verify Evidence	Prove (Disprove) Hypothesis
Discovery	Need to Find an Explanation of Facts	Find and Defend a Suitable Hypothesis	Choose Best Hypothesis for Testing
Negotiation	Conflict of Interests	Get What You Most Want	Reasonable Settlement Both Can Live With
Information-Seeking	Need Information	Acquire or Give Information	Exchange Information
Deliberation	Dilemma or Practical Choice	Co-ordinate Goals and Actions	Decide Best Available Course of Action
Eristic	Personal Conflict	Verbally Hit Out at Opponent	Reveal Deeper Basis of Conflict

eristic dialogue. Discovery dialogue has been added in new list of the properties of the basic types of dialogue in Table 2.

Each type of dialogue is a normative model that is meant to provide standards for judging how a given argument should be correctly used in a given case. Using any of the normative models to analyze or evaluate a real argument found in a given text of discourse is not in itself a trivial step. Evidence needs to be gathered from the given text of discourse to indicate which normative model is applicable to the case. The normal assumption is that the text of discourse in the given case will provide enough evidence to warrant viewing the argument as supposedly being meant to be a contribution to a particular type of dialogue. Each type of dialogue has three stages, an opening stage, an argumentation stage and a closing stage, where the outcome is determined, in some instances by a third party, like a judge (Walton, 2007).

Of the types of dialogue, the one that has been most thoroughly studied in the argumentation literature so far is the persuasion dialogue. Persuasion dialogue is a term now also commonly used in argumentation studies (see Walton & Krabbe, 1995; Walton, 2007) and artificial intelligence (Bench-Capon, 2003; Prakken, 2006; Bench-Capon & Dunne, 2007). As stated by Prakken (2006, p. 1), persuasion is a model of dialogue in which two or more participants try to resolve a difference of opinion by means of arguments, in order for each party to lead the other participants to change his/her commitments. 'Persuasion' in this normative sense means rational persuasion, where a proponent gets a respondent to accept a proposition by presenting an argument that has a valid form (that of an argumentation scheme) and premises that the respondent has already accepted. Persuasion, in this technical sense of the term, refers to rule-governed changes in a party's commitments as a dialogue proceeds (Walton & Krabbe, 1995). The simplest way to explain the basic structure of a persuasion dialogue is to use the diagram of (Walton, 1989, p. 6) shown in Figure 1.

The proponent's task is to rationally persuade the respondent to come to accept her (the proponent's) thesis. The respondent's task is to rationally persuade the proponent to come to accept his (the respondent's) thesis. The thesis to be proved is a specific statement designated at the opening stage. Each party has a thesis to be proved, or in some instances, only the proponent has a thesis to prove and the job of the respondent is to cast enough doubt on the proponent's attempts to prove her thesis, that her attempts are shown to have failed.

It may seem initially that we can distinguish between the two types of dialogue by saying that deliberation dialogue is about actions, and persuasion dialogue is about the truth and falsity of propositions. Deliberation dialogue is differ-

Figure 1. Structure of rational persuasion in persuasion dialogue

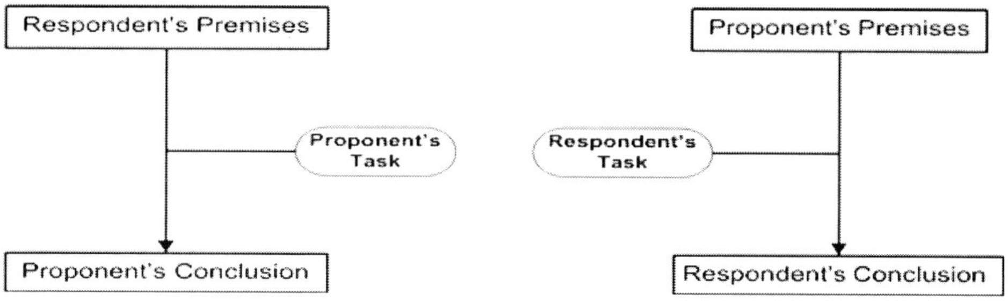

ent from persuasion dialogue, because the goal is to solve a problem posed by a choice of actions that needs to be made in a given set of circumstances (Atkinson, Bench-Capon & McBurney, 2005). However, both deliberation and persuasion dialogues can be about actions, and hence the dividing line between the two types of dialogue is not so simple.

Unlike persuasion dialogue, in deliberation dialogue the burden of proof is not set at the opening stage. During the argumentation stage of a deliberation, each party makes a proposal for a course of action, and these proposals may be modified, accepted or rejected as the argumentation stage proceeds (Walton, 2006). The best proposal, selected at the closing stage, is the one that has the strongest arguments supporting it and that meets the burden of proof. Information seeking-dialogue both seeks information and tries to judge how reliable it is as information to be accepted as factual. Inquiry, in contrast, has a high standard of proof, and uses argumentation so that, ideally, there should never be any need to retract a statement once it has been verified and accepted.

Persuasion dialogue can be classified as a truth-directed type of dialogue, as opposed to deliberation dialogue, which is not aimed at finding the truth of the matter being discussed, but at arriving at a decision on what to do, given a need to take action. While persuasion dialogue is centrally adversarial, deliberation is a collaborative type of dialogue in which parties collectively steer actions towards a common goal by agreeing on a proposal that can solve a problem affecting all of the parties concerned, taking all their interests into account. To determine in a particular case whether an argument in a text of discourse can better be seen as part of a persuasion dialogue or a deliberation type of dialogue, one has to arrive at a determination of what the goals of the dialogue and the goals of the participants are supposed to be. Deliberation dialogue presupposes much more cooperation than persuasion dialogue, which is strongly adversarial.

Before deliberation dialogue is possible, there must be sharing of common goals and common knowledge by all members of the group who are to take part in the deliberation. Dryzek and Niemeyer (2006, p. 639) require that there be a normative metaconsensus, meaning that members of the group agree on a recognition of values that will play a role in the decision-making. They do not have to agree on which values should have priority, but they do at least have to agree on which values are legitimate. We return to some discussion of values in practical reasoning below. The structure of such a deliberation dialogue is comparable to the model of reasoning from generic argument structures in deliberative discourse deliberation developed by Yearwood and Stranieri (2009).

Argumentation in deliberation is primarily a matter of identifying proposals and arguments supporting them and finding critiques of other proposals (Walton et al., 2009). Deliberation dialogue is different from negotiation dialogue, because the negotiation deals with competing interests, and its central role is to resolve a conflict of interests by arriving in a compromise that both parties can live with (Tang and Parsons, 2006). In contrast, in a deliberation dialogue the participants evaluate proposed courses of action according to standards that may be contrary to their personal interests. An important property of deliberation dialogue is that an action-option that is optimal for the group considered as a whole may not be optimal from the perspective of an individual participant (McBurney, Hitchcock & Parsons, 2007, p. 98). In a deliberation dialogue, a participant must be willing to share both his/her preferences and also information with the other participants. The initial situation of deliberation is the need for action arising out of a choice between two or more competing courses of action that are possible in a given situation. The ultimate goal of deliberation dialogue is for the participants to collectively decide on what is the best available course of action for them to take.

Discovery dialogue has been added to the six basic types of dialogue found in (Walton, 2007). On the account given by (McBurney & Parsons, 2001, p. 4), the properties of discovery dialogue and inquiry dialogue are different. In inquiry dialogue, the proposition that is to be proved true is designated at the opening stage, whereas in discovery dialogue the hypotheses to be tested are only formulated during the argumentation stage (Black & Hunter, 2007). A discovery dialogue moves through ten stages (McBurney & Parsons, 2001, p. 5) called open dialogue, discuss purpose, share knowledge, discuss mechanisms, infer consequences, discuss criteria, assess consequences, discuss tests, propose conclusions, and close dialogue. In inquiry dialogue, the proposition that is to be proved true is designated prior to the argumentation stage, but in discovery dialogue the hypothesis only emerges during the course of the dialogue itself.

When confronting an argument used in a text of discourse, a rational critic needs to decide whether it belongs to one type of dialogue or another, using the evidence from the text as the basis for the decision. Figure 2 (Walton & Krabbe, 1995, p. 81) offers a key for arriving at this sort of decision.

Figure 2. Key for determining type of dialogue in a case

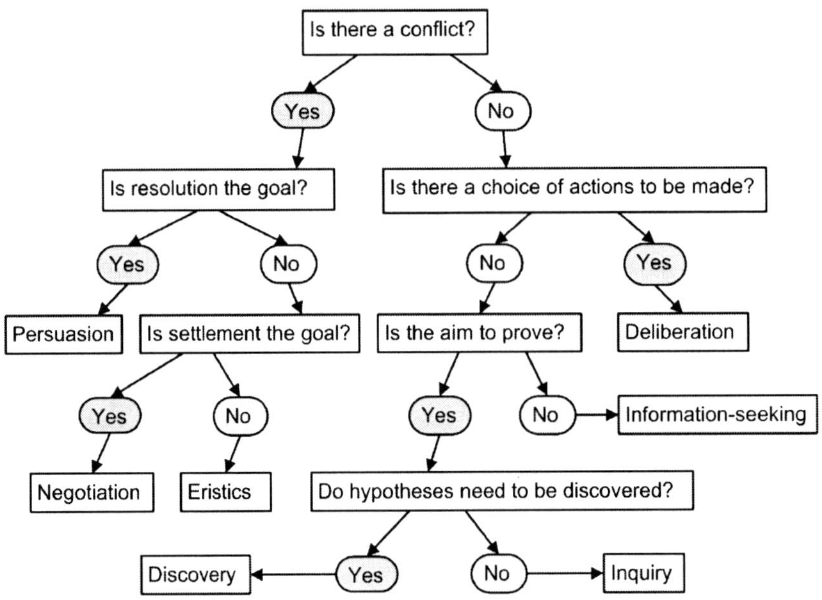

In everyday conversational argumentation, participants may be unclear, when putting an argument or criticizing it, what type of conversation they are supposed to be engaging in.

There can be shifts from one type of dialogue to another during the same sequence of argumentation (Walton & Krabbe, 1995, p. 100-116), for example, there can be a shift from a deliberation dialogue to an information-seeking dialogue. Consider the case where a debate in a legislative assembly is on the question of whether to pass a bill to install a new dam. Before the participants arrive at their decision, they need find out many facts about the plan for the dam, including considerations of what its ecological consequences are likely to be, and what the estimated costs are. They will call in experts, and possibly even people affected by the dam, to collect relevant information. In this case the dialectical shift would be classified as an embedding (Walton & Krabbe, 1995, p. 102), meaning that the goal of the deliberation dialogue is supported by the information-seeking dialogue. In other instances, shifts can be illicit, and even deceptive.

COMPUTATIONAL LITERATURE ON ARGUMENTATION MODELS OF DELIBERATION

Research in multi-agent systems in AI is based around the idea of agents performing tasks using their capability to communicate in order to carry out ordinary actions. This need to structure agent communication has proved to be especially vital in the field of AI called planning. Planning utilizes a sequence of reasoning from an initial state to a goal state. Early formal systems for planning used formal deductive inferences similar to the models of deductive reasoning used in theorem-proving. The operation of planning was described as the process of searching for the sequence of transformations starting from the initial state that will end in the goal state (Carberry, 1990). The classic example is blocks world. In blocks world the initial situation is a pile of numbered blocks on a table. The goal state is to put them in several stacks, so that the blocks in each stack are in a certain order. This kind of reasoning has now been recognized as different from deductive reasoning, and requiring a different model of rationality based on defeasible reasoning of a kind that leads to or justifies actions (Searle, 2001).

The planning system called TRACK (Carberry, 1990) uses a framework of dialogues to verify or falsify inferences about plan recognition: "TRACK assimilates utterances from an ongoing dialogue and incrementally updates and expands the system's beliefs about the underlying task-related plan motivating the information-seeker's queries." Carberry (1990, p. 75). TRACK works as a method of testing and evaluating the defeasible inferences drawn in plan recognition by means of a dialogue between the two agents. Partly it is an information-seeking type of dialogue in which one participant is "seeking information" and the other is "attempting to provide that information" (Carberry 1990, p. 3). However, the one agent is trying to get information from the other agent in relation to some actions. In addition to information-seeking dialogue, deliberation is also involved. The deliberation dialogue is important, because the plan recognizer must have the capability of grasping how sequences of actions make sense as routine ways of carrying out a goal. The plan recognizer must be able to ask the right questions and to fill in the gaps in a sequence of reasoning.

Arguments with implicit premises or conclusions are called enthymemes in logic. They are often based on common knowledge of a kind that has become an important subject for investigation in artificial intelligence. Common knowledge is represented in computing by what is called a frame, a data structure for representing a stereotyped situation, like going to a child's birthday party (Minsky, 1975, p. 2). Common knowledge is based on ordinary ways of doing things familiar to all of us in everyday life. According to Schank and Abelson (1977), common knowledge is based on what they call a *script,* a body of knowledge shared by language users concerning what typically happens in certain kinds of stereotypical situations. Scripts enable a language user to fill in gaps in inferences not explicitly stated in a text of discourse. Schank and Abelson (1977) used the restaurant story as an example. In this story, we are told explicitly that John went to a restaurant, the hostess seated John and gave him a menu, and John ordered a hamburger. Later, John got up from his seat, paid the bill, and left the restaurant. Given the elements of this story as an explicit text of discourse, we can fill in many implicit statements needed to make sense of it as a script. Below, we will show using an example how this process requires the use of scripts to fill in missing steps in a sequence of reasoning.

As will be shown in the example below, one type of reasoning that is very common in deliberation, and even characteristic of it, is practical reasoning, or goal-directed, knowledge-based action-guiding reasoning carried out by an intelligent agent (Bratman, Israel & Pollack, 1988). This type of reasoning is not only fundamental to planning technology, but is now proving to be very important in AI generally (Atkinson, Bench-Capon & McBurney, 2006).

Computational tools for electronic democracy to use the Internet to encourage public participation in deliberative decision-making have recently been built and implemented. One example is the Zeno argumentation framework (Gordon & Karacapilidis, 1997), described by its authors as a tool "designed to be used in mediation systems, an advanced kind of electronic discussion forum with special support for argumentation, negotiation and other structured forms of group decision making". The Zeno system is built on a formal model of argumentation that allows pro and contra arguments used to be weighed against each other in a dialogue format.

Tolchinsky et al. (2003) have built a formal system that combines case-based reasoning (see Ashley, 2006) with argumentation schemes for evaluating argumentation used in deliberative decision-making dialogues. Their system, called *ProCLAIM*, provides a deliberation setting for agents to argue about a decision to be made, and uses a mediator agent that directs the deliberation and evaluates the submitted arguments. *ProCLAIM* uses a case-based reasoning engine to match the current case against previously decided cases. It has an argumentation scheme repository containing both argumentation schemes and their matching critical questions. The example used to illustrate the system is a case of decision-making on human organ transplantation, where a team assembles to arrive at a decision on what to do in a specific instance. The argumentation is organized in a pro-contra structure on a graph.

Another argumentation-based tool to assist deliberative decision-making is the Parmenides system (Cartwright & Atkinson, 2008), specifically designed for a deliberative democracy setting in which the public can submit their opinions on a government policy being considered. This system explicitly adopts an argumentation framework in which argumentation schemes are used to structure the arguments, questions and objections put forward by both sides. When a policy proposal is presented through a webpage interface, an argument to the users supporting the adoption of the proposal can be attached. The user is then guided through a series of web pages that pose appropriate critical questions matching an argumentation scheme to determine which aspects of the proposal the users agree or disagree with. The scheme and critical questions themselves are in the background so that the structure is not visible to the user. Once the user's opinion is sent to Parmenides it is downloaded to a database, where the arguments can be recorded and analyzed, enabling the administrator of the system to judge which parts of the policy justification the public participants agree or disagree with. The Parmenides system has been tested on political debates, including the justification for the 2003 war in Iraq, the UK debate over the legality of fox hunting, and a debate concerning the use of speed cameras on UK roads (Cartwright & Atkinson, 2008).

In the formal model of deliberation dialogue presented by McBurney, Hitchcock and Parsons (2007, p. 100), a deliberation dialogue consists of eight stages.

Open: In this stage a governing question is raised about what is to be done. A governing question, like 'Where shall we go for dinner this evening?' is posed.
Inform: This stage includes discussion of desirable goals, values, constraints on possible actions, evaluation criteria for proposals, and determination of relevant facts.
Propose: Proposals cite possible action-options relevant to the governing question
Consider: This stage concerns commenting on the proposals from various perspectives.
Revise: Goals, constraints, perspectives, and action-options can be revised in light of comments presented and information gathering as well as fact-checking.
Recommend: A proposal for action can be recommended for acceptance or non-acceptance by each participant.
Confirm: The participants can confirm acceptance of the recommended proposal according to some procedure.
Close: The termination of the dialogue.

The model set out in this chapter will follow this eight-stage model of McBurney et al., but three main stages will be distinguished: an opening stage, a closing stage, the six other substages in between making up the argumentation stage. In contrast to the situation in persuasion dialogue, a burden of persuasion is not set at the opening stage. Instead, proposals are put forward during the argumentation stage, and the proposal best supported by the argumentation put forward during the argumentation stage is the one that is accepted at the closing stage.

Yearwood and Stranieri (2004, 2009) have built a dialectical model for practical reasoning within a community and discussed the application of their model to deliberative dialogue. They deploy their generic/actual argument model (GAAM) so that knowledge can effectively be organized within a community of reasoning to carry out transparent decision making. They describe a community of reasoning as a group of individuals who engage in dialogue in order to reason their way toward action (Yearwood & Stranieri, 2009, p. 354).

The initial situation of deliberation is the need for action arising out of a choice between two or more competing courses of action that are possible in a given set of circumstances. The goal of the deliberation dialogue is for the participants to collectively decide on what is the best available proposal of action for the group, based on the facts collected, the goals they share, and the pro and contra arguments that have been brought to bear on each proposal in turn. Each proposal is subjected to critical questioning by all the participants (or all those who speak up), and the arguments both for and against the proposal are articulated as fully as possible. The strengths and weaknesses of each proposal are brought out in the discussion, and it is this evidence that is used to judge which proposal is the best from the point of view of the deliberation. Argumentation in a deliberation is generally defeasible, because new information can come in as the circumstances change, perhaps calling for a reopening of the deliberation.

APPLYING THE METHOD OF ARGUMENTATION TO THE EXAMPLE

There are plenty of examples of arguments put forward and criticized in the conversation between Alice and Bob. Indeed, there are too many to analyze them all, as would be required by a proper case study of the example. Still, some of the arguments can be analyzed in a way that would illustrate that the method of argumentation works. The second argument used in the example conversation can be expressed in the following format, with its premise and conclusion.

Premise: If a house has a nice a view that is good for resale value.
Premise: The two-story house has a nice view.
Conclusion (and premise in the next argument): The two-story house will keep its value.
Premise: Goal: to buy a house that will keep its value.
Conclusion: These are reasons in favor of buying the two-story house.

The first argument made up of the three statements at the top, and its conclusion is re-used as a premise in the second argument. This feature illustrates *chaining of arguments*, where two arguments are combined by sharing a common statement that is a conclusion in one and a premise in the other.

The second argument can be used to illustrate one of the most important types of argument used in deliberation, called practical reasoning. In practical reasoning one premise states a goal, and the other cites a course of action that could be used to fulfill the goal, or provide a means to obtain it. The conclusion states that it would be practically reasonable to carry out this course of action. This form of reasoning is defeasible, meaning that it offers a reason for accepting the conclusion provisionally, but that might be defeated by new information that comes in. For example, if a better means of fulfilling the goal becomes evident, such new evidence would defeat the original argument. The argumentation scheme for practical reasoning has a list of critical questions matching it, and asking a critical question, or backing it with evidence, can defeat the argument.

The simplest form of practical reasoning, called instrumental practical reasoning, is represented by the following scheme (Walton, Reed & Macagno, 2008, p. 323).

Major Premise: I (an agent) have a goal *G*.
Minor Premise: Carrying out this action *A* is a means to realize *G*.
Conclusion: Therefore, I ought (practically speaking) to carry out this action *A*.

Below is the set of critical questions matching the scheme for instrumental practical reasoning (Walton, Reed & Macagno, 2008, p. 323).

CQ_1: What other goals do I have that should be considered that might conflict with *G*?
CQ_2: What alternative actions to my bringing about *A* that would also bring about *G* should be considered?
CQ_3: Among bringing about *A* and these alternative actions, which is arguably the most efficient?
CQ_4: What grounds are there for arguing that it is practically possible for me to bring about *A*?
CQ_5: What consequences of my bringing about *A* should also be taken into account?

The last critical question, $CQ_{5,}$ is very often called the side effects question. It concerns potential negative consequences of a proposed course of actions. Just asking about consequences of a course of action being contemplated could be enough to cast an argument based on practical reasoning into doubt. Another type of argument widely used in the ads is the variant of practical reasoning called value-based practical reasoning (Bench-Capon, 2003). The version of this scheme below is from (Walton, Reed & Macagno, 2008, p. 324).

Premise 1: I have a goal G.
Premise 2: G is supported by my set of values, V.
Premise 3: Bringing about A is necessary (or sufficient) for me to bring about G.
Conclusion: Therefore, I should (practically ought to) bring about A.

This form of reasoning brings us back to the consideration of values in deliberation, bringing us back to the notion of normative meta-consensus of (Dryzek, 2006). This scheme too has a matching set of critical questions, and in this set values play important roles. The device of critical questions when applied in a deliberation provides opportunities for group participants to interact with each other and influence the line of argumentation.

It could be helpful to visualize the structure of the argument above, using a tool called Araucaria, based on an Argumentation Markup Language (Reed & Rowe, 2004). Available free on the internet, Araucaria is a software tool for analyzing arguments that helps a user to reconstruct an argument map of a given argument using a point-and-click interface. The user moves the text of discourse containing an argument as a text file with a set of statements into a box in a left window, and then the user can put each statement in a text box and draw an arrow representing each inference from a set of premises to a conclusion. The outcome is a chain of argumentation that appears on the right as an argument diagram, as shown in the example below.

Araucaria was the first argument visualization tool to incorporate the use of argumentation schemes. The scheme for practical reasoning is shown in Figure 3 surrounding the part of the argument that fits that scheme. The user can search through a menu of argumentation schemes, and apply a selected scheme to an argument, displaying how the premises are connected to the conclusion by an inference fitting the scheme. Araucaria supports a distinction between two basic types of arguments. Both the arguments mapped in Figure 3 are shown as linked. In a linked argument, each premise goes together with another, or with others, to support a conclusion. All syllogistic arguments, or arguments that fit argumentation schemes, are linked, because the two premises work together to support the conclusion. In a convergent argument, each argument provides a separate reason to support the conclusion.

Ten lines before the end of the conversation Bob puts forward a proposal, saying that they should buy the condo. Alice replies by putting forward a counter-proposal, saying that they should make an offer on the bungalow. This is a crucial point in the deliberation. After this point each puts forward arguments supporting his/her proposal, but once each proposal has been made, we can see that all of the various arguments before that point in the conversation become either pro or contra each proposal. Thus once the two proposals have been articulated, we could make a large argument map with a complex chain of argumentation leading into each proposal, or at-

Figure 3. Argument map of the example

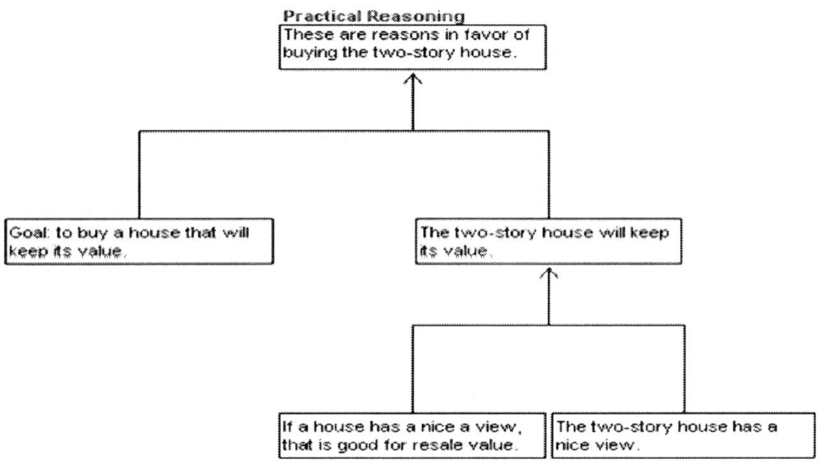

tacking each proposal. In order to determine which of these arguments is stronger, we will first have to do a lot of work to reconstruct these two chains of argumentation, draw an argument map for each, and annotate them to show which premises are acceptable, or have been accepted. We do not have space for this project here, but to give the reader a better idea of the method that needs to be used to carry it out, in Figure 4 we present an argument reconstruction of one of the arguments put forward after the two proposals were made.

What Figure 4 shows is that in order to reconstruct an argument like this from a deliberation that took place in a natural language conversation, we may have to fill in some implicit premises and conclusions. These are shown in the two text boxes at the top that are shaded and have dotted borders. The reader will recall that before the conversation in the example, Alice and Bob had a prior conversation in which they agreed on some shared goals with regard to the project of buying a home. The argument could be analyzed in even finer detail by inserting other implicit premises. For example, alongside the premise on the far left, an implicit premise, 'If a condo is right on the main street, there might be noise in it', forming a linked argument.

CONCLUSION

In the traditional method of decision-making in classical decision theory, the expected utility of an outcome state is calculated on the basis of the likelihood of reaching states by applying a utility function to the transitions between states. The outcome is decided quantitatively by attaching numerical weights to the states by the utility function. The argumentation method of deliberative decision-making compares the arguments on both sides, using argumentation schemes to identify their premises and conclusions and to find missing assumptions. The arguments are evaluated by testing them against critical questions and posing counter-arguments. Practical reasoning is the main scheme most often used in deliberation dialogues. The acceptability of the premises and the structure of the scheme can be used in a given to determine the acceptability of the conclusion.

The argumentation approach is to construct an argument map that represents the chained argumentation on each side, and shows how critical questions and attacks are related to each argument. A new tool that illustrates the methodology is worth mention.

There is a new argumentation system called Carneades (named after the Greek skeptic), that represents critical questions as premises of an

Figure 4. Argument map of the second argument

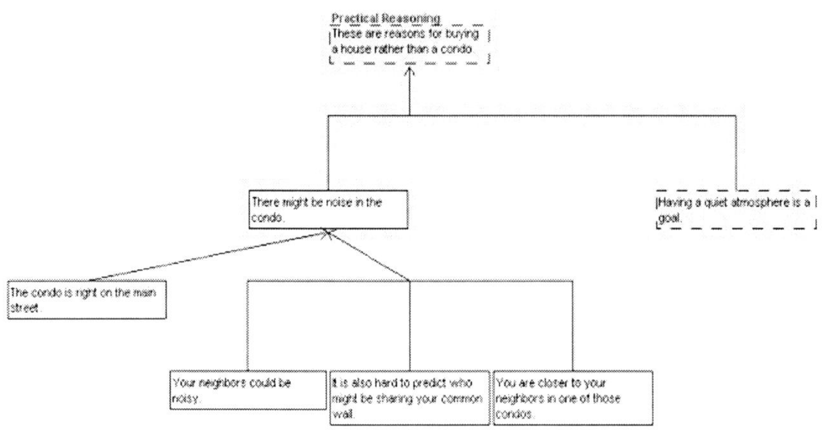

argument. An argument is said to have three kinds of premise: ordinary premises, assumptions and exceptions. Assumptions, like ordinary premises, are assumed to hold, while exceptions are assumed not to hold (Gordon, Prakken & Walton, 2007). Carneades is not only a formal model of argumentation but an argument visualization tool for it is under development as well. Premises in an argument can have various kinds of weight. They can be stated, acceptable or accepted, and Carneades has standards of proof an argument can be weighed against (Gordon & Walton, 2009). So far Carneades has only been applied to argumentation in persuasion dialogue, but a possible future research project is to apply it to modeling argumentation in deliberative decision-making.

In this chapter, it has been suggested that argumentation can provide a method of summing up the reasoning process of the participants in a deliberation dialogue that has some advantages over drawing up a list of factors on either side or calculating expected utilities. This approach takes advantage of a feature of practical reasoning identified in (Searle 2001) to the effect that that our preferences are typically determined as a product of practical reasoning rather than given as an input to it (Walton et al., 2009). A choice can be made at the opening stage on the questions of which goals are shared and which goals have priority, but during the argumentation stage, such preferences can be modified as the application of practical reasoning brings out the consequences of carrying goals, reveals conflicts between goals, and so forth. As critical questions are raised, as new information comes in, and as counter-arguments are put forward to attack a given argument, reasons for or against a proposal can be supported or undercut.

REFERENCES

Ashley, K. (2006). Case-based reasoning . In Lodder, A., & Oskamp, A. (Eds.), *Information technology and lawyers* (pp. 23–60). Berlin: Springer. doi:10.1007/1-4020-4146-2_2

Atkinson, K., Bench-Capon, T. J. M., & McBurney, P. (2005). A dialogue game protocol for multi-agent argument over proposals for action. *Autonomous Agents and Multi-Agent Systems*, *11*(2), 153–171. doi:10.1007/s10458-005-1166-x

Atkinson, K., Bench-Capon, T. J. M., & McBurney, P. (2006). Computational representation of practical argument. *Synthese*, *2006*, 157–206. doi:10.1007/s11229-005-3488-2

Bench-Capon, T. J. M. (2003). Persuasion in practical argument using value-based argumentation frameworks. *Journal of Logic and Computation*, *13*, 429–448. doi:10.1093/logcom/13.3.429

Bench-Capon, T. J. M., & Dunne, P. E. (2007). Argumentation in artificial intelligence. *Artificial Intelligence*, *171*, 619–641. doi:10.1016/j.artint.2007.05.001

Black, E., & Hunter, A. (2007). A generative inquiry dialogue system. In *Proceedings of the Sixth International Joint Conference on Autonomous Agents and Multi-Agent Systems* (AAMAS '07), (pp. 1010-1017). New York: ACM Press

Bratman, M. E., Israel, D. J., & Pollack, M. E. (1988). Plans and resource-bounded practical reasoning. *Computational Intelligence*, *4*, 349–355. doi:10.1111/j.1467-8640.1988.tb00284.x

Carberry, S. (1990). *Plan recognition in natural language dialogue*. Cambridge, MA: MIT Press.

Cartwright, D., & Atkinson, K. (2008). Political engagement through tools for argumentation. In P. Besnard, S. Doutre & A. Hunter (Eds.), *Computational Models of Argument, Proceedings of COMMA-2008,* (Frontiers in Artificial Intelligence and Applications, Vol. 172, pp. 116-127). Amsterdam: Amsterdam.

Dryzek, J. S. (2006). Reconciling pluralism and consensus as political ideals. *American Journal of Political Science*, *50*, 634–649. doi:10.1111/j.1540-5907.2006.00206.x

Gordon, T. F., & Karacapilidis, N. I. (1997). The Zeno argumentation framework. In J. Zeleznikow, D. Hunter & K. L. Branting, (Eds.), *Proceedings of the 6th International Conference on AI and Law* (pp. 10–18). New York: ACM Press.

Gordon, T. F., Prakken, H., & Walton, D. (2007). The Carneades model of argument and burden of proof. *Artificial Intelligence*, *171*, 875–896. doi:10.1016/j.artint.2007.04.010

Gordon, T. F., & Walton, D. (2009). Proof burdens and standards . In Rahwan, I., & Simari, G. (Eds.), *Argumentation and Artificial Intelligence* (pp. 239–260). Berlin: Springer. doi:10.1007/978-0-387-98197-0_12

McBurney, P., Hitchcock, D., & Parsons, S. (2007). The eightfold way of deliberation dialogue. *International Journal of Intelligent Systems*, *22*, 95–132. doi:10.1002/int.20191

McBurney, P., & Parsons, S. (2001). Chance discovery using dialectical argumentation. In T. Terano, T. Nishida, A. Namatame, S. Tsumoto, Y. Ohsawa, & T. Washio (Eds.), *New frontiers in artificial intelligence (*LNAI vol. 2253, pp. 414-424). Berlin: Springer Verlag.

Minsky, M. (1975). A framework for representing knowledge (Reprint). In P. Winston (Ed.), *The psychology of computer vision.* New York: McGraw Hill. Retrieved from http://web.media.mit.edu/~minsky/papers/Frames/frames.html

Prakken, H. (2006). Formal systems for persuasion dialogue. *The Knowledge Engineering Review*, *21*, 163–188. doi:10.1017/S0269888906000865

Reed, C., & Rowe, G. (2004). Araucaria: Software for argument analysis, diagramming and representation. *International Journal of Artificial Intelligence Tools*, *14*(3–4), 961–980. doi:10.1142/S0218213004001922

Schank, R. C., & Abelson, R. P. (1977). *Scripts, plans, goals and understanding*. Hillsdale, NJ: Erlbaum.

Searle, J. R. (2001). *Rationality in action.* Cambridge, MA: MIT Press.

Tang, Y., & Parsons, S. (2006). Argumentation-based multi-agent dialogues for deliberation. In S. Parsons, N. Maudet, P. Moraitis & I. Rahwan (Eds.), *Proceedings of the Second International Workshop on Argumentation in Multi-Agent Systems (ArgMAS '05)* (LNAI vol. 4049, pp. 229-244). Berlin: Springer.

Tolchinsky, P., Modgil, S., Cortes, U., & Sanchez-Marre, M. (2006). CBR and argument schemes for collaborative decision making. In P. E. Dunne & T. J. M. Bench-Capon (Eds.), *Computational Models of Argument: Proceedings of COMMA 2006,* (pp. 71-82). Amsterdam: IOS Press.

Van Gelder, T. (2009, August 10). *Deliberative decisions,* (pp. 5-6). Retrieved August 18, 2009, from http://timvangelder.com

Walton, D. (2006). How to make and defend a proposal in deliberation dialogue. *Artificial Intelligence and Law*, *14*, 177–239. doi:10.1007/s10506-006-9025-x

Walton, D. (2007). *Dialogue theory for critical argumentation.* Amsterdam: John Benjamins.

Walton, D., Atkinson, K., Bench-Capon, T. J. M., Wyner, A., & Cartwright, D. (2009). Argumentation in the framework of deliberation dialogue . In Bjola, C., & Kornprobst, M. (Eds.), *Arguing Global Governance*. London: Routledge.

Walton, D., & Krabbe, E. C. W. (1995). *Commitment in dialogue*. Albany, NY: State University of New York Press.

Walton, D., Reed, C., & Macagno, F. (2008). *Argumentation schemes*. Cambridge: Cambridge University Press.

Yearwood, J. L., & Stranieri, A. (2006). The generic/actual argument model of practical reasoning. *Decision Support Systems*, *41*, 358–379. doi:10.1016/j.dss.2004.07.004

Yearwood, J. L., & Stranieri, A. (2009). Deliberative discourse and reasoning from generic argument Structures. *AI & Society*, *23*, 353–377. doi:10.1007/s00146-006-0069-y

Chapter 2
Intersubjective Reasoning and the Formation of Metaconsensus

Simon Niemeyer
The Australian National University, Australia

ABSTRACT

Group deliberation, according to deliberative theorists, is supposed to produce better outcomes, but there is relatively little specification on the nature of improvement beyond appeals to consensus and improved reasoning. This chapter identifies two inter-related concepts of metaconsensus and intersubjective rationality as outcomes that an authentic deliberative process ought to produce. Importantly, these deliberative ends are consistent with ideal deliberative procedure. They are also empirically tractable, where preference transformation can be described in terms of underlying values, and judgments. Methods for assessing deliberative ends are provided and demonstrated using the Bloomfield Track case study.

INTRODUCTION

To date, the outcomes that group deliberation ought to achieve have been the subject of some confusion and contestation. The most widely recognised benchmark is that proposed by Habermas. In very concise terms this involves a dialogic procedure of reason giving amongst a receptive deliberative group of all those affected by a decision, where the 'force of reason' reveals agreement in the form of complete rational consensus (Habermas, 1984). Habermas's original theory of communicative action is heavily procedural in orientation and much of the debate in deliberative democracy has followed this lead, including consideration of what sort of dialogue should or should not be treated as 'deliberative'. The original ideal is strongly 'rationalist' and has been met with rebuttals from critics as unrealistic and fraught, yielding undesirable side-effects. These critics seek a more inclusive form of deliberation, involving non-rational speech acts, such as storytelling and rhetoric.[1]

While the debate about procedural norms about deliberation continues, perhaps more problematic is that, the Habermasian example of

DOI: 10.4018/978-1-60960-091-4.ch002

rational consensus aside, deliberative theory is vague about deliberative ends. It is a lacuna that is both a cause and a consequence of an empirical blindspot in deliberative democracy, and is borne out of its strongly normative foundations. The counter-factual orientation in much of the early foundations of deliberative democracy has rightly led to criticism regarding its feasibility in the real world. Deliberation in its ideal form is viewed as desirable, but patently unobtainable. Moreover, it is difficult to judge deliberative outcomes because there are few clear and normatively defensible benchmarks that can be used (Burkhalter et al., 2002). Such conceptual uncertainty about deliberative ends has opened up multiple avenues for critiquing examples that claim to represent deliberation, but may not actually be deliberation in a normative sense. The literature is littered with experimental and real world 'deliberation' (criminal juries, social psychology experiments, opinion surveys) as well as theoretical explorations the results of which have been used to critique the deliberative project. In many cases the connection with normative deliberative democracy is tenuous at best (Steiner, 2008).

It seems clear from the above that clarification of the sort of outcomes that deliberation ought to produce is needed. But how should this be done? Simply following the same lines of reasoning regarding ideal normative procedures into the domain of ideal outcomes is risky for two reasons. Firstly, political philosophers particularly good at seeking out dead ends using reason alone: deducing situations in which ideals that are impossible or irreconcilable.[2] Secondly, such ideals — and the related problems identified — may bear little relationship with how real-life deliberation works. One the face of it there is a dilemma between deducing normatively appealing ideals regarding deliberation (as in the case of Habermas) or observing actual outcomes in practice, which might otherwise be undesirable.

However, there is a good deal of room, and need for interaction between theory and empiricism (Bohman, 1998). This is the space occupied by this chapter. Here I seek to reduce the conceptual uncertainty by identifying two types of deliberative end (metaconsensus and intersubjective rationality). Both ends have been identified through a process of intensive exploration of real world deliberation — designed, as far as possible, to replicate the basic normative ideals of deliberative democracy — with the results contrasted to deliberative theory. These ends are phenomenological to the extent that they represent updates of deliberative theory in light of the empirical results. The emergence of metaconsensus and intersubjective rationality is consistent with the idea that the deliberative project should be open to revisiting its normative commitments in light of empirical findings (Rosenberg, 2005). They are observable and have been observed in the analysis of numerous deliberative processes.[3] Not all are recounted here. For simplicity and space I will resort to a single case study for illustration: The Bloomfield Track Citizens' Jury (see Niemeyer, 2002; Niemeyer, 2004).

In the following chapter I begin the discussion by introducing the conceptual framework giving rise to both metaconsensus and intersubjectivity, which I then go on to demonstrate empirically. This treatment is in reverse to the development of these ideas — both have emerged out of empirical research. This is because I seek to present them primarily as normative ideals — albeit ones that have emerged out of a theoretic-empirical interplay. This chapter also demonstrates empirical methods that can be used to evaluate deliberative outcomes and their relationship to these ideal ends.

The following section begins by connecting the task of identifying desirable outcomes with the ideal deliberative processes that are widely recognised in normative deliberative theory. The subsequent two sections detail the two main kinds of desirable outcome: metaconsensus and intersubjective rationality. I then turn to methods that can be used to examine these phenomena empirically, presenting results from the Bloomfield Track case study.

BACKGROUND

Deliberative democracy has been associated with the promise of many things, a number of which remain highly contested. Some claims relate to the nature of outcomes, pertaining for example to the epistemic superiority of deliberation in producing outcomes closer to some independent standard of good collective choice (Cohen, 1986; Estlund, 1997), or its ability to overcome the arbitrariness, instability, and vulnerability to manipulation in collective decision that social choice theory has warned against (McLean et al., 1999; Dryzek & List, 2003).

The most commonly cited outcome is also the most controversial: Habermas's ideal of complete and rational consensus. Elster (1986) interprets this ideal as follows: 'a rational discussion would tend to produce unanimous preferences' The criticisms that this claim attracts reflect wider arguments against the gold standard of consensus. Rescher (1993, p. 3) attributes this adherence to consensus in both Rawlsian and Habermasian interpretations in deliberative democracy as a kind of 'pre-democratic *dirigisme*'. For critics of deliberative democracy, it is often the goal of consensus that explains why deliberative democracy is either unachievable (Femia, 1996; Van Mill, 1996) or undesirable (Sanders, 1997; Young, 2000), achieved by silencing particular voices.

The problem with standards such as rational consensus and epistemic populism is that they privilege exogenous political ideals without drawing directly from the procedural norms that lie at heart of the deliberative project. Nor are they empirically grounded: complete rational consensus is as rare in real life as it is rarefied in its theoretical treatment.

Focusing on normative deliberative ideals, Bohman (1998, p. 403) argues that procedural and epistemic legitimacy are on a collision course — it being impossible to simultaneously guarantee good process and good outcomes. But it is questionable whether this dilemma is intrinsic to deliberative democracy. Thinking through the process of deliberation in broad terms, preference positions should be amenable to change in light of the reasons that are encountered.

Authentic deliberative engagement requires an open mind in a spirit of reciprocity (Benhabib, 1996; Gutmann & Thompson, 1996), where deliberators do not privilege their own perspective above those of others. For Arendt (1961, p. 220), 'enlarged thinking' means transcending 'private subjective conditions' and taking into account the perspectives of others during the process of judgement. Deliberating citizens can ideally investigate generalisable interests (Habermas, 1971) informed by 'communicative rationality' that emphasises mutual understanding. The procedural ideal here involves reflecting on the issue at hand from a shared perspective and articulating good reasons in a public context to co-deliberators. Reasoning from the standpoint of all involved forces "a certain coherence upon one's own views" (Benhabib, 1996, p. 72) — a coherence which is embodied in outcomes.

It is this relationship between a process of deliberative reasoning and 'coherence' that I wish to focus on here. Deliberative processes analysed using the method I outline below have consistently produced two kinds of outcomes. First, deliberation tends to produce *metaconsensus*, or agreement about the nature of the issue at hand, not necessarily on the actual outcome. This occurs to the extent that individuals adhere to the deliberative ideal of transcending private concerns and engage with competing views, taking them into account as part of their evaluations. To the extent this occurs, deliberation should produce metaconsensus.

In brief, metaconsensus works on different levels. At the level of reasons a group can agree on the domain of relevant considerations (involving both beliefs and values) that ought to be taken into account in making a decision — even if they do not agree on the particular facts or ranking of values. At the preference level they can then agree on the character of the choices to be made, although they may not agree with each others' conclusions.

The second type of outcome refers to the form of rationality that emerges from deliberation. *Intersubjective rationality* results from deliberative procedure in which both agreement and disagreement are possible, but are constrained by a condition of consistency regarding the reasons that produce a particular decision. An intersubjectively rational situation emerges when individuals who agree on preferences also concur on the relevant reasons, and vice versa for disagreement.

My argument is that although deliberation should not make concrete claims about making the 'right' decision, it can legitimately claim superiority to the extent that individuals have taken into account all the relevant considerations determined by metaconsensus and intersubjective consistency.

The Deliberating Individual

The general account of deliberative democracy underlying the ideas of metaconsensus and intersubjective rationality is relatively parsimonious. Above I have outlined in general terms the sorts of procedural ideals that have given rise to ideal deliberative ends — involving intersubjective reasoning within a group receptive to and respectful of competing arguments. The account of the deliberative individual is similarly minimalist. The model concerns the formation of individual preferences, which for the purposes here are simply stated choices between viable options. These choices are influenced by normative and cognitive content relevant to the issue at hand — that is, values and beliefs. In 'real world' deliberation there is no neat distinction between beliefs and desires, which are logically interdependent (Dennett, 1971, pp. 92–3).[4] Values and beliefs are thus incorporated into a wider notion of individual subjectivity in which understandings about the world are developed and transmitted.

Following from this, there are a number of ways in which deliberation can transform the positions of individuals. The first results from changes to subjectivity, yielding changes in expressed positions (preferences). Secondly, and more importantly, deliberation shapes the contours of the relevant reasons used in this process. A particular deliberative engagement should result in recognition of those reasons that can lay a legitimate claim on the sorts of choices that need to be made — described below as metaconsensus. As part of this cognitive process, deliberation should also produce more consistent connections between particular subjective positions and corresponding preferences at the group level. The result of this should be a more collectively 'rational' outcome in the sense that there is intersubjective understanding of competing perspectives and consistency between differences in subjective positions and preference positions (intersubjective rationality). In the following I consider these types of outcomes before moving on to explore empirical methods that can be used to assess them before application to the Bloomfield Track example.

METACONSENSUS: OUTCOMES IN THE FORM OF AGREEMENT

Metaconsensus, concerns the type of agreement that both can and ought to be produced by deliberation (Dryzek & Niemeyer, 2006). It represents a move away from the principle of complete consensus as the benchmark for political legitimacy. Although it does not preclude such an outcome, it embraces the principle of plurality within the context of a broader agreement that there are other legitimate points of view that should be admitted to the deliberative table.

As previously discussed, this move to the meta level is connected to key processes at work in deliberation. It also permits nuanced evaluation of the content of any agreement in a way that is consistent with the theory of deliberative democracy. To demonstrate how this works, we need to first work through the components of metaconsensus and their relationship to the deliberative process. Many existing conceptions of consensus proceed in terms of the preference side of the model, such that consensus means agreement on what is to be done. However, consensus can also refer to the subjective component of the model as well. This can be demonstrated by breaking back down into its component parts of values and beliefs. Based on the three component elements of values, beliefs and preferences, there are three kinds of consensus (see Table 1). *Normative consensus* is agreement on values. *Epistemic consensus* is agreement on how actions affect values in cause and effect terms. *Preference consensus* is agreement on what should be done.

The three types of consensus can be demonstrated using the example of the Bloomfield Track. The issue at stake here concerns the future of a crudely constructed road in the tropical far north-east of Australia in an area that has since been listed as of World Heritage significance.

The future of the Bloomfield Track was the focus of a four-day Citizens' Jury conducted as part of research funded by Land and Water Australia in 2000 involving 12 'participants' (Niemeyer, 2002). Four different issue positions regarding the Bloomfield Track were identified through reference to their basic viewpoint: Pragmatist, Preservationist, Optimist and Symbolist. These appear in the first column of Table 2. The second column identifies the value position held most strongly by the group regarding the main issue dimensions of environmental damage and vehicular access for the isolated communities at the northern end of the road. The third and fourth columns indicate the beliefs about connections between means and value positions. The final column indicates the preference ranking generated by each combination of values and beliefs.

Pragmatists prioritise the access needs of the local community as the main value. They do not necessarily believe the road meets this need; nevertheless their preferred option is to keep it open (partly based on an inherent conservatism). Preservationists prioritise environmental values, believe the road impedes these values, and so favour closure. Optimists put community values first, while putatively recognising environmental concerns. Their optimism is underwritten by a belief that the road can promote both values (a kind of 'technocentric' view). Symbolists believe the road benefits the community, but also that its upgrading could actually address their significant concerns about its negative environmental impacts regarding runoff and siltation.

Table 2 reveals no encompassing (complete rational) consensus, but some agreement across aspects of group positions. There are some normative overlaps (Symbolism with Preservation on the priority of environmental values, Optimism with Pragmatism on importance of the commu-

Table 1. Types of consensus

	Element of Preference Construction		
	Value	**Belief**	**Expressed Preference**
Type of consensus	*Normative consensus*	*Epistemic consensus*	*Preference consensus*
	(Agreement on the values that should predominate)	(Agreement on belief about the impact of a policy)	(Agreement on expressed preference for a policy)
Meta counterpart	Recognition of legitimacy of disputed values	Acceptance of credibility of disputed beliefs	Agreement on the nature of disputed choices

From Dryzek and Niemeyer (2006)

Table 2. Preference formation, the Bloomfield Track

POSITION (description)	SUBJECTIVITY (discourse)			PREFERENCE
	Value *Normative level*	Belief *Epistemic Level*		Expressed Preference
	Normative Priority	*Benefit Community*	*Damage Environment?*	*Preference Level*
Pragmatist	Community	No	No	***Keep Road*** *(Leave)*
Preservationist	Environment	No	Yes	***Close road***
Optimist	Community	Yes	No	***Keep Road*** *(Upgrade)*
Symbolist	Environment	Yes	Yes	***Keep road****(Upgrade)*

Adapted from Dryzek & Niemeyer (2006)

nity aspect). At the epistemic level, disagreement across the groups means that agreement on values can coincide with disagreement on preferences. Conversely, Optimists and Symbolists disagree on reasons, but have similar preferences.

Metaconsensus in Operation

It is known from Table 1 that normative, epistemic, and preference consensus all have a "meta" version which good deliberation ought to produce. *Normative metaconsensus* can be defined as shared recognition of the legitimacy of a set of values, while not requiring agreement on the ranking these values. In the Bloomfield Track example, both community and environmental values are recognized as legitimate by all groups. That is, no group is actively opposed to the presence of community and environmental concerns in deliberation — although, as will be shown, the nature of environmental concern did differ at the beginning of deliberation. Ideally, to the extent that metaconsensus is a deliberative ideal, all deliberators should recognize the legitimacy of the values held by other deliberators.

Epistemic metaconsensus refers to agreement on the credibility of beliefs and their relevance to the question under deliberation. Pluralists have long pointed to the need for multiple perspectives to be brought to bear on policy issues. Epistemic metaconsensus means that the engagement of perspectives is productive, as opposed to a clash of competing attempts to eradicate perceived falsehoods. The absence of epistemic metaconsensus can be found in for example debate over the teaching of evolution and creationism in the United States. On environmental issues, epistemic metaconsensus would for example allow that ordinary citizens' claims about health damage caused by pollution would be accepted as equally valid as epidemiological evidence, even when the latter contradicts individual accounts (Tesh, 2000).

Preference metaconsensus refers to the character of choices across options, and most straightforwardly connotes agreement on the range of acceptable alternatives. The whole point of judicial review of legislation rests on the idea that some policy alternatives are unacceptable — such as those that violate basic rights specified in a constitution.

Preference metaconsensus can also refer to the way choices among alternatives are structured. One particular example of such a metaconsensus, well known to political theory, involves single peakendess of preferences (List, 2002) in which structure comes in terms of a single issue dimension along which preferences are aligned. Referring again to the Bloomfield Track, there are five options ranked by deliberators, shown in Table 3.

An example of a single peaked ordering (to use the language of social choice theory) relates to agreement that degree of access is the most

Table 3. Preference options for the Bloomfield Track

Option	Description
Bitumenize	Bituminisation, which would make the track into an ordinary road
Upgrade	Upgrading by covering the road in gravel
Stabilize	Stabilizing steep slopes
Status quo	Leaving the road as it is, accessible only to four wheel drive vehicles
Close	Closure of the road and rehabilitation

important dimension of the issue. In Figure 1 there are three single peaked preference rankings. The broken line is not single peaked, and so this particular preference profile (whose first choice is closure, second choice bitumenization) is outside the preference meta-consensus.

As I will demonstrate below, single peakedness is not the only kind of structure that preference metaconsensus can produce. It occurs to the extent that there is metaconsensus on the relevance of all the options to the issue—because to 'structure' them is to imply a willingness to weigh all the alternatives in respect to the others. If you are willing to structure your choices in the same way as someone else, despite differences in the actual choice, it requires that you be at least willing to understand the nature of the choices that they are making, which applies to both metaconsensus and intersubjective consistency — an outcome that I will return to below.

Alternatively, preference metaconsensus can also imply a change in the range of acceptable options, well as the domain of most important options (a structure within a structure). For example, as will be seen in the Bloomfield case, there was agreement at the end of the process that the effective choice boiled down to one between closure and all other options. This reduction of relevant options is another kind of preference metaconsensus.

Intersubjective Rationality: Coherence in Deliberative Outcomes

Metaconsensus provides one type of coherence that, I argue, ought to result from deliberation in the form of agreement on the relevant domain of reasons and the relevant domain of preferences. It gets us some way toward an idea of the sorts of outcomes deliberation ought to produce, but it is a domain within which all sorts of outcomes of

Figure 1. Single peaked preference orderings of Bloomfield Track policies (from Dryzek and Niemeyer (2006))

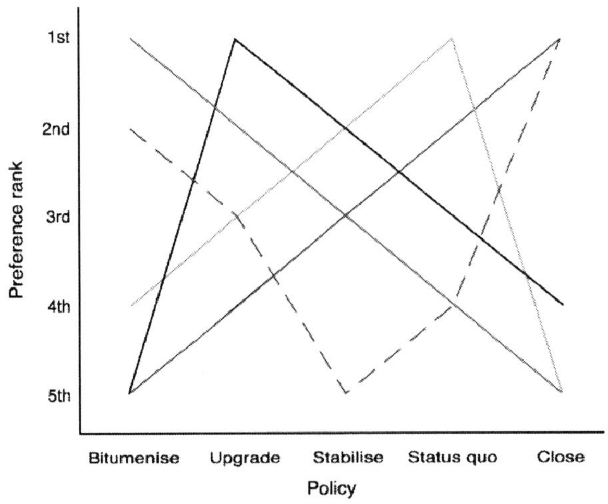

varying claims to deliberative authenticity remain possible. It does not tell us much about the quality of the outcome of any particular deliberative engagement. Also missing is a formal link between values and beliefs on the one hand and preferences on the other that ought to underpin a deliberative form of rationality in both individual and collective choice. In short, deliberation should produce outcomes that reflect the will of its participants.

It is possible to connect metaconsensus and choice in a manner consistent with the ideals that define deliberation. This relationship can be formulated as follows: if authentic deliberation has been achieved, whether or not all individuals end up agreeing at the level of preferences, they ought to at least be capable of identifying and adducing all relevant perspectives dictated by metaconsensus into their reasoning. A good outcome should then reflect the extent to which the individual positions resulting from deliberation reflect the integration of all the concerns present in meta-consensus. The term *intersubjective rationality* is used to describe this ideal (Niemeyer & Dryzek, 2007).

At the heart of intersubjective rationality is the idea that any pair of deliberators with similar subjective positions — in that they agree on values and beliefs — ought also to agree on preferences. Conversely, if they disagree on values and beliefs, they are reasonably expected to disagree on actions. Deliberation should, then, improve the standardization of what should be done in light of any particular individual subjective standpoint.

In its purest form, the condition of intersubjective rationality precludes incompletely theorized agreements, involving working agreements or a *modus vivendi* (Sunstein, 1995). Agreement on outcomes without agreement on reasons is of course a feature of real world deliberation. Indeed, it describes the convergence in the preference positions of Pragmatists and Symbolists in Table 2. However, in proposing intersubjective rationality I seek to establish an ideal outcome based on well-established procedural norms of deliberation. Doing so should contradistinguish processes that approach ideal deliberation in which intersubjective rationality is achieved from those that do not, where other outcomes, such a *modus vivendi* occur. Elsewhere I have suggested that these latter outcomes, although workable, are neither likely to be deliberative in an ideal sense, nor vulnerable, being subject to perturbations in light of changes to the issue in terms of both substance and framing (see Niemeyer, 2007).

Moreover, intersubjective rationality is no more far-fetched than *modus vivendi* — but it is more desirable from a normative deliberative perspective. As I will demonstrate below, it is not a counterfactual ambit, but an empirical observation. Its emergence as a result of the Bloomfield Track CJ dissolved the pre-existing Pragmatist-Symbolist *modus vivendi* and, as argued more forcefully elsewhere, produced an outcome better reflective of democratic will (Niemeyer, 2004).

Implications for Preference Transformation and Will Formation

But what of this democratic will? What does the condition of intersubjective rationality have to say about collective decisions? Complete rational consensus, despite its drawbacks, has the advantage of a definitive collective choice outcome. I make no such claim here in relation to either intersubjective rationality or metaconsensus. These two deliberative ends leave open the kinds of collective decisions that deliberators will come to, although it does constrain them in important ways. Any particular outcome may involve the construction of agreements on courses of action that are not the first preference of any subset of deliberators. These deliberators may assent to the outcome for different reasons (including fear of what may happen in the absence of agreement). Whatever the reason, to the extent that the outcomes at the individual level reflect the two ideals I have laid out, the assent of the collective outcome is more likely because: (1) it is part of a collectively formulated domain of possible outcomes and (2) it is fully understood and accepted as at least legitimate by all deliberators.

The process of reasoning that gives rise to metaconsensus and intersubjective consistency will give rise to preference transformation among individuals to the extent that they end up taking into account a wider (or narrower) range of issues determined by metaconsensus. Prior to deliberation, deliberators in the Bloomfield Track example tended to be focused on a narrow range of concerns that were a product of active attempts to manipulate public opinion (Niemeyer, 2004). The formation of metaconsensus permitted the group to both widen their purview and adjust their positions to take into account the 'new reality' constructed during deliberation. The actual reasoning itself is represented by the formation of intersubjective consistency. Group deliberation produced shared 'lines of reasoning' whereby individuals did not necessarily agree on outcomes, but could readily understand why others had arrived at different conclusions because they were dealing with a shared set of considerations. In practice, the outcome of the Bloomfield Track Citizens' Jury was workable even though there were two distinct groups — those in favour of closure of the road and those in favour of the status quo. Both groups understood one another, and more particularly, could see what it would take to 'come around' to the position of the opposing group.

These outcomes reflect a similar, but more constrained version of deliberative legitimacy proffered by Gutmann and Thompson (1996), based on procedural opportunity for all sides to make claims about preferred outcomes. I would also assert that the deliberative ends that are described in this chapter should also produce collective outcomes that are more stable, in the sense of achieving widespread assent, without risking the alienating effects ascribed to complete rational consensus.

EXPLORING DELIBERATIVE OUTCOMES EMPIRICALLY

I now sketch methods that can be used to evaluate a given deliberative exchange in terms of metaconsensus and intersubjective rationality using data obtained from the Bloomfield Track case study. These subjective positions are explored using Q methodology,[5] which has the advantage of permitting the discovery of relevant subjective positions in a manner that is consistent with a discursive approach. Blaug (1997, p. 107) points out that Q methodology is one of the few empirical methods available for the systematic study of intersubjectivity that 'has been informed by discursive and domination-free notions of opinion formation'. The method also avoids problems of overspecification, lending itself to the discovery of relevant subjective types, rather than establishing them a priori.

Metaconsensus is evaluated using factors produced using Q method in the case of subjectivity and structure found in preferences for preference metaconsensus. Preferences are elicited from participants using a simple ranking of options. Intersubjective consistency is used as an indicator of intersubjective rationality, which combines subjective and preference positions.

Understanding Subjective Positions

Briefly, Q methodology, as applied to the Bloomfield Track case study, entailed presenting each deliberator with a set of 42 statements about the Bloomfield Track issue drawn from interviews and content analysis of media and parliamentary reports. Participants in the citizens' jury on the issue were asked to order these statements in a quasi-normal distribution from 'most agree' to ' most disagree' on a nine-point scale, resulting in a 'Q sort' that is a picture of the individual's subjective orientation to the issue in question. The resulting sorts were subjected to inverse factor analysis.[6]

Participants performed Q sorts and filled in the preference ranking of policy options (based on the list provided in Table 3) both before and after deliberation. The Q sorts were collated and analysed using inverse factor analysis (Centroid extraction, Varimax rotation). Further 'judgmental' rotation was performed to maximise the relationship between the resulting subjective factors and the preferred outcomes of participants. The resulting factors represent archetypal positions pertaining to the Bloomfield Track issue that also correspond to different preference positions, which has been stylistically represented earlier in Table 2. These were interpreted based on the idealized Q sort, which represents the response to each statement of a hypothetical individual in 100% agreement with that factor. The results were 'triangulated' by comparing with dialogue during the deliberative process and follow-up interviews with individual participants.

The four basic subjective positions or factors that were elucidated from the Q sorts (Preservation, Pragmatism, Optimism and Symbolism) reflect their namesakes that have already been described in Table 2 — indeed, the groups in the table were identified using Q methodology. To elaborate slightly on the descriptions provided there, Preservation reflects a strongly pro-conservation orientation and sentiment against development in the region based on a long-term perspective of interconnected impacts. Pragmatism was distinguished by scepticism about evidence supporting particular claims of environmental damage and community access associated with the Bloomfield Track. Optimism resonates with a culturally influenced technocentric belief that developments, such as road construction, represent progress that yields benefits for both humans and the environment. Symbolism reflects a political socialization into the Bloomfield Track issue under the influence of claims made by competing sides of the issue resulting in a grappling with claims concerning community access and reef damage. Both of these were presented in public by opposing sides as part of a strategy of invoking a strong emotional response in a manner consistent with symbolic politics (Edelman, 1985). Consequently, Symbolism is associated with high levels of anxiety in relation to the issue, seeking mechanisms through which to resolve dissonance between extreme views.

The differences between these factors are illustrated numerically in Table 4, which shows the factor 'score' — the typical response for a hypothetical person in complete agreement with that factor — for five of the 42 statements used in the study. I show these five because they best reflect the major differences between Preservation, Pragmatism, Optimism and Symbolism. A score of '4' represents the strongest possible agreement

Table 4. Factor score and average rank for selected statements

		Score				Average Rank	
No.	**Statement**	*Pres*	*Prag*	*Opt*	*Prop*	*Pre*	*Post*
2	I don't know if improving the Bloomfield Track would lead to a rapid acceleration of development in the area to the detriment of the environment.	-3	0	2	0	-0.1	-1.1
7	Erosion from the Bloomfield Track is permanently damaging the coral reefs that fringe the beaches below	1	-2	-2	3	1.4	-0.8
10	No development should be permitted in World Heritage areas such as the Daintree.	3	4	-3	1	0.92	1.92
18	The Bloomfield Track is important because it allows quick access to remote areas of the North.	0	0	2	2	0.5	-0.67
38	The Bloomfield Track may not have been the best idea, but I guess there is probably little point in closing it now that it has been built.	-1	4	-2	2	0.58	-1.58

with a statement, '–4' the strongest possible disagreement. Table 4 also shows the average rank given to each statement by all participants, pre and post-deliberation for these statements.

The table shows a clear difference between Preservation and the remaining factors on whether the road will cause greater levels of (environmentally damaging) development (statement 2). Both it and Pragmatists are against this development (10). Both Pragmatists and Symbolists are sanguine about the issue of leaving the road, given that it already exists (38), but the latter is strongly concerned with its impact on the reef. Moreover, Symbolists share a belief with Optimists that the road provides and important access route.

Table 4 also provides a clue as to the fate of each of these positions as a result of deliberation, using the average rank score for each statement shown in the last two columns for pre- and post-deliberation respectively. Concern about the impact from the road by way of increased development has risen (statements 2 and 10). By contrast, concern about the potential impact on the reef has declined (7). So too has the belief in the importance of the road as an access route (18) as has the conservative disposition to keep the road open (38).

These changes are reflected in the overall predominance of each of the factors in the subjectivity of participants. This can be measured in three ways: average factor loading, number of participants with a significant loading, and number of participants whose loading on a factor is highest.[7] Table 5 provides this information for each factor. It is clear from the table that both Optimism and Symbolism have been greatly reduced in influence during the deliberative process. By contrast, the number of participants principally loaded on Preservation has increased, as has Pragmatism, though remaining a much smaller factor overall.

The decline of Symbolism during deliberation is worth highlighting. The average factor loading has declined from 0.26 to –0.10; the number of significant loadings from six individuals to zero; and the number of principally loaded individuals from two to zero. In short, Symbolism went from being the second most predominant position using all three measures at the outset of deliberation to the smallest, ending up having an effectively nil impact on the subjectivity of participants. As will be shown, the decline of this position in particular has a significant impact on the nature of metaconsensus.

COMPARING METACONSENSUS BEFORE AND AFTER DELIBERATION

Epistemic and Normative Metaconsensus

The disappearance of Symbolism during deliberation changed the nature of epistemic metaconsensus regarding the Bloomfield Track. Prior to the citizens' jury, there was epistemic metaconsensus

Table 5. Level of overall agreement with four subjective positions pre- and post-deliberation

	Pre-deliberation				Post-deliberation			
	Pres	*Prag*	*Opt*	*Sym*	*Pres*	*Prag*	*Opt*	*Sym*
Average factor loading (×100)	59	25	20	26	59	34	15	-1
Number of participants in significant agreement*	11	4	4	6	11	8	2	0
Principal factor	7	2	1	2	9	3	0	0

* Row sums to more than 12 because any one participant can have a significant correlation with more than one subjective position

about the credibility and relevance of two claims. The first claim was that the existence of the track damaged the environment because it caused soil erosion, which in turn led to significant damage to onshore coral reefs (see statement 7 in Table 4). The second was that the track benefited the community by providing residents with easier access to centres of commerce, employment, and social interaction (18).

Evidence for this post-deliberation epistemic metaconsensus comes mainly from analysis of dialogue during the deliberative process by assessing participant dispositions toward the main issues in their speech acts in a manner analogous to, but less advanced than, the measurement of the 'respect' criteria used as part of the Discourse Quality Index (Steenbergen et al., 2003). Exit interviews with individual participants were also used to gauge how beliefs and values changed during the deliberative process from the participant's own perspective (see Niemeyer, 2002).

Direct evidence of the changing nature of metaconsensus during the citizens' jury can be obtained using a juxtaposition of Table 4 and Table 5. They show that some participants disagreed with the two specific claims of reef damage (statement 7) and community access (18) prior to deliberation. After deliberation rejection of them was much more widely shared. Although there was no *simple* consensus on these two claims prior to deliberation, there was *metaconsensus* on their credibility and relevance. This metaconsensus can be divined by comparing the 'forced' Q sorts (in which statements are ranked according to a quota for each response category) that are the basis for the tables provided above to the 'unforced' response (which is unconstrained by quotas, analogous to use of a Likert scale). In these unforced reactions, all participants responded positively to statements 7 and 18 prior to deliberation, but negatively afterward.

After the deliberative process and the disappearance of Symbolism there was epistemic metaconsensus concerning two somewhat different sets of claims. Damage to the onshore reef was no longer viewed as an important environmental concern. It was instead replaced by concern about the knock-on effect in exposing large areas of rainforest to residential and commercial development that would accompany the road (statement 10).

The importance of community access per se did not decline in importance as a valid normative issue. However, as for environmental concern, there was a transformation in the content of metaconsensus under the broader rubric of community access. Epistemic metaconsensus changed such that an existing alternative route to the Bloomfield Track (running inland alongside the World Heritage area) was added to the domain of possible means to this end. Moreover, the Bloomfield Track itself came to be seen not seen as primarily useful for access, but for tourism. Further upgrading the road surface was seen as detracting from the 'wilderness experience' and thus would have questionable economic benefit.

Although the content of epistemic metaconsensus changed, basic normative commitments to community access as a matter of justness and fairness and environmental quality were not substantially altered. This is not surprising. When it comes to normative meta-consensus, one of the more robust findings of the psychology literature on values is that at the abstract level there is a high degree of agreement on the legitimacy of basic values (Rokeach, 1979). Individuals differ mainly on the relative priority of values, and how they apply in particular cases. Normative metaconsensus is pervasive. The Bloomfield Track results are consistent with the psychology literature on this point: the Q factor analysis shows that no normative positions held by any participant was hostile to environmental or community values before or after deliberation. However, prior to deliberation community and environmental values were widely perceived to exist in a zero-sum relationship, such that a tradeoff would need to be made. After deliberation, appreciation of the potential complementarity of community and environmental values had grown. So environmental values were no longer conceptualised in terms that entailed their incompatibility with community values.

Preference Metaconsensus

One type of preference meta-consensus, agreement on a single dimension along which preferences are structured — illustrated above as part of discussion in relation to Figure 1 — can be measured using List's index of preference structuration. This index represents the proportion of individuals in a situation whose preferences are single-peaked on any conceivable dimension (McLean et al., 1999). In the case of the Bloomfield track, this index can be calculated using pre and post-deliberative responses to the five policy options I described earlier.

Table 6 shows the results for the index of structuration for four different preference structures. On the left hand side of the table are the raw preference ranking results pre- and post-deliberation for each participant. On the right hand side of the table are the results of the structuration analysis. Four preference structures are labelled from *X* to *Z*. The first of these (*X*) is the same 'level of access' structure as that represented in Figure 1. Participants who share the same preference structure are indicated by a tick in the column corresponding to that structure for pre and post-deliberation respectively. The bottom row tabulates the total number of participants who share that preference structure and the corresponding index of structuration.

Structuration on the most common pre-deliberative preference dimension (*X*) has not changed strongly as a result of deliberation. Of the twelve participants, three had non-single-peaked prefer-

Table 6. Policy preferences and structuration

	Preference Ranks										*Preference Structures*					
											X		*Y*		*Z*	
	BIT	UPG	STA	MAI	CLO	BIT	UPG	STA	MAI	CLO	{BIT, UPG, STA, MAI, CLO}		{UPG, STA, MAI, CLO, BIT}		{UPG, MAI, CLO, STA, BIT}	
Part.	*Pre-deliberation*					*Post-deliberation*					**Pre**	**Post**	**Pre**	**Post**	**Pre**	**Post**
A	5	4	3	2	1	5	4	3	1	2	✓	✓	✓	✓	✓	✓
B	4	3	1	2	5	5	4	3	1	2	✓	✓		✓		✓
C	5	3	1	2	4	5	4	2	3	1	✓		✓			✓
D	1	4	3	2	5	5	4	3	2	1		✓		✓		✓
E	2	1	3	4	5	5	3	1	2	4	✓	✓		✓		
F	4	3	2	1	5	5	3	1	2	4	✓	✓		✓		
G	2	5	1	4	3	4	5	2	3	1					✓	✓
H	5	4	2	1	3	5	4	3	2	1	✓	✓	✓	✓		✓
I	4	3	2	1	5	5	4	3	2	1	✓	✓		✓		✓
J	4	3	1	2	5	5	4	3	2	1	✓	✓		✓		✓
K	1	2	3	4	5	5	4	3	1	2	✓	✓		✓		✓
L	2	5	3	4	1	5	4	3	2	1		✓		✓		✓
No of single peaks											**9**	**10**	**3**	**10**	**2**	**10**
Index of Structuration											**0.75**	**0.83**	**0.25**	**0.83**	**0.17**	**0.83**

ence orderings on this dominant dimension prior to deliberation, while two had non-single-peaked preference orderings on this dimension after deliberation. This result by itself suggests that deliberation did not substantially increase the degree of this kind of preference metaconsensus across the participants.

However, following deliberation there are actually three preference dimensions that each have ten individuals single-peaked on it. Participant A is the only one with single peakedness on all three structures at both deliberative stages. This despite a change in preference ordering, albeit a slight one, is of particular interest. The only change in preference here involved a switch from Closure as the most preferred option to the second most preferred.

This observation relates to another dynamic at work than that suggested by preference structuration, which generates a different kind of preference metaconsensus in the form of option reduction. What this indicates is a qualitative change in the content of preference meta-consensus, which after deliberation consisted of the idea that the crucial choice was between closure and all other alternatives. The importance of the closure/non-closure choice is also demonstrated by the concurrence between those participants that self-reported a significant preference change and those who experienced a reversal on this binary choice. Thus the full range of options on the degree of access dimension was no longer key (although, as will be seen, still relevant enough to distinguish different positions).

In other words, following deliberation the effective number of options at the heart of the Bloomfield Track issue has been reduced to two, although there exists structuration across all options. Why did this occur? The analysis so far has given us little insight into why the metaconsensus I have observed at the subjective level has translated into metaconsensus at the preference level. To make this connection, I need to now turn to intersubjective consistency.

Intersubjective Consistency

Intersubjective rationality requires that the level of agreement among any pair of individuals in terms of preferences should be proportional to the level of agreement in terms of reasons (subjective agreement). It is possible to measure this relationship using the observations in terms of subjective positions (Q sort) and preference positions (preference survey). This measure, labelled intersubjective consistency (IC) begins by representing on a scatterplot the correlations between pairs of individuals. On the x-axis is plotted the correlation between the subjective positions (based on their Q sorts) of a given pair. On the y-axis is plotted the correlation between the preference positions. Because preference agreement should be proportional to subjective agreement, an intersubjectively rational situation will feature a positively sloped regression line, around which data points are tightly clustered. The regression coefficient is the indicator of IC.

If, as asserted above, authentic deliberation ought to improve intersubjective rationality, then it should also result in an increase in IC (see also Niemeyer, 2007). This is indeed the case for the Bloomfield Track Jury, as can be seen from Figure 2. The figure shows the pre- and post-deliberative plot for all combinations of pairs of participants in the left and right hand plots respectively. Each plot also shows the regression line, with the 95% confidence contour for the regression. The regression r for each plot is also shown.

It can be seen from Figure 2 that intersubjective consistency has increased dramatically during the deliberative process, moving from a non-significant negative relationship prior to deliberation, to a strong one following deliberation. After deliberation, 56% (r^2) of variation in preferences can be explained by variation in subjectivity, up from 1%.[8]

A high IC is possible to the extent that there is a strong metaconsensus on the relevant issue consideration and preference dimensions. At the

Figure 2. Pre- and post-deliberative intersubjective consistency for the Bloomfield Track Case Study

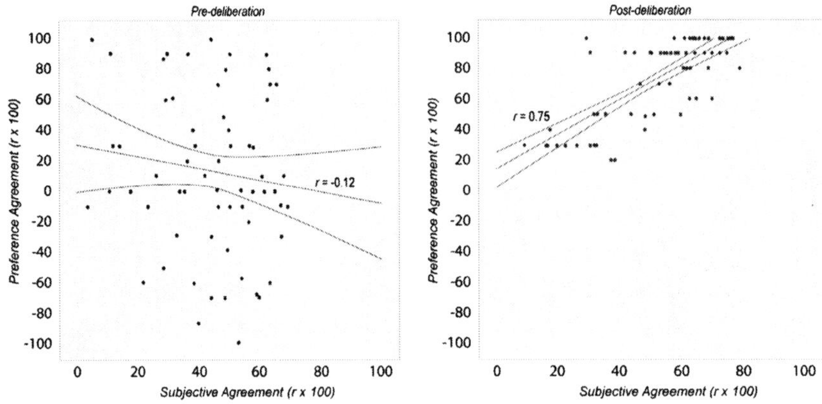

preference level, that there is metaconsensus on the issue of the close/not close road choice post-deliberation can be confirmed by recoding the preference positions according to this binary choice. Doing so results in a post-deliberation IC regression *r* of 0.33, which is statistically significant. It is however less than the IC in Figure 2, where post-deliberative ranks across all the options are included.

Therefore, as I have already suggested, the binary choice is only part of the preference metaconsensus story. In this case, structuration along the level of access dimension accounts for the rest of the high IC. This can be demonstrated by excluding the two individuals in Table 6 who do not conform to the level of access preference structure (*X*, which has a high level of structuration throughout the deliberative process) to see if doing so increases IC, which indeed it does, resulting in a regression *r* of 0.85 compared to 0.75 when all the individuals are included.

Although this evidence cannot be seen as conclusive — larger numbers of participants would be necessary — it does appear possible to demonstrate the relationship between metaconsensus and intersubjective rationality by investigating the extent to which the subset of participants who concur with a particular metaconsensus yield a higher IC than the deliberating group as a whole.

If the relationship between metaconsensus at the preference level and IC can be demonstrated empirically, can the same be done for metaconsensus at the subjective level? The discussion so far suggests that this should be the case. I have posited that intersubjective rationality provides the lynchpin effect between subjective and preference metaconsensus where metaconsensus at the subjective level has a constraining effect on the relevant domain of preferences options and ways in which these should be structured, thus producing a preference metaconsensus.

I have suggested that Symbolism played a key role in distorting what appears to be a strong preexisting consensus at the subjective level, which is mostly expressed in the form of Preservationism. Symbolism was fixated on a particular set of environmental and community concerns that did not survive the deliberative process, hence Symbolism disappearing as a subjective factor. This appears to have improved overall metaconsensus and thus facilitated a higher IC score.

However, it is also possible that the increase in IC is due to reasons other than an emerging metaconsensus. Some of the improvement in intersubjective rationality may be explainable by increased sophistication in political thinking, that is, a simple shift from non-attitudes of the kind often attributed to mass publics (Converse 1964;

Converse 1970) to coherent attitudes as a result of thinking about the issue at hand (Sturgis, Roberts et al. 2005). In this light, the low pre-deliberation IC in Figure 3 may simply reflect non-attitudes on the Bloomfield Track issue, with the deliberation merely producing more coherent post-deliberative positions. However, increasing sophistication along these lines can only be part of the story. High levels of political sophistication (and correspondingly well-formed attitudes) combined with low metaconsensus would still result in a low level of intersubjective rationality.

It is possible to test the relative effects of sophistication and metaconsensus by investigating changes to IC among sub-groups of participants. The small number of participants in the Bloomfield Track study limits the power of the analysis, but it is still illustrative. The result of this analysis is shown in Figure 3. Two rows of plots are provided for Pre- and Post-deliberation IC among three sub-groups. Pairs who are both principally defined as Preservationist are shown in the left hand plots. The middle shows those who are principally defined by a factor other than Preservation. The right-hand plot shows the remaining set of pairs involving a Preservationist and a non-Preservationist.

The figure shows a reasonable level of IC resulting from the Preservationist and non-Preservationist pairs before deliberation. What this suggests is that there is some structure prior to deliberation — rather than a complete absence of attitudes. However, the mixed pairs show a substantial (but not quite significant) negative relationship. After deliberation the IC does not change for Preservationists (though consensus increases dramatically). There are only three non-Preservationists left following deliberation, so it is not possible to get meaningful IC results from this group. The Mixed pairs, by contrast, have improved dramatically in terms of IC.

Figure 3. IC plots by type of subjective-metaconsensus

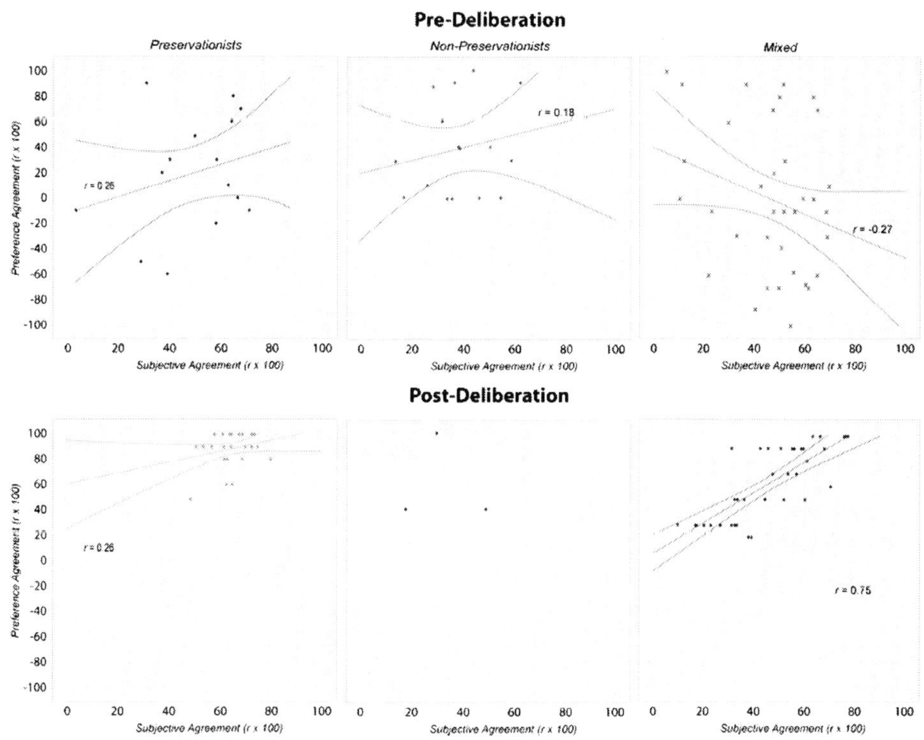

These results suggest that lack of sophistication prior to deliberation was less important than lack of metaconsensus, which emerged during the deliberative process. The key here is the relatively high IC among the pre-deliberative Preservation and Non-Preservation pairs compared to the Mixed pairs. This suggests two different types of relationships between subjective position and preferences, or two different kinds of intersubjective consistency before deliberation, depending on which subjective position is most influential. Thus it appears that there was a good level of political sophistication before deliberation in terms of consistency between particular subjective positions and preferences, which is not as strong when looking at the group as a whole. This is not surprising given the potentially powerful effect of symbolic politics. I have shown in very broad terms how the deliberative process removed this distorting effect and produced a single metaconsensus, resulting in a high IC among all participants.

The methods that I have used to demonstrate this have been largely illustrative. No doubt there is considerable scope for the development of formal methods for identifying the metaconsensus-IC relationship.[9] But it is not the intention of this chapter to provide a definitive empirical methodology. The main aim is to argue for the desirability of metaconsensus and intersubjective rationality as deliberative ends, as well as to showcase methods that can be used to examine them empirically.

CONCLUSION

Deliberative outcomes are amenable to empirical investigation that illuminates the effectiveness of deliberative process. I have stressed two sorts of outcomes, metaconsensus and intersubjective consistency that good deliberation should produce. Empirical analysis that deploys these measures can inform the normative project of deliberative democracy in important ways.

REFERENCES

Arendt, H. (1961). *Between past and future: Six exercises in political thought.* London: Faber and Faber.

Bächtiger, A., Niemeyer, S., Neblo, M., Steenbergen, M. R., & Steiner, J. (2010). Disentangling diversity in deliberative democracy: Competing theories, their empirical blind-spots, and complementarities. *Journal of Political Philosophy*, *18*(1), 32–63. doi:10.1111/j.1467-9760.2009.00342.x

Benhabib, S. (1996). Toward a deliberative model of democratic legitimacy . In Benhabib, S. (Ed.), *Democracy and difference: Contesting the boundaries of the political.* Princeton, NJ: Princeton University Press.

Blaug, R. (1997). Between fear and disappointment: Critical, empirical and political uses of Habermas. *Political Studies*, *45*, 100–117. doi:10.1111/1467-9248.00074

Bohman, J. F. (1998). The coming of age of deliberative democracy. *Journal of Political Philosophy*, *6*(4), 399–423. doi:10.1111/1467-9760.00061

Brown, S. R. (1980). *Political subjectivity: Applications of Q methodology in political science.* New Haven, CT: Yale University Press.

Burkhalter, S., Gastil, J., & Kelshaw, T. (2002). A conceptual definition and theoretical model of public deliberation in small face-to-face groups. *Communication Theory*, *12*(4), 398–422.

Cohen, J. (1986). An epistemic conception of democracy. *Ethics*, *97*(1), 26–38. doi:10.1086/292815

Converse, P. (1964). The nature of belief systems in mass publics . In Apter, D. (Ed.), *Ideology and discontent* (pp. 206–261). New York: The Free Press.

Converse, P. E. (1970). Attitudes and non-attitudes: Continuation of a dialogue . In Tufte, E. R. (Ed.), *The quantitative analysis of social problems*. Reading, MA: Addison-Wesley.

Dennett, D. C. (1971). Intentional systems. *The Journal of Philosophy*, *68*(4), 87–106. doi:10.2307/2025382

Dryzek, J. S., & List, C. (2003). Social choice theory and deliberative democracy: A reconciliation. *British Journal of Political Science*, *33*(1), 1–28. doi:10.1017/S0007123403000012

Dryzek, J. S., & Niemeyer, S. J. (2006). Reconciling pluralism and consensus as political ideals. *American Journal of Political Science*, *50*(3), 634–649. doi:10.1111/j.1540-5907.2006.00206.x

Edelman, M. J. (1985). *The symbolic uses of politics*. Urbana, IL: University of Illinois Press.

Elster, J. (1983). *Sour grapes: Studies in the subversion of rationality*. Cambridge: Cambridge University Press.

Elster, J. (1986). The market and the forum . In Elster, J., & Hylland, A. (Eds.), *Foundations of social choice theory* (pp. 103–122). Cambridge: Cambridge University Press.

Estlund, D. (1997). Beyond fairness of deliberation: The epistemic dimension of democratic authority . In Bohman, J., & Rehg, W. (Eds.), *Deliberative democracy: Essays on Reason and Politics*. Cambridge, MA: MIT Press.

Femia, J. (1996). Complexity and deliberative democracy. *Inquiry*, *39*(3-4), 359–397. doi:10.1080/00201749608602427

Gutmann, A., & Thompson, D. (1996). *Democracy and disagreement*. Cambridge, MA: The Belknap Press.

Habermas, J. (1971). *Knowledge and human interests*. Boston: Beacon Press.

Habermas, J. (1984). *Theory of communicative action*. Boston: Beacon Press.

List, C. (2002). Two conceptions of agreement. *The Good Society*, *11*(1), 72–79. doi:10.1353/gso.2002.0011

Littleboy, A., & Boughen, N. (2006). *Societal uptake of alternative energy futures: Final report*. Newcastle, Australia: CSIRO Energy Transformed Research Flagship.

McLean, I., & List, C. (1999). Can deliberation induce greater preference structuration?: Evidence from deliberative opinion polls . In *Deliberating about deliberative democracy*. Austin, TX: University of Texas.

Niemeyer, S. (2004). Deliberation in the wilderness: Displacing symbolic politics. *Environmental Politics*, *13*(2), 347–372. doi:10.1080/0964401042000209612

Niemeyer, S. J. (2002). Deliberation in the wilderness: Transforming policy preferences through discourse . In *Social and Political Theory, Research School of Social Sciences*. Canberra, Australia: The Australian National University.

Niemeyer, S. J. (2004). Deliberation in the wilderness: Displacing symbolic politics. *Environmental Politics*, *13*(2), 347–372. doi:10.1080/0964401042000209612

Niemeyer, S. J. (2007). *Intersubjective rationality: Measuring deliberative quality*. Canberra, Australia: Deliberative Democracy Research Group, Research School of Social Sciences, The Australian National University.

Niemeyer, S. J. (2007). *Report of the analysis of the UBC Biobank Deliberative Process*. Canberra, Australia: Deliberative Democracy Research Group, Research School of Social Sciences, The Australian National University.

Niemeyer, S. J. (2008). *Deliberation and the public sphere: Macro lessons from micro politics*. Paper presented at the Research School of Social Science theme seminar. Canberra, Australia: The Australian National University.

Niemeyer, S. J. (2008). *New Mexico ForestERA: Analysis of a citizen and stakeholder deliberations on forest restoration issues in northern New Mexico*. Canberra, Australia: Deliberative Democracy Research Group, Research School of Social Sciences, The Australian National University.

Niemeyer, S. J., & Ayirtman, S. (2008). Achieving success in large scale deliberation: Analysis of the Fremantle Bridge community engagement process . In *Micropolitics in Deliberation Reports*. Canberra, Australia: Deliberative Democracy Research Group, Research School of Social Sciences, The Australian National University.

Niemeyer, S. J., & Dryzek, J. S. (2007). *Intersubjective rationality: Using interpersonal consistency as a measure of deliberative quality*. Paper presented at ECPSRR Workshop 23: Advanced Empirical Study of Deliberation, Helsinki, Finland.

Palfrey, T. R., & Poole, K. T. (1987). The relationship between information, ideology, and voting behavior. *American Journal of Political Science*, *31*(3), 511–530. doi:10.2307/2111281

Petit, P. (2001). Deliberative democracy and the discursive dilemma. *Philosophical Issues*, *11*, 268–299. doi:10.1111/j.1758-2237.2001.tb00047.x

Rescher, N. (1993). *Pluralism: Against the demand for consensus*. Oxford: Clarendon Press.

Rokeach, M. (1979). *Understanding human values: Individual and societal*. New York: Free Press.

Rosenberg, S. (2005). The empirical study of deliberative democracy: Setting a research agenda. *Acta Politica*, *40*, 212–224. doi:10.1057/palgrave.ap.5500105

Sanders, L. (1997). Against deliberation. *Political Theory*, *25*, 347–376. doi:10.1177/0090591797025003002

Steenbergen, M. R., & Bächtiger, A. (2003). Measuring political deliberation: A discourse quality index. *Comparative European Studies*, *1*(1), 21–48.

Steiner, J. (2008). Concept stretching: The case of deliberation. *European Political Science*, *7*(2), 186–190. doi:10.1057/palgrave.eps.2210186

Sturgis, P., & Roberts, C. (2005). A different take on the deliberative poll: Information, deliberation, and attitude constraint. *Public Opinion Quarterly*, *69*(1), 30–65. doi:10.1093/poq/nfi005

Sunstein, C. R. (1995). Incompletely theorized agreements. *Harvard Law Review*, *108*, 1733–1772. doi:10.2307/1341816

Tesh, S. N. (2000). *Uncertain hazards: Environmental activists and scientific proof*. Ithaca, NY: Cornell University Press.

Van Mill, D. (1996). The possibility of rational outcomes from democratic discourse. *The Journal of Politics*, *58*, 734–752. doi:10.2307/2960440

Young, I. M. (2000). *Inclusion and democracy*. New York: Oxford University Press.

ENDNOTES

1 For an outline of the nature of this debate, see Bächtiger et al. (2010).

2 Examples include the so-called 'discursive dilemma' (Petit, 2001) and tension between deliberative procedure and deliberative outcomes identified by Bohman (1998). It is notable that that the discursive dilemma has been widely criticized for using a construed example, bearing little relation to how actual deliberation should, or would proceed.

3 For examples see Littleboy et al. (2006), Niemeyer (2002, 2007, 2008, 2008), and Niemeyer, Ayirtman, et al. (2008).

4 This can be illustrated by the adage that 'people believe what they want to', which captures the idea that the dominant values one holds will influence how one interacts with information (Elster, 1983, p. 19). The converse can also hold: the information an individual possesses will affect his or her confidence in beliefs, playing a role in the activation of values. See Palfrey and Poole (1987).

5 The standard exposition of Q methodology is Brown (1980). Introductions to Q can also be found at http://www.qmethod.org.

6 The application of Q methodology to the Bloomfield Track case study is set out in detail in Niemeyer (2002, esp. ch. 6)

7 A factor loading is the measure of correlation between each individual's Q sort and the position described by a particular factor. Significance of correlation was based on a standard error $SE_r = 1/\sqrt{n}$ (Brown 1980), where n (= 42) is the number of statements. At the 95% level a significant loading is $1.95 \times 1/\sqrt{42} = 0.30$, such that a significant affiliation comprised all individuals with a correlation of greater than 0.30.

8 Changes to simple consensus can be also be observed by the distribution of points, where distributions converging toward the top of each axis indicate greater consensus. Subjective consensus has increased only marginally during deliberation from an average correlation of 0.45 to 0.52. Consensus among preferences by contrast has increased dramatically from 0.13 to 0.72.

9 Moreover, the IC measure itself is weakest where there is a high level of consensus at both the subjective and preference levels (as is the case for Preservationists post deliberation in Figure 3). As it stands, it is possible to have a high level of complete (and rational) consensus resulting in a low IC.

Chapter 3
Communication and Group Performance:
Comparing CMC with FTF Decision-Making Groups in Taiwan

Shu-Chu Sarrina Li
National Chiao Tung University, Taiwan

Lin-Mei Huang
Shih Hsin University, Taiwan

Yi-Ching Liu
National Chiao Tung University, Taiwan

ABSTRACT

This study applied functional theory and media-capacity theory to compare face-to-face (FTF) groups with computer-mediated communication (CMC) groups in terms of their group communication and group performance. This study used a field experiment that was integrated into two classes of a communication course at a private university in northern Taiwan, in which 21 groups of 4 to 6 members worked on final group projects. The researchers randomly assigned one class, which had 11 groups (60 persons), to perform the task via CMC, and the other class, which had 10 groups (49 persons), to perform the task via FTF communication. The findings of this study in general support the functional theory and the media-capacity theory. However, some findings are not congruent with those of past studies. For example, previous studies have indicated that the function of criteria establishment was a significant predictor of group performance, while this study found this function to have no effect on the group outcomes. Furthermore, past studies found that the function of social talk had a negative effect on group performance, but this study discovered this function to have a positive effect on group outcomes.

DOI: 10.4018/978-1-60960-091-4.ch003

INTRODUCTION

Contemporary organizations rely heavily on groups to perform various tasks because the handling of many tasks is far beyond the capacity of any individual. Consequently, groups have become an essential part of many organizations. Many studies have sought to identify the factors that influence group effectiveness, and a growing body of evidence indicates that the group communications process by which groups arrive at their decisions influences the effectiveness of their performance. The functional theory proposes that a strong link exists between group interactions and their subsequent group decisions; and it identifies several critical task functions to be performed in the group interaction in order for a group to achieve high-quality decision making (Hirokawa & Salazar, 1999; Hirokawa, 1992; Kuhn & Poole, 2000; Tasa & Whyte, 2005).

As the use of synchronous, text-based messaging systems becomes prevalent in society, more and more organizations are conducting their team work via computer-mediated communication. Therefore, scholars are concerned about the effectiveness of CMC groups when compared with that of FTF groups (Adams, Roch & Ayman, 2005; Baltes et al, 2002; Becker-Beck, Wintermantel & Borg, 2005; Flanagin, Park & Seibold, 2004).

A close examination of the existing literature shows that, with the exception of the study by Li (2007), no studies thus far have applied the functional perspective to examine the decision-making differences between CMC groups and FTF groups. Furthermore, after performing a meta-analysis of 27 studies that compared CMC with FTF group decision making, Baltes et al. (2002) suggested that with the functional perspective having become a major paradigm in the group decision-making literature, more studies should be conducted to examine the application of this theory in the Internet environment. This study has adopted the functional perspective to investigate the differences between CMC and FTF groups in terms of their decision-making process and performance.

LITERATURE REVIEW

Functional Theory on Decision Making

The functional theory assumes that group performance is greatly influenced by the amount of critical thinking that group members employ in their interactions, and that one can assess members' critical thinking by measuring the degree to which a group performs the critical functions required by the task. The critical task functions identified by the theory are:

1. the *problem-analysis function* that helps the group thoroughly and accurately understand the problem in its task,
2. the *criteria establishing function* that allows the group to establish the specific standards that a good decision should have,
3. the *function of generating alternatives* that enables the group to generate as many choices as it can from among which a best decision can be selected,
4. the *function of assessing positive* consequences that helps the group assess the positive aspects of each alternative, and
5. the *function of assessing negative consequences* that allows the group to discuss the negative aspects of each alternative (Gouran & Hirokawa, 1996, 2003; Orlitzky & Hirokawa, 2001).

The existing literature suggests four conclusions regarding the functional theory:

1. Empirical studies confirm that three of the five functions—namely, *problem analysis*, *criteria establishment*, and *assessing negative consequences*, are positively linked to group effectiveness.
2. Past studies show that the *function of assessing positive consequences* has only a moderate effect on group effectiveness. Some studies found that *assessing positive*

consequences has a significant effect on group performance, while others did not.

3. The effect of the alternative-generation function on group performance varied from one study to another, with most studies showing its effect to be limited.
4. A few studies examined the effects of two functions, establishing *task procedures* and *socio-emotional* talk, and found that the two functions had a negative effect on group performance. The possible reason for this finding is that most groups have a limited amount of time in which to accomplish their tasks. If they spend too much time on the two functions, this will reduce the time they have available for more important task functions (Hirokawa & Salazar, 1999; Orlitzky & Hirokawa, 2001).

The existing literature includes some inconsistent findings regarding the functional theory, in that the effects of the five functions on group effectiveness varied from one study to another. Hirokawa (1990), Hirokawa & Salazar (1999), and Orlitzky & Hirokawa (2001) attributed this inconsistency to the tasks employed, because they found that two dimensions of the tasks, complexity and nature of the task, affected the role played by critical functions in group effectiveness. Hirokawa & Salazar (1999) discovered that when the tasks were simple, input variables became more important than process variables. Empirical studies found that some tasks required group members to perform certain functions more frequently than other functions.

CMC and Group Decision Making

Scholars have often used media capacity theories to account for the effect of CMC on group decision making, particularly media-richness theory (Daft & Lengel, 1986; Daft, Lengel, & Trevino, 1987). Media-richness theory stresses that communication effectiveness can be achieved if managers can match task ambiguity to the richness of a communication channel. According to media-richness theory, classification of organizations' communication media corresponds to the degree of richness according to four criteria:

1. the speed of feedback,
2. the capacity of a medium to carry multiple cues such as verbal and nonverbal cues,
3. a medium's ability to use natural language, and
4. the degree of personal focus.

Using the four criteria as a basis, face-to-face communication is the richest medium because it enables communicators to gain instant feedback, to carry verbal and nonverbal cues, to use natural language, and to have the highest degree of personal focus. In contrast, written-communication channels like email have the opposite effect, and hence are classified as lean media. This theory argues that when the task ambiguity is high, multiple interpretations and solutions are possible, and thus a rich medium is necessary for effective handling of the task. However, when the task ambiguity is low, a lean medium is sufficient because the task is simple and predictable.

One capacity theory, called social-influence theory, differs from this view. Social influence theory asserts that managers are not very accurate in differentiating one medium from another with regard to media richness, and that a manager's perception of media richness is more often influenced by personal experiences and employees' attitudes than by the objective characteristics of communication channels. Therefore, the perception of media richness varies between users and with context. Social-influence theory states that managers may perceive lean media (e.g. emails) to be relatively rich if most people in an organization favor using them and are skilled at using them (Fulk & Collins-Jarvis, 2001; Fulk, Schmitz, & Steinfield, 1990). Empirical studies support the assumptions of social-influence theory, but also discovered that respondents' perceptions of media richness were relatively congruent with the objective characteristics of media richness suggested by the media-richness theory (Barkhi, 2005). According to these theories, CMC is a more restricted medium than FTF communication, and hence the effectiveness of CMC groups compared with FTF groups depends a great deal on the nature of the task employed. When the task is complex, FTF groups should perform better than CMC groups because rich media make it more efficient for members to engage in coordination, persuasion, and expression of opinions in group interactions.

The empirical studies regarding CMC group decision making have yielded three conclusive findings:

1. CMC groups took 4 to 10 times longer than FTF groups to finish their tasks because the lack of nonverbal cues made it difficult for the CMC groups to understand each other (Adams, Roch, Ayman, 2005; Althaus, 1997; Baltes et al., 2002; Hancock & Dunham, 2001; Smith & Hayne, 1997).
2. The nature of the task greatly influenced the effectiveness of CMC groups, and CMC groups were less effective than FTF groups when performing the tasks that required high levels of communication. Many studies have used McGrath's task complex (1984) to classify four types of tasks: *generating, choosing, negotiating*, and *executing*. The four types of tasks differ from each other in terms of the amount of communication required to complete the tasks, with choosing and negotiating tasks demanding more communication than the other tasks. Most studies show that CMC groups were less effective than FTF groups when performing the tasks of choosing or negotiating under the non-anonymous condition. However, when CMC groups were anonymous, they were as effective as FTF groups (Adams et al., 2005; Baltes et al., 2002; Becker-Beck et al., 2005).
3. CMC groups were less satisfied than FTF groups with processes and outcomes. Past studies have also shown that CMC groups felt more frustrated with process and with group outcomes than did FTF groups, when performing tasks that required high levels of communication. Furthermore, non-anonymous CMC groups without a time limit were discovered to be as satisfied as FTF groups, whereas anonymous CMC groups were found to be less satisfied than FTF groups (Adams et al., 2005; Baltes et al., 2002; Barkhi, Jacob, & Pirkul, 1999; Ocker & Yaverbaum, 1999; Smith & Hayne, 1997; Straus, 1996).

Research Questions and Hypotheses

The literature strongly suggests that the quality of group decision making is greatly influenced by group interactions. Moreover, past studies show that the successful performance of the three critical functions in group interactions—problem analysis, criteria establishment, and assessment of negative consequences—has a significant effect on group effectiveness.

In this study, groups were tasked with a winter-survival task. Groups were asked to imagine themselves as survivors of a plane crash in Canada's northern forests, and to choose 10 (of 30) items to rely on until a rescue team found them. This task was a type of intellective task that contained problem solving with verifiably correct answers that belonged to the *choosing* category in McGrath's task complex (1984). In *choosing* tasks, considering multiple alternatives and assessing positive alternatives may be more important than establishing criteria to achieve effectiveness. Therefore, in this study, the first hypothesis is:

H1: The groups that can successfully perform the four critical functions—problem analysis, generating alternatives, assessment of positive consequences and assessment of negative consequences—in the group's interactions will arrive at better decisions than the groups that cannot.

The literature on CMC group decision making demonstrates that CMC groups are less effective than FTF groups when dealing with tasks that require high levels of communication and that FTF groups communicate more effectively than CMC groups. This suggests this study's second hypothesis:

H2: When faced with a complex task, FTF groups can perform the four functions - problem analysis, generating alternatives, and assessment of positive and negative consequences - more successfully than can CMC groups.

Only a few studies have examined the effects of *task-procedural function* and of *socio-emotional talk* on group performance, and one such study has found that the two functions had a negative effect on group performance (Orlitzky & Hirokawa, 2001). Furthermore, studies on CMC groups indicate that CMC groups have lower levels of *socio-emotional talk* than FTF groups, due to a lack of nonverbal cues (Walther, Anderson, & Park, 1994). Due to the dearth of conclusive findings regarding the effects of *task-procedural functions* and of *socio-emotional talk* on group performance, this study used two research questions to examine this issue:

RQ1-1: What are the effects of the two functions—*task-procedural functions* and *socio-emotional talk*—on group performance?
RQ1-2: Compared with FTF groups, what are the *task-procedural functions* and the *socio-emotional talk* in CMC groups?

RESEARCH METHODOLOGY

This study adopted a field experimental design to ensure relevance of results to the real situations and to avoid two common shortcomings namely artificiality and zero-history groups— (Wimmer & Dominick, 2003).

The Field Experiment

The researchers integrated the experiment into two classes of a communication course at a private university in northern Taiwan, in which 21 groups of 4 to 6 members worked on final group projects. These groups were created at the beginning of the semester and were required to complete a winter survival task as a class exercise one month before the semester's end. Hence, when group members worked on this class exercise, groups had been working together for almost two-thirds of the semester. Of the 30 items[1] in the task, 10 items were the most important for humans' wintertime survival, another 10 were of secondary importance, and the remaining 10 were useless items. Each of the most important items was worth 10 points, each of the second most important items was worth 5 points, and each of the useless items was worth 0 points.

Subjects included 109 junior college students in communications, divided into two classes. The two classes were conducted in exactly same manner, and varied only by scheduling. The department adopted two measures to ensure that the two classes were equal; firstly, the department used students' entrance examination scores to ensure a balanced distribution of the two classes; and secondly the department required that both classes undergo the same training process, as well as having the same instructors for required courses. Faculty members stated during interviews that the performances of the two classes were almost equal, and nearly indistinguishable in terms of mastery. The researchers randomly required one class, which had 11 groups (60 persons), to perform the task via computer-mediated communication, and the other class, which had 10 groups (49 persons), to perform the task via face-to-face communication. The researchers created a website for the CMC groups, whose members were able to enter their own chat rooms to discuss the task with their group members. Group members set up a time beforehand, and went to the chat rooms simultaneously. Group members were told that there was no time limit, and when they first logged in, they had to read instructions regarding how to complete the task, and then started the text-based discussion, which was automatically recorded for later analysis. The study required that, after the group members finished the task, they should fill out the questionnaire separately and return it via email to a research assistant. To schedule a time, each of the 10 FTF groups were required to sign up with the research assistant, and to come to the laboratory room at the assigned time. Group members were seated around a table in the laboratory, and were then given the task and the task-completion instructions. FTF group members were also told that there was no time limit, and when they finished their discussions, they were to call the research assistant, who would collect the finished task. Their group discussions were video-taped, and after the group members completed their task, they were instructed to fill out the questionnaire separately and to give it to the research assistant. The winter-survival task was translated into Chinese and was checked by three bilingual graduate students who ensured that the translation was correct and understandable to Taiwanese students. The subjects did not have any knowledge of the study purpose, and they regarded this task as one of their course requirements.

Major Variables

One major variable in this study was *communication quality*, which the researchers measured according to the group-interaction performance of the five critical functions. In addition to the five critical functions, this study examined two other functions: *the task-procedure function*, which comprises communicative acts that help group members to understand the procedural aspects of performing the task, and the *socio-emotional function*, which comprises communicative acts that helped group members to entertain one another or to release tension in group interactions.

The second major variable of the study was *group outcomes*, which the study measured by means of two methods: (1) the study assessed the objective group outcomes by examining the scores that groups obtained on the winter-survival task, therefore, groups with better decisions are those groups that obtain higher scores on the winter-survival task, and (2) the study assessed perceived group outcomes by asking group members a set of questions regarding their perceptions of group performance, communication quality, and their personal contribution to the group task.

Data Coding

Interaction analysis is a coding scheme that uses standard procedures to examine verbal behaviors in terms of functions, meanings, and patterns embedded in communication (Putnam & Fairhurst, 2001). This study adopted interaction analysis to analyze the performances of the critical functions in group discussions using the functional coding scheme developed by Hirokawa (1987). Data analysis found that when one member was talking, other group members tended to answer "yes" after each sentence (perhaps as a result of Chinese norms that are concerned with being polite in their communication) (Kim et al., 2001; Ma & Chuang, 2001). Therefore, an utterance with a complete meaning was usually broken into many speech acts. This study used communication utterances that performed a function, instead of a speech act, as a unit of analysis to assess the performance of the seven functions. A communication utterance that performed a function contained three to twenty speech acts, where a speech act was defined as "an uninterrupted verbal utterance; the speech act begins when a speaker starts talking and ends when he or she finishes" (Fisher & Ellis, 1990, p.130).

Two coders were trained with the definitions of the seven communication functions, which they then used to record which of the seven functions a unit performed, and how well (on a scale of 1-7) this unit performed this specific function. Two graduate students were trained as coders for the functional coding, and were kept blind to the purpose of the study. Two types of coder reliability, unitizing and categorizing, were calculated for the functional coding. This study used the procedure provided by Guetzkow (1950) to calculate coder reliability. A conservative level, 0.01, was applied to the calculations of both the unitizing and the categorizing reliability. Guetzkow's procedure for categorizing reliability is regarded as a conservative estimate of reliability because it accounts not only for the observed agreement between the coders, but also for the total number of units coded (N), and for the number of categories (K) used to code those units. The unitizing reliability varied from 90% to 100%, with an average of 98% for the 21 groups. Categorization reliability ranged from 71% to 92% with an average agreement of 85%. (see Table 1)

After the coders decided which of the seven functions was performed by a unit, they had to assess how well this unit performed that specific function (from one to seven with seven indicating being performed well and one not being performed at all). This study used Krippendorff's alpha (α) (1980) to estimate the reliability of the two coders' ratings in terms of how well each unit performed a specific function. Krippendorff's alpha was chosen because it allowed the researchers to calculate the reliability for the ratings on a continuous scale, such as those used in this study. Furthermore, α is regarded as a powerful reliability statistic because it calculates reliability for nominal, ordinal, interval and ratio data, and allows the researchers to compare reliabilities for different kinds of data (Hayes & Krippendorff, 2007). The α for the function of problem analysis was 0.86, for criteria establishment was 0.85, for generating alternatives was 0.93, for the assessment of positive consequences was 0.90, for the assessment of negative consequences was 0.95, for the task procedural function was 0.92, and for the socio-emotional function was 0.95.

Factor Analysis

The study used 22 questions with a seven-point scale to measure group members' perceptions of group outcomes. Using the SPSS package to analyze responses, this study conducted a factor analysis of principal components using the varimax rotation. Four factors were extracted from the 22 questions (see Table 2).

Initially, the first factor had seven items, most of which concerned members' satisfaction with group performance, and was thus referred to as perceived group performance. The second factor had six items, all of which were related to the

Table 1.

Group number	**1**	**2**	**3**	**4**	**5**	**6**	**7**	**8**	**9**	**10**	**11**
Unitizing reliability (%)	99.05	98.38	98.82	100	97.48	96.72	97.06	99.16	99.16	100	100
Categorization reliability (%)	82	92	81	91	91	88	85	85	81	87	91
Group number	**12**	**13**	**14**	**15**	**16**	**17**	**18**	**19**	**20**	**21**	
Unitizing reliability (%)	97.14	90.91	100	100	98.59	100	97.44	98.88	97.52	100	
Categorization reliability (%)	92	71	82	72	81	87	83	84	87	87	

Table 2. Factor analysis of perceived group outcomes

Variables	Factor1	Factor2	Factor3	Factor4
Perceived group performance				
Employing the best way	0.706	0.217	0.046	0.255
Good understanding	0.781	0.079	-0.048	0.031
Satisfied with outcomes	0.847	0.286	0.065	-0.020
Identified with outcomes	0.848	0.250	0.047	-0.016
Communication helps	0.616	0.140	-0.017	0.411
Right direction of discussion	0.485	0.470	0.131	0.048
Free expression	0.596	0.338	-0.017	0.099
Perceived communication quality				
Efficient communication	0.417	0.580	-0.387	0.119
Constructive discussion	0.528	0.569	0.024	0.156
Being cohesive	0.247	0.620	0.035	-0.096
Process satisfaction	0.532	0.633	-0.138	0.030
Communication is helpful	0.199	0.728	0.104	0.021
Communication is necessary	0.136	0.777	0.144	0.109
Task pressure				
Time pressure	-0.075	0.094	0.832	0.180
Having an emergency	0.054	0.301	0.837	0.058
Deciding immediately	0.092	0.360	0.683	0.012
A complete task	0.011	-0.119	0.540	-0.087
Perceived personal performance				
Playing an essential role	0.123	-0.059	0.091	0.794
Consulted by members	0.201	0.220	-0.039	0.630
Contributing to discussion	0.210	0.011	-0.063	0.828
Eigenvalues	4.542	3.596	2.837	2.194
% of Variance Explained	20.646	16.346	12.897	9.971
Alpha	0.8756	0.8265	0.7495	0.7098

quality of communication in group interactions, and was thus referred to as perceived communication quality. The third factor had five items, all of which concerned the task-related work pressure, and was thus referred to as task pressure. The fourth factor had four items, all of which concerned members' personal contribution to group interaction, and it was referred to as perceived personal performance. After the factor analysis, a reliability analysis was performed on each of the factors. The Cronbach alpha values for the first two factors were above 0.80, indicating a relatively high internal consistency. However, the Cronbach alpha values for the last two factors were below 0.40, but after the researchers deleted one item from each of the factors, their alpha values rose above 0.70. Hence, the last two factors were reduced to four and three items, respectively.

RESEARCH FINDINGS

Factors that Influence Group Performance

In this section, we discuss solutions and recommendations for dealing with the issues, controversies, or problems presented in the preceding section.

The data analysis shows that scores for the objective group outcomes varied from 65 to 85, and the average scores for CMC and FTF groups are 76.82 and 73.5, respectively. The result of a t-test indicates that there was no significant difference between the two types of groups in terms of the scores of the objective group outcomes. The researchers questioned the appropriateness of using absolute counts of acts to compare CMC and FTF groups, and thus calculated the proportions of the functions in each of the groups for comparison. To avoid the problem of collinearity, the researchers entered the numbers, the quality and the proportions of the seven functions into three different multiple-regression analyses, with the objective group outcomes as the dependent variable. The results showed that the proportions of the seven functions did not have any significant relationships to group outcomes, but the number and the quality of the seven functions did. The data are summarized in Table 3.

Table 3. Hierarchical multiple regression analysis for group outcomes

Predictor variables	Standardized beta of objective outcome	Significance
Media	-0.351	0.208
Multiple R	0.092	
Adjusted R Square	0.045	
Increased R Square		
Number of Problem Analysis	1.809**	0.003
Number of Criteria Establishment	1.202	0.203
Number of Generating Alternatives	4.999*	0.040
Number of Positive Consequences	1.915*	0.025
Number of Negative Consequences	2.375*	0.022
Number of Procedural Function	1.384	0.129
Number of Social Talk	1.320*	0.044
Number of Total Units	-9.589*	0.042
Multiple R	0.734	
Adjusted R Square	0.516	
Increased R Square	.0.471	
Predictor variables	Standardized beta of objective outcome	Significance
Media	-0.568*	0.031
Multiple R	0.092	
Adjusted R Square	0.045	
Increased R Square		
Quality of Problem Analysis	0.575*	0.020
Quality of Criteria Establishment	-0.327	0.241
Quality of Generating Alternatives	-0.063	0.787
Quality of Positive Consequences	0.693*	0.032
Quality of Negative Consequences	0.031	0.885
Quality of Procedural Function	-0.102	0.709
Quality of Social Talk	-0.415	0.141
Multiple R	0.663	
Adjusted R Square	0.438	
Increased R Square	0.393	

*=p<0.05; **=p<0.01; ***=p<0.001

The data in Table 3 demonstrate that among the nine variables entered into the first multiple-regression analysis, five variables—the number of *problem analysis* (Beta=1.809, t=0.003), the number of *generating alternatives* (Beta=4.999, t=0.040), the number of *positive consequences* (Beta=1.915, t=0.025), the number of *negative consequences* (Beta=2.375, t=0.022), and the number of *social talk* (Beta=1.320, t=0.044)—were positive predictors for the *objective group outcomes*, while the number of *total units* was discovered to be negatively correlated with the *objective group outcomes* (Beta=-9.589, t=0.042). The results from the second multiple-regression analysis show that among the eight variables entered into the regression analysis, the quality of *problem analysis* (Beta=0.575, t=0.020)and the quality of *positive consequences* (Beta=0.693, t=0.032) were positively correlated with the *group outcomes*, while *media* was negatively correlated with the *group outcomes* (Beta=-0.568, t=0.031), indicating that the CMC groups' performance was better than the FTF groups' performance.

The Differences between CMC and FTF Groups

To understand the differences between CMC and FTF groups, this study conducted t-tests to compare the CMC groups with the FTF groups in terms of the number, the quality, and the proportions of the seven functions, the number of total units, and the objective group outcomes. The results are summarized in Table 4.

The data in Table 4 demonstrate that the FTF groups' means were significantly higher than the CMC groups' means for nine variables—the number of *total units* (t=-3.97, p=0.001), the number of *problem analysis* (t=-2.48, p=0.023), the number of *criteria establishment* (t=-2.34, p=0.030), the number of *positive consequences* (t=-3.72, p=0.001), the number of *task-procedural function* (t=-2.54, p=0.020), the quality of

Table 4. Comparisons between CMC and FTF groups

	Group	N	Mean	Standard Deviations	t-value	Sig.	Effect Size Estimates[a]
Number of total units	CMC FTF	11 10	36.55 61.10	12.72 15.60	-3.97***	0.001	-1.73
Number of problem analysis	CMC FTF	11 10	1.45 3.00	1.70 1.05	-2.48*	0.023	-1.10
Number of criteria establishment	CMC FTF	11 10	6.55 10.50	4.30 3.31	-2.34*	0.030	-1.03
Number of positive consequences	CMC FTF	11 10	7.64 12.10	2.62 2.89	-3.72***	0.001	-1.62
Number of task-procedural function	CMC FTF	11 10	6.73 10.60	3.64 3.31	-2.54*	0.020	-1.11
Quality of criteria establishment	CMC FTF	11 10	20.45 33.90	14.59 14.65	-2.11*	0.049	-0.92
Quality of positive consequences	CMC FTF	11 10	17.82 30.10	6.48 9.41	-3.51*	0.002	-1.52
Quality of task-procedural function	CMC FTF	11 10	18.91 31.00	9.03 16.74	-2.09*	0.050	-0.90
Proportion of negative consequences	CMC FTF	11 10	.7273 5.800	1.56 4.78	-3.206*	0.009	-1.43

*=p<0.05; **=p<0.01; ***=p<0.001; [a]=Cohen's d

criteria establishment (t=-2.11, p=0.049), the quality of *positive consequences* (t=-3.51, p=0.002), the quality of *task-procedural function* (t=-2.09, p=0.050), and the proportion of *negative consequences* (t=-3.206, p=0.009). To further understand the differences between the two types of groups, this study conducted four t-tests to compare the CMC groups with the FTF groups in terms of the four perceived group outcomes, and the results are summarized in Table 5.

The data in Table 5 show that the CMC groups' means were significantly higher than the FTF groups' means in terms of *task pressure* (t=2.73, p=0.01) and *personal contribution* (t=2.03, p=0.05), and that the FTF groups perceived a significantly higher degree of *communication quality* (t=-4.29, p=0.00) than did their CMC counterparts.

DISCUSSION CRITICAL THINKING AND GROUP PERFORMANCE

1. The Four Major Functions

The first hypothesis states that, in their interactions, the groups that successfully perform the four critical functions - *problem analysis*, *generating alternatives*, and *assessment of positive* and *negative consequences* - will reach decisions that are better than those reached by groups that do not successfully perform the four critical functions. The data in Table 3 show that the number of *problem analysis*, the number of *generating alternatives*, the number of *assessing positive consequences*, and the number of *assessing negative consequences* are all positively correlated with the *objective group outcomes*. Furthermore, this study found that the quality of *problem analysis* and the quality of *assessing positive consequences* were positively linked with the *group outcomes*. Therefore, the findings support the prediction of the first hypothesis.

In a finding that is consistent with past research, this study discovered that the successful performance of the *problem-analysis* function in a group interaction was a strong predictor of the group performance because both the number and the quality of this function were found to have a significant effect on the group outcomes (Gouran & Hirokawa, 1996, 2003; Orlitzky & Hirokawa, 2001). Previous studies also show that the function of *assessing negative consequences* consistently exerted a significant effect on group performance, but this study found this function to have only a moderate effect on group performance because the number of *assessing negative consequences* was found to be a significant predictor for the group outcomes, while the quality of *assessing negative consequences* was not. A possible explanation for this inconsistency is that the nature of the winter-survival task led group members to focus their

Table 5. Comparisons between CMC and FTF groups for perceived outcomes

	Group	N	Mean	Standard Deviations	t	Effect Size Estimates[a]
Group Performance	CMC FTF	60 49	33.97 34.08	5.74 5.06	-.11 (p=.91)	-0.02
Communication Quality	CMC FTF	59 47	41.80 47.96	7.70 6.88	-4.29*** (p=.00)	-0.84
Task Pressure	CMC FTF	60 48	19.83 17.35	4.45 4.97	2.73** (p=.01)	0.53
Personal Contribution	CMC FTF	59 49	15.08 14.06	2.71 2.49	2.03* (p=.05)	0.39

*=p<0.05; **=p<0.01; ***=p<0.001; [a]=Cohen's d

attention on the positive consequences of the 30 items rather than on the negative consequences. Indeed, this study found that the group members tended to discuss the positive aspects of the 30 items when making decisions for selection. Moreover, this study discovered that the mean number of *assessing positive consequences* (9.8) was much higher than that of the number of *assessing negative consequences* (2.5).

Past studies have indicated that the function of *criteria establishment* was a significant predictor of group performance. This study found that this function had no effect on the group outcomes. The reason for this finding may be that the winter-survival task is a type of choosing task that is less complicated than decision-making tasks or negotiating tasks, and thus it was sufficient for the group members to discuss the positive and negative aspects of these items to make the correct decisions.

The data in this study show that the function of *assessing positive consequences* was a strong predictor of the group outcomes because both the number and the quality of this function were significantly, positively correlated with the group outcomes, while the findings from past studies indicate that this function had only a moderate effect on group performance (Orlitzky & Hirokawa, 2001). One explanation for this finding is that the nature of this task prompted the group members to concentrate their attention on the positive aspects instead of the negative aspects of the 30 items, and hence, when groups were performing a task that belonged to the *choosing tasks* category in McGrath's task complex (1984), the function of *assessing positive consequences* of alternatives played a group-performance role that was more important than that played by *assessing negative consequences* of alternatives. Furthermore, this study found that the function of *generating alternatives* had only a moderate effect on the group outcomes, a finding that is consistent with the findings of Orlitzky & Hirokawa (2001).

2. The Task-Procedural Function and Social Talk

Research question 1-1 explores the effects of the *task-procedural function* and *social talk* on group performance. The results indicate that neither the number nor the quality of the *task-procedural function* had a significant effect on the group outcomes, a finding that is congruent with the findings from the literature. Past studies show that this function had a limited effect on group performance and that if groups spent too much time on this function, it had a negative effect on the group performance (Gouran & Hirokawa, 1996, 2003). However, this study found that the number of *social talk* was positively correlated with the group outcomes, a finding that is contrary to the findings of past studies that found this function to have a negative effect on group performance (Gouran & Hirokawa, 1996, 2003). The reason for this contradictory finding may be that the relationship between *social talk* and *group performance* may be curvilinear, and when there is too much or too little *social talk*, it may be detrimental to *group performance*. Optimal performance may be achieved, however, when a small amount of *social talk* enhances the group-communication atmosphere, but does not significantly detract from a group's productive time. This logic is validated by the findings that the mean for the number of *social talk* is 2.1, which is the lowest among the seven functions, and that this number is small when compared to the mean of the total units (48.2).

THE DIFFERENCE BETWEEN CMC AND FTF GROUPS

1. The Seven Functions in Group Communication

The second hypothesis predicted that FTF groups' performance of the four Functions - *problem analysis*, *generating alternatives*, and *assessing the negative* and *positive consequences* - would be more successful than CMC groups' performance. The data in Table 4 demonstrate that FTF groups had significantly higher means than the CMC groups in terms of the following four variables - the number of *problem analysis*, the number of *assessing positive consequences*, the quality of *assessing positive consequences*, and the proportion of *assessing negative consequences*. However, this study found that the FTF groups did not differ from the CMC groups regarding the means of the quality of the alternative generating function. Therefore, the findings only partially support the prediction of H2.

Research question 1-2 investigate the differences between the CMC groups and the FTF groups in terms of two functions - *the task-procedural function* and *the function of social talk*. The data in Table 4 show that the FTF groups' means were significantly higher than the CMC groups' means in terms of the number of the *task-procedural function* and the quality of the *task-procedural function*. However, the two types of groups exhibited no differences regarding the function of social talk.

In summary, this study found that the FTF groups performed most of the seven functions better than the CMC groups and that the FTF groups had a much higher number of *total units* than the CMC groups. Therefore, these findings are, in general, congruent with the assumptions of both the media richness theory and the social influence theory, which states that FTF communication is the richest medium and can achieve communication effectiveness much better than a lean medium such as CMC channels (Daft, Lengel, & Trevino, 1987; Fulk & Collins-Jarvis, 2001).

2. The Difference in the Group Performance

This study found that the FTF groups performed most of the seven communication functions more successfully than the CMC groups, but the data in Table 3 indicate that the CMC groups' decision outcomes were better than the FTF groups' decision outcomes. This finding is inconsistent with the functional theory because it assumes a positive link between the quality of group communication and group performance. To understand this inconsistent finding, this study closely examined the data and found that five of the seven functions examined in the study - *the function of problem analysis*, *the function of generating alternatives*, *the function of assessing positive consequences*, *the function of assessing negative consequences*, and *the function of social talk* - had a positive effect on the group outcomes. The data analysis shows that the CMC groups had a significantly lower number than the FTF groups in terms of *the function of problem analysis*, but that these two groups did not differ regarding the quality of this function. This finding suggests that, although the CMC groups were unable to generate as many problem-analysis communication units as did the FTF groups, the CMC groups caught up with the FTF groups in terms of the quality of this function. This study found that the functions of *generating alternatives*, *assessing negative consequences*, and *social talk* were significant predictors of the group outcomes, and the data show that the CMC groups did not differ from the FTF groups on these functions. The data also showed that the function of *assessing positive consequences* is positively correlated with the group outcomes and that the FTF groups' means were significantly higher than the CMC groups' corresponding means in terms of this function's numbers and this function's quality. However, the function of *criteria establishment* was found to have no effect on the group outcomes, but the data show that the FTF groups' means were significantly higher than the CMC groups' means in terms of this function's numbers and this function's quality. Furthermore, the data in Table 3 indicate that the number of *total units* was negatively correlated with the group outcomes.

These findings indicate that group outcomes degraded with a greater number of total units of group communication, even if they were exercised as a way to deploy the seven functions. Group members should have identified the functions necessary for achieving group effectiveness and should have then spent more time on those important functions. This study found that the FTF groups were able to produce a greater number of communication units for the seven functions, but it seems that the FTF groups could not discern which functions were more important than other functions. As a consequence, they spent time on both important and unimportant functions. In contrast, the CMC groups were unable to produce as many communication units for the seven functions as the FTF groups did, but the CMC groups knew how to allocate their time for different functions to achieve better performance. Therefore, this study found that the CMC groups' outcomes were significantly better than the FTF groups' outcomes. This finding is congruent with findings from the media-richness theory and from the social-influence theory because past studies show that fewer cues in the CMC environment strengthen group members' task-oriented behavior, while multiple cues in the FTF environment are likely to distract group members from their tasks. Furthermore, the existing literature does not include conclusive findings regarding which of the two groups had a better performance (Baltes et al., 2002; Walther et al., 2001).

3. The Difference in the Perceived Group Outcomes

This study found that the CMC groups felt much less satisfied with the communication quality of their interactions than did the FTF groups, but that the CMC groups had significantly higher means than did the FTF groups in terms of *task pressure* and *perceived personal contribution* to the group outcomes. These findings are congruent with the existing literature and demonstrate that, with fewer cues available, CMC groups were unable to freely express themselves and that CMC groups were thus less satisfied with the group process than were FTF groups. However, past studies show that the lack of nonverbal cues in the CMC environment made the group communication more equally distributed among members, and this study found that the CMC group members perceived a more equal personal contribution to the group outcomes than did the members of FTF groups. Moreover, past studies found that CMC groups took 4 to 10 times longer than FTF groups to finish tasks, and hence the CMC groups felt a much higher degree of task pressure than did the FTF groups (Adams, Roch & Ayman, 2005; Baltes et al., 2002; Barkhi, Jacob & Pirkul, 1999).

SUMMARY OF THE RESEARCH FINDINGS

The findings support the prediction of the first hypothesis because this study found that the four functions—*problem analysis*, *generating alternatives*, *assessing positive consequences*, and *assessing negative consequences* are all positively correlated with the *objective group outcomes*. Research question 1-1 explores the effects of the *task-procedural function* and *social talk* on group performance. The results indicate that the function of *social talk* had a positive effect on the group outcomes, while the *task-procedural function* did not. Research question 1-2 investigate the differences of two functions - *the task-procedural function* and *the function of social talk* between the CMC groups and the FTF groups. This study found that the FTF groups performed better than the CMC groups in terms of the *task-procedural function*, while the two types of groups exhibited no differences regarding the function of social talk. The second hypothesis was partially supported because this study discovered that FTF groups had significantly higher means than the CMC groups in terms of the three functions - *problem analysis*, *assessing positive consequences*, and *assessing negative consequences*. However, this study found that the FTF groups did not differ from the CMC groups regarding the alternative generating function.

CONCLUSION

This study has applied functional theory and media-capacity theory to compare FTF groups with CMC groups in terms of their group communication and group performance. The findings of this study in general support the functional theory and the media-capacity theories.

In need of further clarification are:

1. evidence that the CMC groups were much better than the FTF groups in terms of distinguishing which functions were important from those that were not;
2. that the CMC groups could then allocate their time for group discussion accordingly, and
3. the fact that the CMC groups' performance was better than the FTF groups' performance.

Cultural factors may have influenced these outcomes. The subjects of this study were Chinese students who belonged to a collectivistic culture. Collectivistic cultures have a tendency to be situation-centered. When people are in situation-centered cultures, they are more concerned about their relationships than about the tasks at hand, and thus prefer to use indirect communication to make sure that harmonious relationships are maintained and that people do not lose face. Therefore, a possible explanation for the above findings is that the members in the FTF condition paid too much attention to the ways in which they could maintain harmonious relationships among themselves and that, as a result, the FTF groups did not perform as well as the CMC groups (Kim et al., 2001; Law et al., 2000; Ma & Chuang, 2001; Tsui & Farh, 1997). Future studies should try to clarify this issue by comparing different cultural groups in terms of CMC and FTF communication.

The study used a winter survival task that belongs to one of the choosing tasks in McGrath's task complex for the field experiment. Compared to those tasks in real organizations, this task is too simplistic. It is possible that this task was not difficult enough, and thus CMC groups achieved better performance than their FTF counterparts. Future studies should use more difficult tasks such as decision-making tasks or negotiating tasks to clarify this issue. Furthermore, the researchers found that the subjects of the study knew that the class exercise on group decision making was not relevant to their class. However, the subjects felt this exercise interesting, and thus they reacted positively to this exercise.

REFERENCES

Adams, S. J., Roch, S. G., & Ayman, R. (2005). Communication medium and member familiarity: The effects on decision time, accuracy, and satisfaction. *Small Group Research*, *36*(3), 321–353. doi:10.1177/1046496405275232

Althaus, S. L. (1997). Computer-mediated communication in the university classroom: An experiment with on-line discussions. *Communication Education*, *46*(3), 158–174. doi:10.1080/03634529709379088

Baltes, B. B., Dickson, M. W., Sherman, M. P., Bauer, C. C., & LaGanke, J. S. (2002). Computer-mediated communication and group decision making: A meta-analysis. *Organizational Behavior and Human Decision Processes*, *87*, 156–179. doi:10.1006/obhd.2001.2961

Barkhi, R. (2005). Information exchange and induced cooperation in group decision support systems. *Communication Research*, *32*(5), 646–678. doi:10.1177/0093650205279352

Barkhi, R., Jacob, V. S., & Pirkul, H. (1999). An experimental analysis of face to face versus computer mediated communication channels. *Group Decision and Negotiation, 8*(4), 325–347. doi:10.1023/A:1008621423120

Becker-Beck, U., Wintermantel, M., & Borg, A. (2005). Principles of regulating interaction in teams practicing face-to-face communication versus teams practicing computer-mediated communication. *Small Group Research, 36*(4), 499–536. doi:10.1177/1046496405277182

Daft, R. L., & Lengel, R. H. (1986). Organizational information requirements, media richness and structural design. *Management Science, 32*(5), 554–571. doi:10.1287/mnsc.32.5.554

Daft, R. L., Lengel, R. H., & Trevino, L. K. (1987). Message equivocality, media selection, and manager performance: Implications for information systems. *Management Information Systems Quarterly, 11*(3), 355–366. doi:10.2307/248682

Fisher, B. A., & Ellis, D. G. (1990). *Small group decision making: Communication and the group process*. New York: McGraw-Hill.

Flanagin, A. J., Park, H. S., & Seibold, D. R. (2004). Group performance and collaborative technology: A longitudinal and multilevel analysis of information quality, contribution equity, and members' satisfaction in computer-mediated groups. *Communication Monographs, 71*(3), 352–372. doi:10.1080/0363452042000299902

Fulk, J., & Collins-Jarvis, L. (2001). Wired meeting: Technological mediation of organizational gatherings . In Jablin, F. M., & Putnam, L. L. (Eds.), *The new handbook of organizational communication: Advances in theory, research, and methods* (pp. 624–663). Thousand Oaks, CA: Sage Publications.

Fulk, J., Schmitz, J., & Steinfield, C. W. (1990). A social influence model of technology use . In Fulk, J., & Steinfield, C. W. (Eds.), *Organizations and communication technology* (pp. 117–140). Newbury Park, CA: Sage Publications.

Gouran, D. S., & Hirokawa, R. Y. (1996). Functional theory and communication in decision-making and problem-solving groups: An expanded view . In Hirokawa, R. Y., & Poole, M. S. (Eds.), *Communication and group decision making* (2nd ed., pp. 55–80). Thousand Oaks, CA: SAGE Publications.

Gouran, D. S., & Hirokawa, R. Y. (2003). Effective decision making and problem solving in groups: A functional perspective . In Hirokawa, R. Y., Cathcart, R. S., Samovar, L. A., & Henman, L. D. (Eds.), *Small group communication* (8th ed., pp. 27–38). Los Angeles, CA: Roxbury.

Guetzkow, H. (1950). Unitizing and categorizing problems in coding qualitative data. *Journal of Clinical Psychology, 6*(1), 47–58. doi:10.1002/1097-4679(195001)6:1<47::AID-JCLP2270060111>3.0.CO;2-I

Hancock, J. T., & Dunham, P. J. (2001). Impression formation in computer-mediated communication revisited: An analysis of the breadth and intensity of impressions. *Communication Research, 28*(3), 325–347. doi:10.1177/009365001028003004

Hayes, A. F., & Krippendorff, K. (2007). Answering the call for a standard reliability measure for coding data. *Communication Methods and Measures, 1*(1), 77–89.

Hirokawa, R. Y. (1987). Group communication research: Considerations for the use of interaction analysis . In Tardy, D. H. (Ed.), *A handbook for the study of human communication* (pp. 229–245). Norwood, NJ: Ablex.

Hirokawa, R. Y. (1990). The role of communication in group decision-making efficacy. *Small Group Research*, *21*(2), 190–204. doi:10.1177/1046496490212003

Hirokawa, R. Y. (1992). Communication and group decision-making efficacy . In Cathcart, R. S., & Samovar, L. A. (Eds.), *Small group communication* (6th ed., pp. 165–177). Dubuque, IA: William C. Brown.

Hirokawa, R. Y., & Salazar, A. J. (1999). Task-group communication and decision-making performance . In Frey, L. R., Gouran, D. S., & Poole, M. S. (Eds.), *The handbook of group communication theory and research* (pp. 167–191). Thousand Oaks, CA: Sage Publications, Inc.

Kim, M. S., Aune, K. S., Hunter, J. E., Kim, H. J., & Kim, J. S. (2001). The effect of culture and self-construals on predispositions toward verbal communication. *Human Communication Research*, *27*(3), 382–408. doi:10.1111/j.1468-2958.2001.tb00786.x

Krippendorff, K. (1980). *Content analysis: An introduction to its methodology*. Beverly Hills, CA: Sage Publications.

Kuhn, T., & Poole, M. S. (2000). Do conflict management styles affect group decision making?: Evidence from a longitudinal field study. *Human Communication Research*, *26*(4), 558–590. doi:10.1111/j.1468-2958.2000.tb00769.x

Law, K. S., Wong, Chi-Sun, Wang, D., & Wang, L. (2000). Effect of supervisor-subordinate guanxi on supervisory decisions in China: An empirical investigation. *International Journal of Human Resource Management*, *11*(4), 751–765. doi:10.1080/09585190050075105

Li, S. S. (2007). Computer-mediated communication and group decision making: A functional perspective. *Small Group Research*, *38*(5), 593–614. doi:10.1177/1046496407304335

Ma, R., & Chuang, R. (2001). Persuasion strategies of Chinese college students in interpersonal contexts. *The Southern Communication Journal*, *66*(4), 267–278.

McGrath, J. E. (1984). Groups: Interaction and performance. Englewood Cliffs, NJ: Prentice-Hall, INC.

Ocker, R. J., & Yaverbaum, G. J. (1999). Asynchronous computer-mediated communication versus face-to-face collaboration: Results on student learning, quality and satisfaction. *Group Decision and Negotiation*, *8*(5), 427–440. doi:10.1023/A:1008621827601

Orlitzky, M., & Hirokawa, R. Y. (2001). To err is human, to correct for it divine: A meta-analysis of research testing the functional theory of group decision-making effectiveness. *Small Group Research*, *32*(3), 313–341. doi:10.1177/104649640103200303

Putnam, L. L., & Fairhurst, G. T. (2001). Discourse analysis in organizations: Issues and concerns . In Jablin, F. M., & Putnam, L. L. (Eds.), *The new handbook of organizational communication: Advances in theory, research, and methods* (pp. 78–136). Thousand Oaks, CA: Sage Publications.

Smith, C. A. P., & Hayne, S. C. (1997). Decision making under time pressure: An investigation of decision speed and decision quality of computer-supported groups. *Management Communication Quarterly*, *11*(1), 97–126. doi:10.1177/0893318997111005

Straus, S. G. (1996). Getting a clue: The effects of communication media and information distribution on participation and performance in computer-mediated and face-to-face groups. *Small Group Research*, *27*(1), 115–142. doi:10.1177/1046496496271006

Tasa, K., & Whyte, G. (2005). Collective efficacy and vigilant problem solving in group decision making: A non-linear model. *Organizational Behavior and Human Decision Processes*, *96*(2), 119–129. doi:10.1016/j.obhdp.2005.01.002

Tsui, A. S., & Farh, J. L. L. (1997). Where guanxi matters: Relational demography and guanxi in the Chinese context. *Work and Occupations*, *24*(1), 56–79. doi:10.1177/0730888497024001005

Walther, J. B., Anderson, J. F., & Park, D. W. (1994). Interpersonal effects in computer-mediated interaction: A meta-analysis of social and antisocial communication. *Communication Research*, *21*(4), 460–487. doi:10.1177/009365094021004002

Walther, J. B., Slovacek, C., & Tidwell, L. C. (2001). Is a picture worth a thousand words?: Photographic images in long term and short term virtual teams. *Communication Research*, *28*, 105–134. doi:10.1177/009365001028001004

Wimmer, R. D., & Dominick, J. R. (2003). *Mass media research: An introduction* (7th ed.). Belmont, CA: Wadsworth Publishing Company.

ADDITIONAL READING

Blasio, P. D., & Milani, L. (2008). Computer-mediated communication and persuasion: Peripheral vs. central route to opinion shift. *Computers in Human Behavior*, *24*, 798–815. doi:10.1016/j.chb.2007.02.011

D'Urso, S. C., & Rains, S. A. (2008). Examining the scope of channel expansion: A test of channel expansion theory with new and traditional communication media. *Management Communication Quarterly*, *21*(4), 486–507. doi:10.1177/0893318907313712

Ebel, R. L. (1951). Estimation of the reliability of ratings. *Psychometrika*, *16*(4), 407–424. doi:10.1007/BF02288803

Esser, J. K. (1998). Alive and well after 25 years: A review of groupthink research. *Organizational Behavior and Human Decision Processes*, *73*(2-3), 116. doi:10.1006/obhd.1998.2758

Fuller, S. R., & Aldag, R. J. (1998). Organizational Tonypandy: Lessons from a quarter century of the groupthink phenomenon. *Organizational Behavior and Human Decision Processes*, *73*(2-3), 163. doi:10.1006/obhd.1998.2760

Gouran, D. S., & Hirokawa, R. Y. (1983). The role of communication in group decision making: A functional perspective . In Mander, M. S. (Ed.), *Communication in translation* (pp. 168–185). New York: Praeger.

Hart, P. T. (1998). Preventing groupthink revisited. *Organizational Behavior and Human Decision Processes*, *73*(2-3), 306. doi:10.1006/obhd.1998.2764

Hirokawa, R. Y. (2003). Communication and group decision-making efficacy . In Hirokawa, R. Y., Cathcart, R. S., Samovar, L. A., & Henman, L. D. (Eds.), *Small group communication* (8th ed., pp. 125–133). Los Angeles, CA: Roxbury.

Hirokawa, R. Y., Erbert, L., & Hurst, A. (1996). Communication and group decision making effectiveness . In Hirokawa, R. Y., & Poole, M. S. (Eds.), *Communication and group decision making* (2nd ed., pp. 269–300). Thousand Oaks, CA: SAGE Publications.

Hollingshead, A. B., Wittenbaum, G. M., Paulus, P. B., Hirokawa, R. Y., Ancona, D. G., & Peterson, R. S. (2005). A look at groups from the functional perspective . In Hollingshead, A. B. (Ed.), *Theories of small groups (CH2)*. Thousand Oaks, CA: Sage Publications, Inc.

Janis, I. L. (1989). *Crucial decisions*. New York: The Free Press.

Kramer, R. M. (1998). Revisiting the Bay of Pigs and Vietnam decisions 25 years later: How well has the groupthink hypothesis stood the test of time? *Organizational Behavior and Human Decision Processes*, *73*(2-3), 236. doi:10.1006/obhd.1998.2762

McCauley, C. (1998). Group dynamics in Janis's theory of groupthink: Backward and forward. *Organizational Behavior and Human Decision Processes*, *73*(2-3), 142. doi:10.1006/obhd.1998.2759

Moorhead, G., Neck, C. P., & West, M. S. (1998). The tendency toward defective decision making within self-managing teams: The relevance of groupthink for the 21st century. *Organizational Behavior and Human Decision Processes*, *73*(2-3), 327. doi:10.1006/obhd.1998.2765

Paulus, P. B. (1998). Developing consensus about groupthink after all these years. *Organizational Behavior and Human Decision Processes*, *73*(2-3), 362. doi:10.1006/obhd.1998.2767

Paulus, T. (2007). CMC modes for learning tasks at a distance. *Journal of Computer-Mediated Communication, 12*(4), article 9.

Pesendorfer, E. M., & Koeszegi, S. T. (2007). Social embeddedness in electronic negotiations. *Group Decision and Negotiation*, *16*(4), 399–415. doi:10.1007/s10726-006-9057-3

Peterson, R. S., Owens, P. D., Tetlock, P. E., Fan, E. T., & Martorana, P. (1998). Group dynamics in top management teams: Groupthink, vigilance, and alternative models of organizational failure and success. *Organizational Behavior and Human Decision Processes*, *73*(2-3), 272. doi:10.1006/obhd.1998.2763

Raven, B. H. (1998). Groupthink, Bay of Pigs, and Watergate reconsidered. *Organizational Behavior and Human Decision Processes*, *73*(2-3), 352. doi:10.1006/obhd.1998.2766

Roch, S. G., & Ayman, R. (2005). Group decision making and perceived decision success: The role of communication medium. *Group Dynamics*, *9*(1), 15–31. doi:10.1037/1089-2699.9.1.15

Salazar, A. J., Hirokawa, R. Y., Propp, K. M., Julian, K. M., & Leatham, G. B. (1994). In search of true causes: Examination of the effect of group potential and group interaction on decision performance. *Human Communication Research*, *20*(4), 529–559. doi:10.1111/j.1468-2958.1994.tb00334.x

Timmerman, C. E., & Scott, C. R. (2006). Virtually working: Communicative and structural predictors of media use and key outcomes in virtual work teams. *Communication Monographs*, *73*, 108–136. doi:10.1080/03637750500534396

Turner, M. E., & Pratkanis, A. R. (1998). A social identity maintenance model of groupthink. *Organizational Behavior and Human Decision Processes*, *73*(2-3), 210. doi:10.1006/obhd.1998.2757

Turner, M. E., & Pratkanis, A. R. (1998). Twenty-five years of groupthink theory and research: Lessons from the evaluation of a theory. *Organizational Behavior and Human Decision Processes*, *73*(2-3), 105. doi:10.1006/obhd.1998.2756

Walsh, J. P., & Maloney, N. G. (2007). Collaboration structure, communication media, and problems in scientific work teams. *Journal of Computer-Mediated Communication, 12*(2), article19.

Whyte, G. (1998). Recasting Janis's groupthink model: The key role of collective efficacy in decision fiascoes. *Organizational Behavior and Human Decision Processes*, *73*(2-3), 185. doi:10.1006/obhd.1998.2761

Yang, S. C., Hung, W. C., Sung, K., & Farn, C. K. (2006). Investigating initial trust toward e-tailers from the elaboration likelihood model perspective. *Psychology and Marketing*, *23*(5), 429. doi:10.1002/mar.20120

KEY TERMS AND DEFINITIONS

Communication Quality: Measured by the group-interaction performance of the five critical functions.

Computer-Mediated Communication: Any communication behaviors that occur through the use of two or more networked computers.

Functional Perspective: The functional perspective on group decision making proposes that a strong link exists between group interactions and their subsequent group decisions; and it identifies several critical task functions to be performed in the group interactions in order for a group to achieve high-quality decision making.

Group Communication: Any interaction in a group with more than two people being involved.

Group Performance: Group outcomes including objectives and perceived group outcomes.

Media-Richness Theory: This theory stresses that communication effectiveness can be achieved if managers match task ambiguity and the richness of a communication channel.

Social Influence Theory: Asserts that managers are not very accurate in differentiating one medium from another with regard to media richness, and that managers' perceptions of media richness are more often influenced by personal experiences and employees' attitudes than by the objective characteristics of communication channels.

ENDNOTE

1 Essential items (10 pts each): Lighter, Blankets, Shirt & Pants, Juice, Chocolate candy, Life raft, Knife, Flare gun, Teapot, Plastic trash bags. Useful items (5 pts each): Flashlight, Manila cord, First-aid kit, Wine, Asprin, Shaving kit, Water purification tablets, Crackers, Container of water, Coke. Useless items (0 pts each): Compass, Fire extinguisher, Radio, Newspapers, Air map, Pistol, Clock, Plastic utensils, Flight bags, Toilet paper.

Chapter 4
Information Sharing and Processing in Computer-Mediated Interactions

Jamonn Campbell
Shippensburg University, USA

ABSTRACT

This chapter will examine the effects of internet technologies on the collective information sampling (CIS) bias. The CIS bias is the tendency for task groups to favor the introduction, discussion, and use of shared or commonly known information during group discussion, rather than utilizing unshared information that is uniquely known by individual group members. The oversampling of shared information may lead groups to make suboptimal decisions when the content of their unshared information is critical for solving the task. The inherent structure and socially liberating effects of computer-mediated communication (CMC) may offer some solutions for reducing the CIS bias, and thereby increase the decision quality of electronic task groups.

INTRODUCTION

Groups are commonly called upon to make decisions, because collectively their individual members have access to more task relevant information than any one person. Based on this notion, groups should make better decisions than the average individual group member due to the variety of insights and information that each person brings to discussion. For instance, classic research by Shaw (1932) found that groups vastly outperformed individual participants across a series of complex tasks. Not only did groups make fewer mistakes during their tasks, they were better at error detection and self-correction than were individuals.

Typically decision-making groups have access to two types of task relevant information; items that are shared (common) are known to all group members and unshared (unique) items that are uniquely known by only one team member. Optimal group performance is based on the premise that individual members will in fact share their unique insights with the group, thus helping teams to make more informed decisions (Larson

DOI: 10.4018/978-1-60960-091-4.ch004

& Christensen, 1993; Steiner, 1972). For example, organizations will often delegate hiring decisions to committees instead of leaving the decision up to one individual. If committee members are provided with the opportunity to interact with job candidates one-on-one, then each member will learn unique items of information about each of the potential hires that are not known by the other team members. During group discussion, if the team members pool all of the information that they learned about the candidates during the interview process, both unique and common (e.g. resume items), then the group as a whole should be able to make a more informed hiring decision than any of the individual members.

Prior research has demonstrated that the above scenario is more likely to be the exception rather than the rule. Stasser and Titus (1985) discovered that groups were poor at exchanging unique information during group discussion, instead choosing to focus on shared information that was known to all group members. This biased information processing negates many of the potential benefits of group collaboration and decision-making. By limiting their discussion to information that is commonly held by all group members, discussants don't fully weight the merits of their options and by doing so, increase the likelihood that they will make a poor decision. This biased sampling process is especially damaging when the task relevant information is unequally distributed among the group members, as is the case with most group decision tasks.

The goal of this chapter is to examine the intra-group and environmental processes that contribute to the biased sampling of shared and unshared information during group discussions, as well as the impact of this unbalanced information exchange on decision-making. First, we will discuss factors that affect collective information sharing in traditional face-to-face (FTF) groups. Next, we will focus on how the structure and properties associated with computer-mediated communication (CMC) may serve to facilitate information sharing and potentially improve collective group performance. Lastly, we will examine several interactionist and integrative models that strive to explain some of the conflicting findings of past research in terms of information sharing.

Collective Information Sharing

Research on information sharing in groups has typically utilized a hidden profile paradigm to examine the processing of shared and unshared items during decision making. A hidden profile is created by systematically varying shared and unique information among group members so that the optimal choice is initially obfuscated during pre-discussion analysis. Returning to our hiring example, if a three-person committee discusses all of the shared and unique information about two potential job candidates and realizes that candidate X has seven positive traits and candidate Y has four positive traits, and they have the same number of equally weighted negative and neutral qualities, then X would be the optimal choice. However, if most of the positive information regarding candidate X is discovered during one-on-one interviews (i.e. unique), and most of the positive information regarding candidate Y is shared information (e.g. contained in his/her resume'), then this would result in a hidden profile. Candidate X's superior status will only be discovered if the group brings up all of their unshared knowledge during their discussion of the candidates.

Figure 1 illustrates the above scenario, excluding the neutral and negative items. A hidden profile is created initially if the committee notes four of candidate Y's positive traits but only one of candidate X's positive traits from their initial resumes. In other words, candidate Y looks very good on paper and candidate X needs to brush up on his/her resume' skills. If however, during the one-on-one interview process, each of the three committee members learns two more notable-

Figure 1. The distribution of shared and unshared information in a typical hidden profile

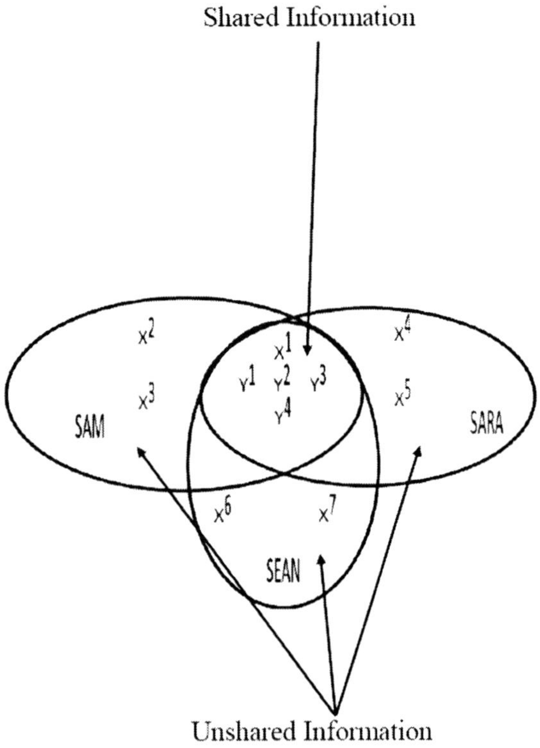

positive facts about candidate X that only they know (unique information), this would mean candidate X actually has seven notable positive traits compared to Y's four. Hence, candidate X is the superior choice, but his/her superior status is initially hidden from the group and can only be revealed through an exhaustive discussion of all the candidates' information.

The dilemma becomes more apparent when you consider that even after the one-on-one interviews, each group member individually knows about four positive traits regarding candidate Y and still only three of X's positive traits, one shared with the other members and two unique traits. So not only does candidate Y appears to be the best choice on paper, even after the interview process, each individual team member would still most likely evaluate candidate Y slightly higher than candidate X. However, if the group members effectively pool all of their shared AND unique information during group discussion it should become apparent that candidate Y is in fact the superior choice, with seven total positive attributes of note compared to candidate X's four. As stated previously, most teams will have a tendency to focus primarily on shared information during group discussion leading them to make a sub-optimal decision.

This tendency for groups to oversample shared information has been dubbed the collective information sampling bias (CIS) and appears to be a fairly robust and pervasive phenomenon affecting decision-making groups (Stasser & Titus 1985; Stasser & Titus, 2003; Wittenbaum & Stasser, 2000). Stasser and Titus (1985) initially demonstrated how biased information sampling could lead four-person groups to make less than optimal choices following discussion of a hidden-profile task regarding political candidates. The researchers found dramatic changes from pre-discussion to post-discussion preferences based on their distribution of shared and unshared information. In their control condition where all the information was shared, the percentage of participants who chose the optimal candidate increased from 67% to 87% following group discussion. However, when information was unequally distributed, utilizing a hidden profile paradigm, the percentage of group members who correctly identified the optimal candidate dropped from 25% to 20% following discussion.

One can see the paradoxical nature of hidden profiles. Although the correct/optimal solutions to these problems can only be uncovered via collective decision making, engaging in collaborative-information processing will typically lead groups to make suboptimal choices. Furthermore, even when unique information was mentioned by Stasser and Titus' (1985) groups, analysis of the discussion content data indicated that the unshared items did not have the same impact as shared information. Participants were much more likely to remember shared information when introduced by their group members over unshared items. The researchers suggest that this lack of overall impact combined with the fact that unique information had a lower probability of being introduced into discussion at all, are two factors that make hidden profiles so problematic.

Stasser and Titus' (1985) CIS model asserts that the probability that a group will discuss a piece of information p(D) is a function of the number of group members who share the information (n) and the likelihood that any of those members will mention the item p(M):

$$p(D) = 1 - [1 - p(M)]^n$$

(Stasser & Titus, 1985; Wittenbaum & Stasser, 2000, p. 6). According to the model, if group members mention roughly 25% of their task relevant information, then the probability of discussing any one shared piece of information is.58 = $1 - [1 - .25]^3$. Shared information would have close to a 60% chance of being discussed. Conversely, the probability of discussing unique items is significantly lower because only one person has access to the item of information. As predicted by the model, unique information would only have a 25% change of being brought up during collaboration ($.25 = 1 - [1 - .25]^1$) (Campbell & Stasser, 2006; Stasser & Titus, 1985, Wittenbaum & Stasser, 2000). Therefore, shared information has a systematic sampling advantage and is more likely to dominate group discussion and influence team members over unshared information (Stasser, Taylor, & Hanna, 1989).

Based on this mathematical premise, larger decision-making groups should be even more susceptible to the CIS bias than smaller teams (Stasser, Taylor, & Hanna, 1989). The disparity between discussion of an item of shared over unique information should exponentially increase as additional members are added to the group. The probability that an item of shared information will be brought up during group discussion increases as the size of the group increases. Conversely, group size does not affect the probability that unshared information will be discussed, because by definition only one person can hold and access this information. Consistent with the model's predictions, Stasser, Taylor and Hanna (1989) found that three and six-person groups working on a decision-making task mentioned on average 45% of their shared information compared to only 18% of their unshared items. Although six-person groups had longer discussions and introduced more overall information, this only served to increase the amount of shared information that was introduced, while unshared was unaffected. The overall sampling advantage was increased for the six-person teams, with these groups mentioning almost 60% of their shared items compared to only 19% of their unique information. The researchers also found that shared information was much more likely to be attended to and repeated once it was introduced into discussion, especially by the six-person teams.

Since this is simply a mathematical model of information processing, it does not take into account many of the cognitive and social factors that influence collective decisions, and therefore should only be viewed as a baseline predictor of group functioning and discussion. The CIS model is contingent upon the assumptions that both shared and unshared items of information are similar in impact, that group members have relatively equal status members, information loads, communication rates, recall independence, and that environmental cues do not favor one set of cues over another (Wittenbaum & Stasser, 2000).

Limits to Information Sampling

In order for groups to make effective decisions, whether working on hidden profiles or otherwise, members must effectively communicate both shared and unshared task relevant information

(Campbell & Stasser, 2006, Stasser & Titus, 1985; Stasser, Stewart, & Wittenbaum, 1995). While in most cases, groups are likely to discuss commonly held information with each other, there are a number of factors that hinder the introduction of unique information into the collective decision-making process. Stasser and Birchmeier (2003) contend that premature consensus seeking, biased sampling processes, and the enhanced credibility of common knowledge serve to limit the introduction and influence of unshared information.

There are a number of heuristic processes that bolster the notion that majority preference or group consensus are equated with correctness (Chaiken & Stangor, 1987; Stasser & Titus, 2003). Often times a simple majority vote at the beginning of group discussion will determine the focus and topic of discussion. In hidden profile situations, focusing on these pre-discussion preferences can lead groups to ignore unshared items and may prevent them from making optimal choices. Premature consensus seeking will also interfere with information exchange, which again is crucial for discovering hidden profiles. Since group consensus has positive implications from intra-group affect and satisfaction, group members who privately disagree with the group, or may have unique information that contradicts the majority may be unwilling to introduce these items for fear of disrupting a harmonious group. Groups who are able to rapidly achieve consensus feel more positive about their decision-making experience, their team members, and they are more confident in their decisions. Quickly achieving consensus on hidden profiles leaves groups feeling very satisfied and confident in their likely incorrect decisions.

Even when heuristic processes are overcome and groups do engage in information exchange, much of this sampling process still tends to favor group members' pre-discussion preferences, rather than viable alternatives. (Stasser & Birchmeier, 2003; Stasser & Titus 1985). Individuals tend to be overconfident in the initial judgments they make regarding a situation. In general, to revise one's belief is to admit that one has made a mistake and most people are generally uncomfortable engaging in self-doubt. Another factor that contributes to our overconfidence is how we engage in information processing. Individuals engage in a biased search process for information that confirms their initial beliefs while ignoring potentially disconfirming cues. This confirmatory bias leads individuals to minimize the impact or ignore completely items of information that call into question our initial assumptions (Swann & Read, 1981). Consistent with these confirmatory search processes, Gigone and Hastie's (1993) research on the common knowledge effect demonstrated how shared information that supports pre-discussion preferences tends to have more influence over group decisions, and similarly biases collaborative decision-making.

Furthermore, the aforementioned CIS model asserts that shared information simply has a higher probability of being mentioned by group members (Stasser & Titus, 1985). Because the probability of introducing an item of information is predicated on the number of group members who have access to this information, unshared information will be at a disadvantage. Consequently, the sampling disparity will be exacerbated as the size of the groups increase (Stasser, Taylor & Hanna, 1989).

Lastly, researchers contend that shared information is more compelling and influential when introduced into discussion and therefore does not suffer the same credibility concerns that plague unshared information (Stasser & Birchmeier, 2003). Experimental evidence indirectly demonstrates this enhanced credibility and impact as group members are more likely to correctly remember, repeat, and include shared information in post-discussion summaries over unshared items (Stasser, Taylor, & Hanna, 1989; Stasser & Stewart, 1995). Common information is viewed as more credible because it can be socially validated by other group members who also have access to this information (Wittenbaum & Bowman, 2004;

Wittenbaum & Park 2001). Unshared information on the other hand may be seen as less reliable because the group has to rely on the memory and comprehension of only one individual. Consequently, this information becomes less impactful during the decision-making process because it appears to be less valid, at least from a shared consensus viewpoint, than common knowledge.

SCOPE Model

Although the shared information bias is a pervasive phenomenon in decision-making groups, Stasser and Birchmeier (2003) suggest that specifically designing groups in order to achieve effective communication can serve to mitigate this bias and promote the introduction and effective use of unique information in collaborative decisions. The researchers identified five factors: social engineering, climate, opportunity, procedures, and expert roles (SCOPE), which can assist in promoting more effective information processing and decision making. Social engineering concerns the overall group composition and design of the interaction, while taking into account the members' areas of expertise. Climate involves fostering norms associated with critical discourse and analysis of information (e.g. priming critical analysis). Opportunity centers on the premise that groups must not only have access to unique information, but they also need sufficient time for that information to enter in and influence discussion. Procedural variations involve altering the ways in which groups engage in task related discussions and decision-making. Lastly, specifically assigning and/or making members aware of each others' areas of expertise can facilitate the emergence of unshared information (For a comprehensive review of the SCOPE model see Stasser & Birchmeier, 2003). Most relevant to this chapter are how changes in procedural behaviors (e.g. the use of technological enhancements) and opportunity (e.g. removing temporal constraints) can foster the introduction and use of unshared information in group decision-making.

Group Discussion in Computer-Mediated Communication

Over the past three decades, computers have dramatically transformed our personal, social, and professional lives. From an organizational perspective, computers and their related internet technologies have been hailed as a technological panacea for many problems that have affected worker productivity and decision-making. There are three main advantages that computer-mediated communication offers over traditional forms of organizational interaction: parallel communication, group memory, and anonymity (Dennis, 1996; Dennis, Hilmer, Taylor, & Polito, 1996).

In synchronous online communication (e.g. internet chat), discussants are able to simultaneously exchange information without blocking each other's communication efforts. This parallel ability to enter and exchange information many substantially increase the amount of information that is exchanged in computer-mediated groups. Face-to-face groups often suffer from "production blocking" effects due to the sequential nature of verbal communication. Group members must wait for the current speaker to finish before they can add their information to the discussion (Dennis, 1996; Dennis et al., 1996). While speakers are waiting their turn they are susceptible to any number of informational disruptions (e.g. disorganization and memory corrosion) that can block their communication efforts. Communicators may also be dissuaded from introducing information once they hear what their fellow teammates have to say.

Computer-mediated communication also has the ability to enhance collective memory in task groups. In electronic discourse, there is typically a running log of the entire group's conversation that each member can refer back to throughout the course of discussion. Group members are able to read and reread items of information at their own individual pace and preference. This enhanced group memory may mitigate many of the information processing problems that affect FTF groups. For example, FTF discussants may experience great difficulty trying to process previously mentioned information while simultaneously attending to newly introduced items (Dennis et al., 1996).

Lastly, the anonymity associated with electronic communication may reduce the influence of social context cues associated with power, status, and prestige that are found in face-to-face interactions, leading to more critical discourse. Due to the anonymous nature of the medium and the reduced influence of the aforementioned cues, low status members, minority opinion holders, and those who are uncomfortable expressing their views, may be more likely to participate in group discussions, leading to more equalized rates of participation and information sharing (Dennis, 1996; Kiesler, Siegel, & McGuire, 1984; Walther, 1996; Walther & Burgoon, 1992).

Although CMC has the potential to facilitate exchange, the findings regarding the overall effects of electronic communication on decision-making have been somewhat mixed. Research supporting a technologically deterministic view of the medium demonstrated that groups using computerized systems to make decisions tended to be more hostile (i.e. insults, derogatory language), took longer to reach consensus, and yielded more polarized decisions than comparable face to face groups (Kiesler et al., 1984; Kiesler & Sproull, 1992). Researchers contend that the increased hostility and conflict in CMC is an inherent property of electronic communication making it an ineffective medium for collaborative decision-making (Kiesler et al., 1984). Kiesler and colleagues (Kiesler et al., 1984; Kiesler & Sproull, 1992) proposed a cues-filtered out model to account for the deleterious effects of electronic communication on group performance. According to the researchers, rather than enhancing information exchange and processing, the absence of social context cues (e.g. nonverbal cues and environmental details) hinders the perception of an adaptation to social roles, structures, and group norms during CMC. It is this attenuation of social cues that deregulates behavior, decreases social inhibitions, and reduces attention for social concern and evaluation. Instead of facilitating task-relevant information exchange, electronic groups may become deregulated and less task-focused, resulting in poor decision-making.

Kiesler and Sproull (1992) reported that in general, computer-mediated task groups take between four and ten times longer to reach consensus, render more unconventional and considerably risky decisions, and have discussions that are more hostile and deregulated than comparable face-to-face groups. Conversely, face-to-face discussion groups make more elaboration attempts and exchange more overall information than CMC groups, which results in less extreme decisions.

Consistent with the cues-filtered out model, McGuire, Kiesler, & Siegel (1987) found that CMC groups failed to adhere to conventional decision-making norms during discussions of potentially risky ventures. Participants in CMC and face-to-face groups acting as fictional corporate managers were given the task of decided between two budget proposals. They were instructed to first make individual choices and then as a group to reach consensus concerning the budget proposals (one gain situation and one loss situation). According to prospect theory (Tversky & Kahneman, 1981), group decisions should be risk-averse for gain situations and risk seeking for loss situations, which accurately predicted the behaviors of the FTF groups. CMC groups on the other hand were less influenced by situational norms and typically defaulted to the first mentioned alternative. However, consistent with Tajfel and Turner's (1986) social identity theory, researchers have found that under conditions of enhanced group identification, anonymous CMC groups will adhere to established group norms (Reicher, Spears, & Postmes, 1995; Postmes, Spears, & Cihangir, 2001).

Information Sharing and Computer-Mediated Communication

Research looking specifically at the impact of CMC on information sharing and decision-making has similarly found that electronic groups generally tend to underperform their FTF counterparts on hidden-profile tasks (Baltes et al., 2002; Dennis, 1996; Campbell & Stasser, 2006; Hollingshead, McGrath, & O'Connor, 1993; Straus & McGrath, 1994). These studies typically examine information exchange, use, recall, and overall decision quality across comparable CMC and FTF groups. As predicted by the CIS bias (Stasser & Titus, 1985), both verbal and electronic teams tend to rely primarily on shared items of information, leading them to make suboptimal choices on the hidden profile tasks. As mentioned above, FTF groups will typically mention a substantially higher proportion of shared information compared to unshared items. However, this was not the case with many electronic groups. The enhanced anonymity and the ability to communicate in parallel, does seem to at times facilitate the exchange of task-relevant information in CMC groups. For example, Dennis (1996) found that groups using a group support system (GSS), a digital system that integrates communication and decision technologies (DeSanctis & Gallupe, 1987), introduced 50% more shared and unshared into discussion than did their FTF counterparts. However, CMC groups were still just as likely to make a suboptimal decision as the FTF groups who were relying upon even less information. Even though the CMC groups discussed more overall information, they still rendered suboptimal decisions.

There are a number of potential explanations for why CMC groups are unable to outperform FTF groups on decision-making tasks, even when they mention more information during discussion. One explanation may be credibility concerns regarding unique items of information, which similarly affects FTF groups. When shared information is introduced into discussions, it can be socially validated because more than one member has access to these items and can vouch for their accuracy and validity. Consequently, a process of mutual enhancement occurs whereby an individual's credibility is enhanced when they introduce socially valid information. Listeners who can validate this information also feel bolstered in their abilities because information that they hold has been introduced into the group's discussion by a co-member (Wittenbaum & Bowman, 2004; Wittenbaum, Hubbell, & Zuckerman, 1999; Wittenbaum & Park, 2001). This mutual enhancement process leads teams to further discuss and repeat shared information thereby strengthening to the overall CIS bias.

In an empirical demonstration of mutual enhancement, Wittenbaum et al. (1999) had dyads participate in a "discussion" regarding two fictional job candidates. The participants wrote down and exchanged 10 items of information about each candidate with each other that the authors secretly altered to reflect a shared or unshared information bias. Partners who received lists from their partners that contained primarily shared information rated their partners and themselves as more knowledgeable and competent than those who received lists containing primarily unshared items of information.

So although CMC groups may at times exchange more information than FTF groups, the increase sampling of unique times may actually harm the communicator's credibility. If an individual mentions a large amount of unshared information, the other group members may rate this person's competence lower than someone who primarily mentions shared information. While this unique information is critical for uncovering a hidden profile, it is unable to be socially validated and thus lessens the perceived competence and knowledge of the communicators, thereby minimizing its impact.

Increased cognitive constraints are another explanation for why CMC groups fail to use unshared information even when it is introduced in discussion. Computer-mediated teams may suffer from information overload especially when dealing with temporal pressures. CMC groups may be at a disadvantage because electronic communications are more cognitively and physically laborious than FTF interactions. For example, non-verbal information such as head-nods, winks, and smiles are often omitted from electronic interactions because they take more time and effort to transmit and process compared to verbal interactions. Without these social context cues, CMC groups often find it difficult to assess the emotional and psychological states of their group members, leading to increased levels of miscommunication, frustration, and cognitive effort (Graetz et al., 1998; Kruger et al., 2005). Due to the features of parallel communication, CMC users also receive more information at a faster rate than they are typically accustom to and may experience difficulty attending to and integrating all the items (Dennis, 1996). The speed of FTF discussion is naturally regulated by sequential turn taking and paralinguistic cues (e.g. a bewildered expression) that is absent in electronic communication. This information onslaught may prevent electronic communicators from utilizing critical items when they are introduced into discussion. Excess information may also limit the utility of enhanced group memory during CMC. Although group members have the ability to examine previous discussion items from the computer logs, the sheer amount of information combined with increased temporal demands may limit their ability to do so (Dennis, 1996). Gratez et al. (1998) reported that computer-mediated groups had significantly higher levels of cognitive load than did comparable face-to-face or teleconference groups while working on a performance task. Computer-mediated groups reported experiencing more temporal and mental demands, exerting increased effort, and more frustration than did groups in either a teleconference or face-to-face condition. The authors argue that even though discussion strategies were similar across groups, coordinating and validating information becomes a decidedly more arduous tack in CMC, which has detrimental effects on groups' decision-making processes.

Since problem-solving during CMC is still a relatively new experience for most individuals, individuals may struggle to adapt to the unfamiliar constraints of the system the first few times they work with the medium. Theorists have emphasized that many of the inconsistent findings across CMC performance research stem from the fact that the majority of studies utilize ad-hoc groups that are unfamiliar with each other and the group support systems (Mennecke & Valacich, 1998). Reliance on ad-hoc "one shot" groups has traditionally been an area of marked criticism in the field of group research (Moreland, Hogg, & Hains, 1994). Dennis (1996) suggests that the unfamiliar way in which GSS systems organize information may present difficulties for novice users in terms of integrating task relevant information. Participants in his study, using the GroupSystems Electronic Meeting System program as a mode of communication, had to open new windows and search through paragraphs of information when considering each of the four decision alternatives. Although the GSS was useful in promoting information exchanges, the way in which the program separated the discussion logs may have made it too cognitively difficult for novice users to effectively integrate the information when formulating their decisions.

Similarly, Hollingshead, McGrath, and O'Connor (1993) asserted that any change experienced by groups in terms of their membership, technology, or task will dramatically impact task performance. Whereas functioning work groups are primarily engaged the processes of goal negotiation or task execution, teams that are dealing with change must shift their attentional focus away from task related behaviors to areas of conflict resolution and the negotiation of technical issues. Furthermore, switching from FTF to CMC may be even more disruptive because electronic discourse is a less familiar mode of task communication, as it strips away previously relied upon social cues.

In a 13-week longitudinal study examining the impact of change on FTF and computer-mediated groups, Hollingshead et al. (1993) found that FTF groups outperformed their electronic counterparts in terms of task performance and satisfaction. Initially the participants in the CMC condition were unfamiliar with their electronic environments. This initial "change" from their normative verbal mode of communication disrupted their decision-making processes and they spent more time dealing with conflict and technological issues than the FTF groups.

However, these findings were reversed when the researchers swapped the groups' communication modes for two weeks. During weeks seven and eight, former CMC groups discussed problems in a face-to-face setting and previous FTF groups held computer-mediated discussions. Across the two weeks, the former CMC groups now outperformed the form FTF groups on both an intellective and a negotiation task. The change in communication media disrupted the communications patterns of the former FTF groups causing them to spend more time dealing with conflict and technological issues. The media switch was relatively easy for the former CMC groups as they reverted back to their traditional verbal mode of communication.

During week nine of the study, when groups were placed back into their initially assigned mode of communication, the two groups performed equally well on an intellective task. In this case, groups in both conditions were now working in communication modes that they were relatively familiar with and therefore could spend most of their time on task related actions. Once the CMC groups became familiar with their electronic work environment they were readily able to take advantage of their electronic environment and make decisions that were equivalent to their FTF counterparts (Hollingshead et al., 1993).

Theoretical Models of Electronic Decision-Making

Over the years, a number of theoretical models have emerged in an attempt to explain the mixed findings regarding information sharing and electronic decision-making. Much like the SCOPE approach (Stasser & Birchmeier, 2003), these theoretical perspectives focus on socially engineering situational, procedural, and opportunity factors to promote the sampling and use of shared and unshared information in electronic groups.

The media richness theories (Daft & Lengel, 1986; Hollingshead et al., 1993) focus on the interaction between task type and medium, in suggesting that certain types of group processing are more suited for information rich media (e.g. FTF, telecommunication), whereas others are more suited for information lean media (e.g. electronic chat). CMC is considered lean media because it doesn't contain the breadth of paralinguistic cues (nonverbal behaviors, vocal tone, pitch, pauses, inflections, etc.) that are found in information rich media. According to this paradigm, tasks that strictly rely on the generation of ideas or the production of information (e.g. additive tasks, brainstorming, etc.) may be more suited to lean types of media. Supporting this notion, Valacich, Dennis, and Connolly (1994) found that CMC groups outperformed nominal groups in idea production and quality while working on brainstorming tasks, whereas the opposite effect is typically found in FTF comparisons (Paulus & Brown, 2007). On the other hand, information rich media may be more appropriate for situations that require a great deal of information integrations such as negotiations, conflict resolution, and judgment tasks (Straus & McGrath, 1994; Munzer & Holmer, 2009). Social context cues become more important to the decision-making process during judgment tasks and therefore a more robust medium is more suited for these pursuits. In support of this interactionist model, Straus and

McGrath (1994) reported that FTF groups were more effective than computer-mediated groups while working on a judgment task, but these differences disappeared when the groups were working on idea generation and intellective-type problems.

Additionally, researchers have suggested that temporal factors may also interact with task type and media in moderating collective decision-making during CMC (Campbell & Stasser, 2006; Straus & McGrath, 1994; Walther, 1996; Walther & Burgoon, 1992). Studies have indicated that computer-mediated groups working on decision-making tasks take longer to reach consensus and reach agreement less often than comparable face-to-face groups. (Kiesler et al., 1984; Kiesler & Sproull, 1992; Straus & McGrath, 1994). However, since CMC is an inherently slower form of communication that FTF interaction, it is possible that many of the interaction process and performance problems experienced by computer-mediated groups may be attributed to situational time constraints rather than determined by the limitations of the technology (Campbell & Stasser, 2006). If computer-mediated groups are provided with sufficient time to interact with their teammates, it is possible that they will be able to more thoroughly process and synthesize the information that is sampled during discussion, thus enabling them to make more optimal decisions (Straus & McGath, 1994).

Walther's Social Information Processing (SIP) theory contends that compared to FTF teams, computer-mediated groups typically operate under increased temporal demands leading them to focus primarily on task relevant behaviors (Walther, 1993; Walther & Burgoon, 1992). This increased task focus decreases their attention socio-emotional and paralinguistic content and may reduce group cohesion, trust, and consensus-seeking efforts that are also crucial when working on decision-making tasks. Consistent with the SIP model, a meta-analysis of 27 electronic decision-making studies by Baltes et al. (2002) indicated that time was a significant moderator of decision effectiveness in computer-mediated groups. The analysis indicated that there were no differences in decision quality between FTF and computer-mediated groups who were given unlimited time to complete their tasks. On the other hand, when groups were temporally constrained, FTF teams made significantly better decisions than their electronic counterparts.

Across dozens of experiments covering more than two decades of research there are very few studies that report giving computer-mediated groups unlimited, or the perception of unlimited time, to complete their tasks (Baltes et al; 2002: Campbell & Stasser, 2006; Dubrovsky et al., 1991; Weisband, 1992). Providing electronic groups with ample time to complete their tasks should alleviated cognitive constraints and allow them to more effectively utilize the properties of the medium to facilitation decision-making. Even though researchers typically provide groups with what they considered sufficient time to reach a decision (Shirani, 2006), imposing any time constraints may still result in electronic group members experiencing elevated temporal demands over their FTF counterparts, which can limit their cognitive resources and impair their overall problem-solving abilities (Graetz et al., 1998).

Reducing temporal constraints during CMC may alleviate some of the aforementioned discussion problems (credibility, cognitive demands, and unfamiliarity with the medium) associated with poorer decision-making in electronic groups. The SIP model predicts that over time, electronic groups will develop a more sophisticated use of paralinguistic cues leading to experience more cohesion and trust among their fellow team members (Walther, 1993). This increased comfort and trust may lead group members to not only introduce more unique information into discussion, but also to give it more credence than they would have otherwise (Wittenbaum & Bowman, 2003; Wittenbaum & Park 2001; Wittenbaum & Stasser 2000) Similarly, if temporal constraints are reduced in CMC, this may alleviate some of the increased cognitive demands and information overload that characterizes electronic decision-making (Graetz et al., 1998). With temporal concerns removed, electronic discussants may be able to reassign their cognitive efforts and increase their attention to information processing and integration. Having more time to engage in electronic discussions may also provide group members with the opportunity to familiarize themselves with the characteristics of the digital medium. Over time, as group members accommodate to their electronic environment, they may be able to better utilize the enhanced collective processing tools in the GSS (Dennis, 1996).

On the other hand, alleviating temporal constraints may not be as effective in increasing the decision quality for groups working on hidden profiles, due to the peculiar nature of these problems. Although the CIS bias becomes less prevalent over time (Larson & Christensen, 1993), simply having more time to introduce information into group discussion does not mean that the group will validate or even consider these additional items (Stasser & Stewart, 1992; Stasser, Taylor & Hanna, 1989). Providing a group with extended time to discuss a problem may enable them to sample more unshared information, but the team will also typically increase their mentioning of shared information as well. Therefore, longer discussions do not necessarily guarantee higher quality decisions because of group members' overreliance on shared information (Mennecke, 1997; Stasser & Stewart, 1992; Wittenbaum & Stasser, 2000).

In order for increased discussion time to influence decision quality, group focus must be directed away from shared information, towards unshared items. One way to accomplish this is by altering the way in which group members view their tasks. Group members who perceive a task as intellective (i.e. having a demonstrably correct solution) as oppose to a judgment situation, have a tendency to limit premature consensus seeking and focus more on information acquisition and validation strategies, including attending to unshared cues (Laughlin & Ellis, 1986; Stasser & Stewart, 1992).

In line with this theory of task and time interaction, Campbell and Stasser (2006) found that computer-mediated groups, who perceived a task as intellective and were given ample time, were twice as likely to solve a hidden profile task compared to restricted time and FTF groups working on either judgment or intellective problems. Even though CMC groups working with a judgment set of instructions took just as long as the intellective groups to reach decisions, and sample just as much information, they were ineffective at solving the hidden profile problem. The authors contend that once the electronic groups perceived the task as having a demonstrably correct solution, they shifted their focus towards unshared information. However, they were only able to process this information when they were provided with ample time. This increased attention is evidenced by the ample-time CMC groups', who perceived they were working on a solvable task, tendency to repeat and recall significantly more items of unique information than the other groups. The authors suggest that this repetition and recall is indicative of the groups' shifted attention to unique information, which was facilitated by their reduced temporal concerns. Even though the same amount of information was present in the restricted-time and judgment discussions, these individuals were unwilling (i.e. mutual validation concerns) and/or unable (i.e. cognitive constraints) to process the unshared information, leading them to make worse decisions.

Integrative Approaches to Information Sharing

Although the temporal (Walther, 1993; Walther & Burgoon, 1992) and media richness perspectives (Daft & Lengel, 1986; Hollingshead et al., 1993) present a foundation for the study of information sharing and processing in electronic groups, these models limit themselves by focusing on one dimensional comparisons between computer-mediated and FTF groups, instead of a more functional integrative model. The aforementioned theories also tend to focus on too broad a spectrum by ignoring many of the differing intra-discussion processes that characterize group decision-making (Munzer & Holmer, 2009). Not only do the task and medium interact to influence performance, but this interaction may be modified at each step of the decision-making process, affecting more than the overall outcome or success of the group. Instead of focusing primarily on outcome variables (correct vs. incorrect) and simple medium comparisons (CMC vs. FTF), researchers should also strive to identify the numerous sub-processes which may differentially impact the relationship between task and medium (Munzer & Holmer, 2009).

For instance, Munzer and Holmer (2009) examined information sampling in CMC while manipulating group members' abilities to communicate in parallel, receive immediate feedback, and their storage capacity across four different types of electronic interactions. Although the groups were uniformly poor in their efforts at exchanging information and uncovering the hidden profile, there were a number of interesting sub-process interactions for each condition that differentially impacted information sharing. For instance, groups that utilized parallel communication but did not receive immediate feedback mentioned more unique information than those communicating over typical synchronous chat. On the other hand, groups that received feedback when communicating in an asynchronous fashion had more coherent discussions that the typical synchronous chat groups. It may be that a more thorough examination of these sub-process throughout the decision making process may yield further insight into the information sampling bias.

Additionally, an integrative media approach that utilize both information rich and information lean media during the decision making process may offer alternative ways to enhance information sampling. Attenuating the collective information sampling bias presents a peculiar problem because on one hand in order to solve a hidden profile a group must sufficiently sample the available pool of shared and unshared information. If this is that case, then an information lean media such as synchronous chat would be appropriate according to media richness theories (Daft & Lengel, 1986; Hollingshead et al., 1993). However, once the group has introduced the information into discussion, the task of the group shift from production to evaluation. In order to solve a hidden profile the group must process, analyze, and weigh the merits of the information that is brought up during discussion. In this case an information rich media would become a more effective mode of communication (Daft & Lengel, 1986; Hollingshead et al., 1993).

Instead of contrasting the effectiveness of CMC and FTF communication modes, perhaps another alternative would be to use a combination of information lean and rich media when working on problems solving tasks. An integrated media approach could enhance the effectiveness of group discussion above and beyond the use of either mode by itself. Depending on the type of task, groups may want to utilize a type of computer-mediated interaction (synchronous, asynchronous, etc.) during the initial stages of discussion where the purpose is to introduce as much task relevant information as possible, and then shift to either a face-to-face or another alternate mode of electronic discourse when the need for analysis and integration arises. Researchers in education have effectively used this integrated methodology to increase participation and quality of face-to-face classroom discussions following computer-mediated discussions (Dietz-Uhler & Bishop-Clark, 2001).

CONCLUSION

Research on collective information sampling has shown that task groups are affected by a systematic bias favoring the introduction and use of shared over unshared information during group discussion (Stasser & Titus 1985; Stasser & Titus, 2003; Wittenbaum & Stasser, 2000). When unshared information is critical for making optimal decisions, oversampling shared information ensures that groups will make poor decisions. Consistent with the SCOPE model, changes in procedural behaviors (e.g. CMC) and opportunity (e.g. reduced temporal constraints) can minimize the CIS bias and improve group decision quality. The anonymity, parallel communication, and group memory inherent in many forms of computer-mediated communication have the potential to assist groups in overcoming this collective information sampling bias, thus helping groups make more informed and higher quality decisions (Dennis, 1996; Dennis et al., 1996). Researchers have reported that under certain conditions, using electronic decision-making can lead to increased sampling (Dennis, 1996) and use (Campbell & Stasser, 2006) of unshared information resulting in improved group performance.

However, researchers' tendency to rely on "one-shot" ad-hoc groups ignores the powerful role that time plays in information sampling and decision-making (McGrath & Tschan, 2004; Moreland et al., 1994). Theorists have posited that the relative unfamiliarity that individuals have with electronic group work may lead to them to experience increased cognitive and temporal demands (Dennis, 1996; Mennecke & Valacich, 1998). While it appears as if temporal factors play a crucial role in decision-making success, especially in electronic groups, (Baltes, et al; 2002; Campbell & Stasser, 2006; Walther, 1993), there are very few adequate models to account for the interaction of time in the collaborative process.

The problem that remains is identifying the specific conditions where CMC will be appropriate and effective in the group decision- making process. Whereas the basic structure of fact-to-face interaction remains stable in most decision-making paradigms, there is considerable variation to the type of computer-mediated interaction that groups can engage in. Media synchronicity theory describes five dimensions that electronic communication can vary across (Hollingshead et al., 1993; Munzer & Holmer, 2009). These computer-mediated interactions can differ in the type of information that they transmit, parallel communication (e.g. synchronous or asynchronous), immediacy of feedback, rehearsability (e.g. ability to edit messages), and reprocessability (e.g. ability to review saved information). Additionally, there are a number of specialized group support systems (GSS) which can include virtual file folders, voting procedures, and the like which can interact with the aforementioned communication dimensions. Due to this increased variability, our attention may want to shift from focusing on oversimplified outcome variables (e.g. correct vs. incorrect solutions), to an examination of the multitude of sub-process interactions that occur during the electronic decision-making process.

For instance, an integrated media approach may be better suited to clearing up some of the past inconsistencies found computer-mediated decision-making research. It is plausible that outside the laboratory, neither FTF nor CMC are optimally suited on their own for most complex decision-making tasks. Using both mediums in tandem during the decision-making process may serve to increase information sampling, reduce the CIS bias and improve overall group performance. An integrated approach argues for an examination of the various dimensions and sub-processes that occur within the decision-making chain, in hopes of finding the most effective order and combination of mediums to enhance group performance (Munzer & Holmer, 2009).

REFERENCES

Baltes, B. B., Dickson, M. W., Sherman, M. P., Bauer, C. C., & LaGanke, J. S. (2002). Computer-mediated communication and group decision making: A meta-analysis. *Organizational Behavior and Human Decision Processes*, *87*, 156–179. doi:10.1006/obhd.2001.2961

Behavior and Organization Design (no date). *Research in Organizational Behavior, 6*, 191-233.

Campbell, J., & Stasser, G. (2006). The influence of time and task demonstrability on decision-making in computer-mediated and face-to-face groups. *Small Group Research*, *37*, 271–294. doi:10.1177/1046496406288976

Chaiken, S., & Stangor, C. (1987). Attitudes and attitude change. *Annual Review of Psychology*, *38*, 575–630. doi:10.1146/annurev.ps.38.020187.003043

Daft, R. L., & Lengel, R. H. (1984). Organizational information requirements, media richness and structural design. *Management Science*, *32*, 554–571. doi:10.1287/mnsc.32.5.554

De Sanctis, G. L., & Gallupe, R. B. (1987). A foundation for the study of group decision support systems. *Management Science*, *33*, 589–609. doi:10.1287/mnsc.33.5.589

Dennis, A. R. (1996). Information exchange and use in small group decision making: You can lead a group to information, but you can't make it think. *Small Group Research*, *27*, 532–550. doi:10.1177/1046496496274003

Dennis, A. R., Hilmer, K. M., Taylor, N. J., & Polito, A. (1996). Information exchange and use in GSS and verbal group decision making: Effects of minority influence. *Management Information Systems Quarterly*, *20*, 433–457. doi:10.2307/249563

Dietz-Uhler, B., & Bishop-Clark, C. (2001). The use of computer-mediated communication to enhance subsequent face-to-face discussions. *Computers in Human Behavior*, *17*, 269–283. doi:10.1016/S0747-5632(01)00006-1

Dubrovsky, V. J., Kiesler, S., & Sethna, B. N. (1991). The equalization phenomenon: Status effects in computer-mediated and face-to-face decision-making groups. *Human-Computer Interaction*, *6*, 119–146. doi:10.1207/s15327051hci0602_2

Gigone, D., & Hastie, R. (1993). The common knowledge effect: Information sharing and group judgment. *Journal of Personality and Social Psychology*, *65*, 959–974. doi:10.1037/0022-3514.65.5.959

Graetz, K. A., Boyle, E. S., Kimble, C. E., Thompson, P., & Garloch, J. L. (1998). Information sharing in face-to-face, teleconferencing, and electronic chat groups. *Small Group Research*, *29*, 714–743. doi:10.1177/1046496498296003

Hollingshead, A. B., Mcgrath, J. E., & O'Connor, K. (1993). Group task performance and communication technology: A longitudinal study of computer-mediated versus face-to-face work groups. *Small Group Research*, *24*, 307–333. doi:10.1177/1046496493243003

Kiesler, S., Siegel, J., & McGuire, T. (1984). Social psychological aspects of computer-mediated communication. *The American Psychologist*, *39*, 1123–1134. doi:10.1037/0003-066X.39.10.1123

Kiesler, S., & Sproull, L. (1992). Group decision making and communication technology. *Organizational Behavior and Human Decision Processes*, *52*, 96–123. doi:10.1016/0749-5978(92)90047-B

Kruger, J., Epley, N., Parker, J., & Ng, Z. (2005). Egocentrism over e-mail: Can we communicate as well as we think? *Journal of Personality and Social Psychology*, *89*, 925–936. doi:10.1037/0022-3514.89.6.925

Larson, J. R. Jr, & Christensen, C. (1993). Groups as problem-solving units: Toward a meaning of social cognition. *The British Journal of Social Psychology*, *32*, 5–30.

Laughlin, P. R., & Ellis, A. L. (1986). Demonstrability and social combination processes on mathematical intellective tasks. *Journal of Experimental Social Psychology*, *22*, 177–189. doi:10.1016/0022-1031(86)90022-3

McGrath, J. E., & Tschan, F. (2004). *Temporal matters in social psychology: Examining the role of time in the lives of groups and individuals*. Washington, DC: APA Publications. doi:10.1037/10659-000

McGuire, T. W., Kiesler, S., & Siegel, J. (1987). Group and computer-mediated discussion effects in risk decision-making. *Journal of Personality and Social Psychology*, *52*, 917–930. doi:10.1037/0022-3514.52.5.917

Mennecke, B. E. (1997). Using group support systems to discover hidden profiles: An examination of the influence of group size and meeting structures on information sharing and decision quality. *International Journal of Human-Computer Studies*, *47*, 387–405. doi:10.1006/ijhc.1997.0136

Mennecke, B. E., & Valacich, J. S. (1998). Information is what you make of it: The influence of group history and computer support on information sharing, decision quality, and member perceptions. *Journal of Management Information Systems*, *15*, 173–197.

Moreland, R. L., Hogg, M. A., & Hains, S. C. (1994). Back to the future: Social psychological research on groups. *Journal of Experimental Social Psychology*, *30*, 527–555. doi:10.1006/jesp.1994.1025

Munzer, S., & Holmer, T. (2009). Bridging the gap between media synchronicity and task performance. *Communication Research*, *36*, 76–103. doi:10.1177/0093650208326464

Paulus, P. B., & Brown, V. R. (2007). Toward more creative and innovative group idea generation: A cognitive-social-motivational perspective of brainstorming. *Social and Personality Psychology Compass*, *1*, 248–265. doi:10.1111/j.1751-9004.2007.00006.x

Postmes, T., Spears, R., & Cihangir, S. (2001). Quality of decision making and group norms. *Journal of Personality and Social Psychology*, *80*, 918–930. doi:10.1037/0022-3514.80.6.918

Reicher, S. D., Spears, R., & Postmes, T. (1995). A social identity model of deindividuation phenomena. *European Review of Social Psychology*, *6*, 161–197. doi:10.1080/14792779443000049

Shaw, M. E. (1932). A comparison of individuals and small groups in the rational solution of complex problems. *The American Journal of Psychology*, *44*, 491–504. doi:10.2307/1415351

Shirani, A. I. (2006). Sampling and pooling of decision-relevant information: Comparing the efficiency of face-to-face and GSS supported groups. *Information & Management*, *43*, 521–529. doi:10.1016/j.im.2006.01.001

Stasser, G., & Birchmeier, Z. (2003). Group creativity and collective choice . In Paulus, P., & Nijstad, B. (Eds.), *Group Creativity* (pp. 85–109). New York: Oxford University Press. doi:10.1093/acprof:oso/9780195147308.003.0005

Stasser, G., & Stewart, D. (1992). Discovery of hidden profiles by decision-making groups: Solving a problem versus making a judgment. *Journal of Personality and Social Psychology*, *63*, 426–434. doi:10.1037/0022-3514.63.3.426

Stasser, G., Stewart, D., & Wittenbaum, G. M. (1995). Expert roles and information exchange during discussion: The importance of knowing who knows what. *Journal of Experimental Social Psychology*, *31*, 244–265. doi:10.1006/jesp.1995.1012

Stasser, G., Taylor, L. A., & Hanna, C. (1989). Information sampling in structured and unstructured discussion of three- and six-person groups. *Journal of Personality and Social Psychology*, *57*, 67–78. doi:10.1037/0022-3514.57.1.67

Stasser, G., & Titus, W. (1985). Pooling of unshared information in group decision making: Biased information sampling during discussion. *Journal of Personality and Social Psychology*, *48*, 1467–1478. doi:10.1037/0022-3514.48.6.1467

Stasser, G., & Titus, W. (2003). Hidden profiles: A brief history. *Psychological Inquiry*, *14*, 304–313. doi:10.1207/S15327965PLI1403&4_21

Steiner, I. D. (1972). *Group process and productivity*. San Diego, CA: Academic Press.

Straus, S. G. (1996). Getting a clue: The effects of communication media and information distribution on participation and performance in computer-mediated and face-to-face groups. *Small Group Research*, *27*, 115–142. doi:10.1177/1046496496271006

Straus, S. G., & McGrath, J. E. (1994). Does the medium matter?: The interaction of task type and technology on group performance and member reactions. *The Journal of Applied Psychology*, *79*, 87–97. doi:10.1037/0021-9010.79.1.87

Swann, W. B. Jr, & Read, S. J. (1981). Acquiring self-knowledge: The search for feedback that fits. *Journal of Personality and Social Psychology*, *85*, 1095–1106. doi:10.1037/0022-3514.85.6.1095

Tajfel, H., & Turner, J. C. (1986). The social identity theory of inter-group behavior. In Worchel, S., & Austin, L. W. (Eds.), *Psychology of Intergroup Relations*. Chicago: Nelson-Ha.

Tversky, A., & Kahneman, D. (1981). The framing of decisions and the psychology of choice. *Science*, *21*, 453–458. doi:10.1126/science.7455683

Valacich, J. S., Dennis, A. R., & Connolly, T. (1994). Group versus individual brainstorming: A new ending to an old story. *Organizational Behavior and Human Decision Processes*, *57*, 448–467. doi:10.1006/obhd.1994.1024

Walther, J. B. (1993). Impression development in computer-mediated interaction. *Western Journal of Communication*, *57*, 381–398.

Walther, J. B. (1996). Computer-mediated communication: Impersonal, interpersonal, and hyperpersonal interaction. *Communication Research*, *23*, 3–43. doi:10.1177/009365096023001001

Walther, J. B., Anderson, J. F., & Park, D. W. (1994). Interpersonal effects in computer-mediated interaction. *Communication Research*, *21*, 460–487. doi:10.1177/009365094021004002

Walther, J. B., & Burgoon, J. K. (1992). Relational communication in computer-mediated interaction. *Human Communication Research*, *19*, 50–88. doi:10.1111/j.1468-2958.1992.tb00295.x

Weisband, S. P. (1992). Group discussion and first advocacy effects in computer-mediated and face-to-face decision making groups. *Organizational Behavior and Human Decision Processes*, *53*, 352–380. doi:10.1016/0749-5978(92)90070-N

Wittenbaum, G. M., & Bowman, J. M. (2004). A social validation explanation for mutual enhancement. *Journal of Experimental Social Psychology*, *40*, 169–184. doi:10.1016/S0022-1031(03)00091-X

Wittenbaum, G. M., Hubbell, A. P., & Zuckerman, C. (1999). Mutual enhancement: Toward an understanding of the collective preference for shared information. *Journal of Personality and Social Psychology*, *77*, 967–978. doi:10.1037/0022-3514.77.5.967

Wittenbaum, G. M., & Park, E. S. (2001). The collective preference for shared information. *Current Directions in Psychological Science*, *10*, 70–73. doi:10.1111/1467-8721.00118

Wittenbaum, G. M., & Stasser, G. (2000). Management of information in small groups . In Nye, J. L., & Brower, A. M. (Eds.), *What's social about social cognition?* (pp. 1–28). London: Sage.

KEY TERMS AND DEFINITIONS

Collective Information Sampling Bias (CIS): The tendency for group members to introduce and rely on shared rather than unshared information during group discussion.

Group Support Systems (GSS): An electronic communication system used to facilitate group performance.

Hidden Profile: A research paradigm where shared and unshared items of information are systematically distributed to group members in a way which initially obscures the optimal solution.

Mutual Enhancement: A tendency to evaluate group members and oneself more favorably when discussing shared information.

Shared Information: Task relevant information that is known by all group members.

Unshared Information: Task relevant information that is known by only one group member.

Chapter 5
Communities of Practice in Knowledge Management and Organisational Learning

Celina M. Olszak
University of Economics in Katowice, Poland

Ewa Ziemba
University of Economics in Katowice, Poland

ABSTRACT

It is believed that communities of practice (CoPs) more and more frequently become a critical success factor for contemporary organisations. This article, following the network organisation paradigm, identifies the most important properties of CoPs, presents CoPs capacities in the socialisation, externalisation, combination, internalisation (SECI) knowledge spiral model, and suggests information technologies that support the work of CoPs. Theoretical considerations are contrasted against empirical research results obtained in 2008 while examining 160 organisations in Upper Silesia, the most industrialised and densely populated region in Poland.

INTRODUCTION

The 20th and 21st centuries witnessed vast changes in management, viewpoints, perception of corporate success and capacities for supporting enterprises by means of information technology tools. There was some transition from rigid boundaries of organisations and sectors towards symbiotic systems where all stakeholders shaped their organisation value. Traditional components of competitive advantage are being now replaced with the level and quality of relationships between organisations and their environments. Organisations have to undertake effective actions internally but they must also co-operate with their partners including customers, suppliers, sector experts and other informal community groups. Simultaneously, it is observed that institutional acting and learning that both involve the implementation of already verified and standardised methods are not sufficiently effective. More and more frequently, informal processes and communities of practice (CoPs) become critical success factors. The communities in question spontaneously exchange knowledge and, what is more, learning is an integral part of their work. They are believed

DOI: 10.4018/978-1-60960-091-4.ch005

to play a major role in creating and transferring knowledge particularly including tacit knowledge along with unique skills into organisations.

The chapter aims at presenting the role of CoPs in supporting knowledge management and processes of organisational learning. Simultaneously, information technologies and tools that support functioning of CoPs are suggested. Theoretical considerations are contrasted with the results of the research that was carried out in 2008. The research employed a technique of a standardised interview accompanied by a questionnaire with organisations that are based in Upper Silesia – a Polish region that is undergoing rapid economic development. The region in question is the most densely populated and the most industrialised area in the country. Results were aimed to show the CoPs usage level in knowledge management and organisational learning. Moreover, potential reasons for the unsatisfactory level of CoPs usage in Polish organisations were sought.

The theoretical and practical research results that are presented in this article provide organisations with valuable guidelines to be followed while using CoPs in order to manage knowledge, while selecting information technologies and tools that may effectively support any work undertaken by CoPs and while identifying barriers that prevent organisations from using CoPs.

BACKGROUND

Literature Review and Hypothesis Development in the Fields of CoPs and Knowledge Management

A Contemporary Model of an Organisation

From a historical perspective an organisation has always been subject to numerous perceptions. Some attention should be paid to three analogies of an organisation understood as mechanistic, biological and – recently – socio-cultural systems.

In the mechanistic approach, an organisation is seen as an instrument that is used while realising different pre-defined goals. An organisation is a complex machine that functions on the basis of some programme. There is no room for friction and tensions. Structure of any mechanistic organisation provides for a strict control system and is hierarchical and inflexible.

An organisation understood as a biological system is a complex entity that establishes and maintains ongoing relationships with its environment (due to the exchange of energy and information). A biological organisation is made up of functionally diversified subsystems that interact and co-operate with one another and that try to keep organisational equilibrium. Although the system described is able to make decisional choices, its parts are deprived of such ability.

The third - socio-cultural - point of view has recently gained in importance. It stipulates that an organisation is a system that is characterised by collective thinking, which is confirmed by the fact that an organisation is a teleological formation of organisation members. The members in question define criteria of decisional choices. Members of a socio-cultural system are inter-connected by mutual goals and collectively accepted methods of realising their goals. They share their values that are deep-rooted in culture. Culture unites all parts in a consistent entity. Simultaneously, each of the parts mentioned "have some say in" the whole organisation and consensus plays a major role. Members of such organisations participate in network relationships that are largely informalised, thus being entitled to enjoy free communication regardless of organisational units' position and boundaries.

It is difficult not to notice that a socio-cultural model is based on network logics. A network is supposed to generate new relationships between enterprises, customers, suppliers and shareholders and to create collective intelligence. Power of a particular network is built, first of all, by small portions of information and knowledge. People have to be ready to share such portions. A network is a place of innovative relationships where e.g. customer's and producer's identities are slowly being merged (Kelly, 1998). The creative potential of a network is frequently based on multi-disciplinary teams that go beyond traditional organisational boundaries (Skyrme, 1999). Such a situation is mainly manifested in developing networks of connections between customers, suppliers, experts (communities of experts), individuals who work on joint tasks (communities of practice) or who are interested in a particular topic (communities of interest), individuals who work on realising some common objective (communities of purpose) or those who learn together (communities of learning) (Sullivan, 2003; Wenger et al., 2002; Lesser & Storck, 2001). The above networks may function both off and on-line and are frequently built on an ad hoc basis (Newell et al., 2002). According to Dube et al. (2006) their structure is not imposed but rather emerges in a natural way to respond to organisation's changing needs.

It is now necessary to highlight that the philosophy of a network model is based on knowledge and processes of organisational learning. Tacit knowledge that does not result directly from data analyses is particularly valuable for the model discussed. Such knowledge is moulded in a long-lasting process of learning, acquiring experience, cumulating information, grouping information in logical cognitive structures, relating information to emotions and ordering information within different systems of values, which all requires innovativeness, experimenting and learning (Olszak & Ziemba, 2008). Much attention is paid to learning on the higher level, i.e. so-called generative learning that - contrary to adaptive learning - calls for new methods to be used while perceiving the world. This type of learning is referred to as double-loop learning and requires continuous experimenting and analysing results, thus generating changes in any organisational knowledge base. As a result, knowledge is planted in processes, products, services, structure and culture of an organisation. What is more, the knowledge in question is also found in relationships that are established and maintained by an organisation with its closer and further environments.

A network model is influenced by modern information technologies whose application creates new quality of corporate relationships (Castellis, 1996). In case of information exchange, a criterion of time and space is somehow nonexistent. Information technologies integrate business communities in conditions of decentralisation and allow independent and frequently competitive entities to communicate.

COPS NATURE AND OBJECTIVES

CoPs refer to social networks as a new method of shaping organisational structures and economic systems. CoPs provide launch pads for learning, undertaking social activities and sharing interests and knowledge.

CoPs are groups of individuals who are interconnected by means of informal relationships and who share their experiences and knowledge of a particular domain (Sena and Shani 1999). CoPs are created as a result of their members' joint passion for a particular topic. The passion leads to deepening knowledge and enriching experiences of the scope selected by a particular group whose members are usually dispersed geographically but united by means of the following dimensions:

- structural, i.e. other individuals who function in the network are also taken into consideration while undertaking actions;
- relational, i.e. actions undertaken in a community are based on trust; and
- cognitive, i.e. members share common objectives.

With reference to other communities, CoPs are distinguished by means of the following three dimensions advanced by Wenger (1998):

- What do they refer to? (What do they deal with) – the scope of activities undertaken and knowledge to be used by CoPs to get organised (domain);
- How do they function? – individuals become members of CoPs by sharing knowledge and experiences and they are inter-connected as a result of some involvement in joint actions undertaken (community); and
- What kind of skills have they developed? – a common set of means that represent material dimensions of CoPs including, first of all, procedures, sensitivities, artefacts, vocabulary, styles, etc. (practice).

Therefore, it is possible to state that CoPs are groups of individuals who are inter-connected by means of common goals and internal motivation. They are not typical task-oriented teams but rather groups of co-operating individuals who focus on, for instance, a particular project, product or service. Such co-operation may bring about the emergence of new standards, tools, projects, good practices and business patterns (Wenger et al 2002). While co-operating with one another, participants of the community acquire knowledge, generate new knowledge, share knowledge, learn and acquire new skills. As a result, an organisation creates competencies that allow for building and retaining its competitive advantage.

Recognition of CoPs characteristics is necessary to succeed in the CoPs application. Table 1 presents different types of CoPs and their most important features along with consequences organisations face while applying CoPs. A detailed description of CoPs includes such characteristics as demographics, organisational context, membership characteristics and technological environment (Dube et al., 2006). Another area of recent study is ecological communities, which relates to networking through tacit knowledge. Hopkins (2009) identifies a new ecology of learning and thinking. It is holistic and radical and provocative, ranging over systems theory, emergent thinking in the arts and sciences, and new ideas about dialogue and collaboration. Throughout, the focus is how people have ideas and how they can use ideas to develop, collaborate and adapt as well as how to bridge the gap between individual talent and group power.

While analysing the above description, it is difficult not to agree that elaboration and management of different CoPs that are characterised by specific contamination of values pose numerous challenges that have to be faced by managers, designers, leaders and members of the communities in question. There are also different methods of evaluating such CoPs.

CoPs in the Development of Organisational Learning Processes

As it has already been highlighted, processes of organisational learning lay the foundations for existence of contemporary organisations. CoPs are created to utilise employees' knowledge and skills (Brown & Duguid, 1991). CoPs are likened to learning groups in which new observations are transformed into knowledge by means of mutual involvement in joint ventures. Therefore, it is possible to assume that CoPs remarkably contribute to creation of organisations' knowledge resources. In order to verify the thesis formulated this way, the subsequent part of this article is devoted to analyses of specifics of organisational learning. Much attention is paid to creating and transferring knowledge in organisations so that the place and role of CoPs could be shown in the processes discussed.

Originally, the very term of 'learning' was associated with human traits. Due to learning, human beings can transform themselves and do everything they were not able to do before. At present, abilities to learn are also attributed to

Table 1. Characteristics of CoPs (Source: on the basis of Dube et al. (2006))

Category/Feature	CoPs characteristics
Demographics	
Orientation	Operational CoPs focus on the daily operations of the organisation, such as answers to customers' problems. Strategic CoPs shape the organisation's response to important environmental changes or define new products or segment markets
Life span	CoPs can be assembled on a temporary basis and may be initially indeterminate. They are supposed to accomplish a specific purpose, e.g. a response to and *ad hoc* environmental changes. In spite of their initially *ad hoc* nature they are created on a permanent basis, i.e. an on-going mechanism for knowledge sharing
Level of maturity	Launching a new CoP is more challenging, as assembling community members, identifying their common interests, choosing technology, developing standards and processes may prove difficult. However, mature CoPs frequently require reinvestment when the face difficult changes
Organisational context	
Creation process	In practice creation processes are spontaneous or intentional: in the intentional (top-down) approach managers define CoPs development framework, e.g. they define goals of a network and key members of the community, etc. In the spontaneous approach (bottom-up) the initiative to create CoPs is taken by individuals whose goal is not only to gather and share ideas and knowledge but also to help one another
Boundary crossing	CoPs are frequently created to break organisational silos and promote co-operation, learning and information sharing. They are usually based on personal or task oriented relationships and may involve crossing boundaries across work groups, organisational units and organisations
Environment	CoPs creation and development are influenced by many factors including e.g. organisational culture, economic environment, management styles or politics. Taking the economic environment into consideration CoPs are more required in industries where turbulence and rapid change are common, which increases the need for knowledge and lifelong learning
Degree of institutionalised formalism	CoPs may integrate with the formal structure of an organisation and the degree of such integration may be different. Institutionalised CoPs are fully integrated and therefore treated as an organisational unit. However, CoPs do not often enjoy acknowledged existence in the organisation
Leadership	An organisation can either create formal CoPs governance structure where individuals are appointed to specific roles or leave roles and authority relationships to emerge through interaction around expertise. However, if CoPs take important roles in organisations in the context of knowledge management and learning, it is recommended to determine roles of particular community members clearly.
Membership characteristics	
Size	CoPs may associate a different number of members. A large community is more probable if CoPs attract individuals who have diversified and distributed interests. In such a case social relationships are more ephemeral. As a result, it may be more challenging to meet the needs of all members of the community to find valuable knowledge
Geographic dispersion	Geographical dispersion refers to the physical location of the CoPs participants. A high level of dispersion brings about additional challenges because physical distance encourages psychological distance. High geographic dispersion may mean that members are in different time zones, which makes synchronous communications much more difficult. It might also be necessary to face cultural diversity related barriers
Members' selection process	A CoPs open membership means that each interested employee in the organisation may become a member of the community. Such an approach is more in line with the idea of organisation-wide knowledge sharing. CoPs may also function on the basis of a closed membership and then in order to participate it is necessary to meet certain criteria
Members' enrolment	CoPs members' enrolment may be voluntary or compulsory. Volunteers are more motivated that those who are made to participate. Therefore, organisations should create an appropriate system of motivation and promotion
Member' prior community experience	Existing networks of individuals may be the instigator of new CoPs. In such a case, members already know one another and they are used to co-operating, sharing knowledge and learning together

continues on following page

Table 1. continued

Category/Feature	CoPs characteristics
Membership stability	CoPs may have permanent members but can also have changing membership. Open CoPs may experience less stability than closed ones. Stability may also be affected if key actors of the CoP are replaced by new members. The values, norms and communication patterns of new members will be confronted with the ones that were adopted by the CoPs
Members' IT literacy	Contemporary CoPs require IT literacy and skills. CoPs members represent different levels of their IT skills and that is why they prefer different technologies, e.g. some prefer to use video-conferencing, groupware systems or whiteboards. Others tend to use electronic mail only. In case of CoPs whose members are not proficient in IT, it may turn out that exchanging and sharing of knowledge is somehow limited
Cultural diversity	CoPs must facilitate integration of different communities that are characterised by various cultural attributes including national, organisational and professional attributes. National cultures influence styles of management and leadership, setting goals and priorities, decision making and inter-personal relationships, language and problem solving in communication and co-operation. Organisational culture frequently influences diversification of learning and adapting processes, knowledge sharing, employees' expectations and preferences or evaluation systems. Integration of different professional cultures within one CoP is another cultural dimension because members of particular professional groups develop their own knowledge bases, languages and specific vocabulary, technical procedures, values and standards, etc.
CoPs relevance to members	In practice CoPs goals set by managers do not always match objectives set by employees, which does not encourage any development. Hence, it is necessary to make sure that all goals are consistent and match one another
Technological environment	
Degree of reliance on IT	IT makes it possible to overcome barriers of time and space. Depriving CoPs of face-to-face contacts, especially at the beginning, may generate problems related to establishing a sense of identity and a common purpose or to development of knowledge sharing and mutual understanding in the group of all community members
IT availability	CoPs functionality largely depends on IT employed. IT that is applied provides virtual spaces where individuals can meet and discuss (synchronously and asynchronously), store documents, manage documents, co-operate, share knowledge and learn

organisations and different community groups. A learning organisation is an organisation that improves its capacity for creating its own future (Senge 1990). The following assumptions have to be taken into consideration while discussing any learning organisation:

- learning, i.e. a basic value of any organisation;
- 'teamability', which means that all employees participate in the process of learning and that is why organisations must emphasise team learning;
- motivating, which means that organisations must motivate individuals to participate in the processes of learning, creating innovations and contributing to the future of the organisations in question; and
- continuity, which means that the very process of learning ought to be continuous and conscious.

Learning may refer to different individuals and levels. In a majority of cases four levels of learning are discussed: individual, group, organisational and inter-organisational. Individual learning involves employees who acquire new skills, competencies and values on their own. Group learning means collective interactions that are manifested in jointly realised objectives and in communication along with knowledge exchange by employees. As a result of group learning some common approaches to solving problems and making decisions in an organisation are elaborated. Organisational learning refers to the so-called 'memory of an organisation' that is used for consolidating, institutionalising and storing different behaviours, activities, mental maps or values, i.e. results of learning undertaken by different individuals. Inter-organisational learning is also connected with the memory of an organisation although the very issue might be discussed in case of a few organisations.

It seems that the levels of organisational learning mentioned above would have to include community learning that refers to different informal groups, i.e. experts, practitioners, fans or hobbyists. Knowledge that is aimed at meeting the needs of a contemporary organisation and processes of decision making is deposited in numerous places. This knowledge is frequently of an informal nature and is obtained interactively. CoPs are valuable sources of knowledge including tacit knowledge in particular. They are also frequently thought to be the driving force of social capital. By means of spontaneous, free and non-formalised exchange of knowledge, thoughts and experiences, CoPs play a major role in the so-called empiric learning, i.e. combining different resources of knowledge that comes from dispersed locations, verifying such knowledge quickly and filtering to eliminate obsolete and useless knowledge. Significance of CoPs may also be confirmed by the so-called cybernetic learning that involves discovering new ways of perceiving and understanding the reality along with questioning and creating new visions, principles, and rules of functioning along with mental constructs of people. CoPs may contribute to creating new knowledge, combining new knowledge with already existing knowledge or replacing already existing knowledge with new knowledge.

CoPs in the SECI Knowledge Spiral Model

Numerous interesting observations of organisational learning and the role of CoPs in the very process may be made while analysing the Japanese model of a knowledge spiral – SECI – that involves the following processes: socialisation (S), externalisation (E), combination (C) and internalisation (I) advanced by Nonaka et al. (2000).

Socialisation involves sharing experiences, i.e. sharing tacit knowledge that is then used for creating better ideas. Exchange of tacit knowledge is performed when individuals stay together in the same environment and when they undertake different activities together. It is much more difficult to transfer this kind of knowledge by means of written or verbal instructions. Therefore, it is quite easy to agree with the statement that CoPs may play a major role in the process discussed.

Externalisation is a process that involves expressing tacit knowledge by means of notions available. Since transformation of tacit knowledge into explicit knowledge by means of a language is a very complex process, different metaphors, analogies and models find their application. CoPs translate tacit knowledge into an understandable form that may be assimilated by others. In the very process, an individual becomes a part of the group. Intentions and ideas spread and become an integral part of the group's mental world.

Combination involves grouping and combining different elements of any available knowledge in order to obtain a particular system of knowledge. Explicit knowledge is usually transformed into other - more complex - forms of explicit knowledge. On this level a key issue is to provide effective communication and knowledge transfer to particular members of the group.

On the other hand, internalisation is a process as a result of which formal knowledge in the form of e.g. a product concept or production procedures is implemented in activities undertaken or put into practice. There is some integration of newly generated knowledge with already existing systems of production, marketing and sales. Internalisation is a process of learning through acting. Therefore, learning through practice, trainings and exercises allows CoPs to gain some access to knowledge that is possessed both by different individuals and by the whole organisation.

All the processes of knowledge conversion mentioned above may be realised in different special contexts as described by the Nonaka and Konno (1998) concept of Ba. They pay particular attention to the following Ba:

- Originating Ba, i.e. some space for the exchange of emotions, experiences and mental models. Due to co-operation in a group, mutual trust is developed. Hence, it is possible to exchange tacit knowledge between particular individuals, which will result in creating new knowledge;
- Ba of interactions, i.e. some space where tacit knowledge is transformed into explicit knowledge. A major role is played here by dialogues and metaphors. Simultaneously, barriers related to own points of view are overcome and reflections are made;
- Cyber Ba, i.e. some virtual space of interactions that is used for filtering and coding new and already existing explicit knowledge. In a majority of cases Cyber Ba takes a form of a computer system; and
- Exercising Ba, i.e. some space that supports internalisation, i.e. learning. Explicit knowledge is transformed into tacit knowledge in the process of practice.

CoPs that focus on individuals and social structures allow for effective expansion of already existing knowledge through the whole SECI model and in all Ba special contexts. This way CoPs provide a new perspective of knowledge management and organisational learning processes related issues.

Research Study: The IT Support for CoPs and Knowledge Management

CoPs may be developed traditionally or they might be of virtual nature. It is even possible to find the term of VCoPs (Virtual Community of Practice) in the professional literature. According to the Authors transferring CoPs into the virtual world is today – in the era of globalisation and geographical dispersion of organisations and their trading partners – a necessity. Moreover, development of CoPs without information and computer technologies is simply not feasible. Therefore, we subsequently suggest IT that may play a major role while supporting the CoPs development. Particular attention is paid to IT that support tacit and explicit knowledge management.

From a perspective of any researcher, it is very important to implement scientific propositions. That is why, the authors carried out some audit of the actual situation of CoPs and knowledge management in Polish organisations. Unfortunately, it turns out that the results of the scientific research undertaken and related efforts do not present business practice in a favourable light. Organisations have to face numerous challenges posed by CoPs, knowledge management and IT implementation. The research résumé is presented below.

Propositions of Systems and Technologies that Support CoPs and Knowledge Management in CoPs

Philosophical, theoretical and empirical research on IT and knowledge management has been deep since the early 1990s in the field of information science and management. In particular Johannessen et al. (2001) has contributed much to this field in the late 1990s and early 2000s. He and his colleagues have noted "...lack of support for the positive economic impact of IT, we have seen that companies increasingly invest in the new technology. As this technology is limited to the transfer of explicit knowledge, this may relegate tacit knowledge to the background, in spite of the strategic importance of tacit knowledge, hence, leading to the mismanagement of knowledge". We identified how to invest in and use IT in tacit knowledge management to create sustainable competitive advantages. Our purpose was to improve the understanding of the role of tacit knowledge and to reflect on and give guidance on how to handle the relationship between tacit knowledge and IT.

Our proposition on what kind of technologies can be used for supporting of tacit and explicit knowledge management is presented as follows.

Processes of CoPs knowledge socialisation may employ technologies that allow for creating and sharing knowledge while participating in processes, performing tasks and observing phenomena (see Figure 1). It is now necessary to list all tools that allow for communicating and discussing ideas and that support holding conferences, organising group work and making collective decisions. Such tools let CoPs participate in synchronous and asynchronous forms of co-operation and are useful in structuralising and visualising co-operation and in assessing group decision making. Tools of this type may - among others – include the following: Parmenides Eidos Suite (www.parmenides-foundation.org), GroupSystems (www.groupsystems.com), IBM Lotus (IBM www-01.ibm.com/software/lotus), eGroupWare and MS Exchange(Microsoft www.microsoft.com/exchange).

It is recommended that the process of converting tacit knowledge into explicit knowledge should employ technologies that are able to grasp personal knowledge of individual members of a particular group in order to transform such knowledge into knowledge that would be available for the whole organisation (see Figure 1). Tools that are used to codify tacit knowledge are supposed to save knowledge in the structural way in form of text, graphics, video or sound. The tools in question predominantly include systems of content management and tools that are used to support repository creation along with systems of documentation management, e.g. Documentum Enterprise Document Management System (www.documentum.com), IBM Lotus Notes (IBM www-01.ibm.com/software/lotus), Hummingbird DOCS Open (connectivity.opentext.com) or Logotec DDM9000 (www.logotec.com.pl)

A separate class of tools that support processes of externalisation refers to semantic networks and ontology. Semantic networks are a project that is

Figure 1. Technologies for supporting of tacit and explicit knowledge

	Tacit knowledge	Explicit knowledge
Tacit knowledge	**Socialisation** Tacit knowledge → Tacit knowledge - E-mail - Newsletter - Chat - Forum - Messaging - Videoconferencing & audioconferencing	**Externalisation** Tacit knowledge → Explicit knowledge - Semantic Web - Knowledge ontology - Categorization systems - Taxonomy systems - Content management systems - Expert systems
Explicit knowledge	**Internalisation** Explicit knowledge → Tacit knowledge - E-learning - Workflow systems - Document Management Systems - Knowledge Base - Best practices repositories - Search machine - Text mining, Web mining	**Combination** Explicit knowledge → Explicit knowledge - Data base - Data warehouse - Knowledge Base - Document Management Systems - Workflow systems - OLAP - Data mining - Business Intelligence - Decision support systems

supposed to contribute to creation and dissemination of standards to be applied while describing content on the Internet, which should allow for automated searching, processing and transferring the very contents by matching documents with particular meaning. Ontology is an abstract and formalised description of some fragment of the reality that is used to systemise knowledge in organisations and to support searching, standardising and categorising of knowledge. To define ontology the following languages are used: XML (Extensible Markup Language), RDF (Resource Description Framework) or OWL (Web Ontology Language).

In order to support internalisation processes, it is recommended to employ technologies that allow for using acquired explicit knowledge in practice, which simultaneously results in generating tacit knowledge (see Figure 1). At present electronic training that aims at knowledge assimilation and accommodation, analyses of solutions, performance of a task or problem solving more and more frequently is used to create tacit knowledge. Creating tacit knowledge is facilitated by utilising explicit knowledge that is found in systems of documentation management, systems of work flows and systems of group work. Finding necessary knowledge frequently requires time consuming and complicated searches. That is why, some research is being carried out in order to analyse semantic networks and ontologies that were mentioned above. Interesting applications that analyse contents coming from different external and internal sources and then filter and personalise such contents for particular users include Compassware InfoMagnet, Excalibur' Retrieval and Verity Information Server along with text mining and web mining tools.

Technologies and systems that support converting explicit knowledge into explicit knowledge make up a very large set (see Figure 1). They include technologies that allow for localising and acquiring knowledge (e.g. expert networks, Business Intelligence systems, expert systems, ontologies, search engines, OLAP (On-Line Analytical Processing), and data mining (Olszak & Ziemba, 2006), codifying and organising knowledge (e.g. taxonomies, data bases, knowledge bases and data warehouses), disseminating and sharing knowledge (e.g. group work systems, visualisation systems, documentation management systems, electronic mail, etc.) and employing knowledge (e.g. Case Base Reasoning systems and multi-agent systems).

On the market, it is now possible to find numerous tools that support knowledge management. However, organisations have to face some challenges resulting from the selection of appropriate tools and technologies that would be adequate to organisations' requirements. The selection in question depends on many factors including, first of all, solutions that are already existing and being in use, new functionalities required, and new forms of knowledge to be acquired, created and disseminated (structural or non-structural knowledge, knowledge that is process-based, internal or external knowledge, etc.). Implementation and personalisation of software that supports CoPs and knowledge management must be preceded by recognising and understanding actual work of target groups and teams in any organisation and its environment. Therefore, it is usually necessary to carry out a thorough analysis of CoPs participants, formal and informal processes and the exchange of tacit knowledge into explicit knowledge.

At the end of this section of considerations, it is necessary to highlight that a particular technology is only an artefact that may allow for or improve undertaking human actions. Whether application of a particular technology generates any benefits or not, depends on the methodology that defines the way the technology in question is used and that describes the way individuals interact with their technologies. While creating such a methodology, it is necessary to pay attention to the fact that tools may be of different relevance and may play different roles depending on the context they are used in. Additionally, some tools may require a restrictive approach to their utilisation depending on the socio-technical environment they are used in because such environments may be characterised by different values, behaviours and even knowledge. It is particularly necessary to approach - carefully - integration of technologies that support management of knowledge from different sources – even when the same tools are used in all the sources.

Corporate Portals as an Integrated Platform for CoPs

Implementation of single technologies in case of different CoPs may be hazardous and it is likely not to generate expected outcomes. Some help in this respect may be provided by corporate portals. Corporate portals usually perform a function of integrated and homogeneous platforms that are used to implement diversified technologies that are then used in all processes of knowledge conversion, i.e. to acquire, codify and disseminate knowledge, search for knowledge and share knowledge (Collins 2003; Firestone 2003; Sullivan 2003; Terra and Gordon 2002; Woods 2005). It is possible to state that such platforms may be used by employees, particular groups of employees and communities. Creating integrated environments aimed at CoPs requires extensive business and technology related knowledge accompanied by experience and intuition. A quite important role is played here by, among others, identifying CoPs members, business processes observed in CoPs and CoPs knowledge resources. It is also important to identify communication and co-operation inside and outside a particular organisation. Only then, is it possible to select and integrate technologies that are adequate to meet CoPs needs.

Using IT to Develop CoPs and Manage Knowledge

The research carried out allowed for diagnosing the implementation of CoPs and knowledge management in Polish organisations. Using IT to develop and manage knowledge was analysed as well. The research was carried out in 2008. A standardised interview accompanied by a questionnaire was employed. The research was conducted among small and medium enterprises, which represented the private sector. Organisations based in Upper Silesia were subject to research. The region of Upper Silesia was chosen because it is rapidly being developed economically, which has to be taken into consideration while interpreting the research results and referring the results in question to organisations that are active in other regions of Poland.

The research sample consisted of 160 organisations dealing with services, commerce and production. Interpretation of the results requires paying some attention to the fact that a larger group of organisations was supposed to participate in the research. Unfortunately, approximately a quarter of organisations involved responded positively and then agreed to participate in the research. Refusals were not justified appropriately and resulted mainly from time related constraints of individuals who were competent enough to provide valuable information. Enigmatic refusals made some impression that potential respondents wished to avoid discussion about crucial business topics that were not thoroughly recognised by their organisations.

Question 1: Are any Concepts of Knowledge Management and CoPs Known to Your Organisation?

The research results show that in case of less than 50% of organisations subject to research, employees perceive knowledge to be some important asset that decides competitiveness (see Table 2). The knowledge management concept is commonly known in approximately 28% of organisations. However, knowledge of CoPs is manifested by only less than 10% of organisations. Organisations that confirmed their knowledge of CoPs in a ma-

Table 2. Familiarity of the knowledge management and CoPs concepts

Level of the imperative knowledge	Knowledge as a strategic resource	Knowledge management as a strategic process	Major role of CoPs in knowledge management
Commonly known to employees on all levels of management	44.4%	28.54%	9.54%
Not commonly known in an organisation but some activities are being now undertaken to promote the concept among organisation employees	15.04%	17.74%	8.20%
Not commonly known in an organisation but its promotion is thought to be important	14.44%	18.44%	9.20%
Not commonly known in an organisation but some activities are planned to promote the concept among organisation employees	6.94%	11.44%	20.40%
Not commonly known in an organisation and its promotion is thought to be unjustified	6.34%	9.54%	15.20%
It is difficult to say	13.14%	14.64%	37.46%

jority of cases stated that they used CoPs to reach short-term goals and to get involved in ongoing processes (e.g. improvement or development of new products and services). CoPs were also used for widely understood exchange of information. In case of organisations subject to research, CoPs were rather initiated by managers than regular employees. CoPs were strictly integrated with organisations. The fact that some organisations are carrying out or planning activities that promote knowledge, knowledge management and CoPs instils some optimism.

Even some superficial assessment of the research results confirms the thesis that CoPs and knowledge management related theory and practice are not sufficiently popularised in business environments.

Question 2: Which Knowledge Sources are Used in Your Organisation?

Organisations still do not know where they can find required knowledge. Particular sources of knowledge are in a majority of cases mediocre or they are not explored sufficiently (see Table 3). Internal sources of knowledge are recognised and used in the best way. A much more difficult situation is experienced in case of external sources. Organisations frequently have to deal with many difficulties resulting from identification of sources of explicit knowledge (recorded in electronic and paper forms) and tacit knowledge (stored in 'minds' of employees, experts, customers, etc.).

Table 3. Assessment of the level of using sources of knowledge in organisations

Sources of knowledge	Level of knowledge of sources / Level of using sources of knowledge			
	totally sufficient	Mediocre	insufficient	it is difficult to say
Internal knowledge	55.0% / 60.0%	36.3% / 31.9%	6.3% / 5.0%	2.5% / 3.1%
External knowledge	25.0% / 32.1%	51.9% / 49.1%	14.4% / 11.9%	8.8% / 6.9%
Explicit knowledge	39.0% / 35.6%	38/4% / 37.5%	8.2% / 13.8%	14.5% / 13.1%
Tacit knowledge	30.2% / 30.8%	32.1% / 39.0%	11.3% / 8.8%	26.4% / 21.4%

Question 3: What are the Barriers that Prevent Organisations from Developing CoPs and Knowledge Management?

It is possible to state that knowledge management and CoPs utilisation in organisations calls for improvement. It is indispensable to promote CoPs and knowledge management related thinking among employees of any organisation. Managers have to cope with some necessity to overcome barriers that prevent their organisations from using CoPs and knowledge management. The barriers in question predominantly include nonexistent knowledge about CoPs and knowledge management related opportunities and benefits (29.9%), no experience in knowledge management and CoPs particularly including lack of good examples of CoPs, poor knowledge of network participation rules and network etiquettes, lack of up-to-date expert maps and no experience in using adequate IT while supporting CoPs (28.0%), lack of general knowledge of knowledge management and CoPs (27.0%), no ideas and motivation required while launching and implementing knowledge management and CoPs (18.6%) and cultural barriers observed in organisations (15.9%).

Question 4: Does Your Organisation Use IT to Manage Sources of Knowledge?

Organisations still lack information and computer technology solutions that satisfactorily allow for supporting management of external and internal tacit and explicit knowledge (see Table 4). Only in case of using IT to manage internal sources of knowledge, a majority of respondents expressed their total satisfaction (64.3% of replies). The worst situation is observed in case of using technologies to manage tacit knowledge.

Question 5: Which IT are Used by Organisations to Develop CoPs and Knowledge Management?

The research results clearly show unsatisfactory organisation awareness of complexity of problems to be dealt with while implementing information technologies to manage knowledge and develop CoPs. It turns out that in a majority of cases organisations associated information technologies that supported knowledge management with those technologies that had been in use before knowledge management and CoPs related works and experiments were initiated (see Table 5). The following items were given in this context: electronic mail (97.7%), data bases (91.5%), spreadsheets (85.4%) and word processors along with search engines (80.8% each). Web mining and HTML documents were suggested less frequently by respondents (71.1% and 60.8% respectively). A half of enterprises involved declared that they used group work systems including Lotus Notes, Outlook/Exchange, Oracle and WiseGroup to manage knowledge. Communicators of the Gadu-Gadu or Skype type were also used. Unfortunately, technologies that are said to be knowledge management

Table 4. Assessment of the level of using information technologies to manage tacit and explicit knowledge that comes from organisations and their environments

Sources of knowledge	Level of using IT (N=130)			
	totally sufficient	mediocre	insufficient	it is difficult to say
Internal knowledge	64.3%	25.6%	7.8%	2.3%
External knowledge	41.8%	45.0%	9.3%	3.9%
Explicit knowledge	38.7%	41.1%	13.2%	7.0%
Tacit knowledge	20.2%	37.2%	14.7%	27.9%

dedicated are rarely used. Only 20% of organisations have data warehouses and even they do not use all capacities provided by data warehouse technologies. Only 4.7% of respondents confirmed that they used OLAP multidimensional analyses and Business Intelligence systems. Data mining (e.g. Statistica, SPSS and Oracle Data Mining systems) is used only by 11.7% of organisations.

Table 5. Assessment of the level of using particular categories of information technologies to develop CoPs and knowledge management

Information technologies	The level of IT usage (N=130)	
	Answers provided	Percentage of answers provided
Electronic mail	127	97.7%
Data bases	119	91.5%
Spreadsheets	111	85.4%
Word processors	105	80.8%
Search engines	105	80.8%
Web mining	91	71.1%
HTML documents	79	60.8%
Workgroup systems	66	50.8%
Communicators	66	50.8%
Information technology management systems	63	49.2%
Text mining	59	46.1%
Electronic bulletins	47	36.2%
Synchronic trainings	44	33.8%
Asynchronic trainings	41	31.5%
Documentation management systems	38	29.7%
Teleconferences (video and audio teleconferences)	30	23.1%
XML documents	27	20.8%
Data warehouses	26	20.0%
Knowledge data bases	25	19.2%
Discussion lists	18	13.8%
Discussion forums	18	13.8%
Systems of competence management	16	12.5%
Exploration of data from data bases	15	11.7%
Systems of work flow management	14	10.9%
Expert systems	13	10.2%
Chat rooms	12	9.2%
Categorisation and taxonomy systems	11	8.6%
Content management systems	10	7.7%
Business Intelligence Systems	6	4.7%
OLAP multidimensional data analyses	6	4.7%
Ontology systems	2	1.6%

Use of categorisation and taxonomy systems is equally poor (8.6% of the replies given). In case of ontology, only 1.6% of respondents declared its usage, which means that the above solutions are hardly known to organisations. A slightly better situation is observed in case of documentation management systems (29.7%), competence management systems (12.5%) and work flow systems (10.9%). Detailed results of the research carried are presented below (see Table 5).

Question 6: Does Your Organisation Use a Corporate Portal to Develop CoPs and Knowledge Management?

It might seem that as a result of the major role that may be played by corporate portals while managing knowledge and developing CoPs, the corporate portals in question are commonly used in organisations. However, the research carried out contradicts the above thesis (see Table 6). It turns out that only 14.6% of organisations that declared usage of information technologies to support knowledge management and CoPs use corporate portals, thus integrating a majority of information technologies employed. 13.1% of organisations integrate only selected technologies in their corporate portals. A vast number of organisations do not know a solution that is called a corporate portal (36.9%). The fact that 10.8% of organisations do not use corporate portals and do not recognise any need to implement such portals does not bode well for the future.

Question 7: What is a Barrier that Prevents Your Organisation from Using IT to Develop CoPs and Manage Knowledge?

Organisations have by necessity to overcome barriers that prevent them from implementing and using information and computer technologies to support knowledge management and CoPs. The barriers in question predominantly include lack of funds (33.4%), lack of knowledge of information and computer technologies and their capacities to support knowledge management and CoPs, and lack of experience in implementing and using information and computer technologies to support knowledge management and CoPs (32.8% each), no idea how to implement and use information and computer technologies to support knowledge management and CoPs (22.1%), no integration of information and computer technologies with organisation strategies and knowledge management strategies (16.8%), and – finally - technical problems (14.0%). The other answers provided referred to lack of integration of information and computer technologies, legal problems and security of information and computer technologies.

CONCLUSION

The research carried out confirmed the fact that in case of the CoPs usage the level of knowledge and experience of Polish organisations is very low and unsatisfactory. Only few organisations

Table 6. Using corporate portals in organisations

ParticularsWe do not know this platform	Total number (%)48 / 6.9%
We do not use the portal and we deem its implementation to be unjustified	14 / 0.8%
We do not use the portal but we believe that it would be justified to implement the portal in question	25 / 9.2%
We do not use the portal but the portal in question is currently being built	7 / 5.4%
We use the portal that integrates selected technologies of knowledge management only	17 / 3.1%
We use the portal that integrates a majority of knowledge management technologies in use	19 / 4.6%

build their CoPs and undertake some knowledge management related attempts. Organisations that undertook some activities to develop CoPs and manage knowledge were able to see some benefits. CoPs contributed to improvement in organisational knowledge and organisational learning. They allowed for more effective problem solving, sharing experiences and best practices, finding information quickly, identifying gaps in knowledge repositories, holding discussions, exchanging opinions and developing contacts and co-operation inside organisations and with their business partners. Benefits that were most frequently given by organisations may be grouped in two categories referring to organisations and employees (see Table 7).

Taking into consideration CoPs related benefits and remembering about rare practical implementation of CoPs, it is necessary to undertake promotional activities that would popularise the idea of CoPs in organisations, which encouraged us to present the most important information on CoPs in this article. Additionally, the philosophy of a network organisation along with importance of knowledge and organisational learning in a contemporary organisation are presented. The role of CoPs in organisational learning and knowledge management is shown and IT that allow for effective work of CoPs are suggested in the context of tacit and explicit knowledge management. Some barriers that prevent organisations from implementing CoPs and knowledge management are identified along with barriers related to using IT in the above initiatives.

Summing, it is necessary to underline that CoPs are not a universal silver bullet (Wenger et al., 2002).They are not meant to replace teams or business units as structures for serving markets and delivering products and services. Learning in model of CoPs requires an atmosphere of openness. Each CoPs develops a unique atmosphere – intense or laid back, formal or informal, hierarchical or democratic. Whatever norms members establish, thy key is to build a foundation for collective inquire. An effective CoPs offers a place of exploration where it is safe to speak the truth and ask hard questions. Trust is key to this process. Effective CoPs are not necessarily without conflict. In fact, the stronger a community, the better it is able to handle dissension and make it productive. In good CoPs strong bands withstand disagreement, and members can even use conflict as a way to deepen their relationships and their learning.

It is supposed that further effective development of CoPs in Poland will be closely correlated with the development of knowledge management in organisations and economy. In comparison to the last 10 years, Poland has shown significant progress in all knowledge economy index (KEI) pillars in comparison to other West European countries. However, Polish enterprises are still relatively insufficient in terms of innovation system development (KEI for Poland in 2007

Table 7. Benefits from CoPs

Benefits for employees	Benefits for organisations
• Learning from experts and peers • Developing a sense of identity and belonging • Improving ties with colleagues from other locations and organisations • Developing broader perspectives of the organisation and environment • Developing long-term personal networks • Getting recognition for specific skills and knowledge not directly related to their main job description • Improving self-esteem	• Codifying chunks of knowledge carried by employees • Promoting focused organisational learning. knowledge management and innovation • Solving problems quickly • Transferring best practices • Developing professional skills • Helping the organisation recruit and retain talent • Better serving existing clients

was 6,89, for Germany 8,93), quality of institution (for Poland 7,07, for Germany 8,38) and IT (for Poland 6,87, for Germany 8,79). Only for education pillar KEI in Poland in 2007 was higher than in Germany (for Poland 811, for Germany 8,08) according to the World Bank (2009). These facts should be considered in the interpretation of presented results of CoPs.

Methodological aspects related to building CoPs in the context of capacities, functions and properties presented are going to be dealt with by the Authors in subsequent publications. Comparing our study with other Eastern and Western European Economies is also very interesting research field.

REFERENCES

Blackler, F. (1995). Knowledge, knowledge work and organisations: An overview and interpretation. *Organization Studies*, *16*(6). doi:10.1177/017084069501600605

Brown, J. S., & Duguid, P. (1991). Organizational learning and communities of practice: Toward a unified view of working, learning and innovation. *Organization Science*, 2.

Castells, M. (1996). *The rise of network society* (The information age: Economy, society and culture, vol. 1). Oxford: Blackwell Publishers.

Collins, H. (2003). *Enterprise knowledge portals: Next generation portal solution for dynamic information access, better decision making and maximum results*. AMACOM.

Dube, L., Bourhis, A., & Jacob, R. (2006). Towards a typology of virtual communities of practice. *Interdisciplinary Journal of Information, Knowledge, and Management, 1*.

Firestone, J. M. (2003). *Enterprise information portals and knowledge management*. Elsevier Butterworth-Heinemann.

Hopkins, J. (2009). *Creative ecologies: Where thinking is a proper job*. UQP.

Johannessen, J. A., Olaisen, J., & Olsen, B. (2001). Mismanagement of tacit knowledge: The importance of tacit knowledge, the danger of information technology, and what to do about it. *International Journal of Information Management*, *21*(1). doi:10.1016/S0268-4012(00)00047-5

Kelly, K. (1998). *New rules for the new economy: 10 radical strategies for a connected world*. Viking.

Lesser, E. L., & Storck, J. (2001). Communities of practice and organizational performance. *IBM Systems Journal*, 4.

McKenzie, J., & van Winkelen, C. (2004). *Understanding the knowledgeable organization: Nurturing knowledge competence*. Thomson Learning.

Newell, S., Robertson, M., Scarbrough, H., & Swan, J. (2002). *Managing knowledge work*. Palgrave.

Nonaka, I., & Konno, N. (1998). The concept of "ba": Building a foundation for knowledge creation. *California Management Review*, 3.

Nonaka, I., Toyama, R., & Konno, N. (2000). SECI, ba and leadership: A unified model of dynamic knowledge creation. *Long Range Planning*, 33.

Olszak, C. M., & Ziemba, E. (2006). Business intelligence systems in the holistic infrastructure development: supporting decision making in organisations. *Interdisciplinary Journal of Information, Knowledge, and Management, 1*.

Olszak, C. M., & Ziemba, E. (2008). A conceptual model of ICT for intellectual capital management . In Korczak, J., Dudycz, H., & Dyczkowski, M. (Eds.), *Management: Advanced information technologies for management. Wroclaw, Poland: Publishing House of the University of Economics*. Wroclaw.

Sena, J. A., & Shani, A. B. (1999). Intellectual capital and knowledge creation: Towards an alternative framework . In Liebowitz, J. (Ed.), *Knowledge management handbook*. CRC Press.

Senge, P. M. (1990). *The fifth discipline: The art and practice of the learning organization*. New York: Doubleday.

Skyrme, D. (1999). *Knowledge networking: Creating the collaborative enterprise*. Butterworth-Heinemann.

Sullivan, D. (2003). *Proven portals: Best practices for planning, designing, and developing enterprise portals*. Reading, MA: Addison-Wesley.

Terra, J. C., & Gordon, C. (2002). *Realizing the promise of corporate portals: Leveraging Knowledge for Business Success*. Butterworth-Heinemann.

Wenger, E. C. (1998). Communities of practice: Learning as a social system. *The Systems Thinker, 9*.

Wenger, E. C., McDermott, R., & Snyder, W. M. (2002). *Cultivating communities of practice: A guide to managing knowledge*. Boston: Harvard Business School Press.

Wenger, E. C., & Snyder, W. M. (2000). *Communities of practice: The organizational frontier. Harvard Business Review*. January-February.

Woods, E. (2005). Collaboration software: Evolution and revolution . In Rao, M. (Ed.), *Knowledge management tools and techniques: Practitioners and experts evaluate knowledge management solutions*. Butterworth-Heinemann.

World Bank. (2009). Knowledge economy index, 2009. Retrieved November 10, 2009, from http://www.worldbank.org/kam

Chapter 6
A Reasoning Community Perspective on Deliberative Democracy

John Yearwood
University of Ballarat, Australia

Andrew Stranieri
University of Ballarat, Australia

ABSTRACT

This chapter describes some of the current approaches to deliberative democracy and then considers them from the perspective of a reasoning community framework. This approach highlights important tasks, processes and structures that can be used to enhance the processes of groups engaging in deliberative democracy approaches. In particular it focuses attention on the potential for technologies to support groups in achieving broad agreed structured reasoning bases that capture the scope of an issue from multiple perspectives.

INTRODUCTION

Initiatives toward the advancement of electronic democracy do more than provide fora for deliberative democracy. Organizations such as The World Legal Information Institute (WorldLII) is a free, independent and non-profit collaboration of a number of institutes dedicated to the provision of free access to public legal information throughout the world Worldlii (2007). They receive the text of judgments and statutes on a daily basis from thousands of courts and parliaments worldwide. Within a very short time frame, the text is automatically processed and uploaded to the relevant databases making the documents freely available to any Internet user. Combine this with initiatives to ensure that advanced communication technologies are readily accessible to all citizens and there is the capacity for a well-informed citizenry with the capacity to deliberate on issues in a way that has not been historically possible. This chapter looks at some of the structural, organizational and group elements that need to be considered in enabling effective deliberative democracy and the reasoning community framework perspective on these.

Deliberative democracy, first coined by Bessette (1994) combines elements of representative

DOI: 10.4018/978-1-60960-091-4.ch006

democracy with direct democracy by advocating some form of public deliberation about issues in decision-making debates. An example of this kind of democracy can be seen in the deliberative opinion poll model advanced by Fishkin, (1991). James Fishkin developed the concept of Deliberative Polling, which uses television and public opinion research to help members of a reasoning community in understanding an issue that has direct relevance in their lives. Whilst this method does not involve computing technology at all, it does provide evidence to support the notion that reasoning by a group differs from individual reasoning and that benefits for the public can be accrued by better informing citizens and involving them in the decision-making process.

The first step in the Deliberative Polling process is to form a representative sample of the wider community, usually by random sampling. Members of this group are polled on the targeted issues. Following the initial poll, the members are invited to gather at a single venue for a weekend in order to discuss the issues. A set of briefing materials is sent to the participants and is also made publicly available. The participants form small discussion groups and, with the assistance of a trained moderator, engage in dialogue with competing experts and political leaders. These events are either taped and edited or are broadcast on live television. After the deliberations, the sample revisits the original questions and any changes in opinion are noted.

Deliberative polling experiments have been conducted in the U.S., Canada, the United Kingdom, Australia and Denmark, with samples ranging in size from 200 to 466. Such sample sizes have two main benefits. Firstly, they are large enough to include a representative sample of the wider population and hence to return reliable results from statistical analysis. Secondly, they are small enough to enable the process of face-to-face, synchronous discussions. The results from these experiments indicate that the process proved to be especially suitable for issues where the public had little background knowledge of an issue. More importantly, the process had a significant impact on people's understanding of the issues, which resulted in significant proportions of the samples changing their decisions. Thus the process is very useful for the fact-finding phase of reasoning. However, the process requires synchronous and collocated participation, and it is limited in the number of participants that it can accommodate. These drawbacks are overcome by use of web-based technologies.

A key feature of Deliberative Polling is that a representative sample of participants is used for deliberating and ultimately voting on an issue. The deliberations and voting outcomes provide valuable insights, but do not supplant representative decision making. Many initiatives similar to or inspired by deliberative opinion polls have been established in a number of countries. The sites vary according to their support for deliberation. For example, electronic petitions sites are a type of opinion poll where participants can signal their views on a topic though the capacity to access information and engage in deliberation is limited.

A central tenant of deliberative democracy is the ideal that decision making ought to be informed by reasoned debate (Stromer-Galley, 2007). Electronic petition systems go some way toward this by providing easy access to information about an issue and the views of others. In reasoning community terms, this enables participants to perform individual coalescing and form their own views. Those views are then submitted to the community through the petition. The decision making remains the province of the parliament or recipient of the petition. Some sites such as the Australian *GetUp* (www.getup.org.au) do not directly aim their petitions at a decision-making body.

According to Habermas (1984), decisions made by consensus following informed deliberation leads to outcomes that are just and invite compliance. This cannot be said of outcomes determined by a vote. In practice, decision making is often not deliberative because participants do not start on a level playing field. Some participants have a greater voice, are more powerful, less inclined to hear others or are motivated by rewards outside the discussion. The perception that parliaments, local government councils, company boards, school boards and other decision-making bodies are not deliberative has possibly contributed to an unprecedented degree of public cynicism and mistrust of social institutions.

Although deliberation as an ideal may be intuitively appealing, the ascertainment of the extent to which deliberation occurs in a group is a difficult exercise. As judgments about the extent to which groups are deliberative are being made it becomes more important to understand what constitutes deliberation and how it might be measured or qualified.

The process for defining the participants in a deliberative poll corresponds to a decision about the engagement phase of a reasoning community as conceptualized by Yearwood & Stranieri in Chapter xx. In this chapter the extent to which the current deliberative democracy efforts fit with the framework for reasoning communities is examined. The reasoning community framework is also used to understand the elements of the deliberative democracy processes currently in use and examine, in particular structural aspects of the processes. This analysis is used to highlight possibilities for technological support to both permit broader participation and enhance the level of participation and deliberation.

DELIBERATIVE DEMOCRACY

John Dryzek argued for transformation of political processes to include the active and critical engagement of citizenry in deliberative processes in the modern, global world (Dryzek, 1990; Dryzek, 2002). In Australia, GetUp is a not-for-profit organization not affiliated with any political party. The organization elicits suggestions for issues from subscribers, collates information regarding the issue and provides an online bulletin board and voting mechanism. In a relatively short period of time, the subscription to GetUp has exceeded that of all political parties in that country. The Australian Council for Trade Unions (http://Rightsatwork.com.au)campaign provides information regarding changes to industrial relations laws and encouraged online opposition. In three months, the email list grew from 4000 to 170,000 lodgements of opposition.

e-participation initiatives in Europe include the Gov2u project (http://www.gov2u.org/) which is a generic but customizable, project that includes open source programs that facilitate the establishment of electronic participatory petitions. In Spain, over sixty municipalities are electronically implementing Gov2Demoss in the framework of the eConsensus project, which aims to foster public participation and consultation over the Internet; to improve the communication between councils and the citizens they serve; and to provide virtual fora to help citizens organize communities of interest and thus participate more effectively and constructively in public life. In Japan, the Voicebank initiative (Yong, 2005) provides another example of an effective electronic petition site. Voicebank elicits opinions and concerns about government programs from citizens. The opinions and complaints are made available for all citizens and an online discussion of underlying issues often occurs. An electronic petition system in Thailand provides a link for citizens to lodge suggestions and complaints directly to the prime minister (Yong, 2005).

In June 2007, the first e-community of Europeans was launched. It saw the creation of the first multilingual social network, blogging platform and fora for European citizens. With over 300,000 visitors per month and 30,000 subscribers on its newsletter, cafebabel.com has the capacity to change the perspective of European citizens on Europe. It is having a significant impact in the field of Internet journalism. More recently, mobile technologies have been deployed to enable citizens, particularly in remote communities to voice their views and provide input into the decision-making process (Gross, 2000; Brucher, 2003; Cabri, 2005). Further, a mobile system integrated into a geographical information system has been trialled in Portugal by Carvalho (2004).

Electronic petition approaches typically enable very little scope for participants to engage further with any opinions or to conduct deliberative discussions outside the petition forum. Electronic petition systems enable widespread access to information and spaces for discussion do not, in themselves, ensure that deliberation occurs. However, the identification of the nature of deliberation has been controversial.

In 2002, the European Science Foundation established the programme TED (Towards Electronic Democracy) with the objective of developing interactive decision support tools to help develop inclusive e-democratic systems. The project included members from Spain, Canada, Austria, Ireland, Italy, Germany, UK, Netherlands, Russia, Switzerland, Sweden, Czech Republic, Finland and Belgium. TED was based at the Universidad Rey Juan Carlos in Madrid, Spain. The project's aim was to use the internet to enable citizens to participate in societal decision making and voting, rather than simply being involved in electronic referenda. That is, the project team sees stakeholders being involved in all aspects of decision making including "helping to shape options, sharing perspectives, discussing and evaluating pros and cons, addressing uncertainty and moving towards a balanced conclusion". To address this, several areas for development were identified, including: problem structuring, uncertainty modelling, preference modelling, negotiation, risk assessment and communication of issues and the development of a web-based software infrastructure. Many tools were developed (see French, 2007; Insua, 2003; Insua, 2005; Prosser, 2004).

Government based on democratic principles has become increasingly prevalent in the last two centuries to the point that most nation states now are democratic. Key features of these states include the notion that a government, freely and fairly elected by the majority of people, rules on their behalf and by their consent. The protection of human rights, a respect for the rule of law, and the enshrinement of freedom of expression are ideals closely associated with democracy.

The form of government in the Greek city states over two thousand years ago involved a direct democracy where citizens participated directly in debates that resulted in legislation and other decisions that affected the city. Although, participation in meetings of the Assembly was restricted to adult male citizens in Athens, according to Manville (2003), thousands of participants would dutifully attend monthly Assembly meetings. Debates often raged and ultimately decisions were made by a simple majority vote of a show of hands or counting of coloured balls.

In contrast, representative democracy typical of most modern states, involves the appointment of representatives to engage in debates and make decisions on behalf of citizens. Direct democracy is regarded as infeasible as the population sizes increase whereas representative democracies have provided stable and effective government for millions. However critics of representative democracy (including Rousseau, 1968) suggest that citizens are disempowered if removed from the means to direct the process of law making. Indeed, many democratic states report that participation by citizens is decreasing to alarming levels leading to widespread disenfranchisement, particularly for minority groups such as youth, the elderly, rural communities and certain disability groups (Macintosh, 2002; Chutimaskul, 2004).

Åström (2004) distinguishes between three models of digital democracy: Direct democracy, Interactive democracy and Indirect democracy. Direct democracy corresponds to citizens' direct participation in the political process and decision-making. Interactive democracy involves strengthening civic engagement and activism through online discussion and deliberation. The indirect model works to strengthen the core institutions of representative democracy through the provision and dissemination of information and providing transparency. He compares these models in term of their aims, grounds for legitimacy, citizens' role, mandate of the elected and ICT-use focus. For example, intractive democracy has respectively the following characteristics: autonomy, public debate, opinion former, interactive, discussion.

Gibson et al. (2004) consider the effects and impact that the Internet and Information and Communication Technologies have had on the political process and the practice of democracy to date. They find that the hopes for digital information and communications technology to enable movement from representative democracy to more direct forms of democracy have not been realized. The situation is described as better than a null effect with many enhancements to the practices used in representative democracy, in particular some greater engagement with the younger more computer literate generations. In general there seems to be a greater use of the Internet to provide communities with information but there is some disappointment with the extent to which the elected representatives are found to listen to the feedback of the individuals in the community.

THE ELEMENTS OF A REASONING COMMUNITY

The term reasoning community (Yearwood & Stranieri, 2009) is broader and more encompassing than communities of commitment, practice, or purpose. A reasoning community is a group of participants that reason individually, communicate with each other, and attempt to coalesce their reasoning in order to reason collectively to perform an action or solve a problem.

The four key phases inherent in a reasoning community are (Figure 1):

- **Engagement** - involves the selection and recruitment of participants, the identification of the intended audience for the reasoning on the selected issue, the definition of a communication protocol and a decision-making protocol. Participants in a reasoning community are the people that agree on the issue and directly engage in reasoning to solve the problem or perform an action. A protocol for communication is the set of rules that govern exchanges between participants. A decision protocol specifies how the community will ultimately reach a decision; by voting, consensus or other mechanisms.
- **Individual reasoning**. Each participant ascertains facts, makes inferences from facts to draw conclusions and, by so doing contributes reasons to a pool of reasons for the community. A key part of individual reasoning involves an individual's coalescing of reasoning. This is the process of juxtaposing background knowledge with reasons advanced by other participants in order to understand the issue and position his or her claims amidst the others. A participant's coalescing of reasoning involves making sense of reasons in order to as-

Figure 1. Reasoning community phases

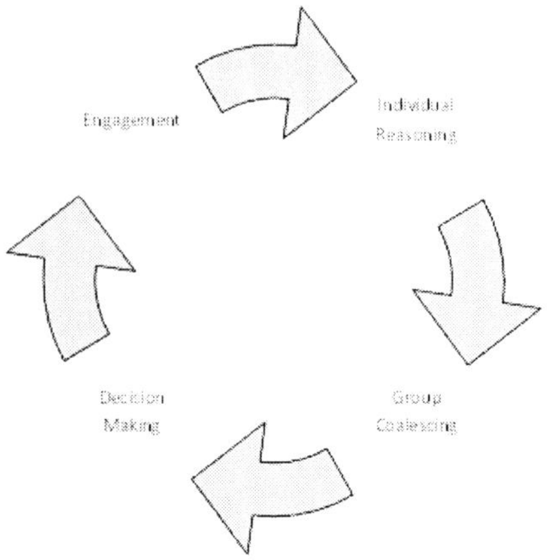

sert their own claims or to understand the claims of others.

- **Group coalescing of reasoning**. The coalescing of reasoning for the entire community involves organizing the terms, concepts and reasoning advanced by participants to the community into an explicit, coherent representation. This is important for shared and democratic decision-making where decisions are made on the basis of reasoned debate. Further, group coalescing enables communities in the future to adopt coalesced reasoning as a starting point for their own deliberations. Most current reasoning communities perform individual coalescing but do not systematically perform group coalescing.
- **Decision-making**. Making a decision requires making a choice between alternatives, actions or solutions considered. In a practical sense it involves the performance of an action or solution of a problem. The resolution of the problem requires the implementation of the decision-making protocols in order to reach a final decision.

MEASURING DELIBERATION

As deliberation is critical to deliberative democracy, it is important that we have tools to help further understand it and that allow us to identify and if possibly quantify it. The identification and measurement of deliberation is difficult despite its appeal as a key part of a group's process that is desirable in underpinning democratic decisions. A clear definition of deliberation in a democratic context by Stromer-Galley (2007) provides a starting point:

"deliberation as a process whereby groups of people, often ordinary citizens, engage in reasoned opinion expression on a social or political issue in an attempt to identify solutions to a common problem and to evaluate those solutions."

Three different approaches for measuring the quality of a deliberation are provided by Niemeyer (2007), Stromer-Galley (2007) and Steenbergen (2003). The Discourse Quality Index (DQI) of Steenbergen (2003) and Steiner (2004) provides an instrument that codifies interactions in order to rate the quality of discourse. The DQI was evaluated on parliamentary discourse in Britain and uses four criteria: the level of justification supporting claims, the extent to which justifications were egalitarian, the degree of respect paid to counter-arguments and the extent to which proposals were based on a goal of reaching genuine consensus.

The approach taken by Stromer-Galley (2007) codifies dialogue using a different scheme that includes indicators for the extent to which reasons underpin opinions, the extent to which participants are engaged in the process and the extent to which discussion was on topic. She defined six key elements of deliberation:

- Reasoned opinion expression - where participants provide reasons for their opinion
- References to external sources such as mass media and other materials - useful to ensure that all participants have access to the same background information. When there is disagreement, claims based on the documents can be more readily understood
- Expressions of disagreement. The free expression of disagreement is important to avoid undue influence of one view, the phenomena called ``groupthink" by Janis (1972). Open expression of disagreement also ensures that participants justify their positions
- Equal levels of participation - ensures that the broadest range of candidate suggestions and solutions are identified and all parties have equal opportunity to participate

- Coherent topic and structure - is concerned with how much of the time is spent on-topic and with the structure of the deliberation
- Engagement between participants - is necessary for discussions to evolve.

Niemeyer (2007) notes that measures of deliberation based solely on the quality of discourse unnecessarily ignore outcomes and that a more effective measure includes an assessment of the extent of the observed transformations resulting from deliberation. This is grounded in the notion of intersubjective rationality,that requires consistency between predispositions in terms of beliefs and values and the preferences or desired outcomes within the deliberation group. Intersubjective rationality is measured by taking every pairing of participants and performing correlation statistics on their beliefs and their desired outcomes. The correlations are then plotted as illustrated in Figure 2 which might be indicative of low intersubjective rationality. Points in the upper right quadrant represent those correlations where the pair of participants adopts similar positions taken on the basis of similar beliefs. In contrast, the correlations in the lower right quadrant represent correlations where pairs of participants desire different outcomes though their underlying beliefs are similar. Figure 3 illustrates what the correlations might look like in the case of high intersubjective rationality. The deliberation amongst a group is more advanced if more pairings are in the top right or lower left quadrants. This demonstrates an alignment between beliefs and outcomes; an intersubjective rationality.

Yearwood & Stranieri (2007) used characteristics related to the structure of reasoning in examining the impact of different approaches to argument mapping on secondary school students reasoning in groups on a government policy issue. The researchers provided all students with the same background information regarding the Aus-

Figure 2. Low intersubjective rationality

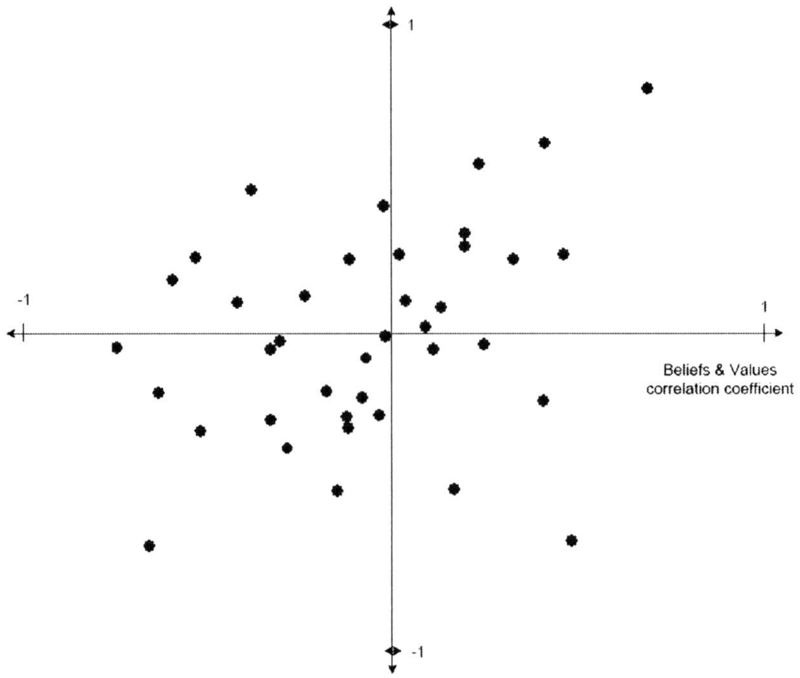

Figure 3. High intersubjective rationality

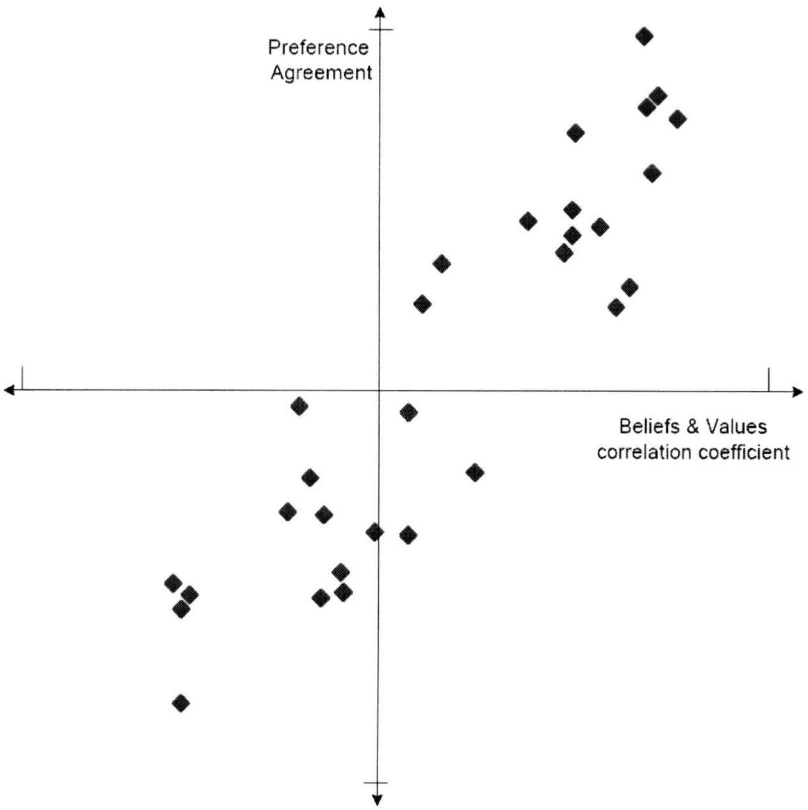

tralian government's policy on the processing of asylum seekers and then divided them into seven groups to discuss the issue. Each group was provided with a different reasoning/knowledge scheme to support their deliberations. These were the original Toulmin argument structure, IBIS (Issue-Based Information System (Rittel & Weber, 1984)) (with and without the positions and arguments laid out), the GAAM (Generic/Actual Argument Model (Yerwood & Stranieri, 2006) (with and without the claims, relevance reasons and inference reasons laid out), and narrative reasoning. There was also a control group who received no support for their reasoning. Learning outcomes were measured with a test on the issue and group deliberation was subjectively measured by experimenter observation of the following features:

- The extent to which there was agreement on claims without the advancement of reasons or consideration of alternatives,
- The extent to which agreement was reached by disagreement with the alternative claims,
- The extent to which agreement was reached because of some other evidence or reason, and;
- The extent to which agreement was reached by a combination of other evidence data or reasons.

Some indicative results from the preliminary study are presented in Table 1. These results are not conclusive but suggest that there are benefits for deliberation to be gained by using techniques for structuring reasoning. Further work is being undertaken with greater emphasis on the groups

Table 1.

GROUP	Rating of reasoning	Rating of group discussion
Unstructured	Reasoning was very poor with most students suggesting one of the solutions and modifying it to account for a health or language concern. Only two students mentioned reasons for their conclusion that were not simply a dislike of the other possible solutions	High. Discussion was based on single points of view put by individuals
GAAM	All students gave some reasons for their solution. Clear and balanced account of relevant factors	High. General discussion of issues
TAS	3 out of the 4 gave reasons and 1 actually rated the importance of a particular factor as a reason. Little breadth of issues	High
IBIS	3 students gave reasons for their choice the other 2 largely said that they liked their choice better than the other solutions without other reasons. Clear articulation	High, but focused on popularized discussion

acting according to the principles of a reasoning community.

In our view, efforts to measure the quality of deliberation, while important, are aligned with the goal of identifying effective technologies to support deliberation. The ability for large numbers of participants to engage in quality deliberative discussion, however this is measured, requires two key technological features:

- Advanced communication technologies. Existing Internet technologies provide the medium for unprecedented numbers of individuals across the globe to exchange views and information
- Schemes for organizing and structuring knowledge so that very complex issues and arguments can be organized in compact ways. Without this, participation in any discussion is restricted to a small number of highly specialized individuals because others will not have time to assimilate the large quantities of information that underpin most current issues.

The first, advanced communication technologies, requires that there is widespread and affordable access to Internet and related technologies. Concerns that a digital divide between those with access to these technologies and consequent access to information and the opportunity to contribute to decisions, and those without this access will only result in a new underclass have been raised by numerous authors including Wilhelm (2000), Kampen (2003), Jaeger (2004) and Wright (2006). Although the ability to connect to the Internet is increasingly prevalent and affordable throughout many countries, access is still far from universal. Further, affordable access to information and communication technologies alone does not ensure empowerment and engagement. Educational, cultural and other socio-political factors can impact significantly on the adoption of digital technologies to enhance participation Wilhelm (2000).

Digital divide issues aside, the second requisite for large numbers of participants to engage in reasoning involves schemes for organizing and structuring reasoning. This is needed because issues that attract the interest of large numbers of participants are invariably complex and involve many claims, facts and documents detailing evidence. For instance, an Environmental Impact Statement documenting arguments associated with the deepening of a shipping channel near the city of Melbourne totals more than 3000 pages. The executive summary alone exceeds 35 pages.

In another participatory decision making initiative in Australia, a forum known as the Energy Futures Forum engaged a variety of stakeholders including energy utilities and mining companies to identify plausible scenarios for generating energy needed by Australian industry and consumers (Morozow, 2006). A key feature of this forum was the inclusion of a deliberative democracy initiative called the Energy Futures Citizens Panels (EFCPs). Twenty participants representing a cross-section of views on sources of energy formed each of three EFCPs held in different major Australian cities. Deliberations were conducted with a facilitator over a four-day period. A great deal of work was done in small groups who reported to the larger group. Facilitators helped groups by summarizing inputs and assisting with the production of each Panel's final PowerPoint presentation. The deliberations provided valuable insight for the drafters of the final report (Littleboy, 2006), though the issues were quite complex and intricate.

A detailed analysis of the EFCP panels led Dryzek & Niemeyer (2006) to observe that deliberation initially involves agreement on those values, beliefs and preferences that are relevant and legitimate to the issues at hand. It does not require consensus about which of them is right or how they should be prioritized, but it does require that all relevant perspectives be actively considered and understood. They call this agreement, metaconsensus. Metaconsensus requires that participants understand each other and the pertinent factors without a need for consensus. Dryzek and Niemeyer (2006) claim that metaconsensus produces outcomes that are tractable and more legitimate than those where metaconsensus is low. It also eliminates irrelevant and strategic arguments.

Argument mapping methods such as Toulmin and its variants, IBIS and GAAM can be seen to provide a diagrammatic representation that work towards representing metaconsensus. In laying out positions and pro and con arguments, the IBIS map advanced by Rittel and Webber (1984) captures aspects of metaconsensus without prescribing a preference ranking or preferred solution. However it gives undue representation and emphasis to polarities. The version of the Toulmin structure deployed by vanGelder (2003) performs a similar role by laying out assertions and reasons for or against them. The generic argument structure proposed by Yearwood and Stranieri (2006) captures the generally agreed structure of reasoning on the issue without explicitly representing polarities at the generic stage.

A central feature underpinning argument mapping methods and measures of deliberation is the use of reasons to underpin opinion. Somewhat surprisingly, Polletta (2006) found that discussion was indeed largely based on reasons however, if a view being advanced was unpopular or unorthodox, its advocate was far more likely to use personal story telling rather than reasons to advance their view. She attributes this to the power of story-telling for engaging with others in an inherently collaborative way. Listeners can identify, sympathize and understand personal accounts in a qualitatively different way than they can when confronted with reasons.

Polletta (2006) cites a respondent in one of the online fora set up to foster participatory debate on the World Trade Center site who used a story to make the point that the new design must respect a sense of reverence the public has for the site:

"Recently, someone I know came back for a visit. She and her family went to Manhattan and took the Circle Line for a ride and she said that when they reached the WTC site the entire ship got silent. It's this reverence that needs to be addressed."

It is difficult to imagine how an assertion depicting the need for a reverent site design backed by well supported reasons could inspire the same sense of drama and impact that the short and simple personal account provides.

In addition to the use of stories for describing a view or making a point in a way that invites community unity rather than cold debate, Polletta (2006) notes that quite often participants in a discussion do not enter deliberations with fixed and certain opinions but rather do so with vaguely held notions and an openness to the views of others so that their own view may be cemented with some certitude.

The central lesson to draw from the varied approaches to measuring deliberation involves, first of all, the acceptance that deliberation can be measured, it is desirable to measure it and that there may be more than one way to measure it. More research is required in order to determine the extent to which the different instruments correlate with each other though measuring different phenomena. It is quite possible, for instance, that deliberation measured using DQI results in deliberation ratings that are similar to those observed with intersubjective rationality. That said, the incorporation of quantitative measures of deliberation that include personal story telling metrics may be far more difficult to achieve.

The measurement of deliberation is of great interest to advocates of deliberative democracy though there is no suggestion that deliberation is restricted to public policy discussions involving large numbers of participants. We adopt the view that approaches to measure deliberation can be interpreted as approaches to measure the extent to which a reasoning community operates deliberatively. The nature of the processes at each phase of a reasoning community's work (engagement, individual reasoning, group coalescing and decision making) can be identified to distinguish a deliberative reasoning community from less deliberative communities. Adopting this view facilitates the discernment of appropriate technologies to advance deliberation and, ultimately, the democratic ideal.

A REASONING COMMUNITY PERSPECTIVE ON DELIBERATIVE DEMOCRACY

If we accept that a central tenant of deliberative democracy is the ideal that decision making ought to be informed by reasoned debate (Stromer-Galley, 2007) then a forum for deliberative democracy is an instance of a reasoning community. Like all reasoning communities, its processes comprise engagement, individual reasoning, group coalescing and decision making. This perspective on deliberative democracies enables a clear specification of elements in each phase that are essential for a community to operate as a deliberative democracy community.

Table 2 summarizes essential elements. For instance, participants are best recruited to be representative of the constituency and so should represent as broad a range of views and beliefs as possible for a deliberative democracy community. The communication protocol should enable all views to be heard and encourages open expression. That all participants have easy access to relevant background information and work with schemes for the representation of their reasoning is particularly important for the individual reasoning phase of a deliberative community but also so that each individual can contribute informed and explicit communication of their views and reasons. Frequent and thorough group communication and coalescing enables agreement on relevant features of a problem solution that is important to avoid individuals becoming overwhelmed by the complexity and size of policy issues. The spirit of deliberative democracy is more aptly captured with a decision-making protocol based on consensus than a resort to voting, regardless of how fair the voting algorithm is. Each of these points is elaborated upon in the next sections.

THE ENGAGEMENT PHASE

The engagement phase of a reasoning community involves the clear identification of the issue, the recruitment of participants, agreement on a protocol for communication and a protocol for decision-making. The engagement phase also involves the articulation of the future audience that may re-use the community's reasoning. Features within each

Table 2. Reasoning community phases and deliberative democracy communities

Reasoning Community Phase	Important Features for Deliberative democracy communities
Engagement	Representative sample. Recruitment of participants to ensure breadth of views. Communication protocol characterized by respect, tolerance and listening
Individual reasoning	Ready access to all relevant background information. Schemes for the explicit representation of reasoning.
Group Coalescing	Frequent communication and capturing of views and reasoning - group coalescing endeavours. Inclusive and non-combative schemes for the explicit representation of meta-consensus.
Decision Making	Full participation in decision-making. Decision-making protocol based more on consensus than voting

of these elements of the engagement phase can be identified that contribute to deliberative democracy to a greater extent than others. For instance, the recruitment of participants for a deliberative democracy forum is ideally performed to include a broad range of stakeholder views, to ensure that all key views are represented and to ensure that participant demographics are representative of those of the broader community.

Recruiting participants to ensure a broad range of views is important so that no opinion is left unheard and also to ensure that the problem is viewed from many perspectives. For the World Trade Center discussion described by (Polletta, 2006), over 800 participants were selected to represent a cross section of views. They were assigned to 26 groups to discuss architectural designs, transportation, housing, and economic development options. This was done so that the solutions were guaranteed to be viewed from a range of pertinent perspectives.

The imperative to recruit participants so that the sample is representative of the broader community is more important in a deliberative democracy setting than in other reasoning communities. This taps into the democratic ideal that decisions are made by people who are affected by the decisions. In the Energy Citizen Panels study by Dryzek (2006), participants were selected by random sampling from telephone directories and further stratified for age, gender, highest level of education and employment status. In the Bloomfield Track deliberations involving the controversial construction of an outback Australian road (Niemeyer, 2003), indigenous participants were prevented from attending by floods.

Although not appearing as a criterion for the measurement of deliberation in existing schemes, the extent to which participants are selected to advance a broad range of perspectives is arguably as important as a criterion for measuring discourse quality once dialogue commences. In its extreme, discourse is conceivably extremely un-deliberative if adherents to controversial views are not invited to the table in the first place.

Communication protocols need to be determined and agreed upon during the engagement phase. A communication protocol includes:

- The medium for communication. Options may include online communication, face to face, email, telephone, SMS text and video conferencing
- The procedural regulation of discussion. This includes rules for turn taking, the appointment or not of facilitators, moderators or chairpersons and time or quantity limits on contributions
- The regulation of content and behaviour. This includes constraints on the content

matter of contributions in addition to the manner in which the contribution is conveyed.

Workshop methods such as the *Search Conference* advanced by Emery and Purser (1996) detail many aspects that facilitate the communication that they claim to be essential for deliberation: respect, involvement of all participants and collaborative attitudes. For example, Search Conference interactions should take place face to face over one or more days to encourage participants to get to know each other. Seating arrangements should be chosen so that participants face each other. Facilitators are required to be well trained to bring quiet participants into discussion and regulate the behaviour of strongly assertive participants. In contrast, the Consult approach described by Afshar et al. (2006) suggests that deliberation can occur with an exclusively online medium provided the system mandates the supply of reasons to underpin suggestions, anonymity and an unbiased voting scheme. Similarly, an email-based Delphi process is commonly believed to engage participants deliberatively.

The reasoning community life cycle model proposes that the decision-making protocol is also determined in the engagement phase. This is so that participants agree and understand on how to put together their reasoning. Deliberation is sometimes associated with consensual decision-making however, this is not necessarily the case. The final decision in the World Trade Center project was made by the City of New York. In the Energy Forum Citizens Panels managed by Dryzek (2006), the panel views were aggregated with the use of the Hare, Condorcet and Borda preferential voting schemes. However, the voting scheme was not determined by the panellists but by the facilitators. The deliberation discussion facilitated by vanGelder (2003) in a workplace setting was made by consensus. However his claim that the facilitated elicitation of a Toulmin-based diagram from a group of participants virtually ensures a consensual decision may be overstated.

A reasoning community identifies the intended audience for the future re-use of their reasoning in the engagement phase. The intended audience for future re-use of the reasoning performed within each reasoning panel can easily be imagined to include new citizen panels discussing an issue like 'energy generation', anywhere in the world, though this was not advanced in the Dryzek (2006) study. Future panels could benefit by having access to the claims, reasons and evidence from the Australian discussions, as a starting point. This would facilitate both individual reasoning and group coalescing.

By adopting a reasoning community view we more clearly see that there are several elements in the Engagement phase that can impact on the quality of deliberation and the outcomes of deliberative democracy fora. Furthermore, most existing measures of deliberation only focus on some of the elements that can feasibly have an impact on the quality of deliberation. Stromer-Galley (2007) discourse based approaches to measuring deliberation use elements of procedural regulation of communication by including indicators such as equal levels of participation while being silent on the impact of the communication medium. Emery and Purser (1996) and Steenbergen (2003) emphasize the importance of respect.

INDIVIDUAL REASONING PHASE

In the individual reasoning phase of the reasoning community life cycle there are two major roles: the individual drawing together or coalescing of knowledge and the formulation of beliefs, opinions, judgments and assertions from that knowledge. As Polletta (2006) notes, many participants commence discussions without clear views of relevant factors or of their own beliefs or assertions. She points out the importance of personal story telling in the task of making sense of the issue. In contrast, the intersubjective rationality approach (Niemeyer, 2007) for measuring deliberation presumes that participants have clearly articulated beliefs and desired outcomes. Furthermore, it seems implicit in intersubjective rationality that participants form their beliefs and desired outcomes somewhat independently of group processing.

Each individual in a (deliberative) reasoning community brings their background, context and perspective on an issue to contribute to the group. However, the feature of deliberative reasoning communities most pertinent to individual coalescing is the provision of reasons to underpin assertions. Whilst communication of opinions and beliefs can be achieved in a variety of ways, it is the extent to which reasons are provided that distinguish the dialogue as being deliberative and can be found in all attempts to measure deliberative quality. For intersubjective rationality, the reasons are assumed to be associated with the beliefs a participant holds. Two of the four criteria cited by Steenbergen (2003) as indicative of deliberation are the level of justification and the content of the justification. The variation to the Toulmin structure that vanGelder (2003) has developed includes specific slots for reasons. Yearwood and Stranieri (2006) used external observers to assess the extent to which reasons were provided by school students in discussing the government's refugee policies. Polletta (2000) found that discussion was based heavily on reasons. Even when personal story telling was used rather than reasons to advance a view in her World Trade Center study, the events of the story can be seen to provide a reason to underpin the point of the story even though both point and reasons remain implicit in the narrative.

Although there is broad consensus that reasons are important for deliberation, there is far less clarity on what reasons are, and how a good reason can be discerned from poor ones. In formal logics the quality of supporting promises is related to their truth value. In defeasible logics and arguments the strength of the reasons is related to the number and strength of supporting and attacking propositions. In informal logic the goodness of reasons is assessed by their explicitness and their support. Support may be assessed by weights assigned in various ways or by popularity or by authority.

To reiterate, the specification of reasons is broadly accepted as an integral element of deliberative democracy. However, in our view, the way reasons are defined and conceptualized is intimately associated with the way in which reasoning is represented. The specification of reasons in a story format is quite different from the specification of reasons in the syllogism, in a defeasible logic, in Toulmin argument structures, IBIS or the GAAM. Consequently, measures of deliberative quality ought to include a notion of the representation of reasoning.

COALESCING

In the group coalescing phase, a reasoning community takes time out, as it were, to take stock of the individual reasoning advanced by each participant. The background knowledge discovered in addition to all claims and reasons are coalesced into a format that acts as a summary and a general representation of all views. Dryzek (2006) observes and suggests that deliberation initially involves agreement on those values, beliefs and preferences that are legitimate and relevant to the issues. They call this agreement or this stage of agreement, metaconsensus. Metaconsensus requires that participants understand each other and the pertinent factors without the need for agreement on decisions or outcomes.

In a small reasoning community such as a patient-doctor community, group coalescing is rarely performed at all. In a larger community involved with more complex issues, the specification of an explicit group coalescing is sometimes performed. The Search Conference technique of small breakout groups brainstorming an issue to present to the entire group enables the group to perform group coalescing, albeit indirectly. Claims, reasons and knowledge pertinent to the issue are elicited and distinguished from those that are not relevant. A Delphi facilitator engaged in organizing, abstracting and summarising input from participants is an agent actively engaged in group coalescing.

Dryzek (2006) claims that high metaconsensus produces outcomes that are tractable and more legitimate than when metaconsensus is low and that it also eliminates irrelevant and strategic arguments. vanGelder (2003) claims that the specification of an argument map goes a long way toward engaging participants to find a solution to a dispute. In our view, the group coalescing is a process that can be performed by participants or external personnel and produces, as an artefact, an explicit representation of the elements of the group reasoning. A metaconsensus provides one description of this conceptually but to be useful needs a representation. An argument map can help with individual reasoning but needs to be at another level of abstraction to be useful as an explicit representation for group reasoning. Stories and narrative although usually individual can capture reasoning that the group subscribe to in a potentially general way.

That group coalescing is performed frequently and represented explicitly is critical for deliberative democracy. Without constant coalescing, participants are readily overwhelmed by the complexity of knowledge, reasons and claims in a large issue. Without some form of coalescing, deliberation can more readily degenerate into attacks on individual claims or onto individuals themselves. For instance, in the EFCPs, a facilitator was used to help the groups. The set of scenarios was modelled as a tree with 3 main branches (high carbon tax, late tax, no tax). In the Bloomfield Track study by Niemeyer (2003) panellists were presented with five preferences and seven key factors to take into account as they deliberated. In particular, they agreed upon the set of relevant factors. For example, the impact of silt run-off on a nearby reef was removed from the list of environmental concerns after scientific evidence showed that this would be minimal.

THE DECISION-MAKING PHASE

In Athenian democracy, decisions were ultimately made by a simple majority vote of adult male citizens present at the discussion. Preferential voting schemes enable participants to rank their preferences and allow for more sophisticated aggregation of voting. Arrow (1963) identified four properties in social preference functions as follows:

- Universal Domain. The voting scheme accounts for all individual preferences and provides a complete ranking of preferences that would be the same ranking each time voters' preferences are presented the same way.
- Pareto Efficiency implies that the outcome of the voting system is not over sensitive to any individual's preferences.
- Non-dictatorship. The aggregation of preferences into an outcome should account for the choices of multiple voters. The result cannot simply copy the preferences of a single voter.
- The independence of irrelevant alternatives entails that should provide the same ranking of preferences among a subset of options as it would for a complete set of options.

According to Arrow's theorem (Arrow, 1963) it is impossible to construct a perfect system of voting that can satisfy all four fairness properties when there are more than two alternatives. List (2002) note that the system known as the Borda Count system violates the fourth fairness condition though this is not regarded as necessarily disadvantageous. Niemeyer (2003) compared three schemes, the Condorcet, Hare and Borda Count for the Bloomfield Track deliberation and found the final preferences differed for all three methods. Following further deliberation the three methods were in complete agreement on rankings of the five solutions to the issue of sealing the outback road although there was a 7-5 split between two options.

Although voting schemes provide a fast, convenient mechanism for the determination of a final decision, there is a sense that resorting to a vote is not in keeping with the ideal of a deliberative discursive community as advanced by Habermas (1984). In that ideal, open dialogue unencumbered by power plays, false or missing information will lead naturally to a consensus. Once each participant is heard and rationality is exalted above other considerations, solutions can be agreed upon unanimously even if the solution goes against a particular individual's self interest.

Features that characterize a deliberative democracy reasoning community from other reasoning communities can be identified at each phase of group reasoning. Currently, different measures of deliberative quality relate to some but not all of the features. In our view, this is an omission that can be clearly seen if a deliberative community is seen as an instance of a reasoning community.

CONCLUSION

Initiatives toward the advancement of electronic democracy do more than provide fora for deliberative democracy. The World Legal Information Institute (WorldLII) is a free, independent and non-profit collaboration of a number of institutes dedicated to the provision of free access to public legal information throughout the world WorldLII (2007). They receive the text of judgments and statutes on a daily basis from thousands of courts and parliaments world-wide. Within a very short time frame, the text is automatically processed and uploaded to the relevant databases making the documents freely available to any Internet user.

Initiatives to ensure that advanced communication technologies are readily accessible to all citizens include affordable access to the Internet, and a level of education and awareness to navigate effectively through the maze of information already available. This includes programs such as those envisaged by Macintosh et al. (2002) that aim to enhance participation by all age groups.

This chapter has considered the nature of deliberative democracy from the perspective of a reasoning community as defined by Yearwood and Stranieri, (2008). The elements and phases of a reasoning community make explicit the many structural and procedural matters that need to be considered and implemented in a deliberative democracy setting. This perspective provides considerations and suggestions for the development of technologies and processes to more readily support groups wishing to engage in and develop deliberative democracy. In doing so we have also suggested that consideration of the appropriate knowledge and reasoning structures can also shed light on the measurement of deliberation and provide insight into the effective of deliberative democracy approaches. The approach is consistent with providing support for the functions of Åström's Interactive democracy.

REFERENCES

Afshar, F., Yearwood, J., & Stranieri, A. (2006). A tool for assisting group decision-making for consensus outcomes in organizations. In Voges, K. E., & Pope, N. K. L. (Eds.), *Business applications and computational intelligence* (pp. 316–343). Hershey, PA: Idea Group Publishing.

Arrow, K. J. (1963). *Social choice and individual values*. New York: Wiley.

Åström, J. (2004). Digital democracy: Ideas, intentions and initiatives in Swedish local governments . In Gibson, R., Römelle, A., & Ward, S. (Eds.), *Electronic democracy: Mobilisation, participation and organization via new ICTs*. London: Routledge.

Bessette, J. (1994). *The mild voice of reason: Deliberative democracy & American national government*. Chicago: University of Chicago Press.

Brucher, H., & Baumberger, P. (2003). Using mobile technology to support e-democracy. In *Proceedings of the 36th Annual Hawaii International Conference on System Sciences*. Washington, DC: IEEE Computer Society.

Cabri, G., Ferrari, L., & Leonardi, L. (2005). A role-based mobile-agent approach to support e-democracy. *Applied Soft Computing*, *6*(1), 85–99. doi:10.1016/j.asoc.2004.12.001

Carvalho, A., Rocha, R., & Oliveira, M. A. (2004). SINUP: Using GIS to support E-democracy. In *Electronic government* ([]). Berlin/Heidelberg: Springer. Retrieved from *http://www.inescporto.pt*]. *Lecture Notes in Computer Science*, *2739*, 341–344. doi:10.1007/10929179_61

Chutimaskul, W., & Funilkul, S. (2004). The framework of e-democracy development. In R. Traunmuller (Ed.), *Electronic government* (Lecture Notes in Computer Science, vol. 3183, pp. 27–30). Berlin/Heidelberg: Springer.

Dryzek, J. (1990). *Discursive democracy: Politics, policy and political science*. Cambridge, UK: Cambridge University Press.

Dryzek, J. (2002). *Deliberative democracy and beyond: Liberals, critics, contestations*. Oxford, UK: Oxford University Press. doi:10.1093/019925043X.001.0001

Dryzek, J., & Niemeyer, S. (2006). Reconciling pluralism and consensus as political ideals. *American Journal of Political Science*, *50*(3), 634–649. doi:10.1111/j.1540-5907.2006.00206.x

Emery, M., & Purser, R. E. (1996). *The search conference: A powerful method for planning organisational change and community action*. San Francisco, CA: Jossey-Bass Public Administration Series.

Fishkin, J. (1991). *Democracy and deliberation: New directions for democratic reform*. New Haven, CT: Yale University Press.

French, S. (2007). Web-enabled strategic GDSS, e-Democracy and Arrow's theorem: A Bayesian perspective. *Decision Support Systems*, *43*(4), 1476–1484. doi:10.1016/j.dss.2006.06.003

Gibson, R., Römelle, A., & Ward, S. (Eds.). (2004). *Electronic Democracy: Mobilisation, participation and organization via new ICTs*. London: Routledge.

Gross, T. (2000). Technological support for e-democracy: history and perspectives. In *Proceedings of 11th International Workshop on Database and Expert Systems Applications, Sept 4 – Sept 8, London, UK,* (pp. 391–395).

Habermas, J. (1984). *The theory of communicative action*. Cambridge, UK: Polity Press.

Insua, D., Fernandez, E., Rios, J., & Rivero, A. (2005). Participatory budget formation through the Web. In Michael H. Böhlen (Ed.), *E-government: Towards electronic democracy* [Proceedings of TCGOV conference, March 2-4, Bolzano, Italy]. Berlin/Heidelberg: Springer.

Insua, D., Holgado, J., & Moreno, R. (2003). Multicriteria e-negotiation systems for e-democracy. *Journal of Multi-Criteria Decision Analysis*, *12*(2-3), 213–218. doi:10.1002/mcda.358

Jaeger, P. (2004). The social impact of an accessible e-democracy: Disability rights laws in the development of the federal e-government. *Journal of Disability Policy Studies*, *15*(1), 19–26. doi:10.1177/10442073040150010401

Janis, I. (1972). *Victims of groupthink: A psychological study of foreign-policy decisions and fiascos* (2nd ed.). Boston: Houghton Mifflin.

Kampen, J. K., & Snijkers, K. (2003). E-democracy. *Social Science Computer Review*, *21*(4), 491–496. doi:10.1177/0894439303256095

List, C., & Pettit, P. (2002). Aggregating sets of judgments: Two impossibility results compared. *Economics and Philosophy*, *18*, 89–110.

Littleboy, A., Boughen, N., Niemeyer, S., & Fisher, K. (2006). *Societal uptake of alternative energy futures: Final report* [Technical report]. Newcastle, Australia: CSIRO Energy Transformed Research Flagship.

Macintosh, A., Robson, E., Smith, E., & Whyte, A. (2002). Electronic democracy and young people. *Social Science Computer Review*, *21*(1), 43–54. doi:10.1177/0894439302238970

Manville, B., & Ober, J. (2003). *A company of citizens: What the world's first democracy teaches leaders about creating great organizations*. Boston: Harvard Business School Press.

Morozow, O. (2006). *The heat is on: The future of energy in Australia* [Technical report]. Melbourne, Australia: Commonwealth Scientific and Industrial Research Organisation.

Niemeyer, S., & Blamey, R. (2003). *The far-north Queensland citizens' jury* [Technical Report]. Canberra, Australia: Land and Water Australia.

Niemeyer, S. J. (2007). Intersubjective rationality: Measuring deliberative quality. Paper presented to Political Science Seminar. Canberra, Australia: The Australian National University Research School of Social Sciences.

Polletta, F. (2006). *It was like a fever: Storytelling in protest and politics*. Chicago: University of Chicago Press.

Prosser, A., & Krimmer, R. (2004). Electronic voting in Europe . In Prosser, A., & Krimmer, R. (Eds.), *Electronic Voting in Europe – Technology, Law, Politics and Society Workshop of the ESF TED Programme together with GI and OCG July, 7th–9th, 2004 in Schloß Hofen/Bregenz, Lake of Constance, Austria onstance*. Austria.

Rittel, H. J., & Webber, M. M. (1984). Planning problems are wicked problems . In Cross, N. (Ed.), *Developments in design methodology* (pp. 135–144). Hoboken, NJ: John Wiley and Sons.

Rousseau, J. J. (1968). *The Social Contract*. Harmondsworth: Penguin.

Steenbergen, M., Bachtinger, A., Sporndliand, M., & Steiner, J. (2003). Measuring political deliberation: A discourse quality index. *Comparative European Politics*, *1*(1), 21–48. doi:10.1057/palgrave.cep.6110002

Steiner, J., Bachtiger, A., Sporndli, M., & Steenbergen, M. (2004). *Deliberative politics in action: Analysing parliamentary discourse*. Cambridge, UK: Cambridge University Press.

Stromer-Galley, J. (2007). Measuring deliberations content: A coding scheme. *Journal of Public Deliberation*, *3*(1), 1–37.

van Gelder, T. (2003). Enhancing deliberation through computer supported argument mapping. In Kirschner, P. A., Shum, S. J. B., & Carr, C. S. (Eds.), *Visualizing argumentation: Software tools for collaborative and educational sense-making*. London: Springer Verlag.

Wilhelm, A. G. (2000). *Democracy in the digital age: Challenges to political life in the digital age*. New York: Routledge.

World Legal Information Institute. (2007). Retrieved from http://www.worldlii.org/

Wright, S. (2006). Electrifying democracy?: 10 years of policy and practice. *Parliamentary Affairs*, *59*(2), 236–249. doi:10.1093/pa/gsl002

Yearwood, J., Avery, J., & Stranieri, A. (2006). Interactive narrative by thematic connection of dramatic situations. In *Proceedings of The Third Narrative and Interactive Learning Environments Conference* University of Edinburgh, (pp. 98 – 110).

Yearwood, J., & Stranieri, A. (2008) A study of frameworks for structured reasoning on reasoning quality and knowledge, Learning & Teaching: an international journal in classroom pedagogy, Vol.1, No.1, pp.71-90.

Yearwood, J. L., & Stranieri, A. (2006). The generic/actual argument model of practical reasoning. *Decision Support Systems*, *41*(2), 358–379. doi:10.1016/j.dss.2004.07.004

Yearwood, J. L., & Stranieri, A. (2008). Deliberative discourse and reasoning from generic argument structures. *Artificial Intelligence & Society*, *23*(3), 353–377.

Yong, J. S. (2005). *e-Government in ASIA: Enabling public service innovation in the 21st century*. Singapore: Times Media.

Section 2
Enhancing Collaborative Reasoning

The second group of chapters focuses more heavily on advancing theoretical and practical approaches that directly enhance group reasoning. Like the chapters in the previous section, they derive from very diverse fields of inquiry.

Chapter 7
Group Decision Making:
How to Obtain Representative Group Judgments from Individual Judgments

Thomas Saaty
University of Pittsburgh, USA

ABSTRACT

This chapter introduces a multicriteria decision making method for measuring intangible criteria alongside tangible ones. It relies on the use of judgments when scales are unavailable, and because scales of measurement use arbitrary units uniformly over their entire range, judgment is often needed to determine the significance of reading on such a scale. A brief discussion is given of how individual judgments can be combined to obtain a representative group judgment. An actual example for selecting the best hospice for a terminally ill patient is developed in detail. Numerous examples of application of the Analytic Hierarchy Process (AHP) process around the world are listed along with references to books written on the subject in many languages.

INTRODUCTION

Decision making is not like measuring something on a scale because it involves human values that are intangible. We make judgments about these intangibles in two ways. One is intuitive and spontaneous and may be the result of long and profound experience. The other is reasoned and deliberative. It relies on both intelligence and the ability to think logically about the many factors that are involved in a decision and the outcome of the interaction and feedback among these factors. It is known that intelligence is not the only factor that determines whether a person is a good thinker and decision maker. Either way by itself may lead to wrong estimates of relations, but usually we use them together to synthesize a richer judgment that is likely to be the wiser way to make a decision. We cannot make a decision without knowing how important each person's value is in each particular decision and how to combine them all in the appropriate way to get the best decision. How to measure and prioritize these values and trade them off is a primary concern of

DOI: 10.4018/978-1-60960-091-4.ch007

decision making. In this chapter we delve into the details of the Analytic Hierarchy Process (AHP), including a validation example, and alerting the reader about the stimulus-response derivation of the 1–9 scale, why we should only compare a few elements at a time and the role of inconsistency in that determination. Finally, why the geometric mean, not the arithmetic mean is the only way to combine individual judgments into a group judgment is discussed along with how, when necessary, to include the priorities of the individuals because of their power and expert knowledge.

The Scale for Paired Comparisons

The greatest test of the validity of a theory of measurement for decision making is the way it deals with judgments of intangible factors. Even the measurements we have for tangibles do not always tell us how accurately those measurements represent our values. The same measurements may not serve all values in the same proportion and may need to be assessed by using new judgments that represent the subjective nature of our values and desires. In other words: rarely do measurements in dollars, tons, yards, and even hours serve all the purposes of decision making. But there are numerous cases where our estimation coincides with the physical dimension being measured. Thus the process we apply to measure intangibles must apply equally well to those tangibles. These values can only be obtained in relative terms because the measurements themselves depend on a unit that is chosen arbitrarily.

A credible decision theory should be able to justify its approach in an objective way by producing successful predictions using the judgments of experts – thus demonstrating that the theory does not impose constraints based on its inherent assumptions. We should not argue defensively that decisions are in any case subjective and the numbers obtained by the theory simply have to be accepted in their own right. What the theory does with numbers must meet the reasonable expectations of experts, because meaningful decision making needs to produce meaningful outcomes from meaningful numbers – not just outcomes dictated by the assumptions of the method. The ultimate question is: How do we elicit meaningful numbers and produce meaningful outcomes?

In multicriteria decisions, we need to assign priorities to the criteria that govern the choice of alternatives. These priorities must be derived by making pairwise comparisons with respect to higher objectives by answering this question: Given two criteria, which one contributes more to the objective (and how much more)? We have no other credible way to do it because only we can judge the relative importance of the criteria with respect to the goal. Moreover we must compare the alternatives in pairs. Pairwise comparison is a tried-and-true procedure that elicits judgment about how many times one element is perceived by an expert to dominate another element used as the unit of measurement, with regard to influence. Rating the alternatives of a decision one at a time rather than comparing them in pairs, relies on a perceived ideal in memory used to rank each alternative by comparing it with that ideal– sometimes by going through an elaborate procedure to give the appearance of realism and rigor but still making a guess as to how desirable that alternative may be.

In the end, to obtain an overall rank for each alternative the priorities must be traded off. Thus decision making depends on numerical representation and how valid it is. Otherwise, as the saying goes, "Garbage in/garbage out" – all in the name of subjectivity and utility, interpreted for each situation in an appropriate and justifiable way.

When applied to estimate physical dimensions, a measurement theory should be able to produce approximate values of such physical quantities. If its estimates are found to be sufficiently accurate and reliable, it can then be applied to estimate values that are intangible. Here we offer examples to show that the AHP, when used by knowledgeable people, produces credible outcomes that match known data.

THE SCALE

Note that the human mind is incapable of comparing very small things with very large things but is able to compare close or homogeneous things on a certain attribute they share. To branch out to widely disparate things we need to cluster the elements into homogeneous clusters with a common pivot element from one cluster to the next and then compare the elements in each cluster and combine them by using the measurement of the pivot. The AHP relies on the following scale of absolute numbers (numbers that cannot be changed by a transformation to other numbers like pound to kilograms or yards to meters and still mean the same thing). Absolute numbers are invariant under the identity transformation. For derivation of this scale from elementary stimulus-response principles, the reader should consult Chapter 1 of Saaty (2005).

We have assumed that an element with weight zero is eliminated from comparison because zero can be applied to the whole universe of factors not included in the discussion.

PRIORITIES, CONSISTENCY, AND NUMBER OF ELEMENTS

The AHP has four axioms: reciprocal judgments, homogeneous elements, hierarchic or feedback-dependent structure, and rank-order expectations. The validation examples and numerous applications demonstrate that the axioms provide a valid logical foundation for the morphological structures used to represent complex decisions, the judgments, and how they are elicited to produce a reliable outcome. Assume that we are given n stones, $A_1,..., A_n$, with known weights $w_1,..., w_n$, respectively. Now suppose that a matrix of pairwise ratios is formed whose rows give the ratios of the weights of each stone with respect to all others. Thus we have the equation: $Aw = nw$ where A is multiplied on the right by the vector of weights w shown in Figure 1.

The result of this multiplication is $\boldsymbol{nw}$. Thus to recover the scale from the matrix of ratios, we must solve the problem $\boldsymbol{Aw} = \boldsymbol{nw}$ or $(\boldsymbol{A} - n\mathbf{I})\boldsymbol{w} = \mathbf{0}$. This is a system of homogeneous linear equations. It has a nontrivial solution if and only if the determinant of $\boldsymbol{A} - n\mathbf{I}$ vanishes, that is, n is an eigenvalue of $\boldsymbol{A}$. Now A has unit rank since every row is a constant multiple of the first row. Thus all its eigenvalues except one are zero. The sum of the eigenvalues of a matrix is equal to its trace, the sum of its diagonal elements, and in this case the trace of A is equal to n. Thus n is an eigenvalue of A, and we have a nontrivial solution. The solution consists of positive entries and is unique to within a multiplicative constant.

To make $\boldsymbol{w}$ unique, we can normalize its entries by dividing by their sum. Thus given the comparison matrix, we can recover the scale. In this case, the solution is any column of A normalized. Notice that in A the reciprocal property $a_{ji} = 1/a_{ij}$ holds; thus, also, $a_{ii} = 1$. Another property of A is that it is consistent: its entries satisfy the condition $a_{jk} = a_{ik}/a_{ij}$. Thus the entire matrix can be constructed from a set of n elements that form a chain across the rows and columns.

In the general case, the precise value of w_i/w_j cannot be given – only an estimate of it as a judgment. Because of lack of complete information and of a perfectly precise understanding of that information, the estimates may by inconsistent. In

Figure 1. The consistent eigenvector formulation

$$Aw = \begin{matrix} & \begin{matrix} A_1 & \cdots & A_n \end{matrix} \\ \begin{matrix} A_1 \\ \vdots \\ A_n \end{matrix} & \begin{bmatrix} w_1/w_1 & \cdots & w_1/w_n \\ \vdots & \cdots & \vdots \\ w_n/w_1 & \cdots & w_n/w_n \end{bmatrix} \end{matrix} \begin{bmatrix} w_1 \\ \vdots \\ w_n \end{bmatrix} = n \begin{bmatrix} w_1 \\ \vdots \\ w_n \end{bmatrix} = nw$$

fact that is the state of affairs for all of us when we make decisions, we cannot be perfectly consistent in estimating the value of w_i/w_j. But we don't want to be very inconsistent because that can lead to deriving wrong priorities and making wrong decisions. In addition, our judgments may be a result of clear and insightful reasoning about some of the facts, while others we only have feelings and inclinations about whose validity we may not be certain. Such guesses are often needed. The more experienced the people who make them, the closer they are likely to make their estimate to the real value of w_i/w_j. For the moment, consider an estimate of these values by an expert who is assumed to make small perturbations of the coefficients. This implies small perturbations of the eigenvalues. The problem now becomes $A'w' = \lambda_{max} w'$, where λ_{max} is the largest eigenvalue of A'. To simplify the notation, we will continue to write $Aw = \lambda_{max} w$, where A is the matrix of pairwise comparisons. The problem now is to find how good the estimate of w is. Notice that if w is obtained by solving this problem, the matrix whose entries are w_i/w_j is a consistent matrix. It is a consistent estimate of the matrix A. A itself need not be consistent. In fact, the entries of A do not even need be transitive; that is, stone A_1 may be preferred to stone A_2 and A_2 to A_3, but A_3 may be preferred to A_1. What we would like is a measure of the error due to inconsistency. It turns out that A is consistent if and only if $\lambda_{max} = n$ and that we always have $\lambda_{max} \geq n$.

Since small changes in a_{ij} imply a small change in λ_{max}, the deviation of the latter from n is a deviation from consistency and can be represented by $(\lambda_{max} - n)/(n-1)$, which is called the *consistency*

Table 1. Fundamental scale of absolute numbers

Intensity of Importance	Definition	Explanation
1	Equal Importance	Two activities contribute equally to the objective
2	Weak or slight	
3	Moderate importance	Experience and judgment slightly favor one activity over another
4	Moderate plus	
5	Strong importance	Experience and judgment strongly favor one activity over another
6	Strong plus	
7	Very strong or demonstrated importance	An activity is favored very strongly over another; its dominance demonstrated in practice
8	Very, very strong	
9	Extreme importance	The evidence favoring one activity over another is of the highest possible order of affirmation
1.1-1.9	When activities are very close a decimal is added to 1 to show their difference as appropriate	A better way to assign the small decimals is to compare two close activities with other widely contrasting ones, favoring the larger one a little over the smaller one when using the 1-9 values.
Reciprocals of above	If activity i has one of the above nonzero numbers assigned to it when compared with activity j, then j has the reciprocal value when compared with i	A logical assumption
Measurements from ratio scales		When it is desirable to use such numbers in physical applications. Alternatively, often one estimates the ratios of such magnitudes by using judgment

index (CI). The consistency index associated with a comparison matrix is compared with a consistency index that represents the average of consistencies of random comparisons at the same matrix order. This index is called the *random index* (RI).

The relations $a_{ji} = 1/a_{ij}$ and $a_{ii} = 1$ are preserved in the matrices to improve consistency. The reason for this is that if stone A_1 is estimated to be k times heavier than stone A_2, one should require that stone A_2 be estimated to be $1/k$ times the weight of stone A_1. If the consistency ratio is significantly small, the estimates are accepted; otherwise, we try to improve consistency by obtaining additional information. Three things contribute to the consistency of a judgment: (1) the homogeneity of the elements in a group, that is, not comparing a grain of sand with a mountain; (2) the number of elements in the group, that is a person cannot hold in mind simultaneously the relations of more than a few objects; and (3) the knowledge and diligence of the decision maker.

To get a feel for what the consistency index might be telling us about a positive n-by-n reciprocal matrix A, consider the following simulation. Choose the entries of A above the main diagonal at random from the seventeen values {1/9, 1/8,…, 1, 2,…,8, 9}. Then fill in the entries of A below the diagonal by taking reciprocals. Put 1's down the main diagonal and compute the consistency index. Do this 50,000 times and take the average, which we call the *random index* (R.I.). Table 2 shows the values obtained from one set of such simulations, and also their first-order differences, for matrices of size 1, 2,…,15, although we always recommend comparing 7 ± 2 elements.

Figure 2 plots the first two rows of Table 2. It shows the asymptotic nature of random inconsistency. Since it would be pointless to try to discern any priority ranking from a set of random comparison judgments, we should probably be uncomfortable about proceeding unless the consistency index of a pairwise comparison-matrix is very much smaller than the corresponding random

Table 2. Random Index with first-order differences

Order	**1**	**2**	**3**	**4**	**5**	**6**	**7**	**8**	**9**	**10**	**11**	**12**	**13**	**14**	**15**
RI	0	0	0.52	0.89	1.11	1.25	1.35	1.40	1.45	1.49	1.52	1.54	1.56	1.58	1.59
First-order differences		0	0.52	0.37	0.22	0.14	0.10	0.05	0.05	0.04	0.03	0.02	0.02	0.02	0.01

Figure 2. Plot of random inconsistency

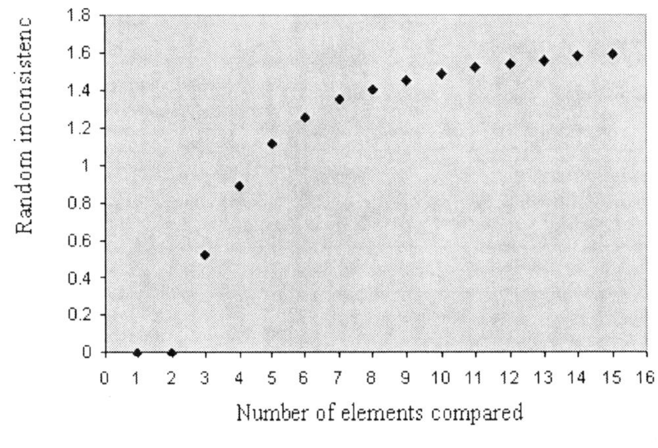

index value in Table 2. The consistency ratio (C.R.) of a pairwise-comparison matrix is the ratio of its consistency index to the corresponding random index value in Table 2.

The notion of order of magnitude is essential in any mathematical consideration of changes in measurement. When we have a numerical value, say between 1 and 10, for some measurements and want to determine whether change in this value is significant or not, we reason as follows: A change of a whole integer value is critical because it changes the magnitude and identity of the original number significantly. If the change in value is in the order of 1 percent or less, it would be so small (by two orders of magnitude) that it would be considered negligible. If this perturbation is a decimal (one order of magnitude smaller), however, we are likely to modify the original value by this decimal without losing the significance and identity of the original number as we first understood it to be. Thus in synthesizing nearly consistent judgment values, changes that are too large can cause a dramatic shift in our understanding and values that are too small cause no shift in our understanding. We are left with only values of one order of magnitude smaller that we can deal with incrementally to change our understanding. It follows that our allowable consistency ratio should be (5% for $n = 3$ and 8% for $n = 4$) and no more than about 0.10 for $n \geq 5$. The requirement of 10 percent cannot be made smaller without trivializing the impact of inconsistency. But inconsistency itself is important – without it, new knowledge that changes a preference cannot be admitted. Assuming that all knowledge should be consistent contradicts our experience that understanding must be continually revised.

If the consistency ratio is larger than desired, we do three things. First find the most inconsistent judgment in the matrix (for example, that judgment for whichis largest). Second, determine the range of values to which the most inconsistent judgment can be changed to improve consistency. Third, ask the judge to consider, changing his judgment to a plausible value in that range. If he cannot, we try with the second most inconsistent judgment and so on. If no judgment is changed, the decision is postponed for better understanding of the stimuli. Judges who understand the theory are always willing to revise their judgments – often not to the full value but partially – and then examine the second most inconsistent judgment and so on.

Before proceeding further, the following observations may be useful. The quality of response to stimuli is determined by three factors: accuracy (or validity), consistency, and efficiency (or amount of information generated). Our judgment is much more sensitive to large perturbations. When we speak of perturbation, we have in mind numerical change from consistent ratios obtained from priorities. The larger the inconsistency – and hence the larger the perturbations in priorities– the greater is our sensitivity to changes in the assigned numerical values. Conversely, the smaller the inconsistency, the harder it is to know where the best changes should be made to produce not only better consistency but also better validity of the outcome. Once near consistency is attained, it becomes uncertain which coefficients should be perturbed by small amounts to transform a nearly consistent matrix to a consistent one. If such perturbations were forced, they could be arbitrary and thus distort the validity of the derived priority vector in representing the underlying decision.

The third row of Table 2 gives the differences between successive numbers in the second row. Figure 4 plots these differences and shows the importance of the number seven as a cutoff point beyond which the differences are less than 0.10 where we are not sufficiently sensitive to make accurate changes in judgment on several elements simultaneously. This result shows that comparing too many elements reduces sensitivity to an inconsistent judgment. Thus in deriving reliable priorities from comparisons, we should compare

Figure 3. Plot of first differences in random inconsistency

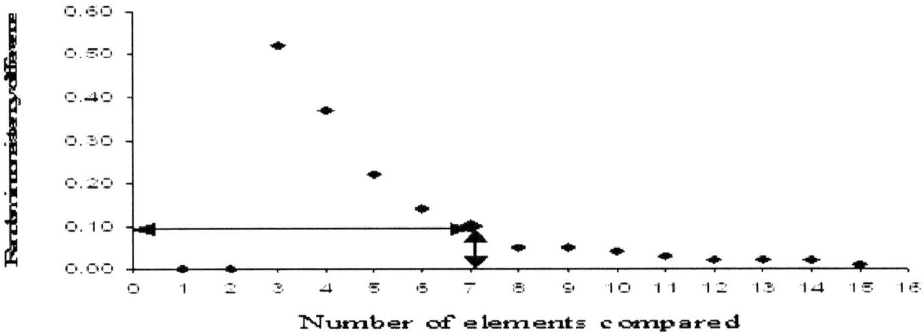

about seven elements, a fact known experimentally to G.A Miller long ago (Miller, 1956).

Sensitivity of the Eigenvector

Stability of the principal eigenvector also imposes a limit on our mental-channel capacity and highlights the importance of homogeneity. To a first-order approximation, perturbation w_1 in the principal eigenvector w_1 due to a perturbation A in the matrix A where A is consistent is given by Wilkinson (1964):

Here T indicates transposition. The eigenvector w1 is insensitive to perturbation in A, if the number of terms n is small; if the principal eigenvalue 1 is separated from the other eigenvalues, here assumed to be distinct (otherwise a slightly more complicated argument can be made); and, if none of the products vjT wj of left and right eigenvectors is small, or if one of them is small, they are all small. But v1T w1, the product of the normalized left and right principal eigenvectors of a consistent matrix, is equal to n, which, being an integer, is never very small. If n is relatively small and the elements being compared are homogeneous, none of the components of w1 is arbitrarily small, and correspondingly, none of the components of v1T is arbitrarily small. Their product cannot be arbitrarily small, w is insensitive to small perturbations of the consistent matrix A. The conclusion is that n must be small, and one must compare homogeneous elements.

When the eigenvalues have greater multiplicity than 1, the corresponding left and right eigenvectors will not be unique. In this case the cosine of the angle between them which is given by, corresponds to a particular choice of and. Even when and correspond to a simple they are arbitrary to within a multiplicative complex constant of unit modulus, but in this case | viT wi| is fully determined. Because both vectors are normalized, we always have | viT wi | < 1.

Figure 4. Five areas

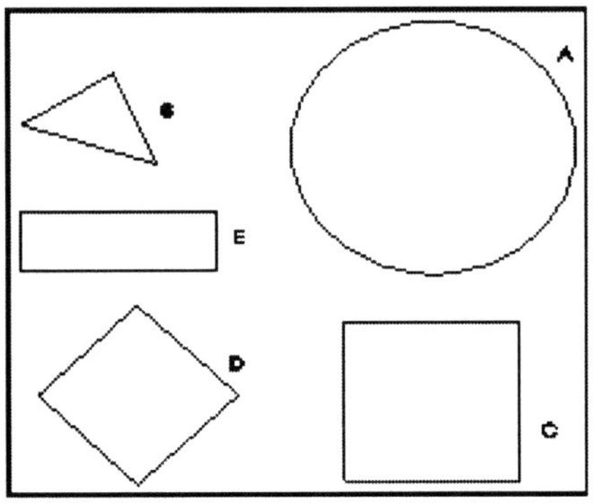

Relative Sizes of Areas

Figure 5 shows five areas. The object is to compare them in pairs to reproduce their relative weights as shown in Table 2. The actual relative values of these areas are approximately A = 0.47, B = 0.05, C = 0.24, D = 0.14, and E = 0.09. Here A is 47 percent of the overall area of the five figures,

Figure 5. Example of a hierarchy for ranking experts

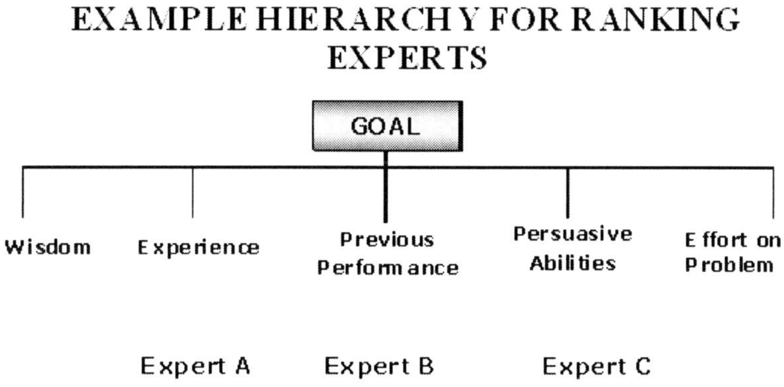

B is 5 percent, and so on. Or, putting it another way, A/B = 0.47/0.05 = 9.4, and A is about nine times as large as B, the relative sizes of the other figures can be determined in the same way.

How much larger (smaller) is an area of the left of the matrix than an area at the top, and by how many times? If it is smaller, one first enters the value of the larger one in the transpose position and uses the reciprocal value above the diagonal.

If there is no computer available, there is a shortcut for computing the approximate priorities: assume the matrix is consistent, normalize each column, and then take the average of the corresponding entries in the columns to obtain the priority vector. The priority vector shown in the right column of the matrix in Table 4 is the actual eigenvalue solution, but using the shortcut calculation gives a similar result for this case.

It is easy to make consistent judgments in this exercise because we are dealing with measures that are familiar to most people. If we were to measure the shapes, we would compute the ratios of their areas exactly, and enter these ratios instead of using judgments. The priority vector would give back the exact relative areas of the figures. The point here is to show that judgments can give accurate results. Sometimes people try to measure the shapes and use formulas dimly remembered from high school geometry to verify the results. The errors associated with physically measuring the figures often give worse results than eyeballing the figures and making judgments.

An interesting thing often happens when the exercise is tried with a group. In the AHP the member's judgments on a pair are combined using the geometric mean. When the combined judgments are used in the matrix, often the answer is closer

Table 3.

Comparisons With Respect to Area Size	Circle	Triangle	Square	Diamond	Rectangle	(priority vector of relative sizes)
Circle	1	9	2.5	3	6	0.488
Triangle	1/9	1	1/5	1/3.5	1/1.5	0.049
Square	1/2.5	5	1	1.7	3	0.233
Diamond	1/3	3.5	1/1.7	1	1.5	0.148
Rectangle	1/6	1.5	1/3	1/1.5	1	0.082

Table 4. The entries in this matrix respond to the question: Which criterion is more important with respect to choosing the best hospice alternative and how strongly?

Choosing Best Hospice	Recipient Benefits	Institutional Benefits	Social Benefits	Priorities
Recipient Benefits	1	3	5	.64
Institutional Benefits	1/3	1	3	.26
Societal Benefits	1/5	1/3	1	.11

to the actual relative sizes of the figures than most of the individuals' answers are.

Aggregating Individual Judgments into a Representative Group Judgment

Let us now talk about how to combine different judgments for comparing an apple and an orange in order to obtain a representative judgment for the group. We also want this representative judgment to satisfy the reciprocal relation. Let us note that we cannot use the arithmetic average. Suppose we have three people estimating how many times the apple is larger than the orange. If the first person says it is twice larger, the second three times larger, and the third four times larger, their average is which is the actual value. But the reciprocal of 3 is not equal to. Now we know that we will never again mention taking the arithmetic average for this purpose.

Since we cannot add and average numbers, we have one other simple alternative to try — that is, to multiply them and see if their average works. Let us observe using multiplication that if all the people give the same value to obtain the representative value for the group, we have to multiply these values and then take their nth root, where n equals the number of people in the group. By perturbing these equal judgments, we would still have to take the root in order to get their representative judgments. The operation of taking the nth root of n numbers multiplied together is known as the geometric mean. We might inquire whether using the geometric mean is the only way to combine group judgments. It turns out that under the fairly general conditions that are needed in group decision making, including the reciprocal property, it is the only way (Aczel & Alsina 1986; Aczel & Saaty, 1983; Saaty, 2000; Saaty, 2005; Saaty, & Peniwati, 2008; Saaty, 2010). To take the geometric average of the numbers 2, 3, and 4 we multiply them and take the cube root because there are three people involved which we can write as (2 x 3 x 4)1/3.Of course everyone knows that (2 x 3 x 4)1/3 = 21/3 x 31/3 x 41/3. Here the root of each judgment is extracted by 1/3. Now let us note that 1/3 + 1/3 + 1/3 = 1. We need this observation in the next step.

Individuals in a group possess different degrees of power as a result of many variables: formal authority within an organizational context, size and strength of an outside constituency, personal charisma, perceived intelligence, expertise on an issue, and the ability to call in favors owed by other group members. These factors, alone and in combination, make some group members stronger than others. If the people themselves have priorities x, y, and z that indicate their respective importance as judges, then their contribution will be different than the accorded value of 1/3; of course we have x + y + z = 1. In this case their judgments would – instead of being raised equally to the power of 1/3 are now raised to the respective powers of x, y, z and their average would be 2x x 3y x 4z (Aczel & Alsina, 1986).

Thus we have a way to obtain a representative group judgment when the people are equally important: a root or power of their judgments is taken that equals 1 over their number n. This means that, in the product of the judgments, each judgment is taken to the power $1/n$, which indicates that all the judges have an equal priority of $1/n$. When the judges are not equally important, the root to take of their judgment is equal to their actual priority. Recall that the sum of the priorities of the judges is equal to 1. The product of their resulting judgments, raised to powers equivalent to their priorities, is then taken for the representative judgment for the group. How to derive priorities for the people can be determined from a hierarchy like Figure 6.

The Hospice Application

Westmoreland County Hospital in Western Pennsylvania, like hospitals in many other counties around the nation, has been concerned with the costs of the facilities and manpower involved in taking care of terminally ill patients. Normally these patients do not need as much medical attention as do other patients. Those who most utilize the limited resources in a hospital are patients who require the medical attention of its specialists and advanced technology equipment — whose utilization depends on the demand of patients admitted to the hospital. The terminally ill need medical attention only episodically. Most of the time, such patients need psychological support. Such support is best given by the patient's family, whose members are able to supply the love and care the patients most need. For the mental health of the patient, home therapy is a benefit. From the medical standpoint, especially during a crisis, the hospital provides a greater benefit. Most patients need the help of medical professionals only during a crisis. Some will also need equipment and surgery. The planning association of the hospital wanted to develop alternatives and to choose the best one considering various criteria from the standpoint of the patient, the hospital, the community, and society at large. In this problem, we need to consider the costs and benefits of the decision. Cost includes economic costs and all

Figure 6. To choose the best hospice plan

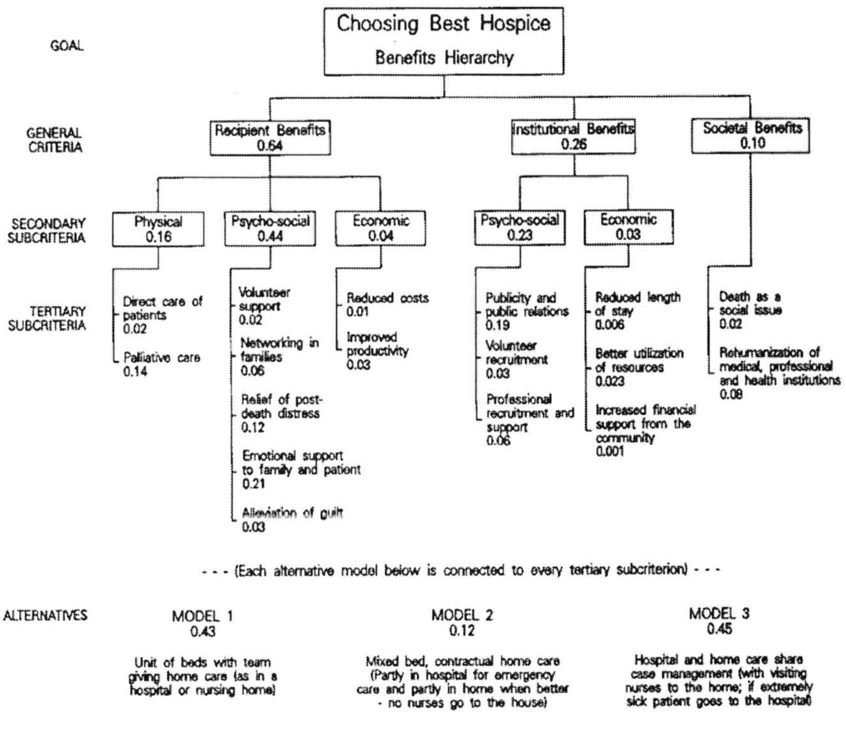

sorts of intangibles, such as inconvenience and pain. Such disbenefits or costs are not directly related to benefits as their mathematical inverses, because patients infinitely prefer the benefits of good health to these intangible costs. To study the problem, one needs to deal with benefits and with costs separately.

Approaching the Problem

I met with representatives of the planning association for several hours to decide on the best alternative. To make a decision by considering benefits and costs, one must first answer the question: In this problem, do the benefits justify the costs? If they do, then either the benefits are so much more important than the costs that the decision is based simply on benefits, or the two are so close in value that both the benefits and the costs should be considered. Then we use two hierarchies for the purpose and make the choice by forming ratios of the priorities of the alternatives (benefits bi/costs ci) from them. One asks which is most beneficial in the benefits hierarchy (Figure 7) and which is most costly in the costs hierarchy (Figure 7). If the benefits do not justify the costs, the costs alone determine the best alternative — that which is the least costly. In this example, we decided that both benefits and costs had to be considered in separate hierarchies. In a risk problem, a third hierarchy is used to determine the most desired alternative with respect to all three: benefits, costs, and risks. In this problem, we assumed risk to be the same for all contingencies. Whereas for most decisions one uses only a single hierarchy, we constructed two hierarchies for the hospice problem, one for benefits or gains (which model of hospice care yields the greater benefit) and one for costs or pains (which model costs more).

The planning association thought the concepts of benefits and costs were too general to enable it to make a decision. Thus, the planners and I further subdivided each (benefits and costs) into detailed subcriteria to enable the group to develop alternatives and to evaluate the finer distinctions the members perceived between the three alternatives. The alternatives were to care for terminally ill patients (1) at the hospital, (2) at home, or (3) partly at the hospital and partly at home.

For each of the two hierarchies, benefits and costs, the goal clearly had to be choosing the best hospice. We placed this goal at the top of each hierarchy. Then the group discussed and identified

Figure 7. To choose the best hospice plan costs hierarchy

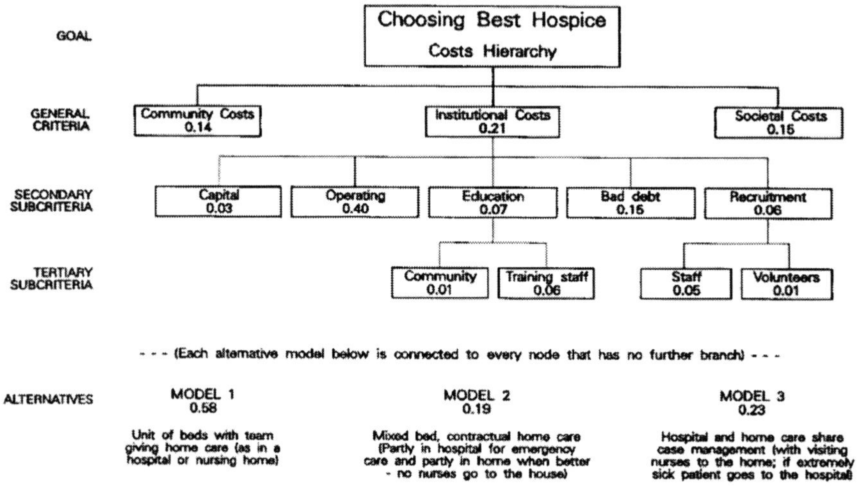

overall criteria for each hierarchy; these criteria need not be the same for the benefits as for the costs.

The two hierarchies are fairly clear and straightforward in their description. They descend from the more general criteria in the second level to secondary subcriteria in the third level and then to tertiary subcriteria in the fourth level on to the alternatives at the bottom or fifth level.

The planning association thought the concepts of benefits and costs were too general to enable it to make a decision. Thus, the planners and I further subdivided each (benefits and costs) into detailed subcriteria to enable the group to develop alternatives and to evaluate the finer distinctions the members perceived between the three alternatives. The alternatives were to care for terminally ill patients at the hospital, at home, or partly at the hospital and partly at home.

At the general criteria level, each of the hierarchies, benefits or costs, involved three major interests. The decision should benefit the recipient, the institution and society as a whole, and their relative importance is the prime determinant as to which outcome is most likely to be preferred. We located these three elements on the second level of the benefits hierarchy. As the decision would benefit each party differently and the importance of the benefits to each recipient affects the outcome, the group thought that it was important to specify the types of benefit for the recipient and the institution. Recipients want physical, psycho-social and economic benefits, while the institution wants only psychosocial and economic benefits. We located these benefits in the third level of the hierarchy. Each of these in turn needed further decomposition into specific items in terms of which the decision alternatives could be evaluated. For example, while the recipient measures economic benefits in terms of reduced costs and improved productivity, the institution needed the more specific measurements of reduced length of stay, better utilization of resources, and increased financial support from the community. There was no reason to decompose the societal benefits into third-level subcriteria, hence societal benefits connects directly to the fourth level. The group considered three models for the decision alternatives, and located them on the bottom or fifth level of the hierarchy: In Model 1, the hospital provided full care to the patients; In Model 2, the family cares for the patient at home, and the hospital provides only emergency treatment (no nurses go to the house); and in Model 3, the hospital and the home share patient care (with visiting nurses going to the home).

To choose the best hospice plan, one constructs a hierarchy modeling the benefits to the patient, to the institution and to society. This is the benefits hierarchy of two separate hierarchies

In the costs hierarchy there were also three major interests in the second level that would incur costs or pains: community, institution, and society. In this decision the costs incurred by the patient were not included as a separate factor. Patient and family could be thought of as part of the community. We thought decomposition was necessary only for institutional costs. We included five such costs in the third level: capital costs, operating costs, education costs, bad debt costs, and recruitment costs. Educational costs apply to educating the community and training the staff. Recruitment costs apply to staff and volunteers. Since both the costs hierarchy and the benefits hierarchy concern the same decision, they both have the same alternatives in their bottom levels, even though the costs hierarchy has fewer levels.

To choose the best hospice plan, one constructs a hierarchy modeling the costs to the patient, to the institution and to society. This is the costs hierarchy of two separate hierarchies

Judgments and Comparlsons

A judgment or comparison is the numerical representation of a relationship between two elements that share a common parent. The set of all such judgments can be represented in a square matrix in which the set of elements is compared with itself. Each judgment represents the dominance of an element in the column on the left over an element in the row on top. It reflects the answers to two questions: which of the two elements is more important with respect to a higher level criterion, and how strongly, using the 1-9 scale shown in Table 1 for the element on the left over the element at the top of the matrix. If the element on the left is less important than that on the top of the matrix, we enter the reciprocal value in the corresponding position in the matrix. It is important to note that the lesser element is always used as the unit and the greater one is estimated as a multiple of that unit. From all the paired comparisons we calculate the priorities and exhibit them on the right of the matrix. For a set of n elements in a matrix one needs n(n-1)/2 comparisons because there are n 1's on the diagonal for comparing elements with themselves and of the remaining judgments, half are reciprocals. Thus we have (n2-n)/2 judgments. In some problems one may elicit only the minimum of n-1 judgments.

As usual with the AHP, in both the cost and the benefits models, we compared the criteria and subcriteria according to their relative importance with respect to the parent element in the adjacent upper level. For example, in the first matrix of comparisons of the three benefits criteria with respect to the goal of choosing the best hospice alternative (Figure 6) recipient benefits are moderately more important than institutional benefits and are assigned the absolute number 3 in the (1,2) or first-row second-column position. Three signifies three times more. The reciprocal value is automatically entered in the (2,1) position, where institutional benefits on the left are compared with recipient benefits at the top. Similarly a 5, corresponding to strong dominance or importance, is assigned to recipient benefits over social benefits in the (1,3) position, and a 3, corresponding to moderate dominance, is assigned to institutional benefits over social benefits in the (2,3) position with corresponding reciprocals in the transpose positions of the matrix.

Judgments in a matrix may not be consistent. In eliciting judgments, one makes redundant comparisons to improve the validity of the answer, given that respondents may be uncertain or may make poor judgments in comparing some of the elements. Redundancy gives rise to multiple comparisons of an element with other elements and hence to numerical inconsistencies. For example, where we compare recipient benefits with institutional benefits and with societal benefits, we have the respective judgments 3 and 5. Now if $x = 3y$ and $x = 5z$ then $3y = 5z$ or $y = 5/3\ z$. If the judges were consistent, institutional benefits would be assigned the value 5/3 instead of the 3 given in the matrix. Thus the judgments are inconsistent. In fact, we are not sure which judgments are more accurate and which are the causes of the inconsistency. Inconsistency is inherent in the judgment process. Inconsistency may be considered a tolerable error in measurement only when it is of a lower order of magnitude (14 percent) than the actual measurement itself; otherwise the inconsistency would bias the result by a sizable error comparable to or exceeding the actual measurement itself.

When the judgments are inconsistent, the decision maker may not know where the greatest inconsistency is. The AHP can show one by one in sequential order which judgments are the most inconsistent, and that suggests the value that best improves consistency. However, this recommendation may not necessarily lead to a more accurate set of priorities that correspond to some underlying preference of the decision makers. Greater consistency does not imply greater accuracy and one should go about improving consistency (if one can given the available knowledge) by making slight changes compatible with one's understanding. If one cannot reach an acceptable level of consistency, one should gather more information or reexamine the framework of the hierarchy.

Under each matrix I have indicated a consistency ratio (CR) comparing the inconsistency of the set of judgments in that matrix with what it would be if the judgments and the corresponding reciprocals were taken at random from the scale. For a 3-by-3 matrix this ratio should be about five percent, for a 4-by-4 about eight percent, and for larger matrices, about 14 percent.

Priorities are numerical ranks measured on a ratio scale. A ratio scale is a set of positive numbers whose ratios remain the same if all the numbers are multiplied by an arbitrary positive number. An example is the scale used to measure weight. The ratio of these weights is the same in pounds and in kilograms. Here one scale is just a constant multiple of the other. The object of evaluation is to elicit judgments concerning relative importance of the elements of the hierarchy to create scales of priority of influence.

Because the benefits priorities of the alternatives at the bottom level belong to a ratio scale and their costs priorities also belong to a ratio scale, and since the product or quotient (but not the sum or the difference) of two ratio scales is also a ratio scale, to derive the answer we divide the benefits priority of each alternative by its costs priority. We then choose the alternative with the largest of these ratios. It is also possible to allocate a resource proportionately among the alternatives.

I will explain how priorities are developed from judgments and how they are synthesized down the hierarchy by a process of weighting and adding to go from local priorities derived from judgments with respect to a single criterion to global priorities derived from multiplication by the priority of the criterion and overall priorities derived by adding the global priorities of the same element. The local priorities are listed on the right of each matrix. If the judgments are perfectly consistent, and hence CR = 0, we obtain the local priorities by adding the values in each row and dividing by the sum of all the judgments, or simply by normalizing the judgments in any column, by dividing each entry by the sum of the entries in that column. If the judgments are inconsistent but have a tolerable level of inconsistency, we obtain the priorities by raising the matrix to large powers, which is known to take into consideration all intransitivities between the elements, such as those I showed above between x, y, and z. Again, we obtain the priorities from this matrix by adding the judgment values in each row and dividing by the sum of all the judgments. To summarize, the global priorities at the level immediately under the goal are equal to the local priorities because the priority of the goal is equal to one. The global priorities at the next level are obtained by weighting the local priorities of this level by the global priority at the level immediately above and so on. The overall priorities of the alternatives are obtained by weighting the local priorities by the global priorities of all the parent criteria or subcriteria in terms of which they are compared and then adding. (If an element in a set is not comparable with the others on some property and should be left out, the local priorities can be augmented by adding a zero in the appropriate position.)

The process is repeated in all the matrices by asking the appropriate dominance or importance question. For example, for the matrix comparing the subcriteria of the parent criterion institutional benefits (Table 5), psycho-social benefits are

Table 5. The entries in this matrix respond to the question, Which subcriterion yields the greater benefit with respect to institutional benefits and how strongly?

Institutional benefits	Psycho-social	Economic	Priorities
Psycho-social	1	7	.875
Economic	1/7	1	.125

regarded as very strongly more important than economic benefits, and 7 is entered in the (1,2) position and 1/7 in the (2,1) position.

In comparing the three models for patient care, we asked members of the planning association which model they preferred with respect to each of the covering or parent secondary criterion in level 3 or with respect to the tertiary criteria in level 4. For example, for the subcriterion direct care (located on the left-most branch in the benefits hierarchy), we obtained a matrix of paired comparisons (Table 6) in which Model 1 is preferred over Models 2 and 3 by 5 and 3 respectively and Model 3 is preferred by 3 over Model 2. The group first made all the comparisons using semantic terms for the fundamental scale and then translated them to the corresponding numbers.

For the costs hierarchy, I again illustrate with three matrices. First the group compared the three major cost criteria and provided judgments in response to the question: which criterion is a more important determinant of the cost of a hospice model? Table 7 shows the judgments obtained.

The group then compared the subcriteria under institutional costs and obtained the importance matrix shown in Table 8.

Finally we compared the three models to find out which incurs the highest cost for each criterion or subcriterion. Table 9 shows the results of comparing them with respect to the costs of recruiting staff.

As shown in Table 10, we divided the benefits priorities by the costs priorities for each alternative to obtain the best alternative, model 3, the one with the largest value for the ratio. Table 10 shows two ways or modes of synthesizing the local priorities of the alternatives using the global priorities of their parent criteria: The dis-

Table 6. The entries in this matrix respond to the question, Which model yields the greater benefit with respect to direct care of patient and how strongly?

Direct care of patient	Model I	Model II	Model III	Priorities
Model I - Unit/Team	1	5	3	.64
Model II - Mixed/Home Care	1/5	1	1/3	.14
Model III - Case Management	1/3	3	1	.26

Table 7. The entries in this matrix respond to the question, Which criterion is a greater determinant of cost with respect to the care method and how strongly?

Choosing best hospice (costs)	Community	Institutional	Societal	Priorities
Community Costs	1	1/5	1	.14
Institutional Costs	5	1	5	.71
Societal Costs	1	1/5	1	.14

Table 8. The entries in this matrix respond to the question: Which criterion incurs greater institutional costs and how strongly?

Institutional costs	Capital	Operating	Education	Bad Debt	Recruitment	Priorities
Capital	1	1/7	1/4	1/7	1	.05
Operating	7	1	9	4	5	.57
Education	4	1/9	1	1/2	1	.14
Bad Debt	7	1/4	2	1	3	.21
Recruitment	1	1/5	1	1/3	1	.07

Table 9. The entries in this matrix respond to the question, Which model incurs greater cost with respect to institutional costs for recruiting staff and how strongly?

Institutional costs for recruiting staff	Model I	Model II	Model III	Priorities
Model I - Unit/Team	1	4	4	.66
Model II - Mixed/Home Care	1/4	1	1	.17
Model III - Case Management	1/4	1	1	.17

Table 10.

		Distributive Mode			Ideal Mode		
Benefits	**Priorities**	**Model 1**	**Model 2**	**Model 3**	**Model 1**	**Model 2**	**Model 3**
Direct Care of Patient	.02	0.64	0.10	0.26	1.000	0.156	0.406
Palliative Care	.14	0.64	0.10	0.26	1.000	0.156	0.406
Volunteer Support	.02	0.09	0.17	0.74	0.122	0.230	1.000
Networking in Families	.06	0.46	0.22	0.32	1.000	0.478	0.696
Relief of Post Death Stress	.12	0.30	0.08	0.62	0.484	0.129	1.000
Emotional Support of Family and Patient	.21	0.30	0.08	0.62	0.484	0.129	1.000
Alleviation of Guilt	.03	0.30	0.08	0.62	0.484	0.129	1.000
Reduced Economic Costs for Patient	.01	0.12	0.65	0.23	0.185	1.000	0.354
Improved Productivity	.03	0.12	0.27	0.61	0.197	0.443	1.000
Publicity and Public Relations	.19	0.63	0.08	0.29	1.000	0.127	0.460
Volunteer Recruitment	.03	0.64	0.10	0.26	1.000	0.156	0.406
Professional Recruitment and Support	.06	0.65	0.23	0.12	1.000	0.354	0.185
Reduced Length of Stay	.006	0.26	0.10	0.64	0.406	0.406	1.000
Better Utilization of Resources	.023	0.09	0.22	0.69	0.130	0.130	1.000
Increased Monetary Support	.001	0.73	0.08	0.19	1.000	0.110	0.260
Death as a Social Issue	.02	0.20	0.20	0.60	0.333	0.333	1.000
Rehumanization of Institutions	.08	0.24	0.14	0.62	0.387	0.226	1.000
Synthesis Normalized		0.428	0.121	0.451	0.424	0.123	0.453
Costs							
Community Costs	.14	0.33	0.33	0.33	1.000	1.000	1.000
Institutional Capital Costs	.03	0.76	0.09	0.15	1.000	0.118	0.197
Institutional Operating Costs	.40	0.73	0.08	0.19	1.000	0.110	0.260
Institutional Costs for Educating the Community	.01	0.65	0.24	0.11	1.000	0.369	0.169
Institutional Costs for Training Staff	.06	0.56	0.32	0.12	1.000	0.571	0.214
Institutional Bad Debt	.15	0.60	0.20	0.20	1.000	0.333	0.333
Institutional Costs of Recruiting Staff	.05	0.66	0.17	0.17	1.000	0.258	0.258
Institutional Costs of Recruiting Volunteers	.01	0.60	0.20	0.20	1.000	0.333	0.333
Societal Costs	.15	0.33	0.33	0.33	1.000	1.000	1.000
Synthesis Normalized		0.583	0.192	0.224	0.523	0.229	0.249
Benefit/Cost Ratio		0.734	0.630	2.013	0.811	0.537	1.819

tributive mode and the ideal mode. In the distributive mode, the weights of the alternatives sum to one. It is used when there is dependence among the alternatives and a unit priority is distributed among them. The ideal mode is used to obtain the single best alternative regardless of what other alternatives there are. In the ideal mode, the local priorities of the alternatives are divided by the largest value among them. This is done for each criterion; for each criterion one alternative becomes an ideal with value one. In both modes, the local priorities are weighted by the global priorities of the parent criteria and synthesized and the benefit-to-cost ratios formed. In this case, both modes lead to the same outcome for hospice, which is model 3. As we shall see below, we need both modes to deal with the effect of adding (or deleting) alternatives on an already ranked set.

Model 3 has the largest ratio scale values of benefits to costs in both the distributive and ideal modes, and the hospital selected it for treating terminal patients. This need not always be the case. In this case, there is dependence between the personnel resources allocated to the three models because some of these resources would be shifted based on the decision. Therefore the distributive mode is the appropriate method of synthesis. If the alternatives were sufficiently distinct with no dependence in their definition, the ideal mode would be the way to synthesize.

The benefit/cost ratios of the three models given in the bottom row of Table 10 are obtained for both the distributive and ideal modes (dividing the priorities of the alternatives by the largest value of an alternative for each criterion thus obtaining the priority one for the alternative with largest value and correspondingly smaller values for the other alternatives). The final ideal values of the alternatives after synthesis are shown in their normalized form. Here one multiplies each of the six columns of priorities of a model by the column of criteria weights on the left and adds to obtain the synthesis of overall priorities, once for the benefits (top half of Table 10) and once for the costs (bottom half of Table 10) and forms the ratios of corresponding synthesis numbers to arrive at the benefit/cost ratio (bottom row of Table 10).

I also performed marginal analysis to determine where the hospital should allocate additional resources for the greatest marginal return. To perform marginal analysis, I first ordered the alternatives by increasing cost priorities and then formed the benefit-to-cost ratios corresponding to the smallest cost, followed by the ratios of the differences of successive benefits to costs. If this difference in benefits is negative, the new alternative is dropped from consideration and the process continued. The alternative with the largest marginal ratio is then chosen. For the costs and corresponding benefits from the synthesis rows in Table 10 I obtained: (see Table 11)

Table 11.

Marginal Ratios:	Benefits:	.12	.45	.43
	Costs:	.20	.21	.59

The third alternative is not a contender for resources because its marginal return is negative. The second alternative is best. In fact, in addition to adopting the third model, the hospital management chose the second model of hospice care for further development.

Absolute Measurement-Ranking Alternatives One at a Time

Cognitive psychologists have recognized for some time that people are able to make two kinds of comparisons - absolute and relative. In absolute comparisons, people compare alternatives with a standard in their memory that they have developed through experience. In relative comparisons, they compared alternatives in pairs according to a common attribute, as we did throughout the hospice example.

Before the development of the AHP, people ranked alternatives only one at a time and did not use comparisons. As a result the ranks of the alternatives were unaffected when new alternatives were added or old ones deleted. Thus it came to be an accepted dogma that if the alternatives are independent from one another, rank should never change. But that is not the case with comparisons. The priority value an alternative receives depends on the quality of all the alternatives with which it is compared. It turns out that preserving rank is not a mathematically proven fact, but a victim of the technique one uses. It is true that from practical considerations one may wish to preserve rank (Buede & Maxwell, 1983; Miller, 1956; Saaty,

2010) Let us now show how rank can be preserved by using the AHP.

People use absolute measurement (sometimes also called rating) to rank independent alternatives one at a time in terms of rating intensities for each of the criteria. An intensity is a range of variation of a criterion that enables one to distinguish the quality of an alternative for that criterion. An intensity can be expressed as a numerical range of values whether the criterion is measurable or qualitative. For example, if ranking students is the objective and one of the criteria on which they are to be ranked is performance in mathematics, the mathematics ratings might be: excellent, good, average, below average, poor; or, using the usual school terminology, A, B, C, D, and F. Relative comparisons are first used to set priorities on the ratings themselves. If desired, one can fit a continuous curve through the derived intensities. This concept may go against our socialization. However, it is perfectly reasonable to ask how much an A is preferred to a B or to a C. The judgment of how much an A is preferred to a B might be different under different criteria. Perhaps for mathematics an A is very strongly preferred to a B, while for physical education an A is only moderately preferred to a B. So the end result might be that the ratings are scaled differently.

Let us illustrate absolute measurement with an example. A firm evaluates its employees for raises. The criteria are dependability, education, experience, and quality. Each criterion is subdivided into intensities, standards, or subcriteria. The managers set priorities for the criteria by comparing them in pairs. They then pairwise compare the intensities according to priority with respect to their parent criterion (as in Table 12) or with respect to a subcriterion if they are using a deeper hierarchy. The priorities of the intensities are divided by the largest intensity for each criterion so the largest one receives the value one and the others proportionately less. Finally, the managers rate each individual (Table 13) by assigning the intensity rating that applies to him or her under each criterion. The scores of these intensities are each weighted by the priority of its criterion and summed to derive a total score for the individual (shown on the right of Table 13). These numbers belong to an absolute scale, and the managers can give salary increases precisely in proportion to the ratios of these numbers. The absolute mode saves time in rating a large number of alternatives.

CONCLUSION

The AHP is a modern tool for decision making that is grounded in our thinking, feelings and one might even say emotions because many people are either not well schooled in logical thinking or they are artists rather than scientists. Even for scientists, the many criteria and alternatives encountered in most complex decisions are hard to reconcile without a formal tool that organizes one's approach and provides a means of eliciting and recording judgments and measurements, synthesizing them to create priorities and then testing the final outcome.

Table 12. Comparisons of the intensities

	Outstanding	Above Average	Average	Below Average	Unsatisfactory	Priorities
Outstanding	1.0	2.0	3.0	4.0	5.0	0.419
Above Average	1/2	1.0	2.0	3.0	4.0	0.263
Average	1/3	1/2	1.0	2.0	3.0	0.160
Below Average	1/4	1/3	1/2	1.0	2.0	0.097
Unsatisfactory	1/5	1/4	1/3	1/2	1.0	0.062

Table 13. Ranking alternatives

	Dependability .4347	Education .2774	Experience .1775	Quality .1123	Total
1. Adams, V	Outstanding	Bachelor	A Little	Outstanding	0.646
2. Becker, L	Average	Bachelor	A Little	Outstanding	0.379
3. Hayat, F	Average	Masters	A Lot	BelowAv	0.418
4. Kesselman,	AboveAv Average	H.S.	None	AboveAv	0.369
5. O'Shea, K	Average	Doctorate	A Lot	AboveAv Average	0.605
6. Peters, T	AboveAv	Doctorate	A Lot	AboveAv	0.583
7. Tobias, K		Bachelor	Average		0.456

The AHP has been generalized to networks with dependence and feedback. Saaty developed the software SuperDecisions that implements the ANP (and therefore also the AHP) with help from expert programmers and has made it available free to the scientific community. It is used widely by the educational community around the world.

There is an international society around the AHP, the International Symposium on the AHP (www.ISAHP.org) that meets biennially. The meeting places have ranged from China to Japan and Indonesia to the US to Chile, Switzerland, and Canada.

Examples of Application since 2005

- Petrobras, Brazil's largest energy company, uses the AHP to help process and guide R&D investments. Petrobras has been recognized for its R&D prowess, pioneering new deep-water exploration and drilling capabilities that led Petrobras to the discovery and operationalization of a massive oil field on the Atlantic seabed.
- National Grid, an international energy delivery company with over 3 million electrical power customers in the U.S., uses AHP to guide its selection of new power transmission and delivery systems. These mission-critical systems are at the center of National Grid's operations, and represent a major selection that has a broad-ranging impact. Bill Tsolias, Manager of National Grid's Energy Management Group, said "We needed a decision process that would receive buy-in from all stakeholders... transmission, distribution, and IS." Decision Lens proven vendor selection process was used to successfully guide multiple groups within National Grid to the "best value" vendor.
- The Commander Navy Installations Command (CNIC) has selected Decision Lens as its decision making and budget allocation software application for enterprise alignment of shore installation support for all Navy Installations globally. CNIC is using AHP to prioritize all military construction projects and base of service activities to align investments and budget requests to strategic priorities and capabilities in support of all Navy operations. This is a $7B decision and resource allocation effort.
- The National Institutes of Health National Cancer Institute (NCI) used AHP to identify specific cancer vaccine target antigens for accelerated research. NCI developed a list of "ideal" cancer antigen criteria/characteristics and evaluated numerous representative antigens against those criteria for potential accelerated funding. The AHP software enabled NCI to capture input from academia, industry and government in an un-biased and structured way. NCI has recognized potential in some developmental cancer vaccines, and is interested in

new approaches for identification, prioritization, and funding of translational cancer research.

Examples of Application before 2002

- Many people around the world are using AHP, not only Americans. Some of the applications mentioned here were done by non-Americans.
- In 1986 the Institute of Strategic Studies in Pretoria, a government-backed organization, used the AHP to analyze the conflict in South Africa and recommended actions ranging from the release of Nelson Mandela to the removal of apartheid and the granting of full citizenship and equal rights to the black majority. All of these recommended actions were then quickly implemented.
- An oil company used it in 1987 to choose the best type of platform to build to drill for oil in the North Atlantic. A platform costs around 3 billion dollars to build, but the demolition cost was an even more significant factor in the decision.
- The process was applied to the U.S. versus China conflict in the intellectual property rights battle of 1995 over Chinese individuals copying music, video, and software tapes and CD's. An AHP analysis involving three hierarchies for benefits, costs, and risks showed that it was much better for the U.S. not to sanction China. The results of the study predicted what happened. Shortly after the study was complete, the U.S. awarded China most-favored nation trading status and did not sanction it.
- British Airways used it in 1998 to choose the entertainment system vendor for its entire fleet of airplanes.
- In 1999, the Ford Motor Company used the AHP to establish priorities for criteria that improve customer satisfaction. Ford gave Expert Choice Inc, an Award for Excellence for helping them achieve greater success with its clients.
- In 2001 it was used to determine the best site to relocate the earthquake devastated Turkish city, Adapazari.
- IBM used the process in 1991 in designing its successful mid-range AS 400 computer. IBM won the prestigious Malcolm Baldrige award for Excellence for that effort. The book The Silverlake Project about the AS 400 project has a chapter devoted to how AHP was used in benchmarking in this effort.

NOTE

In addition, E. Kinoshita has written numerous books in Japanese on the AHP and ANP over the past eighteen years.

REFERENCES

Aczel, J., & Alsina, C. (1986). On synthesis of judgments. *Socio-Economic Planning Sciences*, *20*, 333–339. doi:10.1016/0038-0121(86)90044-3

Aczel, J., & Saaty, T. L. (1983). Procedures for synthesizing ratio judgments. *Journal of Mathematical Psychology*, *27*, 93–102. doi:10.1016/0022-2496(83)90028-7

Buede, D., & Maxwell, D. C. (1995). Rank disagreement: A comparison of multi-criteria methodologies. *Journal of Multi-Criteria Decision Analysis*, *4*, 1–21. doi:10.1002/mcda.4020040102

Miller, G. A. (1956). The magical number seven plus or minus two: Some limits on our capacity for processing information. *Psychological Review*, *63*, 81–97. doi:10.1037/h0043158

Saaty, T. L. (2000). *Fundamentals of decision making with the analytic hierarchy process*. Pittsburgh, PA: RWS Publications.

Saaty, T. L. (2005). *Theory and applications of the analytic network process*. Pittsburgh, PA: RWS Publications.

Saaty, T. L. (2010). *Principia Mathematica Decernendi*. Pittsburgh, PA: RWS Publications.

Saaty, T. L., & Peniwati, K. (2008). *Group decision making: Drawing out and reconciling differences*. Pittsburgh, PA: RWS Publications.

Wilkinson, J. H. (1964). *The algebraic eigenvalue problem*. Oxford: Clarendon Press.

ADDITIONAL READING

Adamus, W. (2008). *The analytic hierarchy & network processes: Application in solving multi-criteria decision problems* (English ed.). Krakow, Poland: Jagiellonian University Press.

Andreichicov, A., & Andreichicova, O. (2000). *Analysis, synthesis and planning in economics* (Russian ed.). Moscow: Finance and Statistics.

Bhushan, N., & Rai, K. (2004). *Strategic decision making: Applying the analytic hierarchy process*. London: Springer-Verlag.

Garuti, C., & Escudey, M. (2002). *Toma de Decisiones en Escenarios Complejos*. Chile: Universidad de Santiago.

Islam, R. (2003). *The analytic hierarchy process: An effective multi-criteria decision making tool*. Kuala Lumpur, Malaysia: International Islamic University Malaysia.

Meixner, O., & Haas, R. (2002). *Computergestützte Entscheidungs – Findung*. Vienna: Redline Wirtschaft.

Rabbani, S. J. R., & Rabbani, S. R. (1996). Decisions in Transportation with the Analytic Hierarchy Process (English ed.). Federal University of Paraiba, Brazil.

Ramík, J. (1999). *Vícekriteriální Rozhodování —Analytický Hierarchický Proces (AHP)* (Czech ed.). Karviná, Czech Republic: Universitas Silesian Opaviensis.

Saaty, T. L. (2007). *Decision making with dependence and feedback: The analytic network process* (Russian ed.). Moscow: O. Andreichicova, Trans.

Schmoldt, D. L., Kangas, J., Mendoza, G. A., & Pesonen, M. (Eds.). (2001). *The analytic hierarchy process in natural resource and environmental decision making (Series in Managing Forest Ecosystems)*. Boston: Kluwer Academic Publishers.

Tansirigongkol, V. (1999). AHP: The most popular decision making process in the world (Thai ed.).

Tone, K. (1986). *Making a decision feels like playing a game: The analytic hierarchy process* (Japanese ed.). Japan: Nitskagiren.

Uberman, R., & Ostręga, A. (2007). Mineral policy in Poland: Mining manual (English ed.). Cracow, Poland: AGH University of Science and Technology Press

Weber, K. (1993). *Mehrkriterielle entscheidungen*. München, Germany: R. Oldenbourg Verlag.

Xu, S. (1986). Applications of AHP in resource allocation, management and conflict analysis (Chinese ed.).

Chapter 8
Distributed Collaborations and the Effect of Sociometric Feedback

Taemie Kim
MIT Media Lab, USA

Alex (Sandy) Pentland
MIT Media Lab, USA

ABSTRACT

Distributed collaborations tend to have different communication patterns and performances compared to co-located collaborations. This chapter proposes using sociometric feedback to understand and help distributed collaborations. The proposed system uses sociometric badges to automatically detect communication behaviors of groups and uses the information to provide real-time visual feedback. The goal of the feedback system is to encourage cooperation and furthermore improve the performance of both individuals and groups. This system could allow distributed collaborations to be more similar to co-located collaborations since it can sense and reintroduce the social signals lost in computer mediated communication. The chapter presents the results of two experimental laboratory studies that examine the effectiveness of the system. Results show that real-time sociometric feedback changes not only the communication patterns of distributed groups but also their performance, making it similar to that of co-located groups.

INTRODUCTION

Recall the last time you were assigned to a new group. You met your group members for the first time, introduced yourself, and then your group members started to talk about the tasks and roles of the group. But as the conversation developed, content of the conversation was not the only thing that you were focusing on: your senses were intensely observing the signals that are underneath the spoken language. Information such as who is sitting next to whom, who talks the most, who sounds the most confident, who asks questions, and who replies are all information that gave you a sense of how the group *works* and how you should behave in the group. Now imagine that this first encounter with the group was through a conference call. Your ears and brain had to work even

DOI: 10.4018/978-1-60960-091-4.ch008

harder to get a sense of the other group members: are they paying attention, are they expecting you to say something, or did you just say something stupid? You put in extra effort but it was much harder for you to get a clear understanding of what was happening and how the group *works*. You noticed that your behavior is somewhat different from what you normally do when you are in the same room with people: you either talked more or less than usual; you didn't know when you are supposed to speak; and sometimes you were less confident about what you just said since you couldn't see the responses of the other party.

As you can recall from your personal experience, distributed collaborations[1] are quite different from co-located collaborations. Researchers in organizational behavior have also confirmed the difference. Hinds and Bailey (2001) have demonstrated that distributed collaborations have very different dynamics compared to co-located collaborations, and that these differences often lead to poorer performance. Rocco (1998) and Bicchieri and Lev-On (2007) found that trust between group members breaks down in computer-mediated communications (CMC). Bouas and Arrow (1995) found that group identity in distributed groups was much lower compared to co-located groups. This makes it extremely difficult for distributed groups to establish and maintain trust and cooperation, which brings down the overall group performance.

Then why are distributed collaborations so different from co-located collaborations? It is surely not the content that is missing, as we have no problem sending and receiving content using voice or text. Then what is the missing piece that makes distributed collaborations so much harder than co-located collaborations? We believe social signals are the answer. Many of the social signal exchanges that provide context to the interactions are lost in distributed collaborations. This makes it harder for group members to understand how the group works, and how they should behave. Additionally, the lack of feedback from one another makes it difficult to reflect on their own behavior. The common use of question marks, exclamation marks, capital letters, and smiley faces are all examples of people trying to augment social signals in text based communication. The advance in communication technology has enabled some level of exchange in social signals. State-of-the-art video conferencing systems allow the exchange of richer visual signals, however there are still many social signals that are lost during the communication. Sense of proximity, back channel conversations, and eye contact are examples of social signals that are still not communicated in CMC.

Pentland describes these social signals as "honest signals" (Pentland, 2008). These are subtle patterns in how we interact with other people. He gives examples of honest signals as influence, mimicry, activity, and consistency. Influence is defined as the amount of influence each person has on another in a social interaction. Influence is measured by the extent to which one person causes the other person's pattern of speaking to match their own pattern. Mimicry is defined as the reflexive copying of one person by another during a conversation, resulting in an unconscious back-and-forth trading of smiles, interjections, and head nodding during a conversation. Increased activity levels normally indicate interest and excitement, as seen in the connection between the activity level and excitement in children. The consistency of emphasis and timing is a signal of mental focus, while greater variability is a sign of additional cognitive load, which is often associated with self-questioning, distracting thoughts, or emotionality. This is because when there are many different thoughts or emotions going on in your mind at the same time, your speech and even your movements become jerky and unevenly accented and paced. These signals create an additional channel of interactions helping group collaboration, which may lead to different decisions or performance.

With the advance of sensing technology, many of these signals can be measured using sensors. We believe that we can obtain a better understanding of group collaborations by measuring these social signals. Moreover there are ambient interface technologies to visualize the social signals without distracting the users (Ishii & Ullmer, 1997). We hypothesize that we can help distributed groups to have similar dynamics and performances of co-located groups by visualizing social signals. Our hypothesis is based on Walther's work (1996) that well designed computer mediation can support interpersonal communication. Moreover he argues that CMC could be extended so that it could facilitate communication that surpasses normal face-to-face interpersonal levels, which he refers to as *hyperpersonal* interaction. In the rest of this article we present the related work that inspired our research and then introduce our sociometric feedback system. In addition, we share findings from two laboratory experiments of our system and its implications.

BACKGROUND

Bos et al. (2002) explored how different modes of communication affect the level of group cooperation. They discovered that text-based communication showed the lowest level of cooperation, followed by audio and video. Face-to-face communication displayed the highest level of cooperation. The cooperation level of groups that communicated via audio and video slowly and vulnerably increased as they repeated the studies over a long period of time. However text-only communication never reached the level of cooperation of face-to-face communication during more than 30 trials. Hence we see that in distributed collaborations mediated by computers, the limited modality of communication delays the development of trust and makes it vulnerable.

Many researchers have tried to overcome the limitations of distributed collaborations by augmenting communication with additional channels such as voice, chat and video (Pedersen et al., 1993; Streitz et al., 1994; Tang et al., 2001). Yankelovich et al. (2006) verified the benefit of high audio quality on users' subjective rating of quality and social presence; however, the gain was offset by larger costs. Additional to the disadvantage of cost, Mark and her colleagues found that these added communication channels often cause distraction to the users (Mark, Grudin, & Poltrock, 1999). They also discovered that even with rich communication channels participants still pay less attention to the group dynamics when distributed compared to when the group was co-located. One of the solutions suggested by Mark et al. was to include an extra member whose explicit role was to facilitate the meeting. However, when human resources are precious, employing an additional member is usually not a viable option. Hence an automated real-time facilitation may be an affordable alternative option.

To measure and change the communication patterns of groups, we propose using automatic measurement and real-time visualization. In the past, researchers have defined measures of group dynamics in order to provide feedback to the group members (Hackman & Helmreich, 1987). However, most of the measures were based on evaluations using surveys or content coding which are inherently subjective and have long turn-around times. Automatic measurement and visualization can provide an objective view for the members to evaluate in a prompt manner. There are many examples of communication patterns being changed using real-time feedback. The most common one is encouraging balanced participation by showing the speaking ratio of the group members (Bergstrom & Karahalios, 2007; DiMicco et al., 2004; Kyulk et al., 2005). A fun example of Basu et al. is the use of an actuator to increase the intensity of lighting on under-participating individuals to emphasize their presence (2001). There have been a few efforts to apply these technologies to distributed collaboration. One example is the work by Leshed et al. where they show language-based feedback on a chatting system using a peripheral visualization (Leshed et al., 2009). This made participants express more agreements and focus more on their language use. Encouraged by these findings we posit that sociometric feedback can change groups' communication patterns to increase their strength of group identity; level of trust; and level of cooperation, which have been show to be strong drivers of their group performance (Fukuyama, 1995; Gambetta, 1988; Kramer & Brewer, 2006). We propose using sociometric sensors that measure various aspects of group communication and provide real-time visual feedback on peripheral displays.

SYSTEM DESCRIPTION

We introduce our system that we call Meeting Mediator (MM). Our system consists of (i) wearable badges that measure communication patterns of groups, (ii) a back end system that synchronizes and analyzes the data, and (iii) a mobile phone display that provides real-time feedback based on the measurements.

Sociometric Badges

The Sociometric badge (Figure 1, right) is a wearable electronic sensing device that collects and analyzes social behavioral data (Olguin Olguin et al., 2009). It is designed to be worn around one's neck like a typical company ID badge. It measures many of the common social signals expressed by the wearer with the exception of video and voice content to avoid the risk of privacy. Its current capabilities include:

- Extracting speech features in real-time to measure nonlinguistic social signals. The badge does not record any speech content, but is capable of identifying social behavior such as speaking time, speaking speed, and speech energy of the user. Turn taking or short affirming phrases reveal social dynamics that can be measured through synchronization of badges of multiple participants.
- Measuring body movement using a single 3-axis accelerometer. This can detect individual activities such as gestures and

Figure 1. The Meeting Mediator system used for the experiment. Sociometric badges (right bottom) capture communication patterns of groups and display it as real-time feedback on mobile phones (left top)

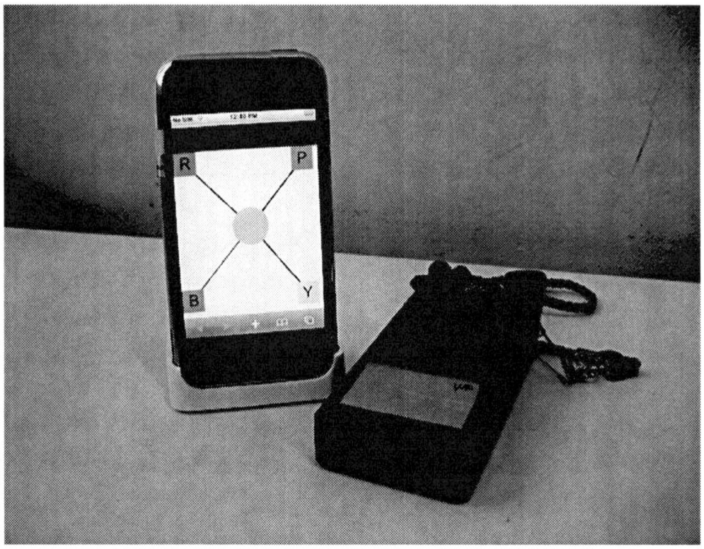

posture as well as social interactions such as body movement mimicry or rhythmic patterns.
- Real-time sending and receiving of information over 2.4GHz radio to and from different users and base stations for real-time communication.
- Detecting proximity data using the 2.4 GHz radio to understand the relational distance and position of multiple wearers. The badge can detect other badges within a 10m radius. This function can be used to detect the distribution of group members.
- Capturing and identifying face-to-face interaction using an IR sensor. By detecting the face-to-face alignment of individuals we are able to detect encounters as well as postural direction.
- Performing indoor user localization by measuring received signal strength from fixed based stations.

Measures

We used the data collected by the sociometric badges to understand group communication patterns. The proximity (radio) and posture alignment (IR) data are excluded from the lab study's analysis as all members' location and posture were constrained to their seats.

Speaking Time (ST)

We define total speaking time as the fraction of time during which an individual speaks, regardless of interruptions or overlap speech from others. Fraction of solo speech time (SST) for individuals is the time that no other person in the group is speaking at the same time the individual is speaking divided by the total time elapsed.

Speech Energy (SE)

We measure the variation in speech energy, in other words, the variation in speech volume. Higher variation in speech energy tends to make the speaker sound more expressive and energetic in speech.

Speaking Speed (SSP)

We measure the speed of speech by approximating the number of voiced segments per speaking time (Choudhury and Basu, 2004). Higher number of voiced segments indicates faster speech.

Average Speech Segment Length (ASL)

We define a speech segment as any one continuous stream of speech from an individual, regardless of interruption or overlap from other participants. A segment ends either by an interruption caused by another participant that resulted in the speaker to stop speaking or a significant length of silence.

Variation in Movement

We define movement energy (ME) as the average amount of body movement, i.e. the amount of gesturing during conversation. Movement energy variance (MEV) is the variation in movement energy over time which indicates the abruptness of movements.

Data Upload and Synchronization

Once the sociometric badges collect data, the mobile badges wirelessly transfer their data to a local base-station badge, which is tethered to a laptop computer. Each mobile badge transmits its data every 5 seconds. Whenever the laptop computer receives data from any of the mobile badges, it transfers the data to a central server where the database is located. The server accesses the data from all four badges and performs synchronization and analysis to calculate features such as speaking balance, overlapping speech, turn taking, and speaking speed. When groups are distributed, each room needs a set of a base-station badge and a laptop computer to upload the data. All other procedures are identical to the co-located case.

Visualization on Mobile Phones

The phone visualization component was designed to promote balanced participation and high level of interactivity. This behavior is not necessarily ideal for all types of group collaboration, but can be optimal for meetings such as brainstorming or problem solving where it is beneficial for all members to share the information that they have (DiMicco et al., 2004). Each of the four participants is represented by a colored square in the corner of the screen (Figure 2). In the user study, the square colors were identical to the color of each participant's badge and seat. The color of the central circle gradually shifts between white and green to indicate interactivity, with green corresponding to a higher level of interactivity. Balance in participation is displayed through the location of the circle: the more a participant talks, the stronger they pull the circle closer to their corner. In addition each member's speaking time is displayed through the thickness of the line connecting the central circle with each member's corner. The visualization is updated every 5 seconds and can be re-initialized every time a new meeting session starts. The data is accumulated throughout the meeting, showing the accumulated communication patterns from the start of the meeting to the current time. The devices are on the table facing each user. The display is intended to be in the periphery of the user's attention, and updates gradually so that it does not require constant attention from the user. Text and small details were also purposefully avoided so that a mere glimpse would be sufficient for information retrieval. The display is designed to be viewed occasionally for most cases. Only an extreme change in the group dynamics will draw the user's immediate attention.

Figure 2. Visualization on the phone emphasizes balance and interactivity in group collaborations: balanced and highly-interactive (left) or un-balanced and less-interactive (right). Circle color denotes group interactivity level, circle position denotes balance in participation, and line thickness denotes speaking time

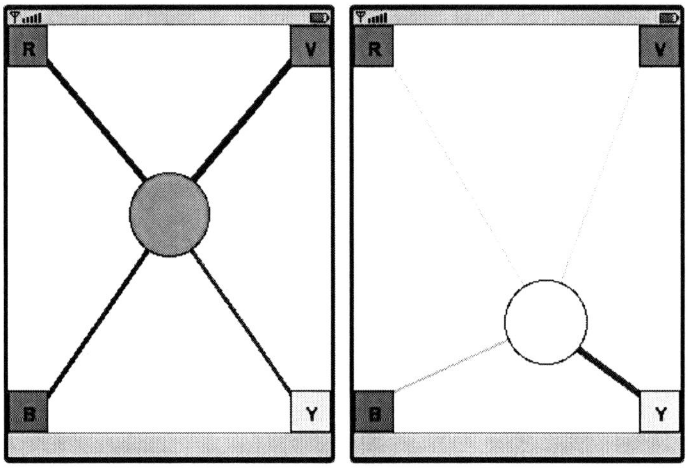

GROUP PROBLEM SOLVING EXPERIMENT

In our first experiment we chose to observe how groups interact with each other in idea generation and problem solving tasks. These types of collaborations are common in real-world meetings and provide insight into the group decision making processes (Romano & Nunamaker, 2004). We conducted a controlled laboratory study where we recruited four subjects per session. The task given to subjects was based on a modification of the game "Twenty-Questions", which integrated both brainstorming and problem-solving scenarios by closely replicating Wilson's experiments (Wilson et al., 2004). At the beginning of a task, each group was given a set of ten yes/no question-and-answer pairs. For the first phase of each task, groups were given 8 minutes to collaboratively brainstorm as many ideas that satisfy the set of question-and-answers. Then, continuing into the second phase, groups were given 10 minutes to ask the remaining ten questions of the Twenty-Question Game to determine the correct solution. Following each task, subjects filled out a post-experiment questionnaire comprised of five-point Likert scale questions regarding their own personality, the group dynamics and each individual's performance for each phase, and if applicable, the utility of the MM system.

Experimental Setup

We recruited 36 four-person groups, resulting in a total of 144 participants (71 male, 73 female, mean age=27.7). All four participants were each asked to wear the sociometric badge throughout the whole duration of the experiment. As the goal of the experiment was to verify the effect of sociometric feedback, a control setting with no feedback was compared to an experimental setting where there were four mobile phones sitting in front of the table, facing each participant. All four mobile phones showed the same feedback to each participant with a minor change of orientation. In the control setting, even though no feedback was given to the participants, their behavior was still measured by the sociometric badges. Also given the consent of all participants, the session was video recorded using two webcams in the corner of the rooms.

Each group participated in two tasks. For one of the tasks, all members were co-located in one room (co-located) and for the other task the members were separated in two pairs and situated in two separate rooms (distributed). The sequence of co-location and distribution were counter-balanced. Hence all groups experienced both co-located and distributed settings, with half of the groups co-located for the first round and then distributed for the second round. Hence all groups experienced both co-located and distributed settings, with half of the groups co-located for the first round and then distributed for the second round. For the other half of the groups, they had to participate in the first round separated into two rooms with no knowledge of who is in the other room. When distributed the participants were not able to see the participants in the other room and were only allowed voice communication.

Methods

To verify the effects of the system, we performed a between-subject experiment comparing 18 groups with mobile feedback on their mobile phones (experimental condition) to 18 groups without mobile phones (control condition). Each team began with one short practice task for which no score was recorded, and then performed two scored tasks. In one task, which we call the co-located case, all four participants were co-located having all audio and video communication available. In the other task, which we call the distributed case, the group was divided into pairs and a conference call setting was simulated by having a curtain between the two pairs. The sequence of co-located and distributed case was counter balanced to eliminate learning effects.

Performance (i.e. scoring) was determined by (1) the number of correct ideas in the brainstorming phase and (2) the number of questions used to arrive at the correct answer in the problem-solving phase. In the brainstorming phase, all participants were asked to write down possible ideas that satisfy the 10 question-and answer pairs. As the goal of brainstorming is to generate as many ideas as possible, we use the total number of ideas generated as a measure for the performance of the brainstorming phase. For the two brainstorming phases (8 minute co-located and 8 minute distributed), individuals generated a mean of 9.4 ideas. In the problem-solving phase, groups asked up to 10 questions to find the correct solution. They received a higher score if they used fewer questions and a zero score if they could not get the answer correct within 10 questions. Hence we use the number of questions each team used as a negated measure of the team's performance. On average, groups used 5.03 out of 10 possible questions to arrive at solution. In the problem-solving phase, groups were encouraged to freely discuss their ideas before deciding on a question to ask the experimenter that would be counted off from the ten possible questions. An additional goal incentive was provided in the form of gift certificates for the top-scoring team.

Hypotheses

The sociometric feedback motivates speech by visualizing the absolute amount of talking per person through the thickness of the lines between the circle and the corner representing each participant. Hence, we hypothesize that feedback will encourage meeting participants to speak more. Secondly, by the color of the center circle, feedback will encourage participants to be more interactive. We define interactivity as the frequency with which the main speaker gives up the stage for interruptions; hence by definition shorter speech segments corresponds to higher interactivity. We believe that the increased speaking time and interactivity will result in a greater number of ideas generated in the brainstorming meeting and higher efficiency in problem solving.

Results

Sociometric feedback had a very strong effect on speaking dynamics in both brainstorming and problem solving session. The primary effect was a dramatic reduction in overlapping conversations. This is in line with our qualitative observation that groups without feedback tended to divide into subgroups and have separate conversations instead of working as one team. The average overlap speaking time is significantly lower for subjects with feedback (mean=31.8% of the total time) than subjects without feedback (mean=49.2% of total time, $F(1,106)=17.8$, $p<.0001$). Due to the large difference in overlapping speaking time, the total speaking time was significantly shorter for subjects with feedback (mean = 41.0% of the total time) than subjects without feedback (mean = 57.1% of the total time, $F(1,106)=15.2$, $p<.001$). Therefore, when subjects were provided with visual feedback through MM, despite speaking less, they were more likely to collaborate with their teammates as one group. Further analysis of speech gives us new insights into the group interactivity level. Subjects with feedback have significantly shorter speech segment lengths (mean = 7.4sec) compared to those without feedback (mean =10.3sec, $F(1,106)=16.8$, $p<.0001$). This relationship is maintained in both brainstorming and problem-solving phases.

Our results suggest that sociometric feedback indeed changes the group dynamics. When feedback is introduced, groups become more polite and collaborative: they work better as one team, and there is more interaction and balance in the participation. However, there was no significant correlation between feedback and performance. In the brainstorming session, though there were changes in the interactivity level of the group, there were no significant changes in the number of ideas generated (Mean = 10.7 without feedback, 8.11 with feedback, $F(1,106)=3.01, p=.08$). This may be interpreted through Wilson's work (2004), where he found that if the given task was easy, more collaboration lead to worse performance. Similarly, in the problem solving phase we did not discover any significant effects with the use of feedback (Mean = 4.84 for groups without feedback, and 5.2 for groups with feedback, $F(1,106)=1.49, p=.23$). Our qualitative observations revealed that in some groups the dominant person would take over the conversation, limiting participation of others, while in other groups the dominant person acted as a facilitator (Mark, Grudin & Poltrock, 1999) who brings out ideas of all participants leading to better performance. Detailed results and analysis can be found in Kim et al. (2008).

SOCIAL DILEMMA EXPERIMENT

As we did not find any correlation with the performance we designed another study so that the collaboration style of the group has a direct influence on the group performance. We used the social dilemma problem to explore how feedback affects communication patterns and performance. A social dilemma can be defined as a situation in which a group of people must decide between maximizing selfish interests and maximizing collective interests (Komorita, 1995). It is generally more profitable for the individual to maximize selfish interests, but if all do so, all are worse off than if everyone had maximized collective interests. The social dilemma paradigm has been used extensively in social psychology and organizational behavior to understand cooperation both in laboratory and real world situations. The paradigm has often been deployed in the field of HCI, where it has been used as a common measure of trust or cooperation (Bradner & Mark, 2002; Kiesler, Sproull & Waters, 1996).

Experimental Setup

We created an identical setup as the group problem solving experiment. We conducted a between subjects experiment of groups of four subjects. A control setting with no feedback was compared to an experimental setting where there were four mobile phones sitting in front of the table, facing each participant. In the control setting, even though no feedback was given to the participants, their behavior was still measured by the sociometric badges. Each group participated in two rounds: once with all group members co-located in one room; and once separated into pairs in two different rooms. The sequence was counter balanced to minimize learning effects.

We recruited 45 groups of four subjects each, total of 180 participants (99 male, 81 female, mean age = 29.4, SD = 10.0). Subjects were recruited on multiple university campuses and through Internet message boards. Every effort was made to keep subjects from having prior knowledge of one another. All subjects were offered a flat fee of $9 for their time and an additional payment between $6-$34, depending on their choice and on the choice of other subjects in the same session. Sociometric badges were provided to all subjects to measure group communication patterns. Due to equipment failure, one of the badges did not collect audio data in the first few experiments. Both badge and performance data was discarded for the feedback-condition sessions using this badge, as the feedback presented to the subjects would not have been accurate. However the performance data of the no-feedback condition sessions were included in the analysis, as the decisions of the subjects were not affected by the badge malfunction. The total number of groups included in the analysis is shown in Table 1.

Methods

To provide a social dilemma situation, we created a conflict between maximizing communal interests

Table 1. Number of four person groups included in the analysis

Performance Data Analysis		
	# of groups with feedback	**# of groups without feedback**
Co-located first	14	9
Distributed first	13	9
Sociometric Data Analysis		
	# of groups with feedback	**# of groups without feedback**
Co-located first	14	9
Distributed first	13	9

and selfish interests: all subjects were told that the goal of the task was to maximize group earning, but the payoff structure was dependent on personal performance. Each subject was given sixteen $1 bills of which they could invest any partial amount to the group fund and keep the rest for themselves. The total money gathered in the group fund was increased 50% by the experimenters and divided equally among the four participants. Hence investing all $16 to the group fund would help increase communal interests, but keeping all $16 would maximize personal earnings given a set of decisions of the other participants. However if all participants were to keep all their money, all participants would earn 50% less than what they would have earned if everyone invested all. The instruction sheet had a sample calculation table and the experimenter confirmed that all subjects comprehended the situation and the consequences of their decisions

Once all participants understood the problem and the goal of the task, each group was given up to 10 minutes to discuss their group strategy. Badge data collected during these discussions were analyzed as we hypothesized that the communication patterns of these discussions would predict the decisions made by the group. After the discussion time, participants were asked to turn away from each other to make their contribution decisions into two envelopes (one labeled group and the other labeled keep) and then handed both envelopes to the experimenter so that individual decisions were not known to the other participants. Once all the envelopes were collected, the experimenter announced the total amount in the group fund and the amount to be distributed. But the individual contribution amounts were never revealed. The same procedure was repeated twice: once co-located and once distributed with the sequence counter-balanced. In both co-located and distributed settings, the final earnings were dependent on the decisions of all four participants, requiring all participants to collaborate to maximize their group or individual earnings. All subjects received the average of their earnings in the two rounds. After each round, the total amount in the group fund was announced to all participants. Hence in the second round, the subjects knew the total group fund amount of the first round.

Level of cooperation is measured by the amount contributed to the group fund. Each individual had two chances to invest $0 to $16 to the group fund, making $32 the maximum amount that a participant can invest. The individual earnings were the sum of what they kept for themselves in the keep envelope plus a quarter of the group fund that was increased 50%. The performance of a group is the sum of individual earnings. In our calculating scheme, group performance has a linear relationship with the total sum of investments in the group fund. The trial was repeated twice and individuals were offered the average of the two trials.

Hypotheses

Studying the literature of cooperation, we discovered that many scholars have found trust to be the strongest driving factor of cooperation (Fukuyama, 1995; Gambetta, 1988). Fukuyama goes further to argue that high trust among citizens account for the superior performance of all institutions in a society. Another important factor that stands out is group identity. Dawes (1988) posits that cooperation is a sign of strong group identity, which indicates itself as the consensus to cooperate. Kramer and Brewer (2006) increased the level of cooperation and group identity by verbally emphasizing the shared nature of the outcome. We measure the subjective impression of group identity using a post-task survey. To measure group identity, we used Hinkle's Group Identity Scale (Hinkle et al., 1998). The scale includes 9 questions that measure the cognitive, evaluative, and affective aspects of group identification. We use all nine questions to create a single measure for group identity. Each question has a 5-point ranging from 1 (strongly disagree) to 5 (strongly agree). The Cronbach's alpha for the group identity scale was 0.88 (N=288, each round counted separately, discarding omitted entries).

Orbell and Kragt (1988) found that having an opportunity to discuss with group members greatly increases the chance of cooperation compared to when they did not have any chance to discuss at all. Building on his findings, the research question we are asking is if the change of style in the discussion would have an effect on the level of cooperation. Sociometric feedback has been shown to make people's communication patterns more interactive. We hypothesize as groups have more communication and higher interaction levels, this will help them to feel stronger ties to one another leading them to make decisions maximizing collective interests rather than selfish interests. More interaction and communication may also lead to inducing commitment of cooperation, which have been shown to have a strong effect on cooperation (Kerr & Kaufman-Gililand, 1994; Orbell 1988). Trust was also measured using a 5-point Likert scale in the post-task survey. The level of trust was asked of each individual member and of the overall group. The questions of the individual members were *"I trust that the person who was on my right/left/across will do what they said in the conversation"*. The question asking for the overall trust level was *"I trust the members of my group"*. The participants had to fill out the post-task survey before the group fund amount was announced.

We believe this effect will exist for both co-located and distributed groups but the effect size will be larger for distributed groups. Our reasoning follows. First if we assume that H1 holds, then we can naturally deduce that bigger changes in communication patterns will lead to bigger changes in cooperative behavior. Second, as Bos and his colleagues (2002) discovered, there is very little headroom for effect for groups collaborating face to face: the group's cooperative level is close to maximum even in their first encounter. However for distributed groups with only audio communication, Bos found the cooperation rate is lower than 70% of the maximum, allowing plenty of headroom for effect. Hence we posit that sociometric feedback will increase the level of group identity, trust, and cooperation, and the size of effect will be greater for distributed groups. Level of cooperation is measured by the amount contributed to the group fund. Each individual had two chances to invest \$0 to \$16 to the group fund, making \$32 the maximum amount that a participant can invest. The individual earnings were the sum of what they kept for themselves in the *keep* envelope plus a quarter of the group fund that was increased 50%. The performance of a group is the sum of individual earnings. In our calculating scheme, group performance has a linear relationship with the total sum of investments in the group fund. The trial was repeated twice and individuals were offered the average of the two trials.

Results

We designed the experiment as two rounds so that we can compare how groups behave differently when they were co-located versus when they were distributed. However as we were conducting the study, we noticed a very apparent pattern: the co-operative level of the second round was strongly driven by the first round ($r = 0.47, p < .00000001$). For groups that were co-located first, the fact that they have already seen each other seem to have an effect even if they were separate in the second round–a result in line with Rocco's finding that trust can be repaired in CMC by having initial face-to-face contact (1998). For groups that were distributed first, the impact of having a report on the first round, which were mostly negative, seem to be much greater than the fact that they got to see the people in the other rooms. Hence it made more sense for us to distinguish the groups as co-located first or distributed first in our observations by merging the first and second rounds. The following results are based on comparing four scenarios of groups: (a) co- located first without feedback; (b) distributed first without feedback; (c) co-located first with feedback; and (d) distributed first with feedback. The former two correspond to the two control settings and the latter two correspond to the experimental settings.

First we observe the effect of sociometric feedback on group identity measured by post-task survey. When compare the four different scenarios, we can see that groups that were distributed first without feedback have a significantly lower group identity than the other three scenarios (Mean = (3.67, 3.25, 3.63, 3.54), $F(3,29)=3.85$, $p<.05$). This means that the feedback had significant effect on groups that were distributed first causing them to have a group identity level similar to that of co-located groups. Similar results were show in the level of trust. Again the groups distributed first show much lower level of trust toward their group members than the other three scenarios (Mean=(4.16, 3.22, 4.08, 3.92), $F(3,29)=3.98$, $p<0.05$). We can infer from the results that the sociometric feedback has a stronger effect on distributed collaboration, making their level of group identity and trust similar to that of co-located groups. As can be seen from the similarity of patterns, group identity and trust have very highly correlations ($r=0.78$).

The final and the most interesting question is if this change in behavior resulted in changes in level of cooperativeness. We look at the total amount of money invested in the group fund as a measure of cooperation. The maximum amount of contribution is $32 which is when the participant fully invested their $16 into the group fund both rounds. We find the same patterns as group identity and trust as we compare the four scenarios. The cooperation level was significantly lower for groups that were distributed first without feedback, but feedback significantly increased the group cooperation level (Mean = $27.85, $23.29, $26.64, $28.69, $X^2(3,176)=10.17$, $p<.05$). Hence the sociometric feedback significantly increased the cooperation level of the distributed groups such that there is now no statistically significant differences between groups that started co-located or distributed. Due to the design of our study, the group performance had a linear relationship with the sum of their individual cooperation levels. As the performance data was non-Gaussian we used a Kruskal-Wallis non-parametric test to examine these effects.

THE POTENTIAL FOR REAL-WORLD APPLICATIONS

Based on our findings our next question is would this system still have an effect in the real world. We have yet to test our system in real-world distributed teams, however we have reasons to believe in its effectiveness in the real world. We deployed the sociometric badges at a leadership forum held in Tokyo, Japan. The student participants wore the badges during all working hours for 7 working days. The Forum involved an experiential group project that involved creative engineering that required cooperation of all group members. The Forum brought together 20 students from the US, mostly from universities in the Greater Boston area, and 20 students from Japan, mostly from universities in the Tokyo area, working in teams of 6 to 8 people. We observed their communication patterns as they were working together in teams. At the end of each day we provided the team members with a sheet of paper including a visualization of the group communication patterns.

Figure 3 shows one of the team's face-to-face interaction pattern captured by the sociometric badges over the course of the week. The size of each circle represents the amount of time a person participated in the conversation. The color of the circle represents how interactive each person was based on their turn-taking pattern (the greener, the more interactive), and the width of the edges represents the turn-taking patterns (the thicker links indicate that the two had a lot of back-and-forth conversations). In this particular team, half of the students in the team were Japanese (4KF, 4MH, 4YI, 4TS) and the other half were American (4HS, 4GO, 4SI, 4BR). We can observe from the width of the links (turn-taking frequency) that the American students interacted mostly among themselves during the first few days and that this pattern changed over the course of the week as the Japanese students became more integrated in the team. We can also see that one of the students (4SI) was taking the most dominant role on the second day but dramatically changes his behavior the next day. His change in behavior was probably due to the visual feedback he received after the second day. By the end of the week the interaction pattern of the team became more balanced, optimized for collaborative idea generation.

Again as this was not a controlled study with statistically sufficient numbers, we cannot verify that our feedback caused this direction of change. The change might have been caused solely by the fact that members got acquainted with each other or that their roles have evolved throughout the week. However given the positive responses of participants and the size of effect, we believe that our feedback did have an impact on the group

Figure 3. Evolution of a team's face-to-face interaction pattern over the course of one week

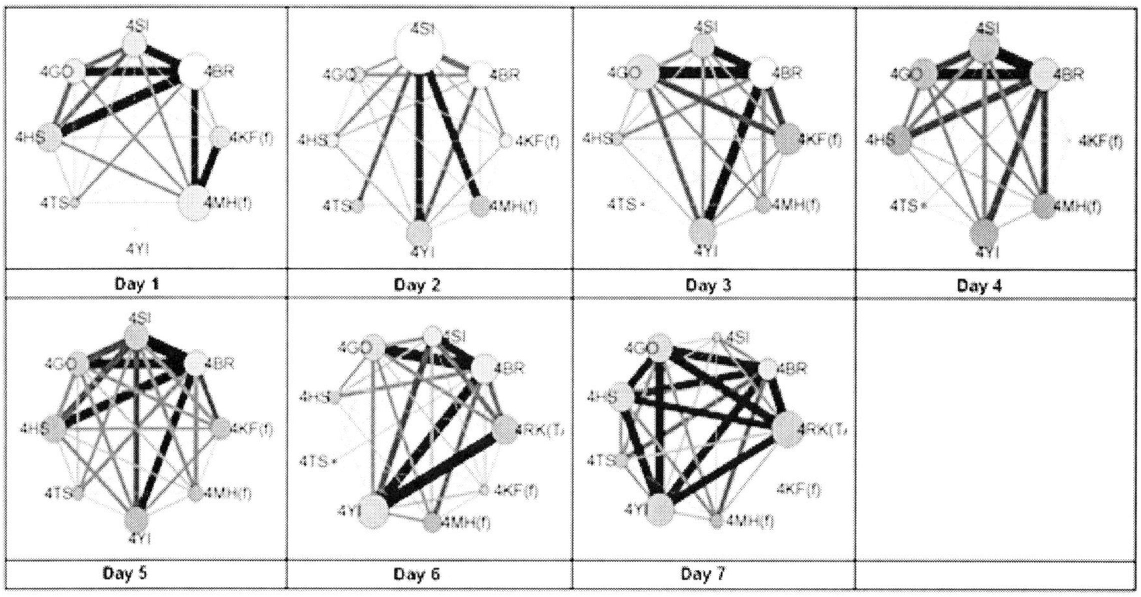

collaboration. This study reveals possibility of using sociometric feedback in the real world along with its potential effect. Moreover based on our findings in our laboratory study, we believe the impact may be deployed to distributed groups and the effect size may be greater for distributed.

CONCLUSION AND FUTURE DIRECTIONS

We showed that a simple visualization of sociometric data can change communication patterns of groups in group problem solving meetings and groups in social dilemma situations. The change in communication patterns were larger for distributed groups than co-located groups. This resulted in increased level of group identity, trust, and cooperation and the increment was significant in groups that were distributed. This in turn increased the overall performance of the distributed groups. We have shown that sociometric feedback is effective in aiding the cooperation of distributed groups, making many of its aspect no longer different from co-located collaboration. Our study provides new possibilities for bridging the gap between co-located and distributed team collaboration.

We propose to continue this direction of research to enhance cooperation and performance of real-world distributed teams. The easiest example could be the system embedded into a conference call system where sociometric feedback can be visualized on monitors. A self reflection system can be designed as a web or mobile application for people, which could be especially helpful for self development or training. Overall the ability to measure group communication patterns and providing guidance for improvement is an emerging area as more and more communication is happening over distance. We believe that the use of sensors and real-time peripheral visualization could help group collaboration, leading to more successful outcomes.

REFERENCES

Basu, S., Chouhdury, T., Clarkson, B., & Pentland, A. (2001). Towards measuring human interactions in conversational settings. *Proceedings of the IEEE International Workshop on Cues in Communication*, IEEE.

Bergstrom, T., & Karahalios, K. (2007). Conversation clock: Visualizing audio patterns in co-located groups. *Proceedings of the Hawaii international Conference on System Sciences*, IEEE, 78.

Bicchieri, C., & Lev-On, A. (2007). Computer-mediated communication and cooperation in social dilemmas: An experimental analysis. *Politics, Philosophy & Economics*, *6*(2), 139–168. doi:10.1177/1470594X07077267

Bos, N., Olson, J., Gergle, D., Olson, G., & Wright, Z. (2002) Effects of four computer-mediated communications Channels on trust development. In *Proceedings of Conference on Human Factors in Computing (CHI)*, 135-140.

Bouas, K., & Arrow, H. (1995). The development of group identity in computer and face-to-face groups with membership change. [CSCW]. *Computer Supported Cooperative Work*, 153–178.

Bradner, E., & Mark, G. (2002). Why distance matters: Effects on cooperation, persuasion and deception. In *Proceedings of ACM Conference on Computer Supported Cooperative Work (CSCW).*

Choudhury, T., & Basu, S. (2004). Modeling conversational dynamics as a mixed memory Markov Process . In *Proceedings of Neural Information Processing Systems (NIPS)*. IEEE.

Dawes, R. M. (1988). Not me or thee but we: The importance of group identity in eliciting cooperation in dilemma situations: Experimental manipulations. *Acta Psychologica*, *68*(1-3), 83–97. doi:10.1016/0001-6918(88)90047-9

DiMicco, J. M., Pandolfo, A., & Bender, W. (2004). Influencing group participation with a shared display. In Proceedings of *ACM Conference on Computer Supported Cooperative Work* (CSCW).

Fukuyama, F. (1995). *Trust*. New York: Free Press.

Gambetta, D. (1988). *Trust: Making and Breaking Cooperative Relations*. Cambridge, UK: Blackwell.

Hackman, R., & Helmreich, R. L. (1987). Assessing the behavior and performance of teams in organizations: The case of air transport crews . In Peterson, D. R., & Fishman, D. B. (Eds.), *Assessment for decision*. New Brunswick, NJ: Rutgers University Press.

Hinds, P. J., & Bailey, D. E. (2001). Out of sight, out of sync: Understanding conflict in distributed teams. *Organization Science*, *14*(6), 615–632. doi:10.1287/orsc.14.6.615.24872

Hinkle, S., Taylor, L. A., Fox-Cardamone, D. L., & Crook, K. F. (1998). Intragroup identification and intergroup differentiation: A multicomponent approach. *The British Journal of Social Psychology*, *28*, 305–317.

Ishii, H., & Ullmer, B. (1997). Tangible bits: Toward seemless interfaces between people, bits, and atoms. In *Proceedings of Conference of Human Factors in Computing*, 234-241.

Kerr, N., & Kaufman-Gililand, C. (1994). Communication, commitment, and cooperation in social dilemmas. *Journal of Personality and Social Psychology*, *66*, 688–698. doi:10.1037/0022-3514.66.3.513

Kiesler, S., Sproull, L., & Waters, K. (1996). A prisoner's dilemma experiment on cooperation with people and human-like computers. *Journal of Personality and Social Psychology.*

Kim, T., Chang, A., & Pentland, A. (2008). Meeting mediator: Enhancing group collaboration with sociometric feedback. In *Proceedings of Conference on Computer Supported Collaborative Work*, 457-466.

Komorita, S. (1995). Interpersonal relations: Mixed-motive interaction. *Annual Review of Psychology*, *46*, 183–207. doi:10.1146/annurev.ps.46.020195.001151

Kramer, R., & Brewer, M. (2006). Effects of group identity on resource use in a simulated commons dilemma. *Rationality and Society*, *18*(4), 443–470.

Kyulk, O., Wang, J., & Terken, J. (2005). Real-time feedback on nonverbal behavior to enhance social dynamics in small group meetings. In *Proceedings of the Conference of Machine Learning for Multi-modal Interaction (MLMI),* 150-161.

Leshed, G., Perez, D., Hancock, J. H., Cosley, D., Birnholtz, J., Lee, S., et al. (2009). Visualizing real-time language-based feedback on teamwork behavior in computer-mediated groups. *Proceedings of Conference of Human Factors in Computing* (CHI).

Mark, G., Grudin, J., & Poltrock, S. (1999). Meeting at the desktop: An empirical study of virtually co-located teams. *Proceedings of the European Conference on Computer Supported Cooperative Work*, 159-178.

Olguin Olguin, D., Waber, B., Kim, T., Mohan, A., Ara, K., & Pentland, A. (2009). Sensible organizations: Technology and methodology for automatically measuring organizational behavior. *IEEE Transactions on Systems, Man, and Cybernetics. Part B: Cybernetics.*

Orbell, J., & Kragt, A. (1988). Explaining discussion-induced cooperation. *Journal of Personality and Social Psychology*, *54*(5), 811–819. doi:10.1037/0022-3514.54.5.811

Pedersen, E., McCall, K., Moran, T., & Halasz, F. (1993). Tivoli: An electronic whiteboard for informal workgroup meetings. *Proceedings conference on Human factors in computing systems*, ACM Press, 391-398.

Pentland, A. (2008). *Honest signals*. Cambridge, MA: MIT Press.

Rocco, E. (1998). Trust breaks down in electronic contexts but can be repaired by some initial face-to-face contact. *Proceedings of Conference of Human Factors in Computing*, 496-502.

Romano, N. C., Jr., & Nunamaker, J. F. (2002). Meeting analysis: Findings from research and practice. *Proceedings of System Sciences*, IEEE Press.

Streitz, A., Geissler, J., Haake, J., & Hol, J. (1994). DOLPHIN: Integrated meeting support across liveBoards, local and remote desktop environments. *Proceedings of the ACM conference on Computer supported Cooperative Work*, 345-358.

Tang, J., Yankelovich, N., Begole, J., Kleek, M., Li, F., & Bhalodia, J. (2001). ConNexus to Awarenex: Extending awareness to mobile users. *Proceedings of Conference on Human Factors in Computer Systems*, 221-228.

Walther, J. (1996). Computer mediated communication: Impersonal, interpersonal, and hyperpersonal interaction. *Journal of Communication Research*, *23*(1), 3–43. doi:10.1177/009365096023001001

Wilson, D. S., Timmel, J. J., & Miller, R. M. (2004). Cognitive cooperation: When the going gets tough, think as a group. *Human Nature (Hawthorne, N.Y.)*, *15*(3), 225–250. doi:10.1007/s12110-004-1007-7

Yankelovich, N., Kaplan, J., Provino, J., Wessler, M., & DiMicco, J. M. (2006). Improving audio conferencing: Are two ears better than one? *Proceedings of the 2006 Conference on Computer Supported Cooperative Work*, 333-342.

ENDNOTE

1 For the purpose of simplicity, geographically distributed groups are referred as distributed groups throughout the manuscript.

Chapter 9
Methods of Group Decisions and Aggregation of Experts' Opinions

Jiří Vaníček
Czech University of Life Sciences in Prague, Czech Republic

Ivan Vrana
Czech University of Life Sciences in Prague, Czech Republic

Shady Aly
Czech University of Life Sciences in Prague, Czech Republic

ABSTRACT

Expert evaluation is often the most reliable way to predict results of a certain process. Estimates from several experts are usually needed in order to achieve correct decision making. This chapter presents and systematically sorts methods for aggregation expert opinions with respect to specific features of the decision making process. Based on information theory, the chapter introduces concepts for understanding expert consensusand also discusses the impact of suppressing/emphasizing extreme estimates or of unequally valuing individual estimates. This chapter addresses the estimation of YES/NO binary results as well as numeric values of a certain attribute and also considers cases when only an estimate or a dispersion of opinions of experts is considered.

INTRODUCTION

In many practical situations, in business and industry, most managerial decision-making problems require a collaborative scheme of decision-making, in which the final decision should rest on the opinions of several experts. This is for instance the case, when we have to predict whether some event will happen or not (e.g. predicting the sale of a newly launched commercial product), or when we must estimate attributes and parameters of a given entity or event, with respect to several decision-making criteria (e.g., assessing an industrial technology with respect to several decision criteria). In the first case, we have to predict a logical value (i.e., binary or YES-or-NO decisions), whereas in the second case, we have to predict a numerical value. Typically, group

DOI: 10.4018/978-1-60960-091-4.ch009

decision-making (GDM) problems are too significant for any single individual to handle. The reliance on group experts' decision is considered a very efficient solution especially in the case of ill-structured and vague decision-making situations. In ill-structured decision situations, it is often difficult to evaluate the correct decision solution beforehand. In such situations, usually not all the relevant variables are known. Secondly, some of the known variables are usually stochastic, vague, and qualitative in nature. In addition, in many decision-making situations, in treating a variety of business, economic, and agricultural decision problems, the problem usually involves several decision assessment dimensions, aspects, or view points. Consequently, multiple, different and specific expertise is often needed in making the decisions. This is why this decision-making environment necessitates the reliance on multiple experts' opinions, in order to enhance the quality of the decision solution obtained. However, most real world GDM takes place in ambiguous decision situations, in which the value of inputs as well as pertinent data and the sequences of the possible actions are vague or not precisely known. Therefore, it is very important to manipulate fuzzy values and concepts of evaluation in order to approximate the vague elements of the decision-making environment and also in order to comprehensively take into account all relevant effects the decision-making solution.

Actually, the most critical issue in the process of experts' GDM is the ability to aggregate or combine these experts' opinion, which may be in a different format or on different judgment scales. The aggregation method should handle all possible cases or situations of experts' judgments, either in the form of binary decisions, or in the form of numerical or linguistic values that commonly arise when attributes are evaluated using multiple criteria. The experts' opinion aggregation or combination methods can be in the form of simple formulas for computing the finally consolidated output of experts' judgments, or in the form of a logical algorithm or procedure for reaching a consensus also leading to a final group decision output. In all cases, these final group opinions must be of a sufficient degree of exactness and free of vagueness.

This chapter presents and systematically categorizes methods of how to aggregate expert opinions with respect to various specifics and objectives of a requestor. Based on information theory this chapter introduces a concept of experts' consensus and discusses the influence of emphasizing consensus or, alternatively, emphasizing the extreme opinions of experts for selection of a method for aggregation of estimates. The chapter also takes into account cases when opinions of experts have different values or weights. The text of this chapter differentiates between situations, when the estimated result is a binary one YES/NO or a numerical value of some attribute. Furthermore, the chapter distinguishes between cases when the requestor is only interested in an estimate itself or when he is also interested in the dispersion of opinions of individual experts.

The evaluation of group decisions and experts' opinions is a broad topic. The objective of this chapter is to show one way to approach this topic in a systematic way and with a precise foundation. It requires the use of some formal mathematical methods, which cannot be fully avoided. The authors have tried to minimize the requirements of previous knowledge of readers from a modern mathematical instruments domain. An explanation of such concepts, as may not be entirely usual, is graphically marked by indented text.

In the next section, the prominent literature on GDM is categorized and reviewed, and the modern research trends in the field are highlighted.

BACKGROUND

Due to the significance and polymorphism or diversity of multi-expert or group decision-making problems, extensive research has been conducted in the field. Since the aggregation or combination of expert's opinions or judgments is a very important element and mechanism for decision-making collaboration, a considerable number of researchers have addressed this issue. Generally, the treatment of GDM problems is performed under one of two main conditions: classical GDM under a crisp or certain environment, and GDM under a fuzzy environment. In the first case, the experts' opinions or judgments are dealt with or considered as crisp or exact, whereas in the second case experts' judgments are viewed as fuzzy or linguistic values. Concerning the first case, GDM under crisp environment, (Al-Ghoneim & Kumar, 1998) differentiated between three types of combination; abstract, rank and measurement. Abstract methods are based on voting procedures that are adopted from GDM theory such as unanimity, majority or plurality. Majority and plurality voting rules are the most widely used. A well-known rank type of combination is the Borda count. In measurement type of combinations, the combination criteria or algorithms have access to a set of numerical scores provided by experts. (Cordella et al. in 1999) stated that combining schemes that exploit information at the measurement level allow us to define combining rules that are more sophisticated and potentially more effective. The combination is based on establishing a unified numerical scale for experts' outputs.

In fact, there are numerous approaches for the combination of scores from various modules into a combined score for a group of modules. A popular approach uses quasi arithmetic means as a family of algebraic combination methods (Smolíková & Wachowiak, 2002; Ben-Arieh, 2005), such as arithmetic mean, geometric means and harmonic mean. In addition to simple combining criteria, some other approaches have been developed that combine experts judgments by consensus reaching (Ngwenyama et al, 1996; Bryson, 1996), or through employing pair-wise comparison judgments also called multiplicative preference relations (Herrera et al., 2001) utilizing the analytical hierarchy process (AHP) (Saaty, 1980; Ramanathan & Ganesh, 1994) and Laarhoven & Pedrycz (1983)), which can be viewed as hybrid approaches involving combination at both the rank and measurement levels.

On the other hand, due to the inherent ill-structured, vague and subjective nature of most real GDM problems, combining experts' judgments under a fuzzy environment is actually the prevalent trend in current GDM. There are several research directions in fuzzy GDM. A considerably large amount of research has been devoted to the application of the different version of Yager's ordered weighted averaging (OWA) operators (Yager, 1988), and linguistic quantifiers in GDM (Ben-Arieh, 2005; Calvo et al., 2004; Torra, 2004; Wang et al, 2007; Chakraborty & Chakraborty, 2007; Chiclana 2007). Other researchers focused on developing aggregation operators on fuzzy sets, including fuzzy numbers and linguistic labels; operations by which several fuzzy sets are combined in some way to produce a single representative either fuzzy or crisp set (Kacprzyk et al., 1992; Bardossy et al., 1993; Chakraborty & Chakraborty, 2007; Vaníček et al., 2009; Tsabadze, 2006; Tastle & Wierman, 2007; Wang, 2009; Zhang & Chu, 2009; Dokas, et al., 2009). Some other research efforts have undertaken consensus-based techniques or procedures as a method for achieving aggregation and arriving at final consolidated outputs (Kacprzyk & Fedrizzi, 1988; Kacprzyk et al., 1992; Ishikawa et al., 1993; Hsu & Chen, 1996; Herrera et al., 1996; Cheng, Y & Lin, 2002; Lee, 2002; Tsabadze, 2006; Tastle & Wierman, 2007). In addition, a considerable number of research attempts have been conducted in utilizing a fuzzified version of multi-criterion decision-making tools such as Fuzzy-AHP (Bozdag et al., 2003, Aly & Vrana, 2006, Duran & Aguilo, 2008; Aly & Vrana, 2008, Huanga et al., 2008; Hsu et al. 2010), fuzzy pairwise comparisons (Boender at al, 1989), multiple preference relations (Donga et al, 2009), and Fuzzy TOPSIS (Triantaphyllou &

Lin, 1996; Chen, 2000; Chu, 2002; Chen & Tzeng, 2004; Mahdavi et al. 2008; Wang & Lee, 2009).

Indeed, GDM under a fuzzy environment is the current modern research trend due to its realistic representation and modeling of most decision-making problems which are inherently ill-defined, subjective and ambiguous. This is exemplified in most recent applications of modern complex group decision-making problems that are mainly ill-defined and vague like those in the field of technology assessment (Bozdag et al., 2003; Duran & Aguilo, 2008; Hsu et al., 2010; Chuu, forthcoming), new product development (Wang, 2009), suppliers selection (Wu, 2009), and R&D projects selection (Huanga et al., 2008).

As, the survey of GDM literature has indicated, the utilization of fuzzy sets or fuzzy logic underpins GDM under a fuzzy environment, while reflecting the current trend in the field of collaborative decision-making.

FUZZY SETS

Not all questions of an outside world can be answered unambiguously. Uncertainty of answers can arise for many reasons: e.g. by stochastic events where individual results can happen with various probabilities, by a subjective assessment which leads to ambiguous conclusions, but also by quantum mechanic situations.

Mathematics uses fuzzy sets or a fuzzy logic to describe some of these uncertain situations. Both these approaches are in principle equivalent and could be converted to each other. Fuzzy systems were first introduced by Zadeh (1965).

The extension of a classical set theory to fuzzy sets is based on the fact that an element from a given universe could belong to a subset only "partially". The measure of this partiality is given as a value of a so-called membership function. The value of this function is a real number from a closed interval <0, 1>. The value 0 indicates that an element definitely does not belong to the set whilst the value 1 indicates that an element definitely belongs to the set. Elements with membership function values from an open interval (0, 1) belong to the set only with a certain measure of certainty, which increases with increasing value of membership function. The fuzzy set theory is then an extension of a classical set theory in the sense that for crisp sets the membership function can only gain values of 0 and 1, and not intermediate values. Basic set operations such as union, intersection and set difference can be introduced between fuzzy sets in such a way as to be an extension of the corresponding operations of classical set theory. This can be accomplished in many ways.

Fuzzy logic is an extension of a classical two-value (binary) logic. The truth-value of a variable or formula in a classical two-value logic can be only FALSE = 0 or TRUE = 1 and also the true valuation of predicates and formulae is a mapping into a two-element set {0, 1} in a classical predicate logic. In extension to a fuzzy logic we admit an uncertainty in a truth of predicates and formulae by admitting that a truth-value can be any number from a closed interval <0, 1>.

Analogically as for set operations, we define logical connectivities in fuzzy logic in such a way that the fuzzy truth-value of a composite formula is determined from fuzzy truth values of its components. Both approaches (fuzzy sets and fuzzy logic) can be converted to each other. There also exists a natural association between operations for fuzzy sets and logical connectivities in a fuzzy logic, where union corresponds to the disjunction; intersection corresponds to the conjunction, and set difference to the negation.

The truth-value of a complex formula is defined in a natural way by the truth values of its components. Both approaches (fuzzy sets and fuzzy logic) can be transformed to the other by the following principle: The value of a membership function $\mu_A(x)$, $0 \leq \mu_A(x) \leq 1$ to the fuzzy set for the element x from the universe to the fuzzy set A can be interpreted as a truth-value of a statement $x \in$

A (the element *x* belongs to *A*). On the other hand, the truth-value of the predicate or the formula *P* can be interpreted as a membership function of this predicate (formula) to the fuzzy set of truth predicates or formulae on the given universe.

More detailed and advanced information about modern fuzzy set and fuzzy logic theory can be found in several monographs, e.g. Hájek (1998) or Klir and Yuan (1995). The application of fuzzy sets theory to decision-making and expert systems is described by Zimmerman (1987).

Instruments of fuzzy sets and fuzzy logic seem to be the relatively most convenient tools for processing of data describing uncertain information, because they enable one to adequately work with uncertainties due to different reasons. We can use them also for the case when expert opinions or sources for a group decision-making lead to uncertain conclusions, and/or for the processing of situations when these sources are uncertain themselves.

It is very important to know not only a prevailing opinion but also a measure of a common agreement for a final decision in these types of uncertainty. Information is often valuable, whether uncertainty follows only from different standpoints of individual respondents while their standpoints are unambiguous, or whether an internal uncertainty of individual participants also exists. We should take into account that a collective opinion of a group or aggregated experts opinion need not be the single input for a final decision.

Based on fuzzy sets and fuzzy logic then also further concepts are used to describe a level of uncertainty. Concepts such as fuzzy number or a fuzzy interval are suitable to estimate the value of some attribute which is not exactly known, but which we want to express its estimate simultaneously with a measure of our uncertainty. We shall explain these concepts immediately before their use.

Fuzzy sets theory moves the aggregation one step away from a simple combination; there is still no representation of why people have disagreed. Although agreement for different reasons is not covered by the model, it seems that fuzzy sets applied to reasons, not to the outcomes may provide a useful extension to our approach.

CONSENSUS

Records of opinions of all experts or group members participating in group-decision making can be unclear. A subject who needs information about a measure of consensus often prefers one single number which adequately expresses this measure of consensus. If opinions of experts were not collected independently but within a common group discussion, it can be helpful to carry out a "preliminary voting" during such a discussion. It is also important for this approach to estimate a measure of achieved consensus clearly, compressed into a single value. Publishing this value instead of individual opinions makes stated points of individual group members anonymous which might be sometimes desirable.

Further useful information for more advanced users can be a level of agreement of experts. Ambiguity of an aggregated experts' opinion can be caused by the following two reasons: individual experts can be unsure of their standpoints, and experts can mutually differ in their opinions, or by a combination of both reasons. It can be useful to provide users with information about what is the main reason of a final summary recommendation or estimate. This could be achieved by a determination of consensus.

The consensus concept used in this chapter has its origin in the work of Tastle and Weirman (2007), where it was introduced for the case of an ordinal Likert scale, in which experts' opinions are expressed by selecting one of a few categories. Moreover, the variables are unsuitably commented and interpreted there. Based on information theory, the entropy of aggregated estimates is calculated, which is a measure of consensus. This chapter presents a more general description and interpretation of the consensus concept. It allows one to also evaluate those estimates where a classification scale need not be necessarily discrete, but an estimate can be expressed by an arbitrary real number within a given span. The proposed approach is also computationally simpler, because it does not require determining the frequencies of classifications in individual categories, but it directly follows from estimates of individual experts.

Let us have n experts ($n \geq 2$) who express their answers YES/NO in an ordinal scale by a number from the closed interval <L, U>, where L means "definitely NO" and U means "definitely YES". The higher selected value, the higher is the certainty in the YES answer. Then L and U are limits, which can, but need not, be used by experts in specific estimates. Let us denote $D = U - L$ a span between extreme options and x_i, $i = 1, \ldots, n$ estimate of the i-th expert. For example, this can be accomplished by a 7-point classification scale 0, 1, 2, …, 6, where $L = 0$, $U = 6$, $D = 6$ or a scale –2, –1, 0, 1, 2, where $D = 4$, but also by a requirement to express certainty level by a membership function to a fuzzy set of where $L = 0$, $U = 1$ and $D = 1$. Let us denote $X = (x_1, x_2, \ldots, x_n)$ the expert's evaluation vector and

$$M = (\sum_{i=1}^{n} x_i) / n$$

an arithmetic mean of experts' estimates. It follows from considerations about entropy of distances of individual estimates from the mean that it makes sense to define consensus by the formula:

$$\text{cons}(X) = 1 + \frac{1}{n} \cdot \sum_{i=1}^{n} \log_2(1 - \frac{\text{abs}(x_i - M)}{D}).$$

It holds:

$0 \leq \text{cons}(X) \leq 1$.
cons(X) = 1 is full agreement of experts in any classification.
cons(X) = 0 is extreme disagreement, when half of the experts select the minimum and the other half the maximum classification.
Dissention is then naturally defined as diss(X) = 1 – cons(X).

The value cons(X) can be considered as a membership function of the given distributions of experts' estimates to the fuzzy set of distributions with full agreement of experts' opinions. Analogically, diss(X) can be considered as a membership function of the given distributions of experts' estimates to the fuzzy set of distributions with full disagreement of experts' opinions.

The following examples illustrate concrete distributions and corresponding values of consensus and dissention. Individual examples describe the behavior of 8 experts in typical situations. They use a scale of seven degrees, where 0 means definitely NO and 6 means definitely YES. (see Table 1)

DISCUSSION OF VARIOUS GDM VARIANTS

This chapter simplifies the most recent GDM aggregation methods developed in modern literature in the field. These aggregation methods should

Table 1.

Expert's evaluations	Arithmetic mean	Consensus
0, 0, 0, 0, 0, 0, 0, 0	0	1
6, 6, 6, 6, 6, 6, 6, 6	6	1
3, 3, 3, 3, 3, 3, 3, 3	3	1
0, 0, 0, 0, 6, 6, 6, 6	3	0
0, 0, 0, 0, 0, 0, 0, 1	0.125	0.945
0, 0, 0, 0, 0, 0, 1, 1	0.25	0.905
0, 0, 0, 0, 0, 0, 0, 2	0.25	0.884
1, 2, 1, 2, 1, 2, 1, 2	1.5	0.874
0, 1, 2, 3, 4, 5, 6, 3	3	0.538
0, 1, 2, 3, 4, 5, 6, 6	3.375	0.417
3, 4, 4, 5, 5, 5, 6, 6	4.75	0.782

allow for consideration of many aspects. We list some of the most important below:

- Opinions and estimates of individual experts need not be of the same weight. Their importance depends on the expert's qualification and experience.
- Provision for dealing with both binary and multi-option GDM situations.
- Handling special cases of decision-making. In some cases we do not care about the dispersion level, in other cases we do. Sometimes from the point of view of preliminary care we should emphasize extreme opinions, in other cases (probably more often) we should suppress them to some extent or even fully eliminate them.

We can obtain a better idea about an expert's opinion, and also a better conclusion, in cases when experts not only declare the conclusion towards which they tend (truth-value or numeric estimate), but they also try to express the fuzziness level of their standpoint. Fuzzy logic shall be used to express the reliability of such a reply as the truthfulness of proposition. Fuzzy intervals or fuzzy numbers shall be used as an estimate of an attribute value.

Approaches will also be described of how to enable experts to express their opinion in the way they are used to (e.g. linguistically, without any knowledge of a fuzzy logic or basics of fuzzy set theory). For instance, it could be expressed in some scale, showing the confidence level of the truth level of their proposition, or graphically by a curve of estimates of possible numerical values. Each expert should have the possibility to formulate his/her estimate in his/her own way, e.g. with an adequately fine scale of his confidence values. Methods of how to unify such approaches for subsequent processing will be also described. However, the above treatment can assume that the experts' decision-makers are asked to provide a homogenous answer from a unified numeric or qualitative scale, or their form of judgments can be transformed and modified. Furthermore, we can deal with the case in which they do not provide homogenous forms of judgments.

Depending on the question formulation, a group decision and expert's opinions can serve two objectives:

1. To select one of two possibilities, for example whether to accomplish a certain action or not, or to judge whether the result of some activity will be a success or a failure.
2. To select a value of a certain attribute of an entity or process, e.g. what gain/loss will be achieved, how long will the process duration be, what will be the costs of some activity, etc. This value can serve as one input for further reasoning.

Depending on whether we are or aren't simultaneously interested about the certainty measure of the answers of individual experts to the given question, we can still distinguish two cases in both the above objectives:

a. We require an unambiguous answer with respect to an expected attribute value in the form of a logical value TRUE/FALSE, or in the form of an integer or real number.
b. We require an answer containing a measure of a subjective uncertainty of a given expert in the form of a value of a membership function to the fuzzy set of positive recommendations, or in the form of a fuzzy interval or a fuzzy number describing an expected attribute value.

Combining both variants we get the following four situations:

1a) The response should be in a binary, two-value logic.

1b) The response in fuzzy logic should be a value of the membership function to the given fuzzy set of positive responses.
2a) The response should be a real number.
2b) The response should be a fuzzy number or a fuzzy interval depending on the experts' confidence on the values of the possible value forecast.

In the following paragraphs we shall analyze and discuss the above stated individual cases 1a, 1b, 2a, 2b.

Various methods of opinion aggregations with respect to which extent they support dispersion among respondents/experts will be mutually compared, whereas an aspect of the computing time complexity will be also considered.

Experts usually express a subjective level of uncertainty of their answer in an ordinal scale. The problem is to transform the ordinal scale into an interval scale where differences between each two consequent values of the scale are equal. In such cases, out of all the aggregation methods, only the calculation of a median and extremes are invariant. Thus an additional problem, which arises, is how to substitute an ordinal scale by an interval scale or ratio scale.

According to the question formulation, two processing alternatives should be distinguished in all the above mentioned cases, because an estimated result can often differ from a recommendation based on the result:

I) In the first condition we only ask for the prediction of some result or a value of either a binary or numeric attribute of a certain entity, whilst we don't expect a recommendation for our behavior from experts. In this situation, we should realize that an error of a binary answer YES instead of NO could have a different real effect than an error of NO instead of YES. Also, a lower estimated numeric value than a real one can have a different effect than a higher estimated value. This means that an expert estimate cannot be equated with a recommendation of how to decide. For decisions based on the expert estimate of a result we should transform obtained aggregated values via the so-called production function, which transforms estimates at their consequences. We shall discuss this for individual estimation cases 1a, 1b, 2a and 2b.

II) In the second condition we directly ask for a recommendation. If it is not directly obvious, we must acquaint experts with consequences of both: of a correct and a wrong estimate of a deviation in attribute value estimate. If this practice was agreed upon, the transformation is not necessary. If this practice was not adopted, the transformation is necessary.

The formulation of a final recommendation after the aggregation of expert opinions must necessarily contain the estimate itself: whether YES or NO, or an expected attribute value. This conclusion could be still supplemented by additional information to instruct users.

In this section, all possible common variants of GDM problem situations will be investigated, and the methods of solutions will be described.

YES/NO Answer at the Given Question

This corresponds to the 1a) situation. Evaluation is very easy when there is a query to directly recommend a certain action.

If the opinions of all requested experts have the same weights, then we determine only scores YES = Y, NO = N. The predominating value obtained by this "voting" represents a final recommendation. The ratio Y: N provides sufficient information about the consensus level. Of course, consensus and dissention of experts' opinions can be also assessed by a membership function value to a fuzzy set of agreement described in the earlier paragraph.

It is suitable to give different weights to individual experts' opinions in some situations. This can be for several reasons, e.g. different qualification of individual experts, our different experience from their previous use. The weights are useful also when experts evaluate different aspects or viewpoints, whilst individual aspects can be differently important. Then we assign different weights to aspects to represent their importance. Then Y is the sum of weights of all experts who answered YES and N is the sum of weights of experts who answered NO. We achieve the final recommendation and consensus estimate again as a ratio Y: N. This approach is universal. It can always be used when we assign a different importance to expert's opinions. The opinion of an expert is simply substituted by the opinions of so many experts with corresponding weight. Further processing is identical in the case when all experts were equivalent. Therefore, we shall not explicitly mention this further in this text.

In the case when we query only the result of some event without determining what importance has a correct estimate and what loss is associated with a wrong estimate, then we should perform another transformation utilizing a benefit function. Such a transformation is possible only in situations when both decision consequences are mutually comparable in the Archimedean structure; see Roberts (1979) or Krantz (1971) for details. If not satisfied, then expert estimates should be considered with great care. For example, in the situation of a large hope for a financial gain versus a small (but not zero) probability of losing life, it is desirable to select a safe approach, which avoids a fatal situation.

In an Archimedean case we have to assess both: gains and losses (positive numbers for a gain, negative for a loss) in the same ratio scale. Uncertainty can be substituted by a probability, which considers the mathematical expectation of benefit. We select the case with a higher mathematical expectation of the benefit.

Both values of expectation of benefit can serve as suitable additional information about the level of certainty with which a given decision is suggested. As for the consensus of respondents, we can proceed from an original frequency of option YES and NO.

Response in a Fuzzy Logic Given as a Value of the Membership Function to the Fuzzy Set of Positive Responses

This corresponds to the 1b) situation. In this case experts are queried not only at their opinion with respect to a result of a given event in a YES/NO form, but also at a level of certainty for this opinion. An answer is usually required in some Likert scale of the type $L, L + 1, L + 2, \ldots, U - 1, U$, where $L < U$. for example $0, 1, \ldots, n$, or $-m, -m + 1, \ldots, 0, 1, \ldots, m$. In the case of an odd number of degrees, a neutral answer NO PREFERENCE is also possible. In the case of an even number, the experts should incline to one of two possibilities YES/NO. The selection of a certain degree reflects the level of experts' certainty. Choice L means an absolute certainty of an answer NO, choice of U means an absolute certainty of an answer YES. Principally, any (real/rational) number from the interval $<L, U>$ can be accepted. Such an arrangement of queries is not usual. It is more convenient and usual to describe individual options by linguistic terms.

The Likert scale and responses of experts can be easily interpreted as values of a membership function to a fuzzy set of positive answers by the use of a linear transformation:

$$\nu(x) = \frac{1}{U - L} \cdot x + \frac{L}{U - L}.$$

The aggregation of opinions from individual experts still needs to be performed. The main challenge of aggregation is caused by the fact that expert evaluation is usually performed using an ordinal scale, because it is difficult to assign an empirical meaning to express a subjective certainty of some opinion.

It is not clear how to measure objective differences between neighboring points of the scale and to ensure that these differences are equal. Even an empirical meaning of these differences need not be clear. As an example, in the school classification it is doubtful, whether the differences between "excellent" and "very good" and between "good" and "not satisfied" have comparable consequences.

Only the calculation of maximum and minimum from given evaluations and the calculation of a median are invariant with respect to such transformations out of further described aggregation methods. All other methods require assessment on an interval scale (where a difference of two evaluations has a sense – then transformation functions in the form $t(x) = a \cdot x + b$, where $a > 0$ are permissible here) or even on a ratio scale, where the value itself is compared to a standard has a meaning (permissible transformation functions are in the form $t(x) = a \cdot x$, where $a > 0$). In the case of ordinal scale only, aggregation can lead to different conclusions when selecting other, equally justified selections of classification degrees. The selection of degrees for an ordinal expert evaluation is a crucial problem. It is extremely desirable to approach a situation when differences between grades are equal. One possibility of how to achieve this is to convert event assessment into assessment of its benefit, e.g. when this benefit can be expressed financially – i.e. financial benefit and loss can be expressed in currency units, which is a ratio scale. All further presented aggregation methods are invariant with respect to permissible transformations on a ratio scale.

Furthermore, we shall briefly introduce some principles discussed in Klir & Yuan (1995). Explanation will be presented for fuzzy sets and corresponding membership functions, i.e. values after the transformation given by the function $v(x)$ Results can be transformed to an arbitrary Likert scale using the inverse transformation to the transformation given by the function $v(x)$

The general aggregation operator is a mapping A, which for each n-tuple of values from a closed interval $\langle 0, 1\rangle$, $n \geq 2$ assigns a number from this interval and satisfies the following conditions:

i. $A(0,..., 0) = 0$.
ii. If $x_j \leq y_j$ for all $j = 1,..., n$, then $A(x_1,..., x_n) \leq A(y_1,..., y_n)$ holds.
iii. Mapping A is continuous. (This condition automatically holds if a scale is discrete, since every function defined at a discrete definition domain is continuous and therefore this requirement drops out).

An aggregation operator is called a fuzzy average if, in addition the following two conditions hold:

- The value of the mapping A is invariant with respect to an arbitrary permutation of its n variables.
- It holds $A(x, x,..., x) = x$, Thus identical estimates lead to the same average.

Any fuzzy average can be used to aggregate the opinion of experts. The arithmetic mean is mostly used.

If all estimates are positive numbers, then the arithmetic mean can be generalized to the quasi-arithmetic mean, also called the power mean or generalized mean, and is defined as

$$A_\lambda(x_1,\dots,x_n) = \left(\frac{1}{n}\cdot\sum_{i=1}^{n} x_i^\lambda\right)^{\frac{1}{\lambda}} \quad \text{for } \lambda \neq 0$$

For a special choice of λ we obtain:

For $\lambda = 1$ ~ arithmetic mean.
For $\lambda = 2$ ~ quadratic mean.
For $\lambda = -1$ ~ harmonic mean.
For $\lambda \to 0$ ~ geometric mean

$$G(x_1,\ldots,x_n) = \sqrt[n]{\prod_{i=1}^{n} x_i} = \exp(\sum_{i=1}^{n} \log(x_i))$$

For $\lambda \to +\infty \sim$ maximum.
For $\lambda \to -\infty \sim$ minimum.

It can be easily proven that the arithmetic mean is always greater than or equal to the geometric mean. The geometric mean is always greater than or equal to the harmonic mean. It generally holds that λ-quasi-arithmetic mean is for fixed data a non-increasing function of the variable λ for $\lambda \in (-\infty, 0) \cup (0, +\infty)$.

Arithmetic, harmonic and geometric means are mutually related by the following relationship:

If x and y are arbitrary real numbers, then both sequences $a_0, a_1, a_2, \ldots$ and $h_0, h_1, h_2, \ldots$, defined as

$$a_0 = x,\ h_0 = y,\ a_{n+1} = (a_n + h_n) / 2,\ h_n = 2 / (1 / a_n + 1 / h_n) \text{ for } i = 0,1, \ldots$$

converge to the geometric mean of x and y.

It is shown in the paper Vaníček 2009, that any aggregation operator can be transformed to another using a transformation by a continuous and increasing function and by applying an inverse function after the aggregation. If A and B are two aggregation operators, then there exists a continuous and increasing function φ, such that

$$B(x_1, x_2,\ldots, x_n) = \varphi^{-1}(A(\varphi(x_1), \varphi(x_2),\ldots, \varphi(x_n))).$$

The specific choice of a quasi-average depends on to the extent that we want to suppress or emphasize extreme values of assessment by individual experts. Results obtained in different alternatives differ mainly in cases when consensus of experts' assessments is low. Values $\lambda > 1$, emphasize influence of extremes, $\lambda < 1$ reduces influence of extremes. A concrete value of λ should be selected according to simulated examples in such a way to correspond to our intuitive perception of how to reduce or emphasize the extreme opinions of experts.

In the case when an ordinal assessment scale has zero as a neutral answer "NO PREFERENCE" and it uses positive numbers for a level of experts' confidence with a YES answer and negative numbers for a level of confidence with NO answer. Then quasi-averages for suppressing or emphasizing extremes can be used as follows:

Exclude all zero assessments and replace each positive assessment λ by the quasi-average of all positive assessments in order to reduce or emphasize extremes. Analogically proceed with negative assessments.

Sum separately all such modified positive assessments, and all such modified negative assessments.

Then sum the two obtained values (results of the above summations) and divide them by the number of experts (including those with zero assessments). Such an arithmetic average of modified values brings a final aggregated estimate which emphasizes extremes for $\lambda > 0$ and suppresses extremes for $\lambda < 0$, and simultaneously offers information about the aggregated confidence of experts to the suggested answer.

The weighted average allows one to respect the various importances of opinions of various experts. The weight vector $(w_1, w_2,\ldots, w_n)$ should satisfy the condition:

$$\sum_{i=1}^{n} w_i = 1.$$

Then the weighted average is determined by a formula:

$$A(x_1,\ldots,x_n) = \sum_{i=1}^{n} w_i \cdot x_i$$

Moreover, this fuzzy average is additive, which means that it holds

$(A\, x_1 + y_1,..., x_n + y_n) = A(x_1,..., x_n) + A(y_1,..., y_n)$ for all $x_1,..., x_n$ and $y_1,..., y_n$.

It can be proven that each additive aggregation operator is a weighted average. Weighted averages can be combined with quasi-averages. The same relations as for averages with various values of λ hold also for weighted averages.

To suppress or emphasize extreme opinions of n experts (where $n \geq 3$) it is appropriate to use ordered weighted averaging operators – OWA, where weights of individual assessments depend on their (ascending) order. Therefore, assessments should be first ordered with respect to their values in an ascending (or descending) sequence, and then we assign them the weights with respect to our wish to suppress or emphasize extreme opinions of experts. In other words, we determine a weight vector $(w_1, w_2,..., w_n)$ satisfying the condition

$$\sum_{i=1}^{n} w_i = 1$$

and find a non-aggregated value from the formula

$$A(x_1,...,x_n) = \sum_{i=1}^{n} w_i \cdot x_{p(i)}$$

where p is such permutation of indices 1,..., n that it holds:

$$x_{p(1)} \leq x_{p(2)} \leq \ ... \ \leq x_{p(n)}.$$

We obtain the following interpretation with respect to selected weight vector w:

$w = (1/n, 1/n..., 1/n)$ ~ arithmetic mean.

$w = (0,..., 0, 1)$ ~ maximum.

$w = (1, 0,..., 0)$ ~ minimum.

$w = (0, 1/(n-2),..., 1/(n-2), 0)$ ~ excludes the "best" and the "worst" evaluation.

$w = (0,..., 0, 1, 0,..., 0)$, where value 1 is only in the middle ~ median.

In this way, a proper selection of the weight vector at OWA allows one to determine the interpretation of aggregation transformation. For example we can select the extent to which extreme opinions are supported or suppressed. An interesting application of the aggregation of fuzzy opinions based on the similarity of distance between fuzzy opinions can be found on Cheng-guuo and Wang (2006).

Response is a Real Number

This corresponds to the 2a) situation. The aggregation of crisp estimates of attribute values expressed by real numbers is in principle governed by the same rules as aggregation of certainty estimates of YES/NO answers, discussed in the previous paragraph. An arbitrary fuzzy average can bring important information. We shall not repeat the description of individual methods.

In the case when the estimated value is in a ratio or (at least) in an interval scale, then the arithmetic mean is important. The maximum, minimum and median of estimates can serve as supplementary information. It can be helpful also to provide information about the distribution of estimates (e.g. in the form of quadrilles, deciles or other percentiles) in the case of a greater number of experts.

If it is clear that the estimated value belongs to a certain interval, the level of certainty of experts can be measured by calculating consensus as described earlier in this chapter.

If estimates are only in an ordinal scale, then the only undoubted (i.e. invariant with respect to admissible transformations) aggregation method is the calculation of maximum, minimum, median and percentiles. Other types of aggregation can provide only supplementary information which is satisfactory as long as we are confident that differences between individual assessments in this scale are proportional to something with an empirical meaning.

Response in a Fuzzy Logic Should be a Value of the Membership Function to the Fuzzy Set of Positive Responses

This corresponds to the 2b) situation. In order to provide experts with the possibility of expressing a level of their confidence that this value will be achieved, in addition to an estimate of the attribute value, we require answers in the form of a fuzzy interval or a fuzzy number.

The following illustrations in Figure 1 briefly remind readers of the basic concepts of fuzzy intervals and fuzzy numbers as special cases of fuzzy sets.

Fuzzy intervals and fuzzy numbers are suitable instruments for expressing individual uncertainty of experts in the estimation of numerical quantities. Individually for each expert they allow one to achieve both the best estimate, and also a possible dispersion and slope with which confidence of gaining values decreases in the vicinity of the best estimate. For fuzzy intervals the trapezoidal shape is considered most often; for fuzzy numbers, a triangular shape is common. Also other shapes of membership function can suit, e.g. a shape similar to a normal distribution.

Arithmetic operations can be introduced for fuzzy intervals and fuzzy numbers, which arise from corresponding operations with fuzzy sets. Out of possible variants of these operations it

Figure 1. Illustrations briefly reminding the basic concepts of fuzzy intervals and fuzzy numbers

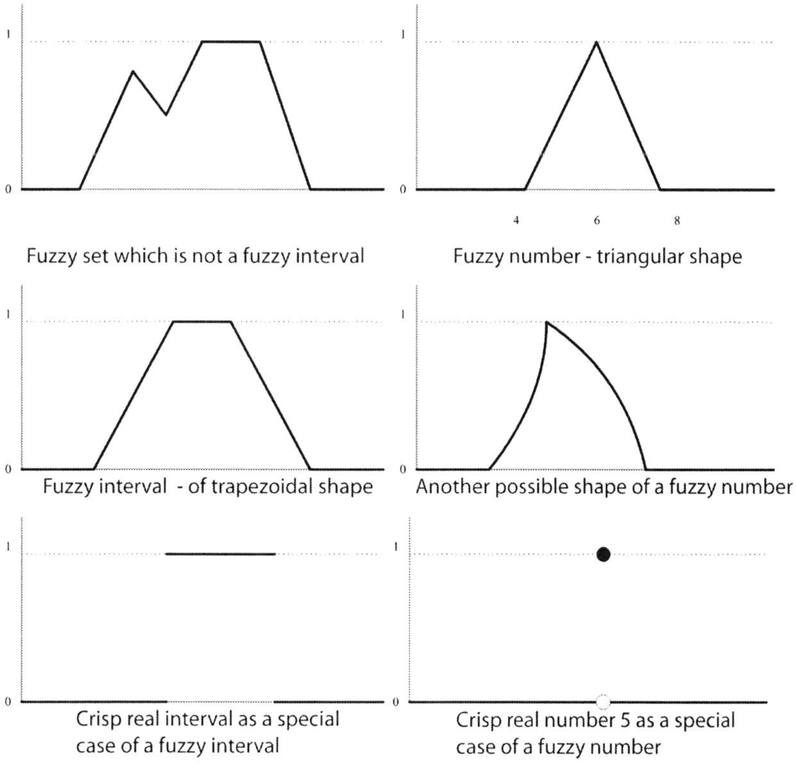

is suitable to select so-called standard fuzzy set operations (sometimes also called Zadeh's or Gödel's operation). These are defined by formulas:

$$\mu_{\bar{A}}(x) = 1 - \mu_A(x);$$
$$\mu_{A\cap B}(x) = \min(\mu_A(x), \mu_B(x));$$
$$\mu_{A\cup B}(x) = \max(\mu_A(x), \mu_B(x)).$$

These are most suitable for work with expert estimates since they suppose that ambiguity reasons of individual experts can be mutually dependent. This better reflects real situations than if we assumed a full independence of events and used the so-called product fuzzy set operations (also called probability operations), which correspond to composition of probabilities of independent events.

Extending arithmetic operations for real numbers at operations with fuzzy intervals and fuzzy numbers can be based at the so-called cylindrical extension of binary relations. This extension leads to the following rules for operations with fuzzy intervals and fuzzy numbers:

Let • be any symbol, from the collection {+, –, ·, / } of operations. Then

$$\mu_{A\bullet B}(x) = \bigcup_{y\in\Re, z\in\Re, y\bullet z=x} \mu_A(y)\cap\mu_B(z)$$
$$= \sup\{\min(\mu_A(y), \mu_B(z)) : y, z \in \Re, y \bullet z = x\}.$$

From $y+z=x$ follows that $y=x-z$. Searching for supreme of the function

$$\min(\mu_A(y), \mu_B(z))$$

for real numbers y, z, such that $y+z=x$ is nothing but searching for supreme of the function

$$\min(\mu_A(x-z), \mu_B(z)$$

for all z. It is analogical also for other operations. Thus we obtain the formulae:

$$\mu_{A+B}(x) = \sup_{z\in\Re}(\min(\mu_A(x-z), \mu_B(z)),$$
$$\mu_{A-B}(x) = \sup_{z\in\Re}(\min(\mu_A(x+z), \mu_B(z)),$$
$$\mu_{A\cdot B}(x) = \sup_{z\in\Re}(\min(\mu_A(x/z), \mu_B(z)),$$
$$\mu_{A/B}(x) = \sup_{z\in\Re}(\min(\mu_A(x\cdot z), \mu_B(z)).$$

The value of $\mu_{A\cdot B}(0)$ should be calculated separately with the use of an original definition because of problems with zero division.

Examples of adding and multiplying two fuzzy numbers are sketched in the following illustrations (see Figures 2, 3 and 4).

Figure 2. Adding and subtracting fuzzy numbers

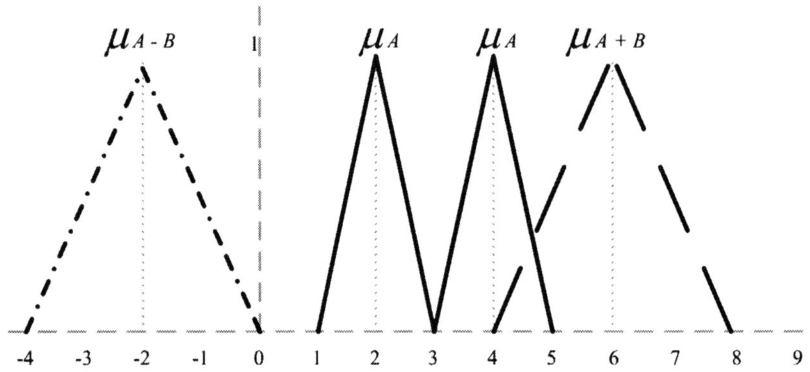

Figure 3. Multiplication of fuzzy numbers. numbers $\mu_{A \cdot B} = \sqrt{x+1} - 2$ *for* $x \in (3, 8)$; $\mu_{A \cdot B} = 4 - \sqrt{x+1}$ *for* $x \in (8, 15)$; $\mu_{A \cdot B} = 0$ *otherwise*

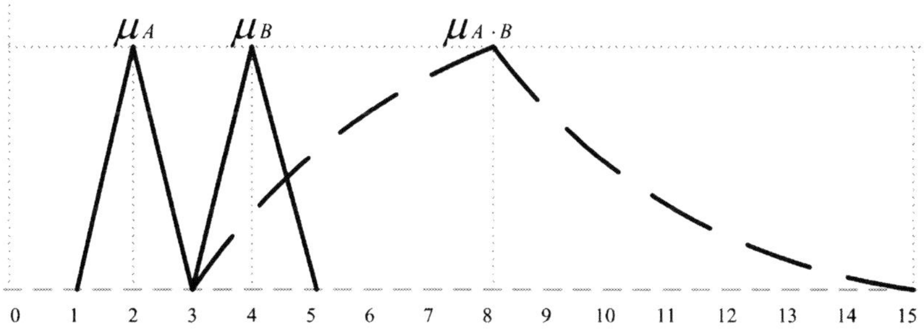

Figure 4. Division of fuzzy numbers. $\mu_{A/B} = (5x - 1) / (1 + x)$ *for* $x \in (1/5, 1/2)$; $\mu_{A/B} = (3 - 3x) / (1 + x)$ *for* $x \in (1/2, 1)$; $\mu_{A/B} = 0$ *otherwise*

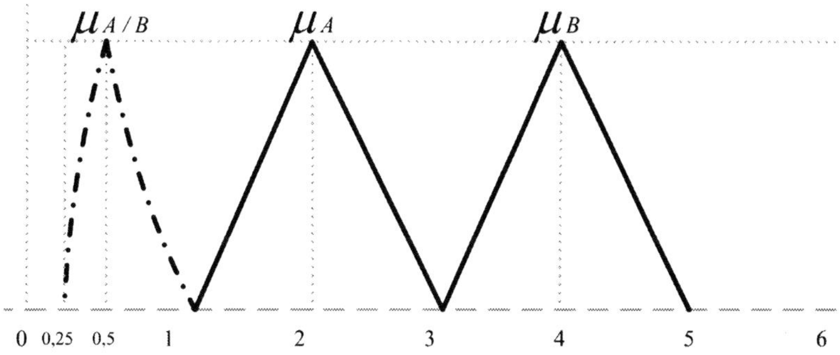

It is clear from the illustrations that analytical processing of more complex calculations with fuzzy numbers is not feasible.

More about fuzzy arithmetic can be found in Kauffman and Gupta (1985).

In utilizing these concepts in expert estimates one can carefully consider the form in which experts should present their estimates. The form of presentation should be easily understandable also for specialists in the given problem domain without their full orientation in the mathematical backgrounds of fuzzy sets and fuzzy logic. We shall mention two alternatives of how to ensure user friendliness for entering experts' opinions.

The first alternative does not necessarily require a graphical interface for communication with an expert. It consists of splitting a range of possible values of estimated attribute at the sequence of intervals, and we ask the estimated possibility that the estimated value falls into each of these intervals.

A graphical interface can further make selection easier when we do not ask for a numerical value but only an interactive input (e.g. by a mouse-click) at that point of scale, which corresponds to confidence of an expert that the value will be from the given interval. In fact, the completion of a table is needed. It is suitable to select the interval <0, 1> for a measure of confidence. Then we deal directly with an estimate of a membership function. Another scale can be transformed to this interval.

The graphical interface is necessary in the second alternative. In this case, the expert is asked to draw (manually or by some computer tool) a curve depicting a function in a Cartesian coordinate system: admissible values of an attribute versus confidence of an expert for that attribute value. Such an input is then digitized for further processing. Both alternatives lead to processing of functions, which are piecewise constant.

In the aggregation of fuzzy estimates (whether numbers or intervals) we can apply the same principles as with an aggregation of crisp numbers. Everything said in the previous paragraph still holds. In numerical calculations we do not use analytical forms of functions but we replace them by piecewise constant functions. We interlace a network of intervals across a support of estimates and the membership function is substituted by a suitable constant in every "small" interval of this network. The membership function drawn by an expert is thus replaced by a piecewise constant function.

Execution of operations with fuzzy intervals and fuzzy numbers is, of course, computationally more demanding and needs computer processing. According to the presented formulae, a time complexity of every arithmetic operation grows quadratically according to the number of intervals in the support. The time complexity is therefore quadratic with respect to granularity of the used scale, and linear with respect to the number of experts.

If we want to provide information about experts' consensus, it is also necessary to calculate consensus separately for each interval of a fuzzy number substitution. Consensus can differ in individual areas of a definition domain of the fuzzy number or interval.

CONCLUSION

This chapter tries compromise between the necessary expression accuracy and clarity for readers, who are not specialized in mathematics. It tries to clarify the essence of the problem and to define and explain basic concepts and their relationships. It does not advance proofs of individual statements. The basic knowledge of readers in fuzzy sets and operations as well as orientation in exact mathematical writing of expressions is assumed. Readers are referred to other resources for more detailed explanation.

In this chapter some selected methods of aggregation of expert opinions in a group expert evaluation are systematically described. The selection of methods does not pretend to be complete. It tries to present individual views on evaluation of expert opinions and to consider further important aspects as e.g. a weight of individual opinions, or whether we want to emphasize or to suppress the influence of extreme opinions. It is desirable to take into account a number of aspects in the selection of a specific method, e.g.: how detailed information is finally needed and how to achieve the necessary detailed information from experts. The following can be considered as the most important:

- *The required level of detail of information and its accuracy*: Whether we need either a simple answer to the given question and, eventually, a numerical estimate of a certain quantity, or whether we need to know also a level of agreement of experts to the answer. Moreover, it can be sometimes interesting also to know an aggregated measure of the confidence of individual experts about their answers. A case of a simple answer can be easier for decisions of a user who is only weakly oriented in a decision problem or who does not want to willingly influence the final decision. On the other hand, more detailed information can be helpful in situations when an expert assessment is only one of several inputs used for adopting a final decision.
- *The level of detail of requested information from experts:* This aspect should not be neglected, because collecting information from experts need not necessarily be easy. Often we cannot expect from an expert, an estimate of his confidence for the given

suggestion or estimate. On the other hand, "sharing doubts" can sometimes make experts' answer more easily. Generally, we cannot automatically expect from experts their knowledge about processing of expert estimates and about the mathematical background of such a processing. Even a basic knowledge of fuzzy set theory cannot be expected. A user friendly interface should be provided to enable easy communication and in order to facilitate non-sharp answers of experts. Transformation of answers into a standardized scale cannot be left to experts themselves. It should be part of the automated processing of their answers.

We can note in conclusion that the group expert evaluation utilizing a suitable aggregation method is one of the most reliable ways for correct decision-making and accurate estimates of quantities, in spite of the great subjectivity of expert estimates and in spite of problems associated with the fact that in the vast majority of situations we deal only with ordinal values.

ACKNOWLEDGMENT

The authors wish to thank the Czech Ministry of Education for support of this work by the Grant No. MSM 6046070904 "Information and knowledge support of strategic control" and Grant No.2C06004 "Intelligent tools for content assessment of relevance of general and specialized data and knowledge resources".

REFERENCES

Al-Ghoneim, K., & Kumar, B. V. K. V. (1998). Unified decision combination framework. *Pattern Recognition*, *31*(12), 2077–2089. doi:10.1016/S0031-3203(98)00030-2

Aly, S., & Vrana, I. (2006). Integrating Multiple Fuzzy Expert System under Restricting Requirements. *Agricultural Economics*, *52*(4), 187–196.

Aly, S., & Vrana, I. (2008). Evaluating the knowledge, relevance and experience of expert decision-makers utilizing the Fuzzy-AHP. *Agricultural Economics*, *54*(11), 529–535.

Bardossy, A., Duckstein, L., & Bogardi, J. (1993). Combination of fuzzy numbers representing expert opinions. *Fuzzy Sets and Systems*, *57*, 173–181. doi:10.1016/0165-0114(93)90157-D

Ben-Arieh, D. (2005). Sensitivity of multi-criteria decision-making to linguistic quantifiers and aggregation means. *Computers & Industrial Engineering*, *48*, 289–309. doi:10.1016/j.cie.2005.01.014

Boender, C. G. E., de Graan, J. G., & Lootsma, F. A. (1989). Multi-criteria decision analysis with fuzzy pairwise comparisons. *Fuzzy Sets and Systems*, *29*, 133–143. doi:10.1016/0165-0114(89)90187-5

Bozdag, C. E., Kahraman, C., & Ruan, D. (2003). Fuzzy group decision-making for selection among computer integrated manufacturing systems. *Computers in Industry*, *51*(1), 13–29. doi:10.1016/S0166-3615(03)00029-0

Bryson, N. (1996). Group decision-making and the analytic hierarchy process: Exploring the consensus-relevant information content. *Computers & Operations Research*, *23*(1), 27–35. doi:10.1016/0305-0548(96)00002-H

Calvo, T., Mesiar, R., & Yager, R. R. (2004). Quantitative weights and aggregation. *IEEE Transactions on Fuzzy Systems*, *12*(1), 62–69. doi:10.1109/TFUZZ.2003.822679

Chakraborty, C., & Chakraborty, D. (2007). A fuzzy clustering methodology for linguistic opinions in group decision-making. *Applied Soft Computing*, *7*, 858–869. doi:10.1016/j.asoc.2006.02.006

Chen, C. T. (2000). Extensions of the TOPSIS for group decision-making under fuzzy environment. *Fuzzy Sets and Systems*, *114*, 1–9. doi:10.1016/S0165-0114(97)00377-1

Chen, M. F., & Tzeng, G. H. (2004). Combining grey relation and TOPSIS concepts for selecting an expatriate host country. *Mathematical and Computer Modelling*, *40*, 1473–1490. doi:10.1016/j.mcm.2005.01.006

Cheng, C. H., & Lin, Y. (2002). Evaluating the best main battle tank using fuzzy decision theory with linguistic criterion evaluation. *European Journal of Operational Research*, *142*, 174–186. doi:10.1016/S0377-2217(01)00280-6

Chengguo, L. L. J., & Wang, Z. (2006). Aggregation of fuzzy opinions under group-decision-making based on similarity and distance. *Journal of Systems Science and Complexity*, *19*, 63–71. doi:10.1007/s11424-006-0063-y

Chiclana, F., Herrera-Viedma, E., Herrera, F., & Alonso, S. (2007). Some induced ordered weighted averaging operators and their use for solving group decision-making problems based on fuzzy preference relations. *European Journal of Operational Research*, *182*, 383–399. doi:10.1016/j.ejor.2006.08.032

Chu, T. C. (2002). Selecting plant location via a fuzzy TOPSIS approach. *International Journal of Advanced Manufacturing Technology*, *20*, 859–864. doi:10.1007/s001700200227

Chuu, S. J. (forthcoming). Selecting the advanced manufacturing technology using fuzzy multiple attributes group decision-making with multiple fuzzy information. *Computers & Industrial Engineering*.

Cordella, L. P., Foggia, P., Sansone, C., Tortella, F., & Vento, M. (1999). Reliability Parameters to Improve Combination Strategies in Multi-expert Systems. *Pattern Analysis & Applications*, *2*, 205–214. doi:10.1007/s100440050029

Dagdeviren, M., & Yuksel, I. (2008). Developing a fuzzy analytic hierarchy process (AHP) model for behavior-based safety management. *Information Sciences*, *178*, 1717–1733. doi:10.1016/j.ins.2007.10.016

Dokas, I. M., Karras, D. A., & Panagiotakopoulos, D. C. (2009). Fault tree analysis and fuzzy expert systems: Early warning and emergency response of landfill operations. *Environmental Modelling & Software*, *24*(1), 8–25. doi:10.1016/j.envsoft.2008.04.011

Donga, Y., Xua, Y., & Yu, S. (2009). Linguistic multiperson decision-making based on the use of multiple preference relations. *Fuzzy Sets and Systems*, *160*, 603–623. doi:10.1016/j.fss.2008.08.011

Duran, O., & Aguilo, J. (2008). Computer-aided machine- tool selection based on a Fuzzy-AHP approach. *Expert Systems with Applications*, *34*(3), 1787–1794. doi:10.1016/j.eswa.2007.01.046

Hájek, P. (1998). Mathematics of fuzzy logic. In *Trends in Logic* (Studia Logica library, vol. 4). Dordrecht: Kluwer Academic Publishers.

Herrera, F., Herrera-Viedma, E., & Chiclana, F. (2001). Multi-person decision-making based on multiplicative preference Relations. *European Journal of Operational Research*, *129*, 372–385. doi:10.1016/S0377-2217(99)00197-6

Herrera, F., Herrera-Viedma, E., & Verdegay, J. L. (1996). A model of consensus in group decision-making under linguistic assessments. *Fuzzy Sets and Systems*, *78*, 73–87. doi:10.1016/0165-0114(95)00107-7

Hsu, H. M., & Chen, C. T. (1996). Aggregation of fuzzy opinions under group decision-making. *Fuzzy Sets and Systems*, *79*, 279–285. doi:10.1016/0165-0114(95)00185-9

Hsu, Y. L., Lee, C. H., & Kreng, V. B. (2010). The application of Fuzzy Delphi Method and Fuzzy AHP in lubricant regenerative technology selection. *Expert Systems with Applications*, *37*(1), 419–425. doi:10.1016/j.eswa.2009.05.068

Huanga, C. C., Chub, P. Y., & Chiang, Y. H. (2008). A fuzzy AHP application in government-sponsored R&D project selection. *Omega*, *36*, 1038–1052. doi:10.1016/j.omega.2006.05.003

Ishikawa, A., Amagasa, M., Shiga, T., Tomizawa, G., Tatsuta, R., & Mieno, H. (1993). The max–min Delphi method and fuzzy Delphi method via fuzzy integration. *Fuzzy Sets and Systems*, *55*, 241–253. doi:10.1016/0165-0114(93)90251-C

Kacprzyk, J., & Fedrizzi, M. (1988). A soft measure of consensus in the setting of partial (fuzzy) preferences. *European Journal of Operational Research*, *34*, 315–325. doi:10.1016/0377-2217(88)90152-X

Kacprzyk, J., Fedrizzi, M., & Nurmi, H. (1992). Group decision-making and consensus under fuzzy preferences and fuzzy majority. *Fuzzy Sets and Systems*, *49*, 21–31. doi:10.1016/0165-0114(92)90107-F

Kauffman, A., & Gupta, M. M. (1985). *Introduction to fuzzy arithmetic: Theory and applications*. New York: Van Nostrand Reinhold.

Klir, G. J., & Yuan, B. (1995). *Fuzzy sets and fuzzy logic: Theory and Application*. Upper Saddle River, NJ: Prentice-Hall PTG.

Krantz, D. H., Luce, D. R., Suppers, P., & Tversky, A. (1971). Foundation of measurement: *Vol. 1. Additive and polynomial representation*. San Diego, CA: Academic Press.

Laarhoven, P. J. M., & Pedrycz, W. (1983). A fuzzy extension of Saaty's priority theory. *Fuzzy Sets and Systems*, *11*, 229–241.

Lee, C. C. (1990). Fuzzy logic in control systems: Fuzzy logic control parts I & II. *IEEE Transactions on Systems, Man, and Cybernetics*, *20*, 404–435. doi:10.1109/21.52551

Lee, H. S. (2002). Optimal consensus of fuzzy opinions under group decision-making environment. *Fuzzy Sets and Systems*, *132*, 303–315. doi:10.1016/S0165-0114(02)00056-8

Mahdavi, I., Mahdavi-Amiri, N., Heidarzade, A., & Nourifar, R. (2008). Designing a model of fuzzy TOPSIS in multiple criteria decision-making. *Applied Mathematics and Computation*, *206*, 607–617. doi:10.1016/j.amc.2008.05.047

Ngwenyama, O. K., Bryson, N., & Mobolurin, A. (1996). Supporting facilitation in group support systems: Techniques for analyzing consensus relevant data. *Decision Support Systems*, *16*, 155–168. doi:10.1016/0167-9236(95)00004-6

Ramanathan, R., & Ganesh, L. S. (1994). Group preference aggregation methods employed in AHP: An evaluation and an intrinsic process for deriving members' weight ages. *European Journal of Operational Research*, *79*, 249–265. doi:10.1016/0377-2217(94)90356-5

Roberts, F. S. (1979). *Measurement theory with applications to decisionmaking, utility and social sciences*. Reading, MA: Addison-Wesley.

Saaty, T. L. (1980). *The analytic hierarchy process*. New York: McGraw-Hill.

Smolíková, R., & Wachowiak, M. P. (2002). Aggregation operators for selection problems. *Fuzzy Sets and Systems*, *131*, 23–34. doi:10.1016/S0165-0114(01)00252-4

Tastle, W. J., & Wierman, M. J. (2007). Consensus and dissention. A measure of ordinal dispersion. *International Journal of Approximating Reasoning*, *45*, 531–545. doi:10.1016/j.ijar.2006.06.024

Torra, V. (2004). OWA operators in data modeling and re-identification. *IEEE Transactions on Fuzzy Systems*, *12*(5), 652–660. doi:10.1109/TFUZZ.2004.834814

Triantaphyllou, E., & Lin, C. T. (1996). Development and evaluation of five fuzzy multi attribute decision-making methods. *International Journal of Approximate Reasoning*, *14*, 281–310. doi:10.1016/0888-613X(95)00119-2

Tsabadze, T. (2006). A method for fuzzy aggregation based on group expert evaluations. *Fuzzy Sets and Systems*, *157*, 1346–1361. doi:10.1016/j.fss.2005.11.015

Vaníček, J., Vrana, I., & Aly, S. (2009). Fuzzy aggregation and averaging for group decision-making: A generalization and survey. *Knowledge-Based Systems*, *22*, 79–84. doi:10.1016/j.knosys.2008.07.002

Wang, T. C., & Lee, H. D. (2009). Developing a fuzzy TOPSIS approach based on subjective weights and objective weights. *Expert Systems with Applications*, *36*, 8980–8985. doi:10.1016/j.eswa.2008.11.035

Wang, W. P. (2009). Evaluating new product development performance by fuzzy linguistic computing. *Expert Systems with Applications*, *36*(6), 9759–9766. doi:10.1016/j.eswa.2009.02.034

Wang, Y. M., Luo, Y., & Hua, Z. (2007). Aggregating preference rankings using OWA operator weights. *Information Sciences*, *177*, 3356–3363. doi:10.1016/j.ins.2007.01.008

Wu, D. (2009). Supplier selection in a fuzzy group setting: A method using grey related analysis and Dempster–Shafer theory. *Expert Systems with Applications*, *36*, 8892–8899. doi:10.1016/j.eswa.2008.11.010

Yager, R. R. (1988). An ordered weighted averaging aggregation operators in multi-criteria decision-making . *IEEE Transactions on Systems, Man, and Cybernetics*, *18*, 183–190. doi:10.1109/21.87068

Zadeh, L. A. (1965). Fuzzy sets. *Information and Control*, *8*, 338–353. doi:10.1016/S0019-9958(65)90241-X

Zhang, Z., & Chu, X. (2009). Fuzzy group decision-making for multi-format and multi-granularity linguistic judgments in quality function deployment. *Expert Systems with Applications*, *36*, 9150–9158. doi:10.1016/j.eswa.2008.12.027

Zimmerman, H. J. (1987). *Fuzzy sets, decision-making, and expert systems*. Boston, MA: Kluwer Academic Publications.

KEY TERMS AND DEFINITIONS

Attribute: An inherent measurable physical or abstract property or a characteristic of entity that can be distinguished quantitatively or qualitatively by a human or automated means.

Averaging Operator: A formula or an algorithm for replacing the set of numbers that represent individual opinions by one number that represents a common opinion.

Consensus: The measure of the expert's opinions matching based on the information theory. The consensus is one when the expert's opinions are completely matching and is zero when the dispersion of expert's opinions is maximal.

Expert Evaluation: Judgment of an activity or event based upon assessment and expertise in an application area, knowledge area, discipline, industry, etc. Such an expertise may be provided by a group or a person with specialized knowledge, skills, experience or training.

Fuzzy Interval: A fuzzy set defined on an open interval (a, b) $a \leq b$, of real numbers, whose membership function gains the value 1 on the interval (x, y), $a \leq x \leq y \leq b$, and is not decreasing on (a, x) and not increasing on (x, b).

Fuzzy Number: A fuzzy set defined on an open interval (a, b) of real numbers, whose membership function gains the value 1 on one and only one point $x \in (a, b)$, $a \leq b$, and is not decreasing on (a, x) and not increasing on (y, b).

Fuzzy Set: A measure of the certainty that a given element belongs to a given set (has a given property), expressed as a membership function with domain on the universe and range on the interval <0, 1>. Its value 1 is interpreted as definitely YES, value 0 as definitely NO.

Measure: A number or a category assigned to an attribute of an entity by making a measurement or an assessment.

Ordered Weighted Averaging Operators (OWA): The weighted averaging operator with weights assigned depending on the position of individual estimation in the ordered set of all estimates.

Weighted Averaging Operators: The averaging operator that computes a common opinion from the individual opinions depending on the weights assigned to individual opinions.

Chapter 10
Cognitive Tools for Group Decision Making:
The Repertory Grid Approach Revisited

Marco Castellani
University of Brescia, Italy

ABSTRACT

This chapter combines the use of cognitive mapping and the repertory grid technique in a socio-organizational perspective and in a problem-solving oriented approach, so as to avoid some recurrent trappings of decision making such as biased goal-oriented behaviour and misleading perceptions of the task environment. The approach requires that a group of people facing a forthcoming choice are randomly split up into three sub-groups of nearly the same number. Subjects in the first sub-group are interviewed about their representation of the problem setting and on potential strategies. In this preliminary step the interviewer, after building up the resulting individual cognitive maps, extracts and codes the main recurrent concepts ("states of the world"). These concepts are used for an evaluation of the task environment by the second sub-group, whose subjects use the repertory grid technique. Individuals are shown how to express their viewpoint in terms of "constructs", which are theoretical abstractions for exploring concepts or real events. The repertory grids elicited provide the building blocks for the final phase of the approach, assigned to the third sub-group. This group is required to generate feasible alternatives for targets derived from grid evidence by exploiting the "province of meaning" through a very simple diagrammatic scheme. The entire approach, which represents a narrow method where each step is linked to the following one, is discussed by making references to the results of a pilot experiment.

INTRODUCTION

The use of different tools for group decision making is widespread in the literature (Gray, 1987; Shakun, 1988). Decision support systems, for example, have been playing an increasingly important role both for routine choices and strategic planning activities inside organizations at all levels (Desanctis & Gallupe, 1987). Despite this, the real effectiveness of tools for the aid of group decision making remains a topic of debate especially for problem setting and problem solv-

DOI: 10.4018/978-1-60960-091-4.ch010

ing issues (Unnikrishnan Nair, 2001). A crucial question under discussion, and one which is probably still unsolved, is the relationship between the exploitation of such tools and the cognitive style of individuals (O'Keefee, 1989). This chapter takes sides inside this debate, not specifically focusing on decision support systems but on more general tools for aiding individuals and group decisions: cognitive mapping (Warren, 1995) and, in particular, the repertory grid (hereafter, RG) technique (Kelly, 1955; Winter, 1985). The main viewpoint is that such tools can help individuals particularly if used in a manner that tries to avoid possible trappings of human decision making. A treatment like this requires the segmentation of the decision-making process and supports the methodological choice of dividing participants in three sub-groups, each one responsible for a stage of the procedure.

A first possible source of bias is behaving in a goal-oriented way during problem setting, a recurrent barrier to the clearer understanding of the situation in its entirety (Walton, 1990). Many studies have emphasised how individuals can be affected by the need to find feasible solutions for the problematic situation they are facing and can be unfocused by so called "motivated reasoning" (Jain & Maheswaran, 2000), or through the perception of goals as always achievable (Kruglanski, 1990). The sharing of such perceptions can lead to major phenomena of uniform thought and static mental models, which represent critical barriers for decision makers (Hodgkinson *et al.*, 1999).

To this end, in order to keep away from possible sources of slackness in the organization life (Christensen & Fjermestad, 1997), the preliminary step of the approach presented here is dedicated to a simple adaptation of cognitive mapping and is addressed to a first sub-group of individuals (see Figure 1). This variation implies a concealment strategy in which cognitive maps, built up through semi-structured interviews, are fragmented in order to highlight just the starting concepts of individuals' vision of the world ("cognitive operational synthesis 1" in the box of Figure 1), and to bypass selective attention that humans sometimes pay just on goals they consider "reachable", even unintentionally.

Another source of a possible misleading perspective that the chapter seeks to analyze concerns an opposite type of obstacle, i.e. the attitude that decision makers often demonstrate of being influenced by their initial perception of the task environment, and unduly too path dependent (Ross *et al.*, 1975). This is the main reason why a second sub-group of subjects is called to make evaluations on elements which are provided by the cognitive maps developed by the sub-group in the preliminary step (see Figure 1). A dedicated use of RG technique is thought to be useful for this aim, since it allows subjects to clarify the nature of the relationship linking given elements and constructs. In this procedure RG is not considered in its common usage, in terms of exploration of the subjects' psychological inner world or interpersonal space, but simply as a tool that can help people to make

Figure 1. A wide-ranging flow-chart of the proposed technique

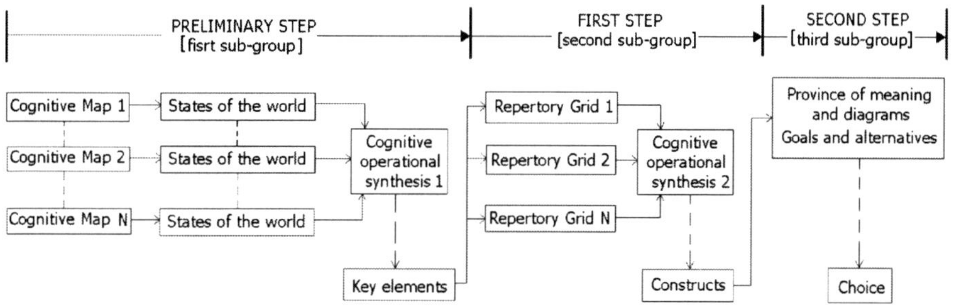

clear the significant meaning assigned to a specific task environment.

A statistical analysis ("cognitive operational synthesis 2" in the box of Figure 1) done on RGs lets the researcher submit to subjects of a third sub-group a set of information they must play upon for the set up of their diagram of the situation, in which the main goals have to be highlighted. The process ends with a narrowing scheme that allows the third sub-group to reach one or more final decisions to undertake. In this second and final step, in fact, individuals must choose the most important goal and suggest an action for achieving it, through an incremental procedure. Moreover, they are supposed not to be biased either by their possible outcome orientation or by their selective attention or over focused attitude (elements and goals are provided to them through the results of the previous sub-group).

The chapter is organized as follows. In the second section the preliminary stage of the procedure is shown, focusing on a problem space representation by the first sub-group. The third section deals with a general exposition of the RG technique (section 3.1) and with its peculiar version applied to the second sub-group of individuals (section 3.2: first step of the procedure). The fourth section's subject is the second and final step of the proposed technique, focused on the diagrammatic reasoning components leading the third sub-group to final decisions. The description of the procedure is coupled at each step with the evidence provided by a "pilot experiment", whose results are used as illustrative examples to indicate the potential of the framework itself. Concluding remarks summarize the main issues addressed in the chapter.

BACKGROUND

The aim of the preliminary step is to obtain the elements upon which the RGs will be set up in the following stage (see Figure 1). As previously explained, the group must be randomly divided in three sub-groups of nearly the same number of people. In the pilot experiment the members (N=12) of a board of directors of a Foundation concerned with *nonprofit* activities lent themselves to be randomly divided into three sub-groups of four people (N_1=4, N_2=4, N_3=4), each one in charge of a stage of the framework. The Foundation, which is operating in the field of cultural and social activities in a city in northern Italy, was facing a change of its organizational vision and had to identify possible strategies to make the change of focus easier and financially beneficial. The Board of directors' members agreed to undergo testing in order to obtain information both on the procedure to be taken for their forthcoming "real" decision-making and on the content of the strategies to be undertaken. They welcomed the proposal to submit themselves to the pilot experiment, firstly because it hinged on a real scenario they had to face, and furthermore for the reason that it could give significant information for the refinement of the problem setting. So, subjects were not bound to comply with the requirements resulting from the pilot experiment, but they had pledged to draw directions for the choices to be made.

The first sub-group (N_1=4) dealing with the preliminary step in Figure 1, was involved only in the problem-setting stage, thus allowing the researcher to build up individuals' cognitive maps and extract the key set of elements. The problem space is considered as simply a sort of simplified representation of the specific task environment in terms of "states of the world", "actions" and "goals" (Wermus, 1997). The working method followed for the drawing up of the cognitive maps included a semi-structured interview in which each subject points out possible choice alternatives to consider for improving the general condition of the organization, both from the organizational and financing viewpoint. The interview that allows us to draw the cognitive map is a semi-structured one, since it lets the interviewees develop arguments without too many constraints. The technique is widely used in research on the mapping of belief-systems (Brown, 1992), because the interviewer simply provides some general basic themes (at most three/four, such as "Explain the whole situation of the operational field of the organization in which you operate", or "Provide some details about the main problems your organization has to face", and so on) about which the interviewees are asked to explain their viewpoint both from a current and a future perspective. Afterwards, responses are coded in order to transform information into a map that brings into focus the ideas, objects, schemata, actions and goals that must be emphasized (Rice, 1978). To this end, items are then placed in a diagrammatic representation, with a simplified adaptation that takes into account just causal and similarity relationships among items (Bougon *et al.*, 1977). Figure 2 shows a subject's cognitive map in the pilot experiment in which various "states of the world" (the ones with an asterisk) placed on the left were present also in the second step of the procedure, owing to the analysis called "cognitive operational synthesis 1" in Figure 1, that is based on the application of an index of similarity[1] that allows us to collect the most recurrent concepts (see Huff *et al.*, 1990; Langfield-Smith & Wirth, 1992 for more information on this technique).

As it was shown before (see the basic plan of Figure 1), the whole set of these states of the world, which identifies the subjects' problem definition or, more generally, the scheme by which

Figure 2. Cognitive map of subject 3 of the first sub-group

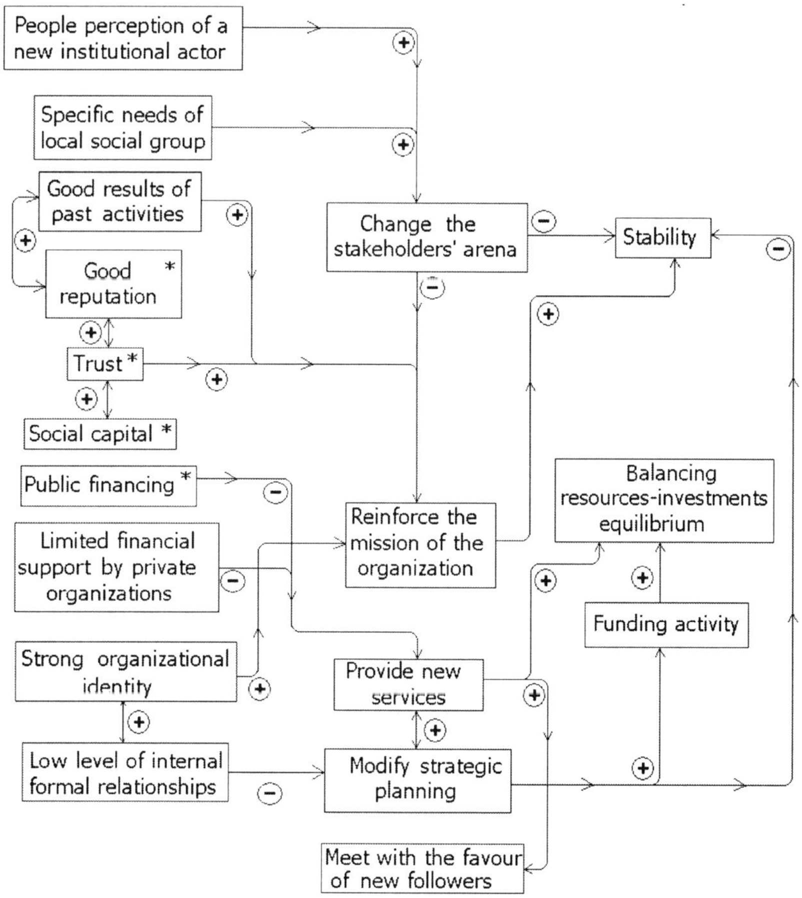

they make sense of reality (Bryant, 1984), represent the main elements for the first step of the RG analysis. Elements of the RGs in the second stage derive from a sort of shared part of the cognitive maps of the first group subjects (see Laukkanen, 1994).

The primary goal of a cognitive map is not to explore the task environment from a global or general perspective, but to highlight relationships between actions and feasible targets. In this chapter a different use of cognitive maps is suggested for the first step of the technique; mainly to explore the problem setting of the task environment or to the structure the situation, more than being focused on the possible solutions. This logic, characterizing many works in the psycho-organizational field, is dedicated to the exploration of the whole collective view of a problem: "The entire set of proximity relationships between items makes it possible to elaborate a graphic image that we call a thematic map. This map represents a collective vision of the themes connected to the problem as it is perceived by all the participants" (Damart, 2008).

Playing upon these findings, the main requirement of the suggested method is that feasible alternatives have to emerge from the process and are not to be identified and fixed ex-ante. Individuals are asked, through an open interview, to illustrate the problem-space (Bohnam & Shapiro, 1984; Hodgkinson & Maule, 2002; Siau e Tan, 2005), starting with questions like "Discuss the present situation and explain what you would do to improve your organization or to make it function better". However, subjects ignore that the interviewer will not make use of a part of their argument, i.e. the one concerning the link between concepts and action. The procedure, in fact, requires the researcher to draw up each cognitive map, following the usual technique or tools for comparing alternatives (Hodgkinson, 2004), and to analyze "states of the worlds" (concepts) by a tool that elaborates their lemmas in order to retain just the most representative (or recurrent) ones[2]. As already outlined, in the pilot experiment a specific index on the adjacency matrix of each single subject's cognitive map was applied (see Langfield-Smith & Wirth, 1992).

It is worth noting that in the case of very small groups, tool application must be carefully regulated or, alternatively, concepts/elements can be simply grouped together and no elaboration is made upon them: they are simply added in a set that does not represent any shared vision of the collective knowledge, the opposite of which happens in many approaches (see Roschelle & Teasley, 1994).

ELICITING CONSTRUCTS

An Overview of the Repertory Grid Technique

The RG technique is mainly based upon the original studies of George Kelly, a scholar who investigated the role of personal experience and consciousness in the frame of individuals' relationship structure (Bannister, 1965).

This investigation technique was traditionally reserved for some disciplines of psychology (particularly clinical psychology), but its use has been gradually developed in the organizational field. RG has been establishing itself since the early Seventies, when researchers turned their attention to this instrument in the managerial sector in order to combine it with the use of cognitive mapping.

In order to fully understand the philosophy of RGs, however, it is indispensable to describe the theoretical framework from which they arise and develop, i.e. the psychology of personal constructs, an approach that dates back to the early 50s (and to some extent differs from the two main "schools" of the period: behaviourism and psychoanalysis. George Kelly, considered the "father" of the psychology of personal constructs, strengthened his treatment in various seminal anticipatory works (Kelly, 1955; 1969), some of which were dedicated to a unified approach of cognition and emotions.

Like Jean Piaget, another kind of progenitor of constructivism in psychology, Kelly embraced the idea that mental processes allow the representation of reality (or better, multiple realities) depending on how subjects anticipate events, according to the metaphor of man as a naïve scientist, in the wake of what Boyd defines a "constitutive metaphor theory" (Boyd, 1979). The evaluation of the "outside" world is made up through a system of personal constructs that can be defined as attributes used by subjects to understand everyday events and human relations, and as theoretical abstractions, but without which no relation to reality would be possible. This philosophical standpoint of Kelly's approach has its main application in the RG technique, which was initially conceived for the study of psychiatric and psychopathological issues, also by Kelly himself for his investigation on depression. The RG requires a preliminary semi-structured interview that enables the researcher to build up a scheme made of rows and columns, in which the cells contain the evaluations made by subjects. In short, a RG is a matrix that relates elements (objects towards which the evaluation is addressed) to a set of constructs (the attributes of the evaluation).

This method has been used in many applications to get information concerning humans' perceptions of different fields of life and it consists of an "elicitation" procedure through which people are asked to explain how two things (elements) are seen as alike and in the same way different from a third. The main reason behind this mechanism can be found in Kelly's assumption, according to which individuals differ in the manner of conceiving their life events, and this understanding is spread over a system of hierarchical relationships between constructs (corollaries), represented by a categorization by similarities and differences of elements belonging to specific areas of reality via a subjective evaluation that orders events. Owing to this dichotomous nature, a construct has two polarities on the basis of which an element of reality (a person, a thing, an event, etc.) may be perceived as similar or dissimilar to another one. Compared to an item, afterwards, a construct shows two polarities, the opposition of which develops not along the lines of a logical but of a more psychological principle (if we assume that the pole of a construct is "intelligent", the other pole cannot be "non-intelligent" because it would extend indefinitely the scope of applicability of the construct and would place it in opposition to the field-corollary, but could be "ignorant" for a person or "unable" or "stupid" for another one). In case more elements are in the neighbourhood of a pole, for them this pole is called 'pole of similarity', otherwise we speak of 'contrast pole' to identify the opposite. When a construct encloses elements that are processed for the first time by subjects, then it is called 'permeable'. Sometimes elements are barred from being included in fields of relevance that are different from the pertaining field of a certain construct; in such a case the construct is called a 'prelative construct'. Considering these brief remarks on the psychology of personal constructs, "Kelly invented the technique of RGs as a means to explore the construct system of people: an attempt to put oneself in the shoes of others, to see the world how they see it, and understand their situations, their problems" (Fransella & Bannister, 1977).

The technique aims at avoiding the possible influence of the researcher, and it uses a test that requires several steps: the initial choice of elements; the presentation of them to the subject; the elicitation of constructs; the placement of elements in the space of constructs (namely the grid itself). In Kelly's original procedure, which primarily operated in a clinical context, elements consisted of 24 descriptions of role for people known by the subject (family, close acquaintances, colleagues, etc.), but they can be defined in different ways, for example in terms of employment (Shubsachs, 1975), feelings (Fransella & Adams, 1965) situations (Fransella, 1972); diseases (Orley & Loff, 1972), tables of Rorschach (Salmon, Arnold & Collyer, 1972); stores (Hudson, 1974) and more recently, commercial banks (Reger, 1990). The choice of elements must respond to the corollary of 'field', in the sense that they must belong to the applicability of the construct and be representative of it. This is a sensitive stage, requiring analytical ability of the observer on the scope of applicability, and for this reason an elicitation of the elements, rather than providing them directly, can be carried out and suggested by subjects themselves.

Application of the Repertory Grid Technique in the Present Research

The main modification to the RG technique here proposed does not concern the developing of the grid itself, with its usual elicitation of constructs, but its starting point. At the beginning, in fact, elements are neither elicited by those interviewed nor presented by the interviewer, but provided by the cognitive maps made on subjects of the first sub-group, in order to leave behind selective attention or possible bias of over-focusing on the initial condition of the task environment, as it is described in session 1 and 2.

Moreover, in the present work, the elicitation of constructs does not occur by considering the name of persons as elements or any aspect connected to the subjects' whole psychological world of relationships, as in Kelly's original technique. Actually, the general aim is to create a "matrix that relates certain characteristics given (by the first sub-group) of the problem domain (alternative events, states or entities) called elements, with the constructs (relevant dimensions to evaluate the similarity and difference between elements) that represent the expert's ability to accurately evaluate the domain problems" (Paramà *et al.*, 1998, p. 250). The basis of such a process is therefore to focus on a specific task environment as it is common in the tradition of those studies that make use of RGs for exploring just a specific domain of human relations (Bradshaw *et al.*, 1993). This is recurrent, for example, in managerial decision making (De Leon & Guild, 2003; Napier *et al.*, 2009), in marketing research (Sampson, 1972) especially focused on customers' perception (Hair *et al.*, 2009), in health sciences for humans' perception of food dishes (Monteleone *et al.*, 1997; Russel & Cox, 2003; 2004), in web commerce (Tan & Tung, 2003), in medical and clinical settings (Böker *et al.*, 2000; Stewart, 1996) and in many other fields of research. As we said earlier in more detail, in the procedure suggested here, items that represent the elements for eliciting personal constructs derive from the elaboration of the cognitive maps of the first sub-group of individuals. In order to make up the grids, individuals of the second sub-group (N_2=4) are asked to make the constructs explicit. Elements obtained in the previous stage are shown in random triads (Siau *et al.*, 2007), on the basis of Kelly's original technique (see Kelly, 1955) and subjects must clarify how two of the three elements are similar (emergent pole, i.e. the construct associated with each element) and the third one is dissimilar (contrast pole). Following the basic hints of other studies making use of RGs, elements are homogeneous considering the area of interest (in our case organizational behavior) and highly representative of the area itself (they derive from a process of integration of various cognitive maps), and simple to be recognized by all subjects (Yorke, 1985). In the example taken from the pilot experiment elements belong to three main sub-areas: the social one, with elements such as "human capital", "social capital", "weak ties", and so on; the organizational one, where elements are "technical support", "organizational skills" and many more; and the operational-market one, with elements as "presence of similar organizations", and others.

The procedure of triadic elicitation, which has been maintained on a minimal level of complexity, starts with the researcher placing elements that have been drawn out from the cognitive maps of the first sub-group (see Feixas *et al.*, 2008) in a column. On the right of the RG, emergent and contrast poles are made explicit by subjects in correspondence with elements (1-5 rating scale).

When two elements are marked by an emergent pole, the rating appears in bold in the grid in square parentheses and it is equal to 1, while it is equal to 5 for the contrast pole. The remaining elements can be associated to the emergent pole (with a rating equal to 2 in round parentheses) or to the contrast pole (with a rating equal to 4 in round parentheses) or categorized by neither pole (rating equal to 3). The example derived from the pilot experiment shows (see Figure 3 above) the RG of a subject belonging to the second sub-group of people (poles, each one with its own exit, have been put below the grid for a better graphical comprehension of the scheme), composed of twenty elements (columns) and fourteen constructs (rows).

It is interesting to point out that contrast poles are not simply built up on the opposite meaning of their own construct, but can be expressed with different implications. Moreover, emergent poles that could be considered as very similar (like in exit 2 and 3) are treated separately because of their singular features. Many statistical analyses can be done upon the different RGs, from sig-

Figure 3. Repertory grid of a second sub-group's subject in the pilot experiment

Exit	1 Human capital	2 Social capital	3 Weak ties	4 Strong ties	5 Good reputation	6 Trust	7 Hihg number of volunteers	8 Ethical values	9 Technical support	10 Organizational skills	11 External competence	12 Managerial skills	13 Presence of similar organizations	14 Many donors	15 Few donors	16 Public financing	17 Local public institutions	18 Troubled political situation	19 Tax exemption	20 knowledge of the law	Exit
1	(3)	(3)	(3)	(2)	(3)	(3)	(3)	(3)	**[1]**	**[1]**	**[5]**	(2)	(3)	(3)	(3)	(4)	(3)	(2)	(3)	(2)	1
2	(3)	(2)	(2)	**[1]**	(2)	(3)	(3)	(3)	(2)	(2)	(2)	(2)	**[5]**	(2)	(3)	(4)	(3)	(4)	**[1]**	(2)	2
3	(3)	**[1]**	(2)	(2)	(2)	(2)	(3)	(2)	(3)	(2)	(3)	**[1]**	(3)	(2)	(3)	**[5]**	(3)	(4)	(2)	(2)	3
4	(3)	(2)	(3)	(3)	**[1]**	(2)	(2)	**[1]**	(3)	(3)	(4)	(3)	(3)	(2)	(3)	**[5]**	(3)	(3)	(4)	(3)	4
5	(1)	(1)	(1)	(1)	(1)	(0)	(0)	(0)	(0)	**[5]**	(0)	(0)	(0)	**[1]**	(0)	(0)	(0)	**[1]**	(3)	(3)	5
6	(3)	(2)	(2)	(3)	(2)	(2)	(3)	**[1]**	(3)	(3)	(3)	(3)	(3)	**[1]**	(3)	(2)	(3)	**[5]**	(3)	(3)	6
7	**[1]**	(2)	(2)	(2)	(2)	(2)	**[1]**	(3)	(3)	(3)	(3)	(4)	(2)	(2)	**[5]**	(3)	(2)	(4)	(3)	(3)	7
8	(4)	(4)	(4)	(4)	**[5]**	(4)	(4)	(4)	(4)	(4)	(3)	(2)	(2)	(4)	**[1]**	(3)	(4)	**[1]**	(4)	(4)	8
9	(2)	(3)	(3)	(3)	(3)	**[1]**	**[1]**	(3)	(3)	(2)	(2)	(2)	(3)	(3)	(3)	(3)	(3)	(3)	**[5]**	(2)	9
10	**[1]**	(2)	(2)	(2)	(2)	(2)	(3)	(2)	(3)	(2)	(2)	**[1]**	(3)	(3)	(3)	(2)	(3)	(3)	**[5]**	(3)	10
11	(2)	(2)	**[1]**	(3)	(2)	**[1]**	(3)	(2)	**[5]**	(2)	(4)	(3)	(3)	(3)	(3)	(3)	(3)	(3)	(3)	(3)	11
12	(3)	(2)	(3)	(3)	**[1]**	(2)	(2)	**[1]**	(3)	(2)	(4)	(3)	(4)	(2)	(4)	(3)	(3)	**[5]**	(3)	(3)	12
13	(3)	(3)	(2)	(2)	(3)	(3)	(3)	(3)	**[1]**	(2)	(2)	(2)	(2)	(3)	(3)	**[5]**	(2)	(4)	(3)	**[1]**	13
14	(3)	(3)	(3)	(3)	(3)	(3)	(3)	(3)	(2)	(3)	**[1]**	(3)	(3)	(3)	(3)	(2)	**[1]**	**[5]**	(2)	(3)	14

Exit	Emergent pole ↔ contrast pole (scale 1 - 5)	Exit
1	are sign of high adaptation capability ↔ is associated to slack	1
2	facilitate fund raising ↔ hampers fund raising	2
3	high possibility of receiving funds ↔ low possibility of receiving funds	3
4	generate good public image ↔ damages public images	4
5	cause rift between donors and recipients ↔ has no influence on this link	5
6	consolidate the mission ↔ can be very negative	6
7	aggregate interests ↔ social immobility	7
8	generate problems for the organization ↔ it is helpful	8
9	are positive for increasing organizational experience ↔ has bad influence on experience	9
10	are powerful for inner growth ↔ it is not good for inner growth	10
11	stimulate interpersonal relationships ↔ it is an obstacle	11
12	create a positive image ↔ is a potential negative sign	12
13	capability to deal with burocracy ↔ slackness	13
14	increase the need of fixed procedures ↔ favours extemporaneous decision making	14

nificant correlation matrixes amongst constructs (see Fransella & Bannister, 1977) to various measures of the cognitive distance (see Bieri *et al.*, 1966). In the example reported from the pilot experiment, significant positive correlations can be found between "trust" and the other elements of the social area (two tailed significance), such as " social capital" (r=0.59; P<0.05), "human capital" (r=0.64; P<0.05), "weak ties" (r=0.55; P<0.05), "high number of volunteers" (r=0.63; P<0.05), "ethical values" (r=0.57; P<0.05), and also between "ethical values" and "social capital" (r=0.67; P<0.01), and between "ethical values" and "good reputation" (r=0.82; P<0.01). This area still shows significant and remarkable correlation between "weak ties" and "human capital (r=0.62; P<0.05), or between "weak ties" and "social capital" (r=0.74; P<0.01). Strong correlation is to be outlined also between "good reputation" and "social capital" (r=0.86; P<0.01). On the other hand it is interesting to point out that "troubled political situation" is significantly negatively correlated with "social capital" (r=-0.64; P<0.05) and with "good reputation" (r=-0.66; P<0.05), while the "knowledge of the law" is positively and significantly correlated with "technical support" (r=0.68; P<0.01), "organizational skill" (r=0.595; P<0.05), and "strong ties" (r=0.64; P<0.05). Several basic considerations can also be made for correlation between constructs. For example, "facilitate fund raising" (exit 2) and "high possibility of receiving funds" (exit 3) are significantly correlated (r=0.64; P<0.01), as "generate problems for the organizations" with "facilitate fund raising" (r=-0.52; P<0.05), "cause rifts between donors and recipients" (r=0.44; P<0.05), "consolidate the mission" (r=-0.54; P<0.05), "aggregate interests" (r=-0.69; P<0.01), "create a positive image" (r=-0.77; P<0.01). Moreover, "generate good public image" is obviously highly positive correlated with "create a positive image" (r=0.67; P<0.01), which shows a lower but significant level with "high possibility of receiving funds" (r=0.5; P<0.05) and a higher one both with "consolidate the mission" (r=0.75; P<0.01) and "generate problems for the organization" (r=-0.77; P<0.01).

A further examination of the RG is the hierarchical perspective of the cluster analysis, which is based upon the Euclidean distance between elements (or constructs) and allows the researcher to focus on similarities and ramifications of the elements in the grid. In the example taken from the pilot experiment, Figures 4 and 5 above in the text show the dendrogram representation of the clustering method (see Clatworthy *et al.*, 2005), and help in focusing on the whole grid and in summarizing proximities between items (elements and constructs) in the cluster tree. An overview of the tree in Figure 4 shows some intuitive associations, such as the one between "social capital" (item 2) and "good reputation" (item 5), an association that creates a group with "ethical value" (item 6), practically independent from other elements. This group, however, is connected with the two principal branches of the cluster tree by means of the element "high number of volunteers" (item 7).

In Figure 5, the hierarchical cluster analysis on constructs shows the role of being "powerful for inner growth" (item 10), in determining the main branch of the tree with "aggregate interests" (item 7), quite isolated from the group formed by the association between "balancing donors and recipients aims" (item 5) and "generating problems for the organizations" (item 8).

All these analyses can be made for each RG of the second sub-group of persons, but in order to overcome the difficulties affecting the confrontation of grids with many different constructs[3], the researcher shows the second subject the grid as it resulted from the first subject analysis, but without any rating inside the cells. The subject is asked to complete the grid by making his/her own pairing between elements and constructs, and to add constructs in case of need, but not to eliminate any of them (if a construct is considered useless, the subject is simply requested to rate it "3" on each element, so as to neutralize the

Figure 4. Hierarchical cluster analysis of elements on RG

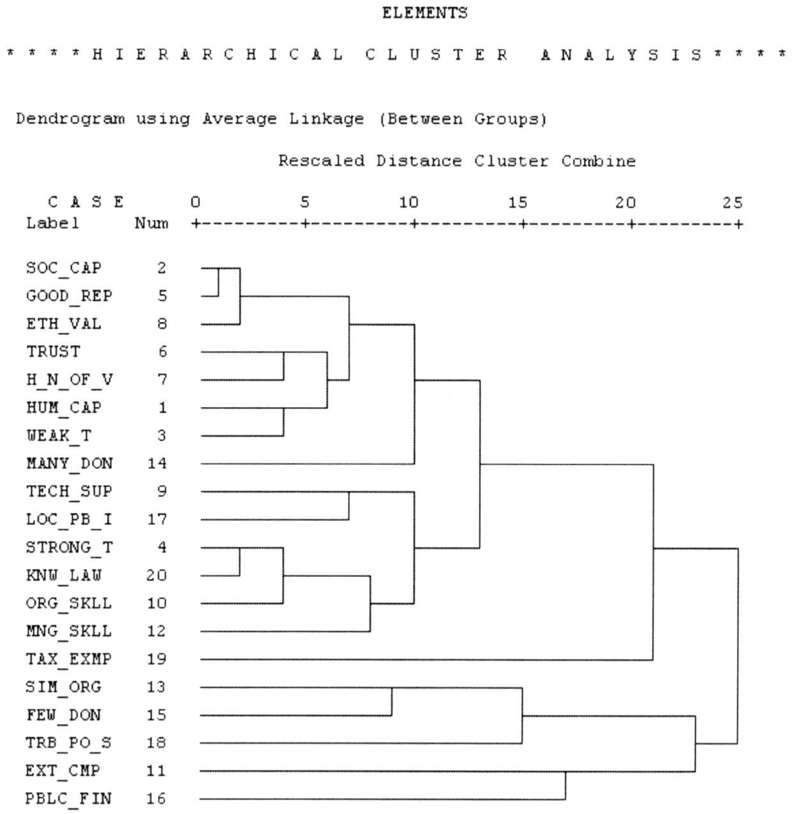

Figure 5. Hierarchical cluster analysis of constructs on RG

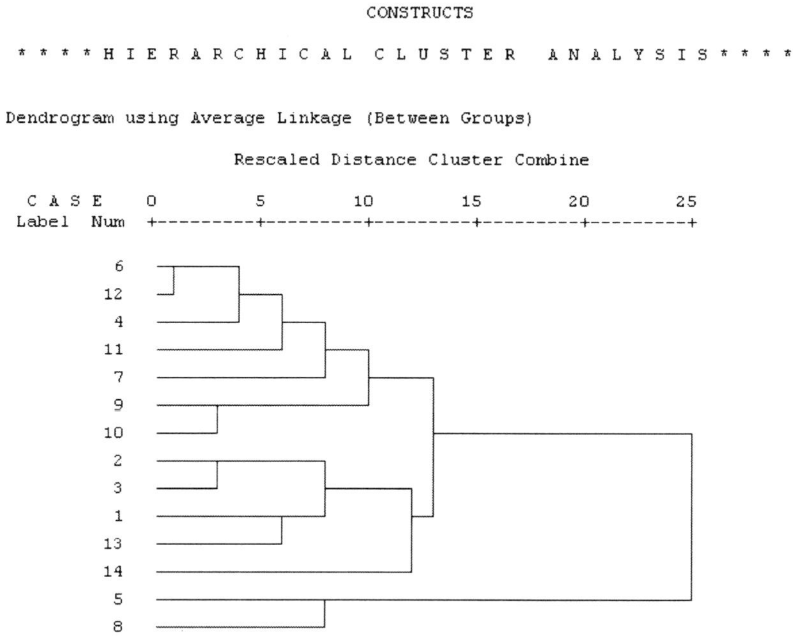

construct). The set of elements and constructs available in the following and final stage of the framework can be derived in various ways, one of which is based on the Generalized Procrustes Analysis (see Gower, 1975). It uses the χ^2 test made up on the quadrants derived from the plotting of the grids on two significant dimensions, in order to evaluate whether two grids (or two groups of grids) are similar or not. In the pilot study a simpler framework was considered, as it is the case in many psychometric studies (Feixas, 1991; Feixas & Cornejo-Alvarez, 1998), that is taking into account just elements and constructs that show a minimum number (at least three for elements and four for constructs) of significant correlations ($r \geq 0.5$ in the pilot experiment) with other items (see Bannister, 1960) and that are not left apart in the hierarchical cluster analysis. This fast and frugal solution allows us to have a set of elements and a set of constructs that have been playing a primary role for subjects of the second sub-group and constitute the basis for the final evaluation by the third sub-group, without undue bias from any individual.

Exploring the "Province of Meaning" and Deciding by Means of Diagrams

This is the second and final stage of the method, and its purpose is to come to an operative and effective procedure of choice emerging from the previous steps, this phase aims at being a sort of natural completion. As shown before, there are different reasons beneath the common use of tools such as cognitive maps and RGs in all the fields of research. In the case of cognitive mapping, the goal is usually to get a very task-specific view of what subjects think about possible results of different ways of action, starting with the existing situations ("real states of the world" or "concepts") or with potential situations ("possible states of the world"). Therefore, the original understanding of the cognitive map approach (Axelrod, 1976; Barr *et al.*, 1992) seems to fit problem solving situations well, but on the other hand it could leave out possible important but hidden links between features of the whole situation that are not directly involved in the sequence "concepts"→ "action" → "results". Hence, in the method suggested here, the main idea was to make use of the cognitive mapping technique not for focusing on possible alternatives, but to build up the general scenario development and then to manipulate the ensuing elements with specific software in order to have a significant set of items (as shown in section 2). Moreover, the RG technique, which is typically used to explore the interpersonal space of individuals (see section 3.1), has been applied to focus on a specific task environment (section 3.2). The *rationale* of the final step can be fully understood only starting with this atypical use that has been made for cognitive maps and RGs in order to overcome possible biases affecting individuals' replies to the interviews as we mentioned above. The general aspiration is that subjects of the third subgroup, forced to use specific items, should not be biased either by being too outcome oriented (and therefore only suggesting alternatives with a supposed high outcome) or by focusing just on the overall evaluation of the situation (and therefore avoiding fixing any possible action to undertake).

Subjects of the third sub-group can count on two sets of information derived from RG analysis (in their turn originated from first sub-group's cognitive maps): elements (concepts) and constructs (evaluation of the elements). Each subject is asked to illustrate a possible scenario and to fix a province of meaning on the grounds of the starting building blocks (the two information sets). A province of meaning[4] is here conceived not in the original strict meaning given by Alfred Schutz, i.e. the style of cognition and the sense that a person gives to his/her meaningful lived experience, but as simply the coupling between the set of elements and the set of constructs that have been found. Therefore, subjects of the third sub-group (N_3=4 in the pilot experiment) are interviewed in order to fix their images of strategy (Wensley, 2003), i.e. a scheme of relations between elements and constructs, and then to suggest an alternative for any link. An example taken from the pilot experiment could help in understanding the development of this stage. The two sets of information, resulted in the RGs analysis, were as follows:

- **Elements:** Human Capital; Social Capital; Weak Ties; Strong Ties; Good Reputation; Trust; Knowledge Of The Law
- **Constructs:** High possibility of receiving funds ↔ low possibility of receiving funds; generate problems for the organization ↔ it is helpful; create a positive image ↔ is a potential negative sign.

Starting from the two sets of information, subjects of the third sub-group are asked to frame the situation in terms of concepts (elements) and goals (constructs) and to show, with a diagrammatic representation, which is the most important target of the set and which action according to the task environment could hit that target. The goal-driven architecture of the scheme is typical of diagrammatic reasoning in its smallest version (Koedinger & Anderson, 1995), which somehow recalls the main plot of a cognitive map, with concepts and goals linked by actions. The components of the (very plain) diagram are the image, i.e. a 2-D space, and a data structure, which is made of objects; these can be labelled as points with their location, lines with axial specification, and regions (Chandrasekaran *et al.*, 2002). In the pilot experiment objects were represented by labels and links, since subjects were simply asked to show where elements, actions and goal are linked to each other, the importance of the actions they would have undertaken in order to reach their goal.

Hence, in the pilot experiment, each individual could count on the same information, which was built up by elements (to be considered as concepts or things) and constructs (to be considered as goals). In order to clarify this passage it is useful to mention an example taken from the pilot experiment. The first subject could play upon concepts as "human capital, social capital, weak ties, strong ties, good reputation, trust, and knowledge of the law", and had to link one or more than one of them with the goal he/she considered as crucial for the future of the organization amongst "high possibility of receiving funds, generate problems for the organization, create a positive image". Since the second construct can be viewed as a goal just for its opposite meaning, due to the bipolar nature of constructs, the subject was asked to consider only the contrast pole ("to be helpful"). The first subject believed that the second goal was too vague and focused on the first one, i.e. "high possibility of receiving funds". After fixing the principal goal, he had to show, in a very simplified way, which action was to be undertaken to hit the target, given the incidence of social capital, human capital, and so on. The answer (action A1) was to "reinforce human capital's capability in organizational skills", an action to which attention was not paid in the first stage of the procedure, in the sense that not enough "states of the world" were linked to it in cognitive maps so as to satisfy the index of minimal degree.

The subject's argument was that the external competencies did not play an essential role for the organization (the subject could view the entire number of elements and verify which one was ejected by the procedure) and the only manner to get a higher level of funds was not to look outside the organization but to strengthen it starting with their components, due to the fact that the organization could also count on significant weak ties, good reputation and knowledge of the law.

At this point, with action A fixed by the first subject, the procedure provided that the second subject, after discussing the solution suggested by the first one, could:

1] Decide whether to accept or not the "primary" one of the goals proposed by the first subject
   - 1.1] If not, to suggest a different goal to be reached (belonging to the set) and an action to undertake (A1)
   - 1.2] If yes, to decide whether to accept or not the action proposed by the first subject
     - 1.2.a] If not, to suggest a different action to undertake (B1)
     - 1.2.b] If yes, to suggest a further instrumental action that could help in undertaking the action proposed by the first subject (A2).

In the pilot experiment the second subject decided to accept the first subject's solution and also his suggested action. Moreover, according to the step [1.2.b], he recommended a further instrumental action in order to render the decision proposed by the first one operative; in detail it was "establishing internal specific courses on financing proceeding". It is worth noting that the justification of an additional instrumental action is crucial with regard to the natural converging of the whole process to a small number of concepts and goals to be considered, which have been filtered in the previous steps. The process goes on with the interviews to the remaining subjects of the third sub-group, who have to follow the same sequence of questions, with a further possible sub-step, that is a supplementary instrumental action for the one suggested by the previous subject. In synthesis, in the case of four people, as in the pilot experiment, the development of the action framework can be horizontal, that is at most four different actions to be undertaken, or vertical, with one major action and three instrumental actions. As the process moves forward and in a narrowing perspective, instrumental actions become more detailed and can exclude the use of the primary action, which could sound too generic. In the example taken from the pilot experiment, the sequence was:

- **Subject 1:**
  Main goal (G1): "high possibility of receiving funds"
  Major action proposed (A1): "reinforce human capital's capability in organizational skills"
- **Subject 2:**
  [1.2.b] further instrumental action proposed (A2): "establish internal specific courses on financing proceeding"
- **Subject 3:**
  [1.2.c] further instrumental action proposed (A3): "performing a check of previous participations in calls for bids"
- **Subject 4:**
  Main goal (G2): "create a positive image"
  Major action proposed (B1): "invest in advertising programme"

To summarize, the first subject's proposal (action A1), was to strengthen organizational capability, but for the second subject, even if A1 was crucial, it merely represented a general purpose. From his viewpoint the organizational capability on which making all efforts was the one more related to the main target ("high possibility of receiving funds" G1). His argument was the following one: only becoming aware of the mechanisms through which private or public institutions identify and decide to finance a certain project, the *nonprofit* organization can participate with higher chances to the calls for bids and therefore of receiving funds. This is because his proposal was to "establish internal specific courses on financing proceeding" (A2). On the other hand, despite being in agreement with colleagues, subject 3 proposed an additional instrumental action, which was to examine the reason of previous failures when the organization had participated in calls for bids, both of public and private institutions (A3: "performing a check of previous calls for bids"). By exploring his diagram it is clear that he is convinced that neither weak and strong ties, nor good reputation or trust (elements that must be considered in the task environment perspective, since they are derived from the RGs analysis of the second group), despite their important function, are enough for improving the situation.

A different viewpoint can be found in the diagrammatic reasoning of the fourth (and last) subject, who believes that the main goal to be considered is only to "create a positive image" (G2) of the organization. In any case it is interesting to outline that the action proposed to hit the target is to "invest in advertising programmes" (B1) and the examination of the diagram shows a particular attention paid by the subject to items of the social area, such as human and social capital, weak ties, good reputation, trust. All of these items are considered essential but not so strong and crucial in their relationship with the goal "creating a positive image", which is thought to be the key for the organizational future.

Carrying out the procedure in the final third stage provides that only actions that do not clash with other ones will be done. In the pilot experiment, on the basis of the outcomes by the four agents, actions to be considered were:

(A2): "establish internal specific courses on financing proceeding"
(A3): "performing a check of previous calls for bids"
(B1): "invest in advertising programmes"

CONCLUSION

This chapter presents an approach for aiding collective decision making, especially applicable for large groups, where subjects who are not willing to take decisions by schemes of bargaining (distributive and/or integrative negotiation) or by voting. The main *ratio* supporting the technique is that it does not emphasize individual differences in the psychological sphere. The approach does not aim to come a solution that supports the deep feature of certain individuals' own view of the world, as is the case in negotiating approaches (Barry & Friedman, 1998) in which a predominant argument (or the so called external sources) often prevails. On the other hand the method aims at assigning different responsibility for the final choice to different sub-groups of individuals in each step, in order to bypass possible biases or heuristics that usually affect group decision making. In particular, two of these are considered: misleading goal-driven behavior in the problem-setting step and ambiguous forms of selective attention paid by decision makers to starting elements of a decision.

The first sub-group is responsible for fixing the main elements, or concepts, of the problem setting (without being aware of doing so, because they are convinced of suggesting strategies and goals). The second one is responsible for identifying the main constructs related to these elements (without being conscious of how it could help the final choice), and the third one for defining actions to be undertaken through the exploration of the provinces of meaning in their minimal representation of the task environment.

The framework structure is indeed time consuming, both in the preliminary step, for the need of analyzing semi-structured interviews for the drawing up of cognitive maps, and in the first step, due to the articulated procedure of RG technique. For this reason it could be of great help if the procedure is managed by a researcher skilled in the usage of such tools. This seems to be a necessary price to pay for trying to avoid the traps discussed in the chapter. Each step of the procedure is linked with the following one, but the essential stage is represented by an atypical use of the RG technique, that is not aimed at building up the interpersonal relationships of subjects but is directed to give rise to a set of information (a number of constructs) which represents the basis for the final choice. In this case, constructs of the RGs, which are the fundamentals of the task environment perspective of subjects and the axis of their view of the world (Klion & Pfenninnger,

1997), are interpreted as goals to be achieved in the next step. Therefore, constructs elicited by individuals (in synthesis the evaluation or simply the attribute that subjects consider as crucial with respect to elements at their disposal), constitute a sort of target to be reached by the third group of subjects. The graphical representation that is requested for subjects of the third sub-group allows the researcher to fix the actions to be undertaken, and is the result of a narrowing process (cognitive maps – RGs – diagrammatic framework) through which final choices emerge and are not simply fixed *ex ante* and discussed by all participants.

Final alternatives to be undertaken are the closing step of a framework that mixes quantitative investigation, by means of statistical analysis on the first group's cognitive maps (index of minimal degree in the adjacency matrix) and on RGs of the second group (significant correlations between elements and between constructs, plus hierarchical cluster analysis with dendrograms), and also a qualitative study on the last evaluation made by subjects of the third group (fixing of goals and actions, plus additional instrumental actions).

The collaborative and coordinated nature of the whole procedure is to be traced back to the interrelationship between steps: each group's contribution plays upon the outcomes of the previous one and adds a specific building block to the framework being proposed. From this perspective, the framework stimulates problem setting and, in its third step, facilitates the emergence of solutions that would not have been taken into account following traditional schemes of decision making process.

REFERENCES

Axelrod, R. (1976). *Structure of decision: The cognitive maps of political elites*. Princeton, NJ: Princeton University Press.

Bannister, D. (1960). Conceptual structure in thought disordered schizophrenics. *The Journal of Mental Science*, *106*, 1230–1249.

Bannister, D. (1965). The rationale and relevance of repertory grid technique. *The British Journal of Psychiatry*, *111*, 977–982. doi:10.1192/bjp.111.479.977

Barr, P. S., Stimpert, J. L., & Huff, A. S. (1992). Cognitive change, strategic action, and organizational renewal. *Strategic Management Journal*, *13*, 15–36. doi:10.1002/smj.4250131004

Barry, B., & Friedman, R. A. (1998). Bargainer characteristics in distributive and integrative negotiation. *Journal of Personality and Social Psychology*, *74*(2), 345–359. doi:10.1037/0022-3514.74.2.345

Bieri, J., Atkins, A. L., Briar, S., Leaman, R. L., Miller, H., & Tripodi, T. (1966). *Clinical and social judgment: The discrimination of behavioural information*. New York: Wiley.

Böker, H., Hell, D., Budischewski, K., Eppelb, A., Härtlingb, F., & Rinnertb, H. (2000). Personality and object relations in patients with affective disorders: idiographic research by means of the repertory grid technique. *Journal of Affective Disorders*, *60*, 53–59. doi:10.1016/S0165-0327(99)00161-5

Bonham, G. M., & Shapiro, M. J. (1984). Mapping structures of thought . In Gallhofer, I., Saris, W. E., & Melman, M. (Eds.), *Different text analysis procedures for the study of decision making* (pp. 29–51). Amsterdam: Sociometric Research Foundation.

Bougon, M. G., Weick, K. E., & Binkhorst, D. (1977). Cognition in organizations: An analysis of the Utrecht Jazz Orchestra. *Administrative Science Quarterly*, *22*, 606–639. doi:10.2307/2392403

Boyd, R. (1979). Metaphor and theory change: What is metaphor for? In Ortony, A. (Ed.), *Metaphor and thought* (pp. 481–532). Cambridge: Cambridge University Press.

Bradshaw, J. M., Chapman, C. R., Sullivan, K. M., Zarley, D., Nims, J., Gavrin, J., et al. (1993). KS-3000: An application of DDUCKS to bone-marrow transplant patient support. *Proceedings of the Sixth Annual Florida AI Res. Symposium (FLAIRS93)*, 18–21.

Brown, S. M. (1992). Cognitive mapping and repertory grids for qualitative survey research: Some comparative observations. *Journal of Management Studies*, *29*(3), 287–307. doi:10.1111/j.1467-6486.1992.tb00666.x

Bryant, J. (1984). Modelling alternative realities in conflict and negotiation. *The Journal of the Operational Research Society*, *35*(11), 985–993.

Chandrasekaran, B., Josephson, J. R., Banerjee, B., & Kurup, U. (2002). Diagrammatic reasoning in support of situation understanding and planning. Proceedings of Army Science Conference, Dec 2-5, Orlando, Florida.

Christensen, E. W., & Fjermestad, J. (1997). Challenging group support systems research: The case for strategic decision making. *Group Decision and Negotiation*, *6*, 351–372. doi:10.1023/A:1008664812184

Clarkson, G. P., & Hodgkinson, G. P. (2005). Introducing Cognizer TM: A comprehensive computer package for the elicitation and analysis of cause maps. *Organizational Research Methods*, *8*(3), 317–341. doi:10.1177/1094428105278022

Clatworthy, J., Buick, D., Hankins, M., Weinman, J., & Horne, R. (2005). The use and reporting of cluster analysis in health psychology: A review. *British Journal of Health Psychology*, *10*, 329–358. doi:10.1348/135910705X25697

Damart, S. (2008). A cognitive mapping approach to organizing the participation of multiple actors in a problem structuring process. *Group Decision and Negotiation*, *19*(5), 505–526. doi:10.1007/s10726-008-9141-y

De Leon, E. D., & Guild, P. D. (2003). Using repertory grid to identify intangibles in business plans. *Venture Capital*, *5*(2), 135–160. doi:10.1080/1369106032000097030

Desanctis, G. R., & Gallupe, B. (1987, May). A foundation for the study of group decision support systems. *Management Science*, *33*(5), 589–609. doi:10.1287/mnsc.33.5.589

Feixas, G. (1991, July). The repertory grid in psychotherapy research. Poster presented at XXII Annual Conference of the Society for Psychotherapy Research, Lyon, France.

Feixas, G., & Cornejo-Alvarez, J. M. (1996). *Manual de la técnica de rejilla mediante el programma RECORD v. 2.0*. Barcelona, Spain: Ediciones Paidos.

Feixas, G., Erazo-Caicedo, M. I., Harter, S. L., & Bach, L. (2008). Construction of self and others in unipolar depressive disorders: A study using repertory grid technique. *Cognitive Therapy and Research*, *32*, 386–400. doi:10.1007/s10608-007-9149-7

Fransella, F. (1972). *Personal change and reconstruction*. London: Academic Press.

Fransella, F., & Adams, B. (1965). An illustration of the use of repertory grid technique in a clinical setting. *The British Journal of Social and Clinical Psychology*, *5*, 51–62.

Fransella, F., & Bannister, D. (1977). *A manual for repertory grid technique*. London: Academic Press.

Gower, J. (1975). Generalised Procrustes analysis. *Psychometrika*, *51*, 33–51. doi:10.1007/BF02291478

Gray, P. (1987). Group decision support systems. *Decision Support Systems*, *3*(3), 233–242. doi:10.1016/0167-9236(87)90178-3

Hair, N., Rose, S., & Clark, M. (2009, March). Using qualitative repertory grid techniques to explore perceptions of business-to-business online customer experience. *Journal of Customer Behaviour*, *8*(1), 51–65. doi:10.1362/147539209X414380

Hodgkinson, G. P. (2004). Causal cognitive mapping in the organizational strategy field: A comparison of alternative elicitation procedures. *Organizational Research Methods*, *7*(1), 3–26. doi:10.1177/1094428103259556

Hodgkinson, G. P., Bown, N. J., Maule, A. J., Glaister, K. W., & Pearman, A. D. (1999, October). Breaking the frame: An analysis of strategic cognition and decision making under uncertainty. *Strategic Management Journal*, *20*(10), 977–985. doi:10.1002/(SICI)1097-0266(199910)20:10<977::AID-SMJ58>3.0.CO;2-X

Hodgkinson, G. P., & Maule, A. J. (2002). The individual in the strategy process: Insights from behavioural decision research and cognitive mapping . In Huff, A. S., & Jenkins, M. (Eds.), *Mapping strategic knowledge* (pp. 196–219). London: Sage.

Hudson, R. (1974). Images of the retailing environment: An example for the use of the repertory grid methodology. *Environment and Behavior*, *12*, 470–494. doi:10.1177/001391657400600403

Huff, A. S., Narapareddy, N. V., & Fletcher, K. E. (1990). Coding the causal association of concepts . In Huff, A. S. (Ed.), *Mapping strategic thought*. Chichester: Wiley.

Jain, S. P., & Maheswaran, D. (2000, March). Motivated reasoning: A depth-of-processing perspective. *The Journal of Consumer Research*, *26*, 358–371. doi:10.1086/209568

Kelly, G. A. (1955). *The psychology of personal constructs*. New York: Norton.

Kelly, G. A. (1969). The role of classification in personality theory . In Maher, B. (Ed.), *Clinical psychology and personality: The selected papers of George Kelly* (pp. 289–300). New York: Wiley.

Klion, R., & Pfenninger, D. T. (1997). Personal construct psychotherapy of addictions. *Journal of Substance Abuse Treatment*, *14*(1), 37–43. doi:10.1016/S0740-5472(96)00130-4

Koedinger, K. R., & Anderson, J. R. (1995). Abstract planning and perceptual chunks . In Chandrasekaran, B., Glasgow, J., & Narayan, N. H. (Eds.), *Diagrammatic reasoning: Cognitive and computational perspectives* (pp. 577–625). Cambridge, MA: The MIT Press.

Kruglanski, A. W. (1990). Lay epistemic theory in social-cognitive psychology. *Psychological Inquiry*, *1*, 181–197. doi:10.1207/s15327965pli0103_1

Langfield-Smith, K., & Wirth, A. (1992). Measuring differences between cognitive maps. *The Journal of the Operational Research Society*, *43*(12), 1135–1150.

Laukkanen, M. (1994, August). Comparative cause mapping of organizational cognitions. *Organization Science*, *5*(3), 322–343. doi:10.1287/orsc.5.3.322

Monteleone, E., Raats, M. M., & Mela, D. J. (1997). Perceptions of starchy food dishes: Application of the repertory grid method. *Appetite*, *28*, 255–265. doi:10.1006/appe.1996.0081

Napier, N. P., Keil, M., & Tan, F. B. (2009). IT project managers' construction of successful project management practice: A repertory grid investigation. *Information Systems Journal*, *19*, 255–282. doi:10.1111/j.1365-2575.2007.00264.x

O'Keefee, R. M. (1989). The implications of cognitive-style findings for operational research. *The Journal of the Operational Research Society*, *40*, 415–422.

Orley, J. H., & Loff, J. P. (1972). The effect of psychiatric education on attitudes to illness among the Ganda. *The British Journal of Psychiatry*, *121*, 137–141. doi:10.1192/bjp.121.2.137

Paramà, J. R., Penabada, M. R., Brisaboaa, N. R., & Betanzosa, A. A. (1998). EECB: A knowledge elicitation tool based on repertory grid and city block metric. *Expert Systems with Applications*, *14*, 249–258. doi:10.1016/S0957-4174(97)00078-X

Procter, H. G. (1985). Repertory grid techniques in family therapy and research . In Beail, N. (Ed.), *Repertory grid technique and personal construct application in clinical and educational setting*. London: Croom Helm.

Reger, R. (1990). Managerial thought structures and competitive positioning . In Huff, A. S. (Ed.), *Mapping strategic thought* (pp. 71–88). Chichester: Wiley.

Rice, T. J. (1978, April). Cognitive mapping: Small group technique in teaching. *Teaching Sociology*, *5*(3), 259–274. doi:10.2307/1317202

Roschelle, J., & Teasley, S. D. (1994). *The construction of shared knowledge in collaborative problem solving*. NATO ASI Series F Computer and Systems Sciences.

Ross, L., Lepper, M. R., & Hubbard, M. (1975). Perseverance in self-perception and social perception: Biased attributional processes in the debriefing paradigm. *Journal of Personality and Social Psychology*, *32*, 880–892. doi:10.1037/0022-3514.32.5.880

Russel, C. G., & Cox, D. N. (2003). A computerised adaptation of the repertory grid methodology as a useful tool to elicit older consumers' perceptions of foods. *Food Quality and Preference*, *14*, 681–691. doi:10.1016/S0950-3293(03)00060-0

Russel, C. G., & Cox, D. N. (2004). Understanding middle-aged consumers' perceptions of meat using repertory grid methodology. *Food Quality and Preference*, *15*, 317–329. doi:10.1016/S0950-3293(03)00073-9

Salmon, P., Arnold, J. M., & Collyer, Y. M. (1972). What do the determinants determine?: The internal validity of the Rorschach. *Journal of Personality*, *36*, 33–38.

Sampson, P. (1972, February). Using the repertory grid test. *JMR, Journal of Marketing Research*, *9*, 78–81. doi:10.2307/3149614

Schutz, A. (1967). *The Phenomenology of the Social World*. Evanston, IL: Northwestern University Press.

Shakun, M. F. (1988). *Evolutionary systems design: Policy making under complexity and group decision support systems*. Oakland, CA: Holden-Day.

Shubsachs, A. P. W. (1975). To repeat or not to repeat? Are frequently used constructs more important to the subject? A study of the effect of allowing repetition of constructs in a modified Kelly repertory test. *The British Journal of Medical Psychology*, *48*, 31–37.

Siau, K., & Sheng, H. X. (2007, November). Important characteristics of software development team members: An empirical investigation using repertory grid. *Information Systems Journal*.

Siau, K., & Tan, X. (2005). Improving the quality of conceptual modelling using cognitive mapping techniques. *Data & Knowledge Engineering*, *55*, 343–365. doi:10.1016/j.datak.2004.12.006

Stewart, T. (1996). Good maintainers and poor maintainers: a personal construct approach to an old problem. *Journal of Fluency Disorders*, *21*, 33–48. doi:10.1016/0094-730X(95)00043-7

Tan, F. B., & Tung, L. L. (2003). Exploring website evaluation criteria using the repertory grid technique: A web designers' perspective. Conference Proceedings of the Second Annual Workshop on HCI Research in MIS, Seattle, WA, 65-9, AUT University.

Unnikrishnan Nair, K. (2001). Cognitive maps of managers and complex problem solving . In Lant, T. K., & Shapira, Z. (Eds.), *Organizational cognition: Computation and interpretation* (pp. 211–240). London: Lawrence Erlbaum Associates Publishers.

Walton, D. N. (1990). *Practical reasoning: Goal-driven, knowledge-based, action-guiding argumentation* (Studies in Epistemology and Cognitive Theory, vol. 2). Lanham, MD: Rowman & Littlefield Publishers.

Warren, K. (1995). Exploring competitive futures using cognitive mapping. *Long Range Planning*, *28*(5), 10–21. doi:10.1016/0024-6301(95)00034-G

Wensley, R. (2003). Strategy as intention and anticipation. In Cummings, S., & Wilson, D. (Eds.), *Images of strategy* (pp. 105–135). Malden, MA: Blackwell.

Wermus, H. (1997, October). Representation, beliefs, and hindrance in understanding. *Cognitive Systems*, *5*(1), 1–16.

Winter, D. A. (1985). Repertory grid technique in the evaluation of therapeutic outcome . In Beail, N. (Ed.), *Repertory grid technique and personal constructs* (pp. 154–170). London: Croom Helm.

Yorke, D. M. (1985). Indexes of stability in repertory grids: A small-scale comparison. *British Educational Research Journal*, *11*(3), 221–225. doi:10.1080/0141192850110303

Zhao, S. (2004). Consociated contemporaries as an emergent realm of the lifeworld: Extending Schutz's phenomenological analysis to cyberspace. *Human Studies*, *27*, 91–105. doi:10.1023/B:HUMA.0000012246.33089.68

ADDITIONAL READING

Baxter, I. A., Jack, F. R., & Schröder, M. J. A. (1998). The use of repertory grid method to elicit perceptual data from primary school children. *Food Quality and Preference*, *9*(1/2), 73–80. doi:10.1016/S0950-3293(97)00031-1

Eden, C., & Jones, S. (1984, September). Using repertory grids for problem construction. *The Journal of the Operational Research Society*, *35*(9), 779–790.

Herman, R. L. (1996). The repertory grid technique: Path to teacher description or teach potential? *Instructional Science*, *24*, 439–459. doi:10.1007/BF00125579

Hodgkinson, G. P., Maule, A. J., Bown, N. J., Pearman, A. D., & Glaister, K. W. (2002, November). Further reflections on the elimination of framing bias in strategic decision making. *Strategic Management Journal*, *23*(11), 1069–1076. doi:10.1002/smj.266

Ilbery, B. W., & Hornby, R. (1983). Repertory grids and agricultural decision-making: A mid-Warwickshire case study. *Geografiska Annaler. Series B, Human Geography*, *65*(2), 77–84. doi:10.2307/490936

Johnson, F., & Crudge, S. E. (2007). Using the repertory grid and laddering technique to determine the user's evaluative model of search engines. *The Journal of Documentation*, *63*(2), 259–280. doi:10.1108/00220410710737213

Kevill, F., Shaw, M., & Goodacre, E. (1982). In-Service diploma course evaluation using repertory grids. *British Educational Research Journal*, *8*(1), 45–56. doi:10.1080/0141192820080105

Klein, J. H. (1994, August). Cognitive processes and operational research: A human information processing perspective. *The Journal of the Operational Research Society*, *45*(8), 855–866.

Larkin, J., & Simon, H. A. (1995). Why a diagram is (sometimes) worth 10000 words . In Glasgow, J., Narayanan, N. H., & Chandrasekaran, B. (Eds.), *Diagrammatic reasoning: Cognitive and computational perspectives* (pp. 69–110). Menlo Park, CA: AAAI Press.

Murray-Prior, R. (1998). Modelling farmer behaviour: A personal construct theory interpretation of hierarchical decision models. *Agricultural Systems*, *57*(4), 541–556. doi:10.1016/S0308-521X(98)00008-0

Nelson, K. M., Nadkarni, S., Narayanan, V. K., & Ghods, M. (2000, September). Understanding software operations support expertise: A revealed causal mapping approach. *Management Information Systems Quarterly*, *24*(3), 475–507. doi:10.2307/3250971

Niu, N., & Easterbrook, S. (2007). So, you think you know others' goals?: A repertory grid study. *IEEE Software*, *24*(2), 53–61. doi:10.1109/MS.2007.52

Solas, J. (1992). Investigating teacher and student thinking about the process of teaching and learning using autobiography and repertory grid. *Review of Educational Research*, *62*(2), 205–225.

Tan, F. B., & Hunter, M. G. (2002, March). The repertory grid technique: A method for the study of cognition in information systems. *Management Information Systems Quarterly*, *26*(1), 39–57. doi:10.2307/4132340

Zuber-Skerritt, O. (1987). A repertory grid study of staff and students' personal constructs of educational research. *Higher Education*, *16*(5), 603–623. doi:10.1007/BF00128424

KEY TERMS AND DEFINITIONS

Cognitive Maps: Graphical images of individuals' viewpoint of the task environment, using items as "states of the word", "actions", and "outcomes".

Diagrammatic Reasoning: Individuals' capability of understanding phenomena by means of graphic (or pictorial) representations and mental images.

Group Decision Making: A process of choice which involves more than one person.

Province of Meaning: A portion of individuals' view of the world in which a stable relation between a goal and an action (with its meaning) is highlighted.

Repertory Grid Technique: Interviewing technique for the analysis of individuals' perception of personal reality, by means of an elicitation process of elements and constructs.

ENDNOTES

1. In the pilot experiment "Decision Explorer" was used for the drawing up of the cognitive maps, while "NVivo" allowed the researcher to group together similar items according to the same headword.

2 The "Cognizer" software could also be mentioned in this direction (see Clarkson & Hodgkinson, 2005).

3 For a possible framework of analysing RGs with many different constructs, developed in the therapy field, see Procter (1985).

4 For an overview of this concept, see the seminal work of Alfred Schutz (1967), who developed its main phenomenological meaning and also Zhao (2004) for a current debate on Schutz's role in the phenomenological analysis.

Chapter 11
Group Decision Making for Advanced Manufacturing Technology Selection Using the Choquet Integral

Cengiz Kahraman
Istanbul Technical University, Turkey

Selçuk Çebi
Karadeniz Technical University, Turkey

İhsan Kaya
Selçuk University, Turkey

ABSTRACT

Advanced manufacturing technology (AMT) is defined as a modern method of production incorporating highly automated and sophisticated computerized design and operational systems. Hence, an investment decision to adopt AMT is a strategic decision. A group decision making process is stressful when group members have different views under multiple and conflicting criteria. Satisfying group members' opinions has a critical impact on a decision. In this chapter, a multiple criteria group decision making problem under a fuzzy environment is used for the selection among AMTs. Choquet integral methodology is used for this selection. A strategic investment problem of a company for a suitable Automated Storage/Retrieval System (AS/RS) is considered and discussed.

INTRODUCTION

The developments of science and technology have led to many new concepts and products, which are replacing the old ones. Flexibility, improvement in productivity and quality, faster response to market shifts, shorter throughput and lead time and savings in inventory and labor costs, enable customer demands to be met in a shorter time. Changing customer preferences and tastes oblige the manufacturer to change his products frequently. Increased consumer awareness has led to the manufacture of high-quality goods. The manufacturing process has to be faster to

DOI: 10.4018/978-1-60960-091-4.ch011

meet market demands at the appropriate time and to overcome competition. All these factors have led to changes in manufacturing processes, which have prompted many manufacturers to adopt computer-integrated manufacturing, namely advanced manufacturing technology (Aravindan & Punniyamoorthy, 2002).

The term "advanced manufacturing technology" (AMT) is broadly defined to include any automated (usually computer oriented) technology used in design, manufacturing/service, and decision support. Components of AMT include computer-aided engineering, factory management and control systems, computer-integrated manufacturing processes, and information integration. Different researchers have defined AMT in various ways. Typical items constituting AMT include: robotics, automated guided vehicles (AGVs), computer numerically controlled (CNC) machines, flexible manufacturing systems (FMS), computer-aided design (CAD), computer-aided manufacturing (CAM), information technology, cellular manufacturing and the use of just-in-time (JIT)/kanban system in the plant (Das and Jayaram, 2003)

AMTs provide many important benefits such as greater manufacturing flexibility, reduced inventory, reduced floor space, faster response to shifts in market demand, lower lead times, and a longer useful life of equipment over successive generations of products. Like many real-world problems, the decision of investing in advanced manufacturing technology frequently involves multiple and conflicting objectives, such as minimizing costs, maximizing flexibility, minimizing machine down times and maximizing efficiency. Application of traditional capital budgeting methods does not fully account for the benefits arising from intangible factors of AMTs (Kahraman et al., 2000).

The rapid growth of the AMT industry is creating problems in new directions. Prospective firms face the situation of having to make a decision among several AMTs, all of which are capable of performing a specific task. Since the development and use of appropriate assessment approaches are crucial to ensuring that the analysis of each AMT project considers all benefits and costs, the selection of a suitable AMT is becoming a more and more complex problem. Therefore the evaluation of the AMT is a multiple criteria problem and group decision making is generally preferred to solve these kinds of problem (Chuu, 2009). Group decision making requires considering multiple perspectives obtained from a group consisting multiple members (Lu et al., 2007). A group decision is required in two situations: (1) when a problem becomes too complex such that the knowledge of a decision maker is inadequate, as in product design, investment decisions and supplier selection and (2) when there are conflicting ideas that influence the decision makers like presidential elections. While the first one is called cooperative group decision making, the second one is called non-cooperative decision making. The common features of both decision making problems is to satisfy multiple decision makers' preferences. Therefore, there are several kinds of group decision making methods in the literature such as authority rule, majority rule, negative minority rule, ranking rule, and consensus rule.

In this chapter, a multiple criteria group decision making problem under fuzzy environment is discussed to select an appropriate AMT by using the Choquet integral since the problem is related to both subjective and objective criteria. For a long time, it has been recognized that an exact description of many real-life physical situations may be virtually impossible. This is due to the high degree of imprecision involved in real-world situations.

Zadeh (1965, 1968) in his seminal papers proposed fuzzy set theory as the means for quantifying the inherent fuzziness that is present in ill-posed problems. Fuzziness is a type of imprecision which may be associated with the sets in which there is no sharp transition from membership to nonmembership. Many problems in the real world deal with uncertain and imprecise data so conventional approaches cannot be effective to find the best solution. To cope with this uncertainty, fuzzy set theory has been developed as an effective mathematical algebra in a vague environment. Although humans are comparatively efficient in qualitative forecasting, they are not good at making quantitative predictions. Since fuzzy linguistic models permit the translation of verbal expressions into numerical ones, thereby dealing quantitatively with imprecision in the expression of the importance of each criterion, some methods based on fuzzy relations are used. When the system involves human subjectivity, fuzzy algebra provides a mathematical framework for integrating imprecision and vagueness into the models (Zimmermann, 1991; Kaya & Çınar, 2008). A major contribution of fuzzy set theory is its capability in representing vague knowledge. In multiple criteria decision making problems, data very often are imprecise and fuzzy. Hence, a decision maker may encounter difficulty in quantifying such linguistic statements so they can be used in deterministic decision making.

The rest of this chapter is organized as follows: In Section 2, the basics of group decision making and a literature review on AMTs are presented. The selection criteria are discussed in Section 3. The fundamentals of the Choquet integral are given in Section 4. A case study is presented in Section 5. Finally, concluding remarks are given in Section 6.

LITERATURE REVIEW: GROUP DECISION MAKING AND AMTS

Group decision making consists of multiple individuals interacting to reach a decision. Each decision maker (expert) may have unique motivations or goals and may approach the decision process from a different angle, but have a common interest in reaching eventual agreement on selecting the "best" option(s). To do this, experts have to express their preferences by means of a set of evaluations over a set of alternatives (Herrera-Viedma et al., 2007). Group decision-making studies reveal that differences in the process by which groups reached decisions may help explain the different findings, generally within the group decision making literature. In some cases, group members privately considered information about a problem and formed individual judgments before meeting as a group. In other cases, groups are exposed to problems for the first time as a group (Moon et al., 2003). The main characteristics of group decisions are described as follows (Lu et al., 2007): (1) A group performs a decision-making task. (2) A group decision covers the whole process of transfer from generating ideas for solving a problem to implementing solutions. (3) Group members may be located in the same or different places. (4) Group members may work at the same or different times. (5) Group members may work for the same or different departments or organizations. (6) The group can be at any managerial levels. (7) There can be conflicting opinions among group members in a group decision process. (8) The decision task might have to be accomplished in a short time. (9) Group members might not have complete information for decision tasks. (10) Some required data, information or knowledge for a decision may be located in many sources and some may be external to the organization. In the following, some studies on AMTs from the literature are briefly summarized. The results of the literature review will be summarized in a comparative manner later.

Demmel and Askin (1992) introduced a multiple objective decision model for the evaluation of AMT, whereby the AMT selection attributes were classified into three categories: (1) pecuniary objective, (2) strategic objective, and (3) tactical objective. The pecuniary objective is based upon discounted cash flow techniques. Shang and Sueyoshi (1995) proposed a selection procedure for an FMS employing the AHP, simulation, and data envelopment analysis (DEA). Sambasivarao and Deshmukh (1997) devised a decision support system for selection of AMT. The system provided the decision-makers with a base of knowledge and approaches to building a justification model, and emphasized that AMT attributes could be classified into two groups, namely, tangible (objective) and intangible (subjective). They suggested AHP, TOPSIS, and a linear additive utility model as alternative multi-attribute analysis methods. Small and Chen (1997) discussed the results of a survey conducted in the US that investigated the use of justification approaches for AMS. According to their findings, manufacturing firms using hybrid strategies, which employ both economic and strategic justification techniques, attain significantly higher levels of success from advanced technology projects.

Kotha and Swamidass (1998) hypothesized that nationality of the firm was an important factor in AMT use. To test the effect of the nationality variable on AMT use they compared the use of 18 AMTs in the US and Japan in an exploratory study using data from 160 US firms and 125 Japanese firms. There was clear evidence that the nationality of the firm was a factor in AMT use; that was AMT use is significantly different in the two countries. O'Kane et al. (2000) reported that simulation modeling could be used for the justification phase of AMT. Efstathiades et al. (2000) investigated the transfer and implementation processes of AMT in developing countries using the Cypriot manufacturing industry as a case study. They also addressed two critical steps: (1) the management processes followed during the transfer of technology into the manufacturing environment, and (2) the steps followed both before and after implementation and productive operation of the technologies. They indicated that despite the distance between the manufacturer and the technology suppliers, no difficulties were experienced in acquiring information about the available technologies and the suitability of these technologies for the specific manufacturing environment.

Arvanitis and Hollenstein (2001) empirically investigated the decisions of firms to adopt AMT based on a comprehensive specification of a rank model of technology adoption using firm-level data for Swiss manufacturing. Chan et al. (2001) presented an overview and guidance for manufacturing companies which planned to invest in AMT. They explained the reasons why the company might encounter problems while adopting AMT and explained the many suggestions offered by the relevant literature for improving the performance of evaluation in AMT investment. According to the four major steps in adopting AMT (i.e. strategic planning, justification, training and installation, and routinization and implementation), they aimed at assisting managers or investors to recognize problems at each step, thus offering appropriate ways to avoid and/or solve those problems. Yusuff et al. (2001) applied the Analytic Hierarchy Process (AHP) to determine the priority weights in predicting the success of AMT implementation because of its capability to structure complicated, multi-decision maker, multi-alternative and multi-attribute problems hierarchically. They applied AHP to forecast the success of AMT implementation by taking into account two possible outcomes, which were predicted according to seven influential factors and called the successful and unsuccessful implementation. Kengpol and O'Brien (2001) applied AHP and cost/benefit analysis to select an advanced technology by incorporating quantitative and qualitative factors. Karsak and Tolga (2001) proposed a fuzzy multiple attribute decision-making (FMADM) procedure for evaluating advanced manufacturing system investments. This approach applied fuzzy discounted cash flow analysis and decision-makers' linguistic assessments to the economic criterion and strategic criteria, respectively. The decision algorithm aggregated the experts' preference ratings for the economic and strategic criteria weights, and the suitability of AMS investment alternatives versus the selection criteria to calculate fuzzy suitability indices. The fuzzy indices were then used to rank the AMS investment alternatives. Abdel-Kader and Dugdale (2001) proposed a FMADM model for assessing AMT investments. This model applied the mathematics of the analytical hierarchy process and fuzzy set theory to aggregate the two major dimensions of financial and non-financial attributes. Gupta and Whitehouse (2001) examined the impact of size of an organization on the manufacturing strategy.

Aravindan and Punniyamoorthy (2002) developed a new model to justify the investment in AMT using the extended Brown–Gibson model which was developed for evaluating alternative plant locations using certain objective and subjective factors. They showed that investment in AMT was attractive if the benefits accruing from the subjective factors were considered. Yurdakul (2004) utilized a combination of AHP and goal programming for selection among computer-integrated manufacturing technologies. The model used AHP to determine attribute weights in goal programming. Sener and Karsak (2007) presented a new decision making approach based on fuzzy regression with non-symmetric coefficients and fuzzy optimization for FMS selection.

Sener and Karsak (2008) proposed a decision model based on fuzzy linear regression and fuzzy multiple objective programming for AMT selection. Thakur and Jain (2008) explored the issues of measurement and comparison of the current state of AMT adoption in India, including important information technology (IT) factors. Comparisons were made between Indian firms, firms in a developed country (Canada), and in a developing country (China) for a worldwide perspective.

Chang and Wang (2009) investigated the potential use of AHP by applying the consistent fuzzy preference relations to predict advanced AMT implementation. They pointed out that the AHP method was not efficient enough because it performed complex computation procedures in paired comparisons and obtaining consistency indicators. To reduce the judgment times and avoid checking inconsistency, they employed the consistent fuzzy preference relations, which inherited some advantages of AHP (distinct hierarchy, effective numerical assessment), to overcome certain drawbacks resulting from this conventional pairwise comparison approach. They also reviewed several methods which are utilized to fulfill the requirements of AMT selection, evaluation, assessment, justification and prediction issues, such as simple estimation methods, breakeven approaches, systems value analysis, scoring models, analytic hierarchy process, simulation and strategic methods. Chuu (2009) developed a fuzzy multiple attribute decision-making method for the group decision-making to improve advanced manufacturing technology selection process. In the proposed approach, a new fusion method of fuzzy information was developed for managing information assessed in different linguistic (multi-granularity linguistic term sets) and numerical scales. Also an application for the Taiwanese bicycle industry was employed. Costa and Lima (2009) presented the rationality for the organizational design development related to AMT adoption. For the development of the organizational design process, it was suggested a specified sequence of activities, in order to consider all the dimensions that define the content of the organizational design.

A result from the literature review is that the Choquet integral has not been used up to date although various MCDM methods have been used for the solution AMT selection problem. In this chapter, it is the first time that Choquet integral is used as a group decision making technique for the solution of the AMT selection problem.

In Table 1, the results of the literature review are summarized in a comparative manner with respect to methods, purpose, and criteria.

Table 1. Categorization and comparison of related works for AMT decision-making

The method or technique	Author(s)	Purpose	Criteria
Analytic hierarchy process (AHP)	Yurdakul (2004)	Selection	Intangible/tangible
	Punniyamoorthy and Ragavan (2003)	Selection	Intangible/tangible
	Yusuff et al. (2001)	Prediction	Intangible
	Kengpol and O'Brien (2001)	Selection	Intangible/tangible
	Mohanty and Deshmukh (1998)	Selection	Intangible/tangible
	Chan et al. (1999)	Selection	Intangible/tangible
	Albayrakoglu (1996)	Justification	
	Oeltjenbruns et al. (1995)	Selection	Intangible/tangible
	Weber (1993)	Selection	Intangible/tangible
	Datta et al. (1992)	Justification	Intangible/tangible
Fuzzy AHP	Chan et al. (2006)	Selection	Intangible/tangible
Fuzzy multiple attribute decision-making	Karsak and Tolga (2001)	Evaluation	Intangible/tangible
	Abdel-Kader and Dugdale (2001)	Selection	Intangible/tangible
Fuzzy Group Decision Making	Chuu (2009)	Selection	Intangible/tangible
	Chuu (2009a)	Selection	Intangible/tangible
Traditional financial justification method	Orr (2002)	Evaluation	Intangible/tangible
Decision support systems	Sambasivarao and Deshmukh (1997)	Selection	Intangible/tangible
Simulation Modeling	O'Kane et al. (2000)	Justification	
Discounted cash flow (DCF)	Aravindan and Punniyamoorthy (2002)	Justification	Tangible
	Demmel and Askin (1992)	Evaluation	Tangible
Data envelop analysis (DEA)	Talluri et al. (2000)	Selection	Tangible
	Talluri and Yoon (2002)	Selection	Tangible
	Shang and Sueyoshi (1995)	Selection	Tangible
Mixed nonlinear programming model	Verter and Dasci (2002)	Planning	
	Bokhorst et al. (2002)	Selection	Tangible
System wide benefits value analysis	Ordoobadi and Mulvaney (2001)	Selection	Tangible
Cluster analysis	Díaz et al. (2003)	Examining	Intangible/tangible
Integrated planning model	Chen and Small (1994)	Planning	Intangible
	Efstathiades et al. (2002)	Planning	Intangible/tangible
Real option valuation technique	MacDougall and Pike (2003)		Intangible/tangible
Fuzzy regression	Sener and Karsak (2007)	Selection	Intangible/tangible
	Sener and Karsak (2008)	Selection	Intangible/tangible

SELECTION CRITERIA FOR ADVANCED MANUFACTURING TECHNOLOGY

Many companies are currently strengthening their competitive positions by updating the technologies used in the manufacturing processes. The selection of the appropriate technology that achieves or matches with the organization objective must be made on the basis of sound decision-making process. The decision to implement AMT is a major decision for many organizations. The success or failure of the organization could be due to this decision, and therefore it is very important that proper consideration be given to all aspects of implementation before a final commitment is made. This is necessary to ensure that all the expected benefits of the decision are realized (Yusuff et al., 2001).

Chan et al. (1999) used the following criteria in Table 2, to evaluate the tangible and intangible benefits for the selection of AMT.

THE CHOQUET INTEGRAL

In this chapter, we are interested in using the Choquet integral which is a generalization of the Lebesgue integral, defined with respect to a non classical measure, often called fuzzy measure, or non-additive measure or also capacity. When the underlying universe is finite, the Lebesgue integral reduces to a (convex) linear combination hence can be assimilated to a particular class of regression models, where the coeffcients are all positive and sum up to one. Hence, the Choquet integral offers a more general model, more precisely, it offers a set of (convex) linear models, each of them being defined in a simplex (Grabisch et al. 2007). The Choquet integral was applied to multi-criteria evaluation by many researchers (Ishii & Sugeno, 1985; Onisawa et al., 1986; Labreuche & Grabisch, 2003; Kojadinovic, 2004; Karsak, 2005; Tsai & Lu, 2006; Grabisch et al., 2007; Berrah, et al., 2008a, 2008b; Schmitt et al., 2008; Saad et al., 2008; Narukawa & Murofushi, 2008; Shieh et al., 2009). The Choquet integral is a method which measures the expected utility of an uncertain event. In imprecise probability theory, the Choquet integral is used to calculate the lower expectation induced by a 2-monotone lower probability, or the upper expectation induced by a 2-alternating upper probability. A fuzzy integral is a sort of general averaging operator

Table 2. Criteria to evaluate AMT Investments

Criteria	Sub criteria
1. Cost	• Product cost
	• Maintenance cost
	• High rate of return
	• Labor cost
	• Material cost
2. Performance	• Compatibility with existing machine
	• Work morale
	• Productivity
	• Utilization
	• Machine breakdown
	• Human integration
3. Quality	• Scraped value
	• Rework
	• Conformance
	• Consistency
4. Delivery	• Transportation
	• Customer services
	• Time scheduling
	• Delivery time
	• Inventory/work in progress
	• Lead time
5. Flexibility	• Design change accommodation
	• Change in product mix
	• Market responsiveness
	• Capacity growth
	• Routing and scheduling flexibility
6. Innovativeness	• Research and development
	• Introduce product variation

that can represent the notions of importance of a criterion and interactions among criteria. The most important feature of a fuzzy integral is its ability to represent a certain kind of interaction among criteria, ranging from redundancy (negative interaction) to synergy (positive interaction). According to Grabisch (1998), there is almost no well established method to deal with interacting criteria, and usually people tend to avoid the problem by constructing independent criteria (or criteria that are supposed to be so). This ability of the fuzzy integral is the reason for its success in various fields of multi-criteria decision-making. The disadvantage of fuzzy integral is complexity of the model, since the number of coefficients involved in a fuzzy integral model grows exponentially with the number of criteria to be aggregated. The main difficulty is to identify all these coefficients, either by some learning data, or by a questionnaire, or both.

To define fuzzy integrals, a set of values of importance is needed. This set is composed of the values of a fuzzy measure. So, a value of importance for each subset of attributes is needed. In the following, some definitions are given to explain the basics of the Choquet integral (Modave & Grabish, 1998):

Definition 1. Let I be the set of attributes (or any set in a general setting). A set function $\mu: P(I) \rightarrow [0, 1]$ is called a fuzzy measure if it satisfies the three following axioms:

- $\mu(\emptyset) = 0$: an empty set has no importance,
- $\mu(I) = 1$: the maximal set has a maximal importance,
- $\mu(B) \leq \mu(C)$ if $B, C \subset I$ and $B \subset C$: a new added criterion cannot make the importance of a coalition (a set of criteria) diminish.

Therefore, in a problem where $card(I) = n$, a value for every element of $P(I)$ including 2^n values is needed. Assuming that the values of the empty set and of the maximal set are fixed, $(2^n$-2) values or coefficients to define a fuzzy measure are needed. So, there is clearly a trade-off between complexity and accuracy. However, the complexity can be significantly reduced in order to guarantee that fuzzy measures are used in practical applications. A fuzzy integral is a sort of weighted mean taking into account the importance of every coalition of criteria.

Definition 2. Let μ be a fuzzy measure on $(I, P(I))$ and an application $f: I \rightarrow \Re^+$. The Choquet integral of f with respect to μ is defined by:

$$(C)\int_I f d\mu = \sum_{i=1}^{n}\left(f\left(\sigma\left(i\right)\right) - f\left(\sigma\left(i-1\right)\right)\right)\mu\left(A_{(i)}\right) \quad (1)$$

where σ is a permutation of the indices in order to have

$$f\left(\sigma\left(1\right)\right) \leq \ldots \leq f\left(\sigma\left(n\right)\right),\ A_{(i)} = \left\{\sigma\left(i\right), \ldots, \sigma\left(n\right)\right\} \quad and \quad f\left(\sigma\left(0\right)\right) = 0$$

by convention.

It is easy to see that the Choquet integral is a Lebesgue integral up to a reordering of the indices. Actually, if the fuzzy measure μ is additive, then the Choquet integral reduces to a Lebesgue integral.

Now we will illustrate how fuzzy measures can be used in lieu of the weighted sum and other more traditional aggregation operators in a Multi-criteria decision making framework. It is shown in Modave and Grabisch (1998) that under rather general assumptions over the set of alternatives X, and over the weak orders $\succ_i$, there exists a unique fuzzy measure μ over I such that:

$$\forall x, y \in X,\ x \succeq y \Leftrightarrow u\left(x\right) \geq u\left(y\right) \quad (2)$$

where

$$u\left(x\right) = \sum_{i=1}^{n}\left[u_{(i)}\left(x_{(i)}\right) - u_{(i-1)}\left(x_{(i-1)}\right)\right]\mu\left(A_{(i)}\right) \quad (3)$$

which is simply the aggregation of the monodimensional utility functions using the Choquet integral with respect to μ.

The global importance of a criterion is given by evaluating what this criterion brings to every coalition it does not belong to, and averaging this input. This is given by the Shapley value or index of importance.

Definition 3. Let μ be a fuzzy measure over I. The Shapley value of index j is defined by:

$$v(j) = \sum_{B \subset I|(j)} \gamma_I(B)\left[\mu\left(B \cup \{j\}\right) - \mu(B)\right] \tag{4}$$

with $\gamma_I(B) = \dfrac{\left(|I| - |B| - 1\right)!.|B|!}{|I|!}$, $|B|$ denotes the cardinal of B.

The Shapley value can be extended to degree two, in order to define the indices of interactions between attributes.

Definition 4. Let μ be a fuzzy measure over I. The interaction index between i and j is defined by:

$$I(i,j) = \sum_{B \subset I|\{i,j\}} \xi_I(B).\left(\mu\left(B \cup \{i,j\}\right) - \mu\left(B \cup \{i\}\right) - \mu\left(B \cup \{i\}\right) + \mu(B)\right) \tag{5}$$

with $\xi_I(B) = \dfrac{\left(|I| - |B| - 2\right)!.|B|!}{\left(|I| - 1\right)!}$.

The interaction indices belong to the interval [−1,+1] and:

- $I(i, j) > 0$ if the attributes i and j are complementary;
- $I(i, j) < 0$ if the attributes i and j are redundant;
- $I(i, j) = 0$ if the attributes i and j are independent.

Interactions of higher orders can also be defined, however we will restrict ourselves to second order interactions which offer a good trade-off between accuracy and complexity.

Definition 5. A fuzzy measure μ is called 2-additive if all its interaction indices of order equal or larger than 3 are null and at least one interaction index of degree two is not null.

In this particular case of 2-additive measures, the following theorem can be given:

Theorem. Let μ be a 2-additive measure. Then Choquet integral can be computed by:

$$(C)\int_I f d\mu = \sum_{I_{ij}>0}\left(f(i) \wedge f(j)\right) I_{ij} + \sum_{I_{ij}<0}\left(f(i) \vee f(j)\right)\left|I_{ij}\right| + \sum_{i=1}^{n} f(i)\left(I_i - \frac{1}{2}\sum_{j \neq i}\left|I_{ij}\right|\right) \tag{6}$$

Note that this expression justifies the above interpretation of interaction indices, as a positive interaction index corresponds to a conjunction (complementary) and a negative interaction index corresponds to a disjunction (redundant).

The success of a Choquet integral depends on an appropriate representation of fuzzy measures, which captures the importance of individual criteria or their combination. In this chapter, the generalized Choquet integral proposed by Auephanwiriyakul et al. (2002) will be used, in which measurable evidence is represented in terms of intervals, whereas fuzzy measures are real numbers, and is an extension of the standard Choquet integral. In contrast to Auephanwiriyakul et al. (2002), Tsai and Lu (2006) proposes another generalization that involves linguistic expressions as well as information fusion between criteria to overcome vagueness and imprecision of linguistic terms in questionnaires.

Steps of the Methodology

The methodology is composed of eight steps (Tsai and Lu, 2006):

Step 1. Given criterion i, respondents' linguistic preferences for the degree of importance, perceived performance levels of alternative locations, and tolerance zone are surveyed.

Step 2. In view of the compatibility between perceived performance levels and the tolerance zone, trapezoidal fuzzy numbers are used to quantify all linguistic terms in this study. Given respondent t and criterion i, linguistic terms for the degree of importance which shows the weight of a criterion is parameterized by $\tilde{A}_i^t = \left(a_{i1}^t, a_{i2}^t, a_{i3}^t, a_{i4}^t\right)$, perceived performance levels which indicates score of an alternative by $\tilde{p}_i^t = \left(p_{i1}^t, p_{i2}^t, p_{i3}^t, p_{i4}^t\right)$, and the tolerance zone which represents specifications of a criterion by $\tilde{e}_i^t = \left(e_{i1L}^t, e_{i2L}^t, e_{i3U}^t, e_{i4U}^t\right)$. In this case study, t=1,2,3,4,5, i=1,2,…,n_j, j=1,2,3,4, n_1=3, n_2=2, n_3=4, n_4=3; where n_j represents the number of criteria in dimension j.

Step 3. Average $\tilde{A}_i^t$, $\tilde{p}_i^t$ and $\tilde{e}_i^t$ into $\tilde{A}_i$, $\tilde{p}_i$, and $\tilde{e}_i$, respectively using Equation (7).

$$\tilde{A}_i = \frac{\sum_{t=1}^k \tilde{A}_i^t}{k} = \left(\frac{\sum_{t=1}^k a_{i1}^t}{k}, \frac{\sum_{t=1}^k a_{i2}^t}{k}, \frac{\sum_{t=1}^k a_{i3}^t}{k}, \frac{\sum_{t=1}^k a_{i4}^t}{k}\right) \quad (7)$$

Step 4. Normalize the location value of each criterion using Equation (8).

$$\tilde{f}_i = \underset{\alpha\in[0,1]}{\|} \bar{f}_i^\alpha = \underset{\alpha\in[0,1]}{\|} [f_{i,\alpha}^-, f_{i,\alpha}^+] \quad (8)$$

where $f_i \in F(S)$ is a fuzzy-valued function. $\tilde{F}(S)$ is the set of all fuzzy-valued functions $f, f_i^\alpha = \left[f_{i,\alpha}^-, f_{i,\alpha}^+\right] = \frac{\bar{p}_i^\alpha - \bar{e}_i^\alpha + [1,1]}{2}$, $\bar{p}_i^\alpha$ and $\bar{e}_i^\alpha$ are α-level cuts of $\tilde{p}_i$ and $\tilde{e}_i$ for all α=[0,1].

Step 5. Find the location value of dimension j using Equation(9).

$$(C)\int \tilde{f}d\tilde{g} = \underset{\alpha=[0,1]}{\|} \left[(C)\int f_\alpha^- dg_\alpha^-, (C)\int f_\alpha^+ dg_\alpha^+\right] \quad (9)$$

where $\bar{g}_i : P(S) \to I(R^+)$, $\bar{g}_i = [g_i^-, g_i^+]$, $\bar{g}_i^\alpha = [g_{i,\alpha}^-, g_{i,\alpha}^+]$, $\bar{f}_i : S \to I(R^+)$, and $f_i = [f_i^-, f_i^+]$ for i=1, 2, 3,..., n_j.

To be able to calculate this location value, a λ value and the fuzzy measures $g(A_{(i)})$, i=1,2,...,n, are needed. These are obtained from Equations (10-12) (Sugeno, 1974; Ishii and Sugeno, 1985):

$$g\left(A_{(n)}\right) = g\left(\left\{s_{(n)}\right\}\right) = g_n \quad (10)$$

$$g\left(A_{(i)}\right) = g_i + g\left(A_{(i+1)}\right) + \lambda g_i g\left(A_{(i+1)}\right), \quad \text{where } 1 \le i < n \quad (11)$$

$$1 = g(S) = \begin{cases} \frac{1}{\lambda}\left\{\prod_{i=1}^n \left[1 + \lambda g\left(A_i\right)\right] - 1\right\} & \text{if } \lambda \neq 0 \\ \sum_{i=1}^n g\left(A_i\right) & \text{if } \lambda = 0 \end{cases} \quad (12)$$

where, $A_i \cap A_j = \varphi$ for all i, j = 1,2,3,…,n and $i \neq j$, and λ∈(-1,∞].

Step 6. Aggregate all dimensional performance levels of the location alternatives into overall performance levels, using a hierarchical process applying the two-stage aggregation process of the generalized Choquet integral. This is represented in Equation (13). The overall performance levels yields a fuzzy number, $\tilde{V}$.

$$\left.\begin{array}{l} \text{main criterion}_{(1)} = (C)\int f dg \\ \vdots \\ \text{main criterion}_{(m)} = (C)\int f dg \end{array}\right\rangle \; V = (C)\int \text{main criterion } dg \quad (13)$$

Step 7. Assume that the membership of $\tilde{V}$ is $\mu_{\tilde{V}}(x)$; defuzzify the fuzzy number $\tilde{V}$ into a crisp value v using Equation (14) and make a comparison of the overall performance levels of alternative locations.

$$F(\tilde{A}) = \frac{a_1 + a_2 + a_3 + a_4}{4} \tag{14}$$

Step 8. Compare weak and advantageous criteria among the locations alternatives using Equation (8).

A CASE STUDY

Flexible material flow, delivery to the point of use, small lots, short production lead times, and high inventory turnover characterize modern manufacturing and distribution operations. Therefore, an automated storage and retrieval system (AS/RS) is used to satisfy the necessities of modern manufacturing systems. AS/RS is a system which consists of a variety of computer-controlled methods for automatically placing and retrieving loads from specific storage locations. The high throughput and short processing times of Automated storage/retrieval systems, promote improved inventory management practices and rapid material handling, and play an essential role in integrated manufacturing and distribution systems. Additional AS/RS benefits generally include high space utilization, labour savings, protection from pilferage and damage, better safety, and real-time material control (Park, 2004). There are five main categories of AS/RS. These are; *(i)* unit load AS/RS, *(ii)* mini load AS/RS, *(iii)* deep-lane AS/RS *(iv)* man-on-board AS/RS, and *(v)* automated item retrieval system. The unit load AS/RS is used to store and retrieve loads that are palletized or stored in standard-sized containers. The mini load AS/RS is designed to handle small loads such as individual parts, tools, and supplies that are contained in bins or drawers in the storage system. A mini load AS/RS is generally smaller than a unit load AS/RS and allows storing more material in less space. The deep-lane AS/RS is a high-density unit load storage system that is appropriate for storing large quantities of stock. The man-on-board AS/RS system allows storage of items in less than unit load quantities. The system permits individual items to be picked directly at their storage locations. The automated item retrieval system is designed for retrieval of individual items or small product cartoons.

In our case study, one of the biggest logistic firms in Turkey plans to install an AS/RS system into its warehouse. The firm will make a selection between a deep-lane AS/RS and a unit load AS/RS by using Choquet integral methodology.

Step 1. The criteria given in Table 3 are selected from the literature to evaluate the alternatives (Chan et al., 1999). Three experts assign their linguistic preferences as seen in Table 4.

Step 2. Trapezoidal fuzzy numbers given in Table 5 are used to quantify the linguistic terms in Table 4.

Table 3. The criteria of AS/RS selection and their symbols

Criteria	Sub criteria
C_1. Cost	C_{11}. Product cost
	C_{12}. Maintenance cost
	C_{13}. Labor cost
	C_{14}. Material cost
	C_{15}. Operating cost
	C_{16}. Breakdown cost
C_2. Performance	C_{21}. Compatibility with existing storehouse
	C_{22}. Productivity
	C_{23}. Utilization
	C_{24}. Rate of breakdown
	C_{25}. Human integration
C_3. Quality	C_{31}.Service Time
	C_{32}. Consistency
C_4. Flexibility	C_{41}. Design change accommodation
	C_{42}. Change in product mix
	C_{43}. Market responsiveness
	C_{44}. Capacity
C_5. Innovativeness	C_{51}.Research and development
	C_{52}. Introduce product variation

Table 4. Individual importances of criteria, the tolerance intervals, and linguistic evaluations for alternatives

		Individual Importances of Criteria			Tolerance Intervals	Deep Line AS/RS			Unit Load AS/RS		
Criteria	**Sub criteria**	**E1**	**E2**	**E3**		**E1**	**E2**	**E3**	**E1**	**E2**	**E3**
C1. Cost		VI	I	VI							
	C11 Product cost	I	I	VI	[M, H]	H	VH	VH	M	H	M
	C12 Maintenance cost	I	M	I	[L, M]	H	VH	VH	M	M	H
	C14 Labor cost	M	I	M	[VL, H]	VL	VL	L	VL	L	L
	C15 Material cost	I	I	I	[M, H]	H	VH	H	M	H	M
	C16 Operating cost	I	VI	I	[L, H]	H	VH	VH	H	VH	H
	C17 Breakdown cost	VI	VI	VI	[L, H]	H	M	H	VH	H	VH
C2. Performance		M	I	M							
	C21. Compatibility with existing storehouse	I	I	VI	[M, VH]	H	VH	H	M	M	H
	C22. Productivity	I	VI	I	[M, VH]	VH	H	VH	H	M	H
	C23. Utilization	M	U	M	[L, H]	H	VH	VH	H	H	M
	C24. Rate of breakdown	M	M	M	[VL, L]	L	L	VL	VL	L	L
	C25. Human integration	M	I	U	[M, M]	M	M	H	VH	VH	H
C3. Quality		M	I	M							
	C31.Service Time	VI	I	I	[L, H]	VH	H	VH	VH	H	H
	C32. Consistency	I	I	I	[M, VH]	H	VH	VH	H	H	H
C4. Flexibility		M	M	I							
	C41. Design change accommodation	U	U	M	[M, H]	H	H	M	L	M	L
	C42. Change in product mix	VU	U	U	[M, H]	L	L	M	M	L	L
	C43. Market responsiveness	I	I	M	[M, H]	H	VH	H	M	M	H
	C44. Capacity	M	I	M	[L, H]	VH	H	VH	M	H	H
C5. Innovativeness		M	M	I							
	C51.Research and development	M	I	I	[L, H]	M	H	M	H	VH	H
	C52. Introduce product variation	M	M	M	[L, H]	L	L	M	M	M	L

Step 3. The aggregation of experts' assessments obtained by Equation (7) is given in Table 6.

Step 4. By using Equation (8), the evaluation values are normalized for each criterion.

Step 5. At different α levels, the values of all criteria in the same dimension are aggregated by using Equation (9). To illustrate the calculation procedure in Steps 4 and 5, an example is presented for deep line AS/RS alternative under innovativeness criterion. Using Equation (8),

$$f, f_i^{\alpha} = \left[f_{1,0}^-, f_{1,0}^+\right] = \frac{[0.667,\ 0.93] - [0.2,\ 0.9] + [1,\ 1]}{2} = [0.283,\ 0.783]$$

and $\left[f_{2,0}^-, f_{2,0}^+\right] = [0.183,\ 0.683]$ are obtained. Their corresponding degrees of importances are $\bar{g}_1^{\,0} = [0.533,\ 0.833]$, $\bar{g}_2^{\,0} = [0.4,\ 0.6]$, respectively. First, the sequence $f_{i,0}^-$ is sorted, where i=1 and 2, as follows: $f_{2,0}^- < f_{1,0}^-$. The corresponding degrees of importances are $\bar{g}_{1,0} = 0.4$ and $\bar{g}_{2,0} = 0.533$, respectively. By solving the equa-

Table 5. The relationship between trapezoidal fuzzy numbers and degrees of linguistic importances in a five-linguistic-term scale

The degrees of importance		Low/High Levels		Trapezoidal fuzzy numbers
Label	Linguistic terms	Label	Linguistic Terms	
VU	Very Unimportant	VL	Very Low	(0,0,0,0.3)
U	Unimportant	L	Low	(0.2,0.3,0.4,0.5)
M	Middle	M	Middle	(0.4,0.5,0.6,0.7)
HI	High Important	H	High	(0.6,0.7,0.8,0.9)
VI	Very Important	VH	Very High	(0.8,1,1,1)

Table 6. Aggregated decision matrix belonging to the group which consists of three experts

Criteria	Individual Importance	Tolerance Zone	Deep Line AS/RS	Unit Load AS/RS
C1	(0.733,0.9,0.933,0.967)			
C11	(0.667,0.8,0.867,0.933)	(0.4,0.5,0.8,0.9)	(0.733,0.9,0.933,0.967)	(0.467,0.567,0.667,0.767)
C12	(0.533,0.667,0.733,0.833)	(0.2,0.3,0.6,0.7)	(0.733,0.9,0.933,0.967)	(0.467,0.567,0.667,0.767)
C13	(0.467,0.633,0.667,0.767)	(0,0,0.8,0.9)	(0.067,0.1,0.133,0.367)	(0.133,0.2,0.267,0.433)
C14	(0.6,0.7,0.8,0.9)	(0.4,0.5,0.8,0.9)	(0.667,0.8,0.867,0.933)	(0.467,0.567,0.667,0.767)
C15	(0.667,0.8,0.867,0.933)	(0.2,0.3,0.6,0.7)	(0.733,0.9,0.933,0.967)	(0.667,0.8,0.867,0.933)
C16	(0.8,1,1,1)	(0.2,0.3,0.6,0.7)	(0.533,0.633,0.733,0.833)	(0.733,0.9,0.933,0.967)
C2	(0.467,0.567,0.667,0.767)			
C21	(0.667,0.8,0.867,0.933)	(0.4,0.5,1,1)	(0.667,0.8,0.867,0.933)	(0.467,0.633,0.667,0.767)
C22	(0.667,0.8,0.867,0.933)	(0.4,0.5,1,1)	(0.733,0.9,0.933,0.967)	(0.533,0.667,0.733,0.833)
C23	(0.267,0.4,0.4,0.567)	(0.2,0.3,0.8,0.9)	(0.733,0.9,0.933,0.967)	(0.533,0.667,0.733,0.833)
C24	(0.4,0.6,0.6,0.7)	(0,0,0.4,0.5)	(0.133,0.267,0.267,0.433)	(0.133,0.267,0.267,0.433)
C25	(0.333,0.433,0.467,0.633)	(0.4,0.5,0.6,0.7)	(0.467,0.633,0.667,0.767)	(0.733,0.9,0.933,0.967)
C3	(0.467,0.633,0.667,0.767)			
C31	(0.667,0.8,0.867,0.933)	(0.2,0.3,0.8,0.9)	(0.733,0.9,0.933,0.967)	(0.667,0.8,0.867,0.933)
C32	(0.6,0.7,0.8,0.9)	(0.4,0.5,1,1)	(0.733,0.9,0.933,0.967)	(0.6,0.7,0.8,0.9)
C4	(0.467,0.633,0.667,0.767)			
C41	(0.133,0.2,0.2,0.433)	(0.4,0.5,0.8,0.9)	(0.533,0.667,0.733,0.833)	(0.267,0.467,0.467,0.567)
C42	(0.067,0.133,0.133,0.367)	(0.4,0.5,0.8,0.9)	(0.267,0.467,0.467,0.567)	(0.267,0.467,0.467,0.567)
C43	(0.533,0.667,0.733,0.833)	(0.4,0.5,0.8,0.9)	(0.667,0.8,0.867,0.933)	(0.467,0.633,0.667,0.767)
C44	(0.467,0.633,0.667,0.767)	(0.2,0.3,0.8,0.9)	(0.733,0.9,0.933,0.967)	(0.533,0.667,0.733,0.833)
C5	(0.467,0.633,0.667,0.767)			
C51	(0.533,0.667,0.733,0.833)	(0.2,0.3,0.8,0.9)	(0.467,0.633,0.667,0.767)	(0.667,0.8,0.867,0.933)
C52	(0.4,0.6,0.6,0.7)	(0.2,0.3,0.8,0.9)	(0.267,0.467,0.467,0.567)	(0.333,0.533,0.533,0.633)

tion $0 = \frac{1}{\lambda}\left\{\prod_{i=1}^{2}[1+\lambda \bar{g}_i]-1\right\}$, λ=0.313 is obtained. Then, their fuzzy measures are derived from as follow;

g(A(2)) = g2 = 0.533

g(A(1)) = g1 + g(A(2)) + λ g1 g(A(2)) = 1

The aggregated Choquet integral values for the main criterion C5 are calculated as $(C)\int \tilde{f} d\tilde{g} = [0.237,\ 0.753]$

Step 6. Similar to Steps 4 and 5, the overall values are obtained for AS/RS systems, as shown in Table 7.

Step 7. From Table 7, the defuzzified overall values of AS/RSs using generalized Chouqet Integral are obtained as 0.710 and 0.713. Hence Unit Load AS/RS is the best alternative for the firm.

Step 8. Weak and advantageous criteria for alternatives are clarified in Table 7 by normal and bold characters, respectively. The numbers in bold characters in Table 7 represent that the alternative has more advantage than the other for that criterion.

Table 7. Defuzzified values of AS/RS alternatives

Criteria				**Deep Line AS/RS**	**Unit Load AS/RS**
		Deep Line AS/RS	**Unit Load AS/RS**	**Rank1**	**Rank2**
Overall Value		(0.468,0.631,0.811,0.882)	(0.458,0.634,0.812,0.882)	**0.698**	0.696
C1		(0.5,0.643,0.813,0.883)	(0.5,0.65,0.817,0.883)	0.71	**0.713**
	C11	(0.417,0.55,0.717,0.783)	(0.284,0.383,0.583,0.683)	**0.617**	0.483
	C12	(0.517,0.65,0.817,0.883)	(0.384,0.483,0.683,0.783)	**0.717**	0.583
	C13	(0.084,0.15,0.567,0.683)	(0.117,0.2,0.633,0.717)	0.371	**0.417**
	C14	(0.384,0.5,0.684,0.767)	(0.284,0.383,0.583,0.683)	0.583	**0.483**
	C15	(0.517,0.65,0.817,0.883)	(0.484,0.6,0.783,0.867)	**0.717**	0.683
	C16	(0.417,0.517,0.717,0.817)	(0.517,0.65,0.817,0.883)	0.617	**0.717**
C2		(0.379,0.506,0.753,0.839)	(0.366,0.513,0.687,0.792)	**0.619**	0.59
	C21	(0.334,0.4,0.684,0.767)	(0.234,0.317,0.583,0.683)	**0.546**	0.454
	C22	(0.367,0.45,0.717,0.783)	(0.267,0.333,0.617,0.717)	**0.579**	0.483
	C23	(0.417,0.55,0.817,0.883)	(0.317,0.433,0.717,0.817)	**0.667**	0.571
	C24	(0.317,0.434,0.634,0.717)	(0.317,0.433,0.633,0.717)	0.525	0.525
	C25	(0.384,0.517,0.584,0.683)	(0.517,0.65,0.717,0.783)	0.542	**0.667**
C3		(0.397,0.52,0.797,0.873)	(0.35,0.455,0.757,0.855)	**0.647**	0.604
	C31	(0.417,0.55,0.817,0.883)	(0.383,0.5,0.783,0.867)	**0.667**	0.633
	C32	(0.367,0.45,0.717,0.783)	(0.3,0.35,0.65,0.75)	**0.579**	0.513
C4		(0.386,0.525,0.767,0.853)	(0.287,0.422,0.668,0.783)	**0.633**	0.54
	C41	(0.317,0.434,0.617,0.717)	(0.184,0.333,0.483,0.583)	**0.521**	0.396
	C42	(0.184,0.334,0.484,0.583)	(0.184,0.333,0.483,0.583)	0.396	0.396
	C43	(0.384,0.5,0.684,0.767)	(0.284,0.417,0.583,0.683)	**0.583**	0.492
	C44	(0.417,0.55,0.817,0.883)	(0.317,0.433,0.717,0.817)	**0.667**	0.571
C5		(0.233,0.383,0.643,0.753)	(0.283,0.447,0.717,0.822)	0.5	**0.567**
	C51	(0.283,0.417,0.683,0.783)	(0.383,0.5,0.783,0.867)	0.542	**0.633**
	C52	(0.183,0.333,0.583,0.683)	(0.217,0.367,0.617,0.717)	0.446	**0.479**

CONCLUSION

In this chapter, a multiple criteria group decision-making problem under a fuzzy environment has been discussed. To illustrate the chapter, a selection problem between AS/RSs has been taken into consideration by using the Choquet integral. The characteristic feature of the problem is that it includes both objective and subjective and conflicting criteria. Hence, three experts' expertise is used in order to cope with the complexity of the problem. To satisfy the group decision, the arithmetic mean method has been used. The Choquet integral has provided an excellent tool for the solution of our advanced manufacturing technology selection problem including conflicting criteria.

For future research, another aggregation method from the literature may be used in order to compare the results and put forward the consistency of the group decision.

REFERENCES

Abdel-Kader, M. G., & Dugdale, D. (2001). Evaluating investments in advanced manufacturing technology: A fuzzy set theory approach. *The British Accounting Review*, *33*, 455–489. doi:10.1006/bare.2001.0177

Albayrakoglu, M. (1996). Justification of new manufacturing technology: A strategic approach using the analytic hierarchy process. *Production and Inventory Management Journal*, *37*(1), 71–76.

Aravindan, P., & Punniyamoorthy, M. (2002). Justification of advanced manufacturing technologies (AMT). *International Journal of Advanced Manufacturing Technology*, *19*, 151–156. doi:10.1007/s001700200008

Arvanitis, S., & Hollenstein, H. (2001). The determinants of the adoption of advanced manufacturing technology. *Economics of Innovation and New Technology*, *10*(5), 377–414. doi:10.1080/10438590100000015

Auephanwiriyakul, S., Keller, J. M., & Gader, P. D. (2002). Generalized Choquet fuzzy integral fusion. *Information Fusion*, *3*, 69–85. doi:10.1016/S1566-2535(01)00054-9

Berrah, L., Mauris, G., Montmain, J., & Cliville, V. (2008a). Efficacy and efficiency indexes for a multi-criteria industrial performance synthesized by Choquet integral aggregation. *International Journal of Computer Integrated Manufacturing*, *21*(4), 415–425. doi:10.1080/09511920701574255

Berrah, L., Mauris, G., Montmain, J., & Cliville, V. (2008b). Monitoring the improvement of an overall industrial performance based on a Choquet integral aggregation. *Omega-International Journal of Management Science*, *36*(3), 340–351. doi:10.1016/j.omega.2006.02.009

Bokhorst, J. A. C., Slomp, J., & Suresh, N. C. (2002). An integrated model for part operation allocation and investments in CNC technology. *International Journal of Production Economics*, *75*(3), 267–285. doi:10.1016/S0925-5273(01)00118-9

Chan, F. T. S., Chan, H. K., Chan, M. H., & Humphreys, P. K. (2006). An integrated fuzzy approach for the selection of manufacturing technologies. *International Journal of Advanced Manufacturing Technology*, *27*(7–8), 747–758. doi:10.1007/s00170-004-2246-9

Chan, F. T. S., Chan, M. H., Lau, H., & Ip, R. W. L. (2001). Investment appraisal techniques for advanced manufacturing technology (AMT): A literature review. *Integrated Manufacturing Systems*, *12*(1), 35–47. doi:10.1108/09576060110361528

Chan, F. T. S., Chan, M. H., & Mak, K. L. (1999). An integrated approach to investment appraisal for advanced manufacturing technology. *Human Factors and Ergonomics in Manufacturing*, *9*(1), 69–86. doi:10.1002/(SICI)1520-6564(199924)9:1<69::AID-HFM4>3.0.CO;2-1

Chang, T. H., & Wang, T. C. (2009). Measuring the success possibility of implementing advanced manufacturing technology by utilizing the consistent fuzzy preference relations. *Expert Systems with Applications*, *36*, 4313–4320. doi:10.1016/j.eswa.2008.03.019

Chen, I. J., & Small, M. H. (1994). Implementing manufacturing technology: An integrated planning model. *Omega*, *22*(1), 91–103. doi:10.1016/0305-0483(94)90010-8

Chuu, S. J. (2009a). Selecting the advanced manufacturing technology using fuzzy multiple attributes group decision making with multiple fuzzy information. *Computers & Industrial Engineering*, *57*(3), 1033–1042. doi:10.1016/j.cie.2009.04.011

Chuu, S. J. (2009b). Group decision-making model using fuzzy multiple attributes analysis for the evaluation of advanced manufacturing technology. *Fuzzy Sets and Systems*, *160*, 586–602. doi:10.1016/j.fss.2008.07.015

Costa, S. E. G., & Lima, E. P. (2009). Advanced manufacturing technology adoption: An integrated approach. *Journal of Manufacturing Technology Management*, *20*(1), 74–96. doi:10.1108/17410380910925415

D’ıaz, M. S., Machuca, J. A. D., & Álvarez-Gil, M. J. (2003). A view of developing patterns of investment in AMT through empirical taxonomies: New evidence. *Journal of Operations Management*, *21*(5), 577–606. doi:10.1016/j.jom.2003.03.002

Das, A., & Jayaram, J. (2003). Relative importance of contingency variables for advanced manufacturing technology. *International Journal of Production Research*, *41*(18), 4429–4452. doi:10.1080/00207540310001595819

Datta, V., Sambasivarao, K. V., Kodali, R., & Deshmukh, S. G. (1992). Multi-attribute decision model using the analytic hierarchy process for justification of manufacturing systems. *International Journal of Production Economics*, *28*(2), 227–234. doi:10.1016/0925-5273(92)90035-6

Demmel, J. G., & Askin, R. G. (1992). A multiple-objective decision model for the evaluation of advanced manufacturing system technologies. *Journal of Manufacturing Systems*, *11*(3), 179–194. doi:10.1016/0278-6125(92)90004-Y

Efstathiades, A., Tassou, S., & Antoniou, A. (2002). Strategic planning, transfer and implementation of advanced manufacturing technologies (AMT): Development of an integrated process plan. *Technovation*, *22*, 201–212. doi:10.1016/S0166-4972(01)00024-4

Grabisch, M., Kojadinovic, I., & Meyer, P. (2007). A review of methods for capacity identification in Choquet integral based multi-attribute utility theory applications of the Kappalab R package. *European Journal of Operational Research*, *186*(2), 766–785. doi:10.1016/j.ejor.2007.02.025

Gupta, A., & Whitehouse, F. R. (2001). Firms using advanced manufacturing technology management: An empirical analysis based on size. *Integrated Manufacturing Systems*, *12*(5), 346–350. doi:10.1108/09576060110398500

Herrera-Viedma, E., Chiclana, F., Herrera, F., & Alonso, S. (2007). Group decision-making model with incomplete fuzzy preference relations based on additive consistency. *IEEE Transactions on Systems, Man, and Cybernetics. Part B, Cybernetics*, *37*(1), 176–189. doi:10.1109/TSMCB.2006.875872

Ishii, K., & Sugeno, M. (1985). A model of human evaluation process using fuzzy integral. *International Journal of Man-Machine Studies*, *22*(1), 19–38. doi:10.1016/S0020-7373(85)80075-4

Jahanshahloo, G. R., Lotfi, F. H., & Izadikhah, M. (2006). Extension of the TOPSIS method for decision-making problems with fuzzy data. *Applied Mathematics and Computation*, *181*, 1544–1551. doi:10.1016/j.amc.2006.02.057

Karsak, E., & Tolga, E. (2001). Fuzzy multi-criteria decision-making procedure for evaluating advanced manufacturing system investments. *International Journal of Production Economics*, *69*, 49–64. doi:10.1016/S0925-5273(00)00081-5

Karsak, E. E. (2005). Choquet integral-based decision making approach for robot selection. *Knowledge-Based Intelligent Information and Engineering Systems*, 635-641.

Kaya, İ., & Çınar, D. (2008). Facility location selection using a fuzzy outranking method. *Journal of Multiple-Valued Logic and Soft Computing*, *14*, 251–263.

Kengpol, A., & O'Brien, C. (2001). The development of a decision support tool for the selection of advanced technology to achieve rapid product development. *International Journal of Production Economics*, *69*, 177–191. doi:10.1016/S0925-5273(00)00016-5

Kojadinovic, I. (2004). Unsupervised aggregation by Choquet integral based on entropy functionals: Application to the evaluation of students. *Modeling Decisions for Artificial Intelligence*, *3131*, 163–174.

Kotha, S., & Swamidass, P. M. (1998). Advanced manufacturing technology use: Exploring the effect of the nationality variable. *International Journal of Production Research*, *36*(11), 3135–3146. doi:10.1080/002075498192337

Labreuche, C., & Grabisch, M. (2003). Choquet integral for the aggregation of interval scales in multicriteria decision making. *Fuzzy Sets and Systems*, *137*(1), 11–26. doi:10.1016/S0165-0114(02)00429-3

Lu, J., Zhang, G., Ruan, D., & Wu, F. (2007). *Multi-objective group decision making methods: Software and applications with fuzzy set techniques*. Singapore: Imperial Collage Press.

MacDougall, S. L., & Pike, R. H. (2003). Consider your options: Changes to strategic value during implementation of advanced manufacturing technology. *Omega*, *31*(1), 1–15. doi:10.1016/S0305-0483(02)00061-0

Modave, F., & Grabisch, M. (1998). Preference representation by Choquet integral: The commensurability hypothesis. *Proc. 7th Int. Conf. on Information Processing and Management of Uncertainty in Knowledge-Based Systems (IPMU)*, 1-8.

Mohanty, R. P., & Deshmukh, S. G. (1998). Advanced manufacturing technology selection: A strategic model for learning and evaluation. *International Journal of Production Economics*, *55*(3), 295–307. doi:10.1016/S0925-5273(98)00075-9

Moon, H., Conlon, D. E., Humphrey, S. E., Quigley, N., Devers, C. E., & Nowakowski, J. M. (2003). Group decision process and incrementalism in organizational decision making. *Organizational Behavior and Human Decision Processes*, *92*(1-2), 67–79. doi:10.1016/S0749-5978(03)00079-7

Narukawa, Y., & Murofushi, T. (2008). Choquet-Stieltjes integral as a tool for decision modeling. *International Journal of Intelligent Systems*, *23*(2), 115–127. doi:10.1002/int.20260

O'Kane, J. F., Spenceley, J. R., & Taylor, R. (2000). Simulation as an essential tool for advanced manufacturing technology problems. *Journal of Materials Processing Technology*, *107*, 412–424. doi:10.1016/S0924-0136(00)00689-0

Oeltjenbruns, H., Kolarik, W. J., & Schnadt–Kirschner, R. (1995). Strategic planning in manufacturing systems: AHP application to an equipment replacement decision. *International Journal of Production Economics*, *38*, 189–197. doi:10.1016/0925-5273(94)00092-O

Onisawa, T., Sugeno, M., Nishiwaki, M. Y., Kawai, H., & Harima, Y. (1986). Fuzzy measure analysis of public attitude towards the use of nuclear energy. *Fuzzy Sets and Systems*, *20*, 259–289. doi:10.1016/S0165-0114(86)90040-0

Ordoobadi, S. M., & Mulvaney, N. J. (2001). Development of a justification tool for advanced manufacturing technologies: System-wide benefits value analysis. *Journal of Engineering and Technology Management*, *18*, 157–184. doi:10.1016/S0923-4748(01)00033-9

Orr, S. (2002). A comparison of AMT strategies in the USA, South Africa and Germany. *International Journal of Manufacturing Technology Management*, *4*(6), 441–454. doi:10.1504/IJMTM.2002.002517

Park, B. C. (2001). An optimal dwell point policy for automated storage/retrieval systems with uniformly distributed, rectangular racks. *International Journal of Production Research*, *39*(7), 1469–1480. doi:10.1080/00207540010023583

Punniyamoorthy, M., & Ragavan, P. V. (2003). A strategic decision model for the justification of technology selection. *International Journal of Advanced Manufacturing Technology*, *21*(1), 72–78. doi:10.1007/s001700300008

Saad, I., & Hammadi, S. (2008). Choquet integral for criteria aggregation in the flexible job-shop scheduling problems. *Mathematics and Computers in Simulation*, *76*(5-6), 447–462. doi:10.1016/j.matcom.2007.04.010

Sambasivarao, K. V., & Deshmukh, S. G. (1997). A decision support system for selection and justification of advanced manufacturing technologies. *Production Planning and Control*, *8*(3), 270–284. doi:10.1080/095372897235325

Schmitt, E., Bombardier, V., & Wendling, L. (2008). Improving fuzzy rule classifier by extracting suitable features from capacities with respect to Choquet integral. *IEEE Transactions on Systems, Man, and Cybernetics. Part B, Cybernetics*, *38*(5), 1195–1206. doi:10.1109/TSMCB.2008.925750

Sener, Z., & Karsak, E. E. (2007). A decision model for advanced manufacturing technology selection using fuzzy regression and fuzzy optimization. *IEEE International Conference on Systems, Man and Cybernetics,* 565-569.

Sener, Z., & Karsak, E. E. (2008). A decision making approach based on fuzzy regression and fuzzy multiple objective programming for advanced manufacturing technology selection. *IEEE International Conference Industrial Engineering and Engineering Management,* 964-968.

Shang, J., & Sueyoshi, T. (1995). A unified framework for the selection of a flexible manufacturing system. *European Journal of Operational Research*, *85*, 297–315. doi:10.1016/0377-2217(94)00041-A

Shieh, J. I., Wu, H. H., & Liu, H. C. (2009). Applying a complexity-based Choquet integral to evaluate students' performance. *Expert Systems with Applications*, *36*(3), 5100–5106. doi:10.1016/j.eswa.2008.06.003

Small, M. H., & Chen, I. J. (1997). Economic and strategic justification of AMT inferences from industrial practices. *International Journal of Production Economics*, *49*, 65–75. doi:10.1016/S0925-5273(96)00120-X

Sugeno, M. (1974). *Theory of fuzzy integrals and its applications*. Unpublished doctoral dissertation. Tokyo Institute of Technology, Japan.

Talluri, S., Whiteside, M. M., & Seipel, S. J. (2000). A nonparametric stochastic procedure for FMS evaluation. *European Journal of Operational Research*, *124*(3), 529–538. doi:10.1016/S0377-2217(99)00188-5

Talluri, S., & Yoon, K. P. (2002). A cone-ratio DEA approach for AMT justification. *International Journal of Production Economics*, *66*(2), 119–129. doi:10.1016/S0925-5273(99)00123-1

Thakur, L. S., & Jain, V. K. (2008). Advanced manufacturing techniques and information technology adoption in India: A current perspective and some comparisons. *International Journal of Advanced Manufacturing Technology*, *36*, 618–631. doi:10.1007/s00170-006-0852-4

Tsai, H. H., & Lu, I. Y. (2006). The evaluation of service quality using generalized Choquet integral. *Information Sciences*, *176*(6), 640–663. doi:10.1016/j.ins.2005.01.015

Verter, V., & Dasci, A. (2002). The plant location and flexible technology acquisition problem. *European Journal of Operational Research*, *136*(2), 366–382. doi:10.1016/S0377-2217(01)00023-6

Weber, S. F. (1993). A modified analytic hierarchy process for automated manufacturing decisions. *Interfaces*, *23*(4), 75–84. doi:10.1287/inte.23.4.75

Yurdakul, M. (2004). Selection of computer-integrated manufacturing technologies using a combined analysis hierarchy process and goal programming model. *Robotics and Computer-integrated Manufacturing*, *20*, 329–340. doi:10.1016/j.rcim.2003.11.002

Yusuff, R. M., Yee, K. P., & Hashmi, M. S. J. (2001). A preliminary study on the potential use of the analytical hierarchical process (AHP) to predict advanced manufacturing technology (AMT) implementation. *Robotics and Computer-integrated Manufacturing*, *17*, 421–427. doi:10.1016/S0736-5845(01)00016-3

Zimmermann, H. J. (1991). *Fuzzy set theory and its applications*. Dordrecht, Germany: Kluwer Academic.

KEY TERMS AND DEFINITIONS

Advanced Manufacturing Technology (AMT): A modern method of production incorporating highly automated and sophisticated computerized design and operational systems.

Automated Storage and Retrieval System (AS/RS): Consists of a variety of computer-controlled methods for automatically placing and retrieving loads from specific storage locations.

Choquet Integral: A generalization of the Lebesgue integral. Choquet integral is a way of measuring the expected utility of an uncertain event.

Decision Making: Can be regarded as an outcome of mental processes leading to the selection of a course of action among several alternatives.

Fuzzy Logic: A form of multi-valued logic derived from fuzzy set theory to deal with reasoning that is approximate rather than precise.

Fuzzy Sets: Are sets whose elements have degrees of membership and are extension of the classical notion of set.

Group Decision Making: Is decision making in groups consisting of multiple members.

Multi-Criteria Decision Making (MCDM): A discipline aimed at supporting decision makers who are faced with making numerous and conflicting evaluations.

Chapter 12
Indexing as an Ontological Support for Legal Reasoning

Erich Schweighofer
University of Vienna, Austria

ABSTRACT

Legal reasoning is still insufficiently understood. Formalisations of legal reasoning have failed so far, due in particular to the non-consideration of societal and legal complexity. Thus, the network of co-operative reasoning and its compilation and analysis should be represented by an index of legal and fact concepts of a legal ontology. The Dynamic Electronic Legal Commentary (DynELCom) constitutes such a semantic description and analysis of the legal information system. Knowledge acquisition tools support the establishment and refinement of the extensive ontology. Since a first ontology exists in European law, international law and Austrian law, this chapter also presents early experiments on EU state aid law which show the feasibility of the approach. Next steps will be a deep refinement of the ontology and the development of a dialogue system.

INTRODUCTION

New research slightly takes the veil of legal work and explains how lawyers cope with the vast body of legal materials, interpret it, apply it and – taking into account its underlying principles – develop the law. Law is handicraft and art but – still in development – also applied science. Since the 1970s, information and communication technology has played a major role of a catalyzer for applied research on law. Results are still premature as investment and acceptance of these endeavours remains insufficient. However, legal information systems have already changed the way lawyers handle legal documents, information and knowledge. Legal reasoning – the holy grail of lawyering – has been so far only slightly touched by automation aims.

Complexity might describe the missing component of research. It sounds liked a buzzword but there is much more behind. Firstly, it describes the present complexity of legal systems, the mil-

DOI: 10.4018/978-1-60960-091-4.ch012

lions of documents, the gigabytes of data, 100,000 and more rules etc. and the resulting information crisis of law. Secondly, is refers to the complexity of today's society and the ongoing globalisation, where information technology and internet provide one of the main fundamentals. Thirdly, it refers to the main reason of existence of law – the reduction of social complexity by governance.

Lawyers have to reduce social and legal complexity in order to achieve manageable containers of complexity. This can be a legal case, a domain of specialization, a co-operation of knowledge workers, some outsourcing etc.

It should be evident that automated legal reasoning should achieve the same level of mastering legal complexity, e.g. not handling only a very small subset of the legal order, a simple regulation like the *Library Regulations* but a particular domain of the law. This sub domain can be defined by quantity (e.g. a high number of cases) or quality (e.g. the law for a particular business or an administrative task). So far, theoretical and practical research has not sufficiently taken into account this requirement. Resulting *"nascituri"* of prototypes have strongly embarrassed lawyers and are responsible for the present low appraisal of this research in lawyer's circles.

Information retrieval can collect all materials and provide hints for relevant information. Legal reasoning requires agents able to handle a tremendous complexity in law and society. Human beings, e.g. lawyers, cope with the complexity with specialization (and longer working hours). Intelligent software agents can handle only standard cases so far. Thus, the complex social reality has to be sufficiently reduced for automating legal reasoning.

What can be the next step? Considering AI research, a sufficiently intelligent legal software agent may see the world in about 50 years. The main obstacle is not the complexity of reasoning but the required very high standards of natural language understanding and interpretation of visual input.

Thus, we argue for a simplified model of reasoning based on advanced indexing. As a well-known tool of reduction of information, indexing is heavily used in books, libraries and retrieval systems. It will provide much more support than just information retrieval. Given "established reasoning", indexing may offer a sufficiently strong link between facts and rules in order to support semi-automated handling of standard cases.

This approach is based on established technologies. Text of the law is available in legal information systems. Legal ontologies provide a specification of a juridical domain in a computer-usable way. World ontologies describe persons, objects, situations and events in the world. Techniques of indexing, summarisation and categorisation support a semi-automatic description and analysis of the text corpus of a legal information system; solving the problem of the bottleneck of knowledge acquisition.

Such an approach relies heavily on collaborative reasoning in a given legal domain. The various positions of legal authorities, in particular parliaments, courts, government offices, and legal authors are compiled and put into a coherent system.

These tools systematically put together, form the core of an indexing tool of legal situations, acts, concepts, rules and principles called *Dynamic Electronic Legal Commentary (DynELCom)*. Similar to its paper counterpart, it describes the law in an analytical structure. However, it is semi-automatic and dynamic, e.g. update is done on a daily basis supported by automatic indexing and categorisation. Further, it includes much more factual descriptions. The main value of a commentary, the index, is properly formalised and thus better understandable. Graphical structures help the lawyers, the formalisation the computers. Thus, a computer-useable structured description will offer a "simplified syllogism" for automated reasoning in standard cases. It is then up to legal practice to implement these "simplified syllogisms" in knowledge systems, dialog systems etc.

The remainder of the paper is organised as follows: section 2 describes the background of research on legal reasoning and ontologies and its shortcomings. Section 3 focuses on indexing and its potential. Section 4 deals with the concepts of the Dynamic Legal Electronic Commentary. Future work and conclusions complete the paper in section 5.

BACKGROUND

Artificial intelligence & law scratches the traditional understanding of law as a handicraft and art discipline. Steinmüller (1993) has called it the ongoing automation of legal intellectual work. That is evident: In order to automate a deep understanding of legal reasoning is required. Since the 1970ies, much work has been devoted to this research aims; however, with humble practical results.

Research on automation of legal reasoning is characterised by a dispute on the right point of departure. Should it be "top" legal reasoning in artificially small domains (e.g. Library Regulations) or "lower" legal reasoning in realistic legal domains? The first approach is strongly based on legal theory and logic programming and starts with the big problem that legal reasoning is still insufficiently understood. The second approach has its roots in legal information retrieval and the needs to make it more intelligent, e.g. advanced information retrieval, "simple" ontologies, text summarisation or text categorisation.

Legal Reasoning

Legal reasoning shares the same fate with law: is still insufficiently analysed. Eminent legal theorists like Hart (1994, p. 1) or Dworkin (1986, pp. 400-) point to the "persistent questions" about the nature of law and the role of lawyers in this system of rules. Alexy (1991, pp.349) focuses on the rational legal discourse as a way of explaining the law as such. Kelsen's pure theory of law (1960) explains major functions of the modern legal systems but excludes many factors by limiting the analysis to a formal interpretation of the body of positive law. We can add to this theoretical discourse our own humble contribution (following Luhmann (1968, 2002) in describing the law as a system for reduction of complexity (Schweighofer, 2008, pp. 42-). Legal theory shows the difficulty in defining the law but also its complexity (Haft, 2009, pp. 10-)[1]. Thus, we have still to rely on a quite practical view of law and legal reasoning for our study.

Legal reasoning focuses on legal practice, e.g. how judges (and administrators) should decide cases. In the field of legal theory, it occupies the more practical end. Legal theorists would classify these work into three categories: "(a) reasoning to establish the existing content of the law on a given issue, (b) reasoning from the existing content of the law to the decision which a court should reach in a case involving that issue which comes before it, and (c) reasoning about the decision which a court should reach in a case, all things considered." (Dickson, 2005).

Paraphrasing it, legal reasoning involves establishing the materials in a given issue, the interpretation of these materials and making a decision on facts and legal questions considering all relevant issues. However, in our view, this list is incomplete as social and legal complexity is missing. Thus, we add another category (d): handling the complexity of society and its governance system.

It is evident that society is getting more complex: more people, more objects, more actions, going global; resulting in many more combinations. Thus, all socioeconomic models only work with significant simplification (*ceteris paribus* clause). Law reflects this complexity by an ever growing number of legal instruments. The fact of an information crisis of law is now part of conventional wisdom; even if the first book of (Simitis, 1970) was not followed by other substantive studies.

Dynamic change of these rules describes the still developing state of many rules. Adaptations to new facts are very common but also struggling for finding the "best rule".

Globalisation adds to complexity in increasing the numbers of constituent factors, e.g. people, objects, actions and combinations. Further, models of governance have to be extended to other means of stabilisation of behaviour, e.g. regimes, customs or morals in order to include the many ways governments and human beings regulate their affairs (Schweighofer, 2008).

These facts aggravate the work of a lawyer, in particular the judge. He has to know the world, its aims of governance (government, law-makers, civil society etc.) with the reflecting ethics and morals, its regulative systems, the available sanctions, the ethical aims and the aims of those in power in order to make a consistent case.

It is not easy to quantify this requirement. Firstly, it is a rebuff of "toy systems" with about 200 rules. A sufficiently complex logic-based knowledge system should have about 10,000 to 20,000 rules. Further, the number of cases that such a system can handle should be sufficiently high. For ontologies, the conceptual structure, the system of knots and arcs, should contain about 10,000 concepts in a still shallow but sufficiently broad system.

Legal theorists concentrate on issues (b) and (c). It is common understanding nowadays that the idea "*le juge n'est pas la bouche de la loi*" is unrealistic. The view that law is to be identified by reference to authoritative social sources (Haft, 2009, pp. 226) does not take into account the argumentative nature of legal practice with its strong and deep disagreements (Dworkin, 1986, chapter 1). In their works on legal reasoning, Hart (1991, 124), Dworkin (1986, 45) and Alexy (1991, 219) devote a good part of their analysis to interpretation.

Thus, the theory of legal reasoning is aware of the complexity and difficulty of legal work. Interpretation and argumentation are tools of this process and may lead to a better understanding of legal thinking. However, it is in no way a convincing start for a formalisation.

As a last point, it is worth remembering that legal practice has a different understanding of legal reasoning. Lawyers do not need a good knowledge of legal philosophy; they can do a good job without knowing or reflecting much about legal theory. Thus, legal practice must be different from the complex theories of legal thinking of Kelsen, Hart, Dworking, Toolmin etc. Law in social reality is strongly determined by a very high number of rules, complex procedures and various authorities. It is complexity, that of the vast amount of rules and the indeterminate number of social situations.

In a sketchy way, legal reasoning can be classified in three types. First order logic or production rules as a practical way of understanding legal rules (e.g. complex IF-THEN EXCEPT (... (EXCEPT...)) rules; decision trees as representations of procedures) and authority marking, and highly efficient reading and interpreting of legal materials. These ways of thinking may be considered as easy; in practice, it is not. Complexity of rules, many exceptions (many exceptions of exceptions of exceptions etc.), diversified decision trees and dynamic authorities, combined with the not completely understandable social reality makes legal reasoning so difficult in practice. In complex cases, legal philosophy and legal theory have to be also taken into account in order to motivate authorities in the proper direction.

Thus, the focus of automation of legal reasoning should be a handling of a very high and complex number of rules, highly structured decision trees, reliable handling of authorities and support (e.g. semi-automation) of legal interpretation. Authoritative reasoning in law is based on taking into account all relevant legal authorities in a legal system. This aspect was also very often not properly taken into account in the automation of legal reasoning.

Automation of Legal Reasoning and its Poor Success

Computers can simulate many forms of human intelligence: memorising, calculating, searching but also some analogy. Legal skills are not limited to legal reasoning; legal research, writing, negotiating etc. are considered equally important in practice. Text processing is now considered as an indispensable tool for lawyers; appropriate information and communication infrastructure and internet broadband access are standard nowadays.

Main streams were logic-based knowledge systems (for a while so-called legal expert systems), argumentation, concept-based knowledge systems, case-based reasoning, knowledge discovery, natural language processing, text analysis and text categorisation. Nowadays, all these efforts have been consolidated under the heading of legal ontologies. It is here not the space to give an overview of these developments; we refer to existing overviews of Zeleznikow and Stranieri (2005), Lodder and Oskamp (2006), Schweighofer (1999), Zeleznikow and Hunter (1994), and the proceedings of the International Conference on Artificial Intelligence and Law, most recently that of 2009 and its related conferences, particularly LOAIT.

In my view, searching was the first and is still the most important area of automation of legal work. Without doubt, indexing and retrieval is considered as an important method in information science (Buder et al., 1997). In law, the same holds true now due to the influence of information and communication technology. In the Gutenberg age, the skill of searching (remembering, reading and associating of related materials based on low-level indices e.g. card indices of libraries) was too close to legal reasoning as to be considered as a special method. Even now, many legal authors on legal methodology consider information retrieval only as a new method of information gathering (see e.g. as a typical example the highly accepted book on Austrian legal methodology by Kerschner, 2007). That is outdated now; information retrieval as an automation of matching of information components and facts constitutes the most important methodological change of jurisprudence for a long time. It is focused on the (a) category of legal reasoning, e.g. finding relevant materials.

Some theoretical work is required for an appropriate legal theory. A legal information system has to be considered as a formalisation of law (Schweighofer, 1999). Establishing and maintaining the huge information systems of today with sufficient efficiency was a big move from a humble start (see the overview of present European legal information systems by Berger, 2009). Information retrieval was and is a *hard rock science* with clear aims and methods. "Information retrieval systems process files of records and requests for information, and identify and retrieve from the files certain records in response to the information requests. The retrieval of particular records depends on the similarity between the records and the queries, which in turn is measured by comparing the values of certain attributes attached to records and information requests" (Salton, 1999). "Information retrieval (IR) is finding materials (usually documents) of an unstructured nature (usually text) that satisfies an information need from within large collections (usually stored on computers)" (Manning, Raghavan, & Schütze, 2008, p. 1). Later, hypertext as a tool for indexing and retrieving relevant information or structuring information in a different (and also flexible way) achieved prominence as a non-linear representation of knowledge. "Hypertext is a form of representation of hyperstructures with strong relations to the basic theory of text retrieval and older efforts." (Bing, 1998) Information retrieval and hypertext are well understood and work with high efficiency and speed in information systems but also the internet. The tool of indexing takes advantage of the very powerful technique of Boolean search logic. In legal information retrieval, the system of Horty is still around (Bing, 1995). Processing speed, storage, handling etc. has tremendously improved but the hard core – matching of Boolean combinations – remained the same. Information retrieval is a good example of a theoretically well founded research with precise goals – improving recall and precision – but very important repercussions for law (for an overview and analysis see Turtle, 1995; Schweighofer, 1999).

Legally speaking, Information retrieval automates the indexing process. Whereas Boolean retrieval still focuses on the user, conceptual information retrieval (Schweighofer, 1999; Smith, 1995; Dick, 1987) wants to replace and support logic but also associative reasoning with intelligent indexing. Conceptual information retrieval systems use knowledge representation techniques in order to encode the semantics of legal concepts (Cross & de Bessonet, 1985; Bing, 1987; Mital & Johnson, 1992; Rose, 1993). With the advent of legal ontologies, the various representation techniques (semantic networks, conceptual graphs, concept frames, object-oriented programming) can be more or less incorporated in the new representation techniques. The main problem of conceptual information retrieval remains the bottleneck of knowledge acquisition. Below, we will discuss the option of semiautomatic text analysis.

Automation of legal reasoning requires a standardisation of a decision process. This may be an algorithm, a decision tree, some matching techniques etc. This programme can be solely determined by man or partly supplemented by learning programmes (e.g. machine learning).

The question of complexity was never openly discussed in logic-based knowledge systems. Haft (2009) rightly mentions that legal methodology never really considered complexity as a problem of law. Complexity reflects also the collaborative aspect of legal reasoning. Legal systems are based on a high level of co-operation between the various legal authorities even if the co-operation is a formal one. Courts must apply the rules of the law-maker, but the law-maker must also take into account legal practice. Legal authors have to compile, analyse or even develop the law in order to make it applicable.

Should the programme be able to handle 20,000 rules or 10,000 concepts in order to cover a sufficiently big legal domain with common complexity or should the reasoning more sophisticated? Bench-Capon & Prakken (2006) have described this as the difference between mechanical and intellectual reasoning. For the main line of AI & law, the answer is clear from the very beginning. It had to be intelligent reasoning, e.g. logic programming, knowledge systems, neural networks, legal ontologies etc. However, legal practice has always challenged this focus as too unrealistic (see e.g. Haley, 2009 or Lauritson & Gordon, 2009) but without much success.

In should be noted that Searle's "Chinese room example" (Searle, 1990, see also the reference in Bund, 1990) shows that the difference between mechanical and intellectual reasoning may be very often blurred. In law, it is difficult to say what looks more sophisticated. Most lawyers would see the handling of complexity as the more important intelligent task.

It remains a mystery why legal theory and practice played such a minor role in building the first legal expert systems. Starting with the famous project of the *British Nationality Act as a Logic Programme* (Kowalski & Sergot, 1991) many projects flourished (for an overview see Susskind, 1987; Zeleznikow & Hunter, 1994; Schweighofer, 1999; Zeleznikow & Stranieri, 2005; Lodder & Oskamp, 2006 and the ICAIL proceedings). No legal theorist ever considered legal reasoning as a similar task as logic; however, exactly that happened. Very few argued that the approach is unrealistic (see in particular Leith, 1986). Contrary to that, legal logic is still more conceptual logic than first order logic (Schweighofer, 1999, p. 28). The other parts of legal reasoning were insufficiently taken into account.

The tool was logic programming as an implementation of rule-based reasoning. Characteristics of these prototypes were first order logic, small number of rules (about 200) and narrow domains of application. Most projects failed due to insufficient understanding of legal reasoning and the inability of handling a bigger quantity of rules (Haft et al., 1992).

In Germany, the failure was named LEX due to a project that tried to automate understanding of legal text (Haft et al., 1987). Most successful applications derive from the work of Softlaw (now Haley) (Haley, 2009) that focused on formalising a big domain with first-order logic and about 20,000 rules. Higher level logics mostly remained theory and was rarely implemented as a prototype.

In recent years, the focus of logic-based reasoning has moved to argumentation (Bench-Capon & Prakken, 2006, 61). Important works are those of McCarty (1977), Gordon (1995, 2009, 137), Prakken (1997, 2009, 58), Bench-Capon (2006, 2009) and Sator (2009, 108). Thus, logic-based knowledge systems are now mostly pushed aside by argumentation endeavours. It is symptomatic that Lodder and Oskamp (2006) do not even include a chapter on legal logic in their book. It has proven as a good way of analysis of legal argumentation. Problems of structuring and formalizing contested legal argument dominate present research; the chosen legal domains are far too small for practice.

We may put the focus again on the fourth component of legal reasoning: complexity. It is obvious that automation of legal reasoning failed to achieve the required complexity. It should be noted that most researchers did not aim to master the complexities of law; they wanted to master the complexities of legal thinking. Whereas this goal is honourable and fascinating, legal practice struggles with the complexities of rules and facts.

As intelligent legal reasoning cannot provide support in legal practice and mechanical reasoning has shown its strength in this domain, we consider mechanical reasoning is a good way of mastering legal complexity.

Case-based reasoning compares a legal problem with previous cases for guiding legal reasoning (Ashley 1990, 2006, 2009). It shares with information retrieval the re-use of previous knowledge. As the knowledge representation is much more complex with factors and dimensions of a case, the conclusions are more precise. The reasoning may be considered as associative and not just mechanical. So far, case-based reasoning has not proven to be a big advantage in moving to in bigger legal applications due to its very high knowledge acquisition tasks. Some research tries to find a way of semi-automatic knowledge acquisition (Brüninghaus & Ashley 2009) but results are promising but not convincing.

Text analysis and text categorisation are focus of legal research since the 1980ies (for an overview see Schweighofer, 1999; Zeleznikow & Stranieri, 2006). Important work in this field constitute the projects KONTERM/LabelSOM/GHSOM (Schweighofer, 1999; Merkl & Schweighofer, 1997; Schweighofer et al., 2002), SALOMON (Moens, 1997), FLEXICON (Smith et al., 1995), SMILE (Ashley & Brüninghaus, 2001), and Support Vector Machines (Gonçalves & Quaresma, 2005; Zeleznikow & Stranieri, 2006). This research has its parallel in computational linguistics (for an overview see Grewendorf & Rathert, 2009). Legal research followed the developments in computational linguistic analysis, in particular the resurgence of interest in empirical and statistical methods of language analysis with human interference (Church, 1993). Legal information systems are huge text corpora; legal thesauri and ontologies offer a structured description of legal systems. Text analysis provides the link between the syntactic and semantic description of the law. Text analysis works with text patterns, e.g. words in its grammatical variations, word combinations, word use and word context. It is not an easy task to build a database for about 5000 concepts and update it on a regular basis. Legal texts are more difficult than other disciplines due to the highly specialised language subject to sudden dynamic changes (e.g. a new law, an overruling judgement). So far, 90-95% of quality can be achieved that is not considered as sufficient for legal environments with the goal of 100% accuracy. We argue to solve this problem with a co-operative model man/machine. Taking advantage of the strength of the partners, the computer is handing the text extraction, classification and filtering work whereas the human being oversees the process and checks the quality of the output. So far, this co-operation has hardly worked in practice. Even legal authors are reluctant to use such semi-automatic techniques for their analysis of legal text corpora.

In the 1990s, ontologies as a conceptualisation of a domain (Gruber, 1993) were considered as the way of reorganisation of AI & law research. Legal ontologies were developed and tested (see for an early analysis Bench-Capon/Visser 1997). Later, the University of Amsterdam has set the standards with LRI-Core and now LKIF (Hoekstra, Breuker, De Bello and Boer 2007). The rules base of ontologies was always broader than that of legal knowledge systems. Much research was reused for a more appropriate knowledge representation. Legal ontologies were implemented for tasks of conceptual information retrieval, knowledge representation, multilingual information retrieval or exchange of information and knowledge (see Casanovas et al., 2007; Casellas et al.; 2009). Besides this variety of uses, in practice there are very few applications. This fate is shared with the semantic web (Berners-Lee et al. 2001). Semantic mark-up makes the text intelligent and active but it has still to be proven that a machine-machine co-operation is more efficient than those of a man-machine one. Legal ontologies are highly developed technological concepts with few applications.

From the point of view of legal complexity, ontologies bring many advantages and flexibility. Depending on the quality of the ontology, shallow ones like advanced dictionaries are thesauri can handle the whole body of law. From the point of view of legal formalisation, legal ontologies are sufficiently open to any form of knowledge representation. Thus, the combination of semi-automatic methods of text analysis for developing and updating a conceptual ontology seems to be very promising. The main advantage of legal ontologies consists in the available technical standards and their flexibility of representation. Advanced thesauri, concepts, links, rules etc. can be easily represented.

Reasons for Failure

The ideas of automated legal reasoning and legal ontologies are somewhat flawed: too ambitious goal, non-consideration of the dynamics of law, misunderstanding of the role of law, an insufficient knowledge base and weak co-operation model between software and person and missing complexity.

Firstly, the goal was much too ambitious. The focus was on a full computerisation of legal reasoning (e.g. the example of the British Nationality Act prototype). But law is not only a logic set of rules but a system of social governance what makes legal reasoning so difficult. A close look to today's commentaries and handbooks demonstrates that legal writers have given up the task of pre-reasoning of all hypothetical cases. In general, only an excellent index is offered on relevant legal materials, ideas, beliefs, rules, principles, facts etc. and its relations. This mixture of a more or less good text with many references and footnotes could be considered as the most-appropriate way of indexing for lawyers. Such knowledge only works with an experienced lawyer that is doing the legal reasoning, e.g. reading, understanding and applying the knowledge. Also the conceptual logic is characteristic for these works that is too difficult to formalise for legal logic.

Secondly, the dynamic character of law, its rules and concepts was not taken into account. Law is changing on a daily basis. Society is developing all the time. Concepts may modify its interpretation over the years of application. Any sophisticated knowledge application is faced with a big maintenance problem.

Thirdly, law is a social tool for shaping politics, economics and society, a mechanism for social change. Rules are set into a social context. Lawyers must have a very good understanding of life and its social complexities. Besides acceptance of rules as an advantage in social life, governmental or social sanctions "motivate" people to comply with rules. Human beings, their life and thus also the social context are changing all the time. Thus, its rules, concepts and procedures are always closely related to a particular society. Application may change according to different social environments. As a social tool, the requirements of complexity of social life, society and maintenance of change have to be taken into account. So far, these needs are not satisfied by the informational reality and computational power. The scaling-up challenge from "toy systems" to real life applications remains particularly difficult in the legal domain.

Fourthly, automation requires a very concise implementation framework for a particular domain. Very often, however, the law-maker is not providing this service. Some difficult questions are left for the administrations and courts to decide. Thus, if the lawmaker is not establishing required clarity the system maker has to do it. However, only few institutions can deliver. Courts could but cannot do it; legal writers often lack the qualification of leading authority. Therefore, administrations are left. They can decide how the law is applied and – more important – also take the resulting legal and financial risks. Therefore, only systems with such a knowledge base are successful in practice.

Fifthly, if software cannot produce the final result human beings and software agents have to work together. Law is from people for people. Thus, all applications and procedures have to achieve a workable co-operation between software and human agent which the latter in the driving seat. The computer may partly replace the lawyer, support the judge but can offer fully automated solutions only in very special cases.

Sixthly, law is a tool of shaping society. In modern legal systems, many authorities have power to add, abolish or modify rules in the legal system. Further, rules are often not sufficiently precise for an application. Only a working co-operation between the various authorities and legal practice provides the basis for an efficient legal system. Therefore, the co-operative aspect is a very relevant one that was not sufficiently taken into account in previous knowledge-based systems.

Taking into account these facts, AI & law research must change in order to move on. Developments in theory of law and logic are still too premature in order to be of practical relevance so far. Ontologies are not a magic solution for knowledge representation but provide a sufficiently powerful environment to move on with simplified representations of societal complexity.

MASTERING SOCIETAL COMPLEXITY WITH INDEXING

The size of a legal text corpus in a complex global word makes the ongoing substantive analysis of rules a very difficult task. Lawyers have to find alternatives for traditional handbooks and commentaries representing so far the hard care of legal dogmatics: describing the components of knowledge, e.g. legal documents, document segments (sections, paragraphs, etc.), concepts etc. and establishing a conceptual structure, e.g. links between the components. The challenge consists in the societal complexity with the enormous amount of rules and facts and the dynamic change of society and the legal system.

Today, the expert community on automated reasoning in law must take a very difficult decision. Should the focus remain on legal theory or production of "toy systems" or should theory and practice accept as a preliminary requirement taking into account sufficient societal complexity, e.g. represented in legal information systems and on the web? In the last 20 years, the AI & law community has moved to new aims and methods but the main approach has been only slightly modified.

We would strongly argue for applications with sufficient societal complexity that is represented in legal information systems and the web. The law as syntactically formalised in legal information systems has to be supplemented by a semantic legal web allowing semiautomatic or automatic applications in the law firm, in government or in business.

Social complexity should be only reduced by specialisation (subdomains) or teamwork (e.g. sharing the work between lawyers, paralegals, librarians, documentalists, and software agents). Automation in law will work only with sufficiently working man machine co-operation that should be properly identified.

It should also be mentioned that group reasoning is an important part of this approach. Legal dogmatics is characterised by the compilation of positions of authorities in law, e.g. parliament, government, courts, legal writers etc. Thus, pieces of knowledge of many persons have to be taken into account. The main quality of a legal commentary consists in the compilation and systematisation of many thousand chunks of data in a particular domain. Thus, a formalisation has to be sufficiently open in order to include many pieces of information from many authorities of legal information.

This ambitious approach of supporting the mastering social and legal complexity in a particular domain with semantic structuring requires a simplified formalism. We limit our approach to the topic (a) of legal reasoning, e.g. establishing the relevant information for a given legal problem. The formalisation consists in graphs and an index, e.g. links between nodes of factual and legal information and decision trees for reasoning based on such a network.

This network of linking relevant information is called indexing in information science. The British indexing standard describes an index as "a systematic arrangement of entries designed to enable users to locate information in a document" (BS3700:1988). Information is reduced for easier identification of relevant materials.

In law, two main types of information entries exist: rules and facts. Complexity derives from the high number of types of facts and rules, the enormous quantity of facts and rules and the resulting very high number of links between these nodes, difficulty of exact and fuzzy linking of these nodes. However, this does not prevent using graphs for formalisation of this knowledge.

A graph is an abstract representation of a set of objects where some pairs of the objects are connected by links (Reisinger, 1987, 96). Facts and rules become objects and the relations are represented by links. In comparison with other approaches of legal reasoning, a graph is a simple formalism.

It is not trivial to find a useable abstraction of the legal domain due to the enormous size of the knowledge domain. A legal database contains millions of documents with an inverted index (dictionary) of hundred thousands of words, millions of cross-references between documents and classifications and thesauri with thousands of entries. In our view, the broadest work on abstraction of the whole body of a legal system was done in concept jurisprudence (Schweighofer, 1999, p. 29). More sophisticated approaches with logic or case-based reasoning were limited to much smaller domains. Luckily, concept jurisprudence and legal ontologies go easy together. Conceptual structure can be represented as a lexical ontology. Rules and facts can be described as concepts with headers and definitions. Relations become links. Contrary to the paper environment, numbers of objects, types of objects and links are not limited. Another very important aspect consists in the automation option. Applications may re-use this formalized information.

We do not disregard the relevance of the other topics of legal reasoning for "hard cases". However, in most legal cases, the so-called "simple cases", the main work of the lawyer concerns establishing the facts and the links to existing solutions, e.g. relevant cases. Then, legal reasoning follows a more or less complex decision tree and can be formalized, including the systems of Haley (2009). It has to be stressed that facts have to be sufficiently established by the user or the lawyer. Attempts of semi-automatic formalising of facts have failed so far, like the LEX project of University of Tübingen (Haft 1987). In modern applications of legal logic in the field of e-government dialog systems or web forms handle this task to a sufficient degree. In other cases, a lawyer has to establish the facts as his main first task.

Four types of indexing are relevant in law: mental index, card indices, information retrieval and hypertext.

The mental map constitutes the main tool of organisation of a lawyer's knowledge. It is located in the human brain and structures legal knowledge according to mastering of law and personal preferences. These maps are based on legal vocabulary, conceptual structures and some logic. They develop through legal training and legal practice. The size of today's text corpora of legal systems cannot be mastered with the mental map only. Nowadays, card indices, retrieval systems and hypertext support the human brain. However, it remains an important tool of indexing, in particular for experienced lawyers.

Card indices of libraries are the best example of the technique in the Gutenberg age. They are still used in smaller environments. The matching is simplified compared to the mental map. It is based on matching of entries of authors, titles, classifications or a thesauri; associative thinking cannot be handled.

The limitations of paper and manual matching are overcome by information retrieval systems. No limits to the size of the text corpus or to the number of indexing entries exist anymore. Boolean search logic has tremendously empowered the search for vocabulary or indexing categories. However, the simplification becomes even more evident. The support of information retrieval as a tool of indexing, searching and retrieving relevant information in unstructured databases is restricted to hints to relevant information. Thus, AI & law experts have considered indexing with information retrieval as a very shallow technique that was and is not justified. However, the human factor is much more important in this man-machine cooperation than in AI & law. Without a sufficiently trained brain information retrieval techniques do not work very well in legal environments.

Hypertext goes beyond formalised matching. It is a tool for indexing and retrieving relevant information or structuring information in a different (and also flexible way) is a non-linear representation of knowledge. In particular, associations can be represented. Semiautomatic generation of hypertext links has not only facilitated the linking of documents in the traditional way (e.g. citations) but also offered options for associative indexing (Schweighofer, 1999).

To sum up, we reduce the information of a legal text corpus by using a conceptual structure describing rules and facts. Relations between facts and rules are represented by links. It is a simple formalisation using graph theory, lexical ontologies and indexing. Lawyers do not have to cope with hundreds of thousand words in a legal information system but "only" with ten thousand words in a conceptual structure. Like the shift from to the semantic web, it is a move from an unstructured vocabulary to a structured one that links also to the mind map of a lawyer. Terminological use and conceptual thinking of lawyers have to be combined. The main advantage of this approach consists in the support of the lawyer in this indexing process. In easy cases, linking of facts with rules may provide the solution of a legal problem. Then, decision trees and dialog systems offer a fine automation option in law. In hard cases, strong hints for relevant rules are provided. It is astonishing to notice that a legal commentary is exactly providing this service. The text corpus is compressed into a systematic representation explaining and clarifying legal rules. Having some knowledge of concepts or the structure of the statute, all relevant information is mentioned and further references are given.

THE DYNAMIC LEGAL ELECTRONIC COMMENTARY AS A NEW MODEL OF INDEXING

The Dynamic Legal Electronic Commentary (DynELCom) consists of a textual, e.g. syntactic representation of the legal order that is supplemented by a semantic representation of the legal system (e.g. conceptual representation of rules), a semantic representation of the world (e.g. conceptual representation of facts) and a legal link structure between these repositories of knowledge. Knowledge acquisition is supported by semi-automatic text summarisation and text classification. A sketchy inference machine allows automated reasoning in "easy cases". The easier formalisation of knowledge and semi-automatic knowledge acquisition allow dynamic semi-automatic updating of the knowledge base.

The goal is an ontological index like that in legal commentaries, however, without the textual components. It is obvious that the readability of such ontological structures is limited and will require some training. However, the exact representation of the underlying conceptual und logical structure of the legal system is much better represented.

The DynELCom is a model of a semantic legal knowledge system. Legal knowledge is formalised with tools of the semantic web and of legal ontologies. The legal text corpus is represented with about 10,000 concepts (with definitions and attributes), represented in a conceptual structure with links. Conceptual jurisprudence but also conceptual information retrieval prove the feasibility of this approach.

The main difference to existing approaches of legal ontologies lies in the fact that world ontologies (e.g. facts) are also included in this conceptual structure. As many quite developed ontological representations of world knowledge already exist, such knowledge can be used for enrichment of an ontological representation of the legal system.

A major part of the DynELCom consists of the link structure between the facts (world ontology) and rules (legal ontology). Thus, legal reasoning is supported that may be sufficient in "easy cases and a valuable support in contradictory situations.

The formalisation of a legal knowledge domain with the DynELCom allows also semi-automatic and automated applications. Conceptual search of links to factual and legal concepts are obvious results of this representation. This search can be supported by dialog systems that support the user in establishing relevant facts of a case. Thus, a sketchy form of automated legal reasoning can be offered, e.g. a "simplified legal syllogism". The facts of a case are properly refined by a dialog system leading to a factual concept but also a legal concept.

The DynELCom faces the dynamics of the legal system. The indispensable indexing and analysis process is supported by semi-automatic categorisation and text analysis. Computational linguistics, text extraction, document categorization and text summarization tools are now sufficiently powerful so that good results can be achieved in very short time.

The analysis of the DynElCom is based on a co-operative work model between the man and the computer. The legal information system provides the basis for the commentary. The knowledge base with the ontology and semi-automatic text analysis provides extensive knowledge of the text corpus of the legal information system. Software tools are information retrieval, hypertext, knowledge management, text summarisation, text categorisation and the inference machine. Manually, ontologies have to be established and maintained, semi-automatic indexing must be constantly fine-tuned and inference engines must be supervised. Such work is presently done by legal authors and practitioners. With the DynELCom, a concentration of such analysis takes place in a semi-automatic way. The main advantage is real time delivery, higher quality and lower costs.

The main advantage of the DynELCom seems to be obvious: in "easy cases", much of the work can be automated. Lawyer services would become cheaper with higher quality. As the demand for more and cheaper legal services can be strongly noticed, such a system would find high acceptance in practice. Existing barriers to law due to financial constraints of many citizens would be reduced.

The Formal Model of DynELCom

The DynELCom consists of the following elements: text corpus, an ontology consisting of an ontology of the legal system, a world ontology, anchors between the legal and world ontology, knowledge acquisition tools, a sketchy inference engine and a dialogue system.

Text corpus: Existing legal information systems are the basis for the text corpus. The textual representation should be supplemented by a visual representation. Only few modifications are envisaged; mostly hypertext links to the ontology. Further, the list of document types may be improved in some cases as well as the broad classification of documents.

Ontology: The ontology consists of a legal ontology, a world ontology and links (anchors) between the legal and the world ontology. Elements of the ontology are 3 types of frames: legal frame, fact frame and anchor frame. A frame contains a header, definitions (with sources), classification codes, and relations (to other frames, e.g. synonym, homonym, polysem, hyponym, hyperonym, antonym etc. but also to an anchor). The anchor frame can best be described as a citation with a header, the identification (abbreviation or number) and links to facts and legal concepts. For the representation, existing standards of the semantic web and legal ontologies are implemented; in particular OWL, RDF and LKIF.

This first step with a frame-like representation of legal concepts, factual concepts and the anchors between facts and rules will be followed by a second step that intends a more sophisticated ontological representation of the legal system. This representation focuses on space, persons, actions, material rules and procedural rules.

Action space for persons will be the real space and the cyberspace. Persons can be natural persons and quasi-persons (robots or software agents subject to sufficient action options in real space or cyberspace). Objects in the space are physical objects (thinks), energy and quasi-physical objects (e.g. web store). Actions can be physical processes (actions or non-actions in real space or quasi-physical processes (actions on the web). Mental objects and mental processes consist of combinations of these elements. Due to social practice, such "virtual" sets are considered as a unified object or process (e.g. organizations, enterprises, associations, families etc.) Law builds on the existing physical and social structure of persons, objects and processes but modifies it or adds particular elements. Persons can be natural or legal (e.g. limited company, international organization, state), objects are physical, mental or legal, and actions are physical, mental, or legal. It is obvious that the differences between the social reality and law (as representation as the world should be) are a high interest in any legal system. The representation is structured in concepts (an ontology), rules (classified according to the systems of Allen (1957) or the more advanced system of Herrestad (1998)), and factual situations. Isomorphism is respected via direct links to norms but also its logical representation. Legally relevant links between the world ontology and the legal ontology provide support for legal reasoning (e.g. possible factual situations or legal consequences of certain facts). In the LKIF terminology, such a function is called anchors (LRI-Core Ontology). They provide an anchor function as links between the social spectrum of actions and legally.

Knowledge acquisition tools: Text extraction and summarisation tools are decisive for the knowledge acquisition. Dynamic updating of the knowledge base cannot be effectuated otherwise. The tools consist of a knowledge base containing the extraction, summarization and classification rules with header, rule, definition and relations and several tools for semi-automatic text analysis providing information on relevant documents, extract important text passages, classify documents, deliver definitions etc.

We have developed prototypes and applications on corpora-based text analysis for about 20 years now. Due to space restrictions, we can provide only an overview of the methods.

A pre-defined list of descriptors can be checked against a text corpus with the KONTERM method (Schweighofer, 1999). The various term occurrences are clustered according to the context allowing a structuring of homonyms and polysems. Thus, the various meanings in the text corpus can be analyzed. Further, the term lists are used for an automated description of documents. The context of a term is used for describing and analyzing the various meanings. These representations were incorporated in the description of homonyms and polysems.

The self-organising map is a general unsupervised tool for ordering high-dimensional data in such a way that alike input items (e.g. documents) are mapped close to each other (Schweighofer et al., 2001). In such a map, similar documents are grouped together. An extension allows the building of various layers and clusters of the map (growing hierarchical self-organising map). Further, common similarities of a cluster can be described with keywords (labelling of self-organising maps). A further extension allows the manual refinement of vector weights of the documents with data enrichment tools. The descriptions were used for the creation and refinement of the thesaurus, in particular concerning synonyms (Schweighofer et al., 2002).

Further, we have also take advantage of the GATE library for text analysis. The GATE ANNIE (A Nearly New Information Extraction System) tool is very helpful for a more detailed analysis: segmentation of documents (tokenizer), words, gazetteer, sentence splitter and semantic tagger. The GATE JAPE tool (Regular Expressions Over Annotations) is implemented for a similar purpose. It is much more powerful in bigger text environments but does not allow so sophisticated representations of meanings as the KONTERM method.

Sketchy inference engine: In first step with a simplified ontology, the inference is not much more than a hint of relevance like in the information retrieval system. Factual concepts are matched with legal concepts and vice versa. In case of a more complex ontology, an inference engine is required. Decision trees are represented as complex IF-THEN-statements with a mechanism for prioritizing rules. Such statements are interpretations of facts, rules, concepts and anchors in the ontology. Such an inference engine allows the representations of a legal syllogism and a quicker handling of relevant information.

Dialogue system: Such a system is intended to converse with a human in a coherent structure (Kolt et al., 2008; Wikipedia: Dialogue Systems, 2010). In the beginning, the dialogue will be text-based with a graphical user interface. A spoken dialogue system is in consideration. Natural language understanding is supported by a robust parser. The purpose of a dialogue system consists in the establishing of facts but also in the clarification of applicable legal rules.

Functions of the DynELCom

The DynELCom can provide many functions: conceptual information retrieval, a commentary for lawyers, a work bench for legal writers and analysts, a support tool for law-makers, and automation of legal reasoning.

Figure 1. Graphical model of the simplified version of the DynELCom

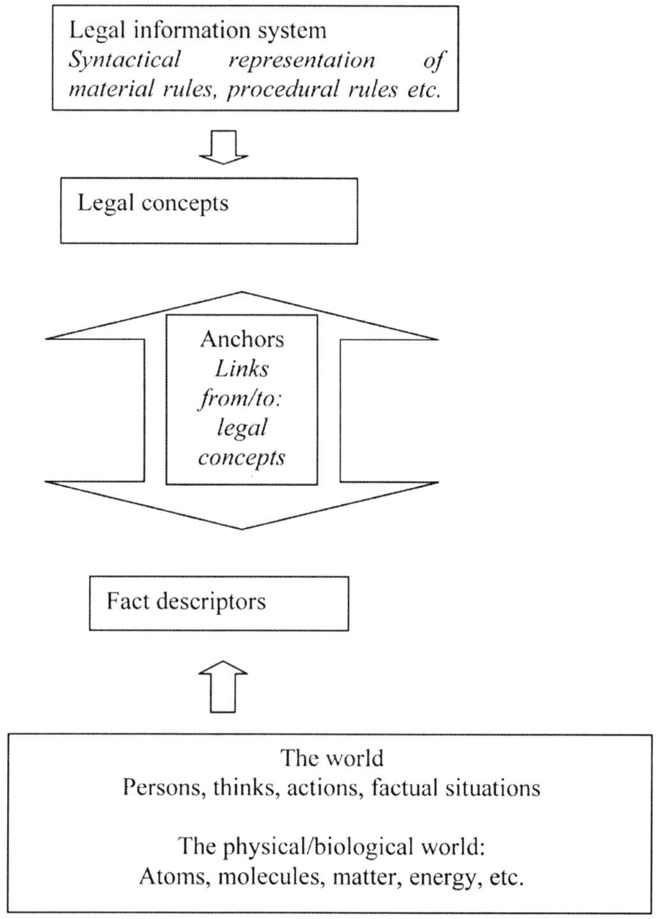

Conceptual information retrieval: The ontology is added and indexed as meta information to the various legal information systems. Thus, a conceptual retrieval based either on factual or legal descriptors is offered. Further, hypertext links can be generated automatically.

Legal commentary (or handbook): The ontology provides a powerful index to the legal system giving factual and legal conceptual structure, citations, and links to relevant documents. Due to the semi-automatic text analysis of the legal system, the commentary can be updated daily. Unlike the legal handbook, a textual description is not offered by the index. In the first stage, the commentary could replace the traditional commentary as it provides the same information with a very fast updating. In the second stage, an atomistic representation of the world and the legal system would be offered. A logical representation of the legal system combined with a powerful world ontology gives a very deep description of the world and its legal systems: persons, objects, actions and its legal characterization and relevance. The world ontology would provide a description of hypothetical facts of legal cases. The legal ontology would give a representation of hypothetical legal relevance and consequences. Thus, much more properly structured information will be available.

Work bench for legal analysis: For legal authors, the DynELCom would be an indispensable tool for their deeper analysis for a legal practitioner but also the legal author. As materials are

Figure 2. Graphical model of the extended version of the DynELCom

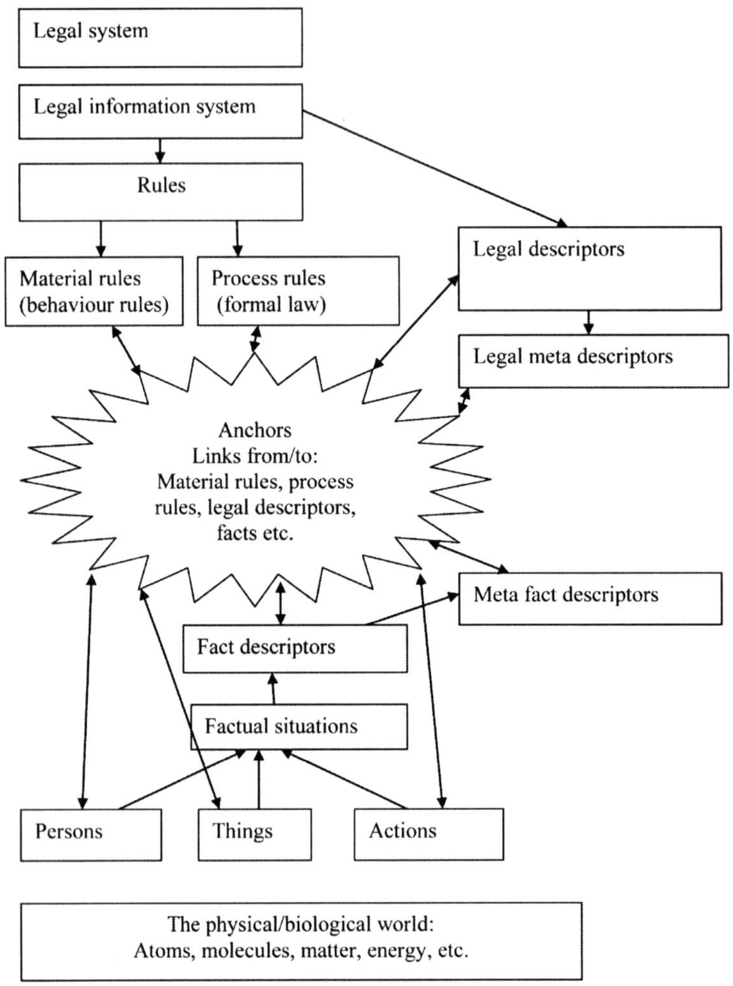

properly classified and pre-analyzed, more time and energy can be used for review of the various legal developments.

Support of law-making: Parliaments have to base their law-making on a reliable analysis of the present law and its application problems. Commentaries provide this service; a DynELCom may move further as much more information on the world is given.

Automation and support of legal reasoning ("simplified legal syllogism"): An important body of legal sources, in particular case law, is devoted to this process of interpretation and clarification of norms. In the DynELCom, "links" between social behaviour and legal rules are included as a tool of conceptual retrieval but also as a support for the legal reasoning process. In standard cases, interpretation and application is sufficiently well known in order to be automated, thus, leaving indexing (and establishing facts) as the main missing component of automated reasoning.

ESTABLISHING THE ONTOLOGIES AND FIRST EXPERIMENTS

Due to time and financial restrictions, the implementation of the DynELCom has mostly remained a concept. However, due to our ongoing practice in European, international and Austrian law we have worked for more than five years on a partial experimental application. The following presentation provides first experiments.

Dynamic Electronic Legal Commentary (DynELCom) on European law, international and Austrian law

Due to our daily work with European, international and Austrian law, much experience exists that provides many ideas for the DynELCom but also a good field for first experiments. The European law ontology is the most developed one.

The text corpora consists of legal information systems that represent more or less the main body of law with sufficient coverage (Austrian law: RIS (Rechtsinformationssystem des Bundes) and the private databases RDB and LexisNexis), European law: EUR-Lex and Europa Server).

So far, we have established an ontology with a sufficient granularity of an ontological representation of a jurisdiction: about 10,000 thesaurus entries, 5,000 citations, up to 200 document types, a classification structure, 100 text extraction and summarization rules. This meta data is stored and updated in a database with different types of knowledge frames:

Fact and legal descriptors: header, definition (with sources), examples (with sources), relations (synonym, homonym, polysem, hyponym, hyperonym, antonym etc.), classification, other information.

- Anchors: header, identification (abbreviation or number), synonyms, classification, author, links, other information.
- Document types: header, identification (abbreviation), use, format, other information.
- Classification: header, code, definition, relations, other information.
- Extraction and summarization rules: header, rule, definition, relations, other information.
- Concepts: header, definition (with sources), related thesaurus entries and citations, relations (synonym, homonym, polysem, hyponym, hyperonym, antonym etc.), classification, legal conceptual structure (ontological model), other information.
- Rules: header, quasi-logical expression, source, type, classification, legal conceptual structure (ontological model), other information.
- Procedures: header, flowchart, source, type, classification, legal conceptual structure (ontological model), other information.

The following examples may show the lexical ontology (note: an (L), (F) or (A) is added to the header in order to distinguish between legal and fact descriptors as well as anchors). The attribute "legal conceptual structure" indicates relevant branches of law.

Legal concept:

```
Header: Measures of equivalent effect (L)
Definition: Discriminatory and non-discriminatory rules of Member States hindering trade between Member States are illegal.
Source: Article 34 TFEU, cases C-267/91 Keck and Mithouard, 120/78 Cassis de Dijon, 8/74 Dassonville
Relations: BT customs, measures of equivalent effect (A), freedom of goods (A)
Classification: 02.40
Legal conceptual structure: customs union, freedom of goods
Other information: none
```

Fact concept:

```
Header: Liqueur in Germany (F)
```

```
Definition: The minimum amount of al-
cohol which should exist in liqueurs
was 25% (up to 1978).
Relations: Measures having equiva-
lent effect
Source: DE Brandtweinmonopolgesetz
(German liquor monopoly act)
Classification: 02.40
Legal conceptual structure: customs
Links: Measures having equivalent
effect
Other information: none
```

Anchor (link):

```
Header: Measures having equivalent
effect (A)
Links: Liqueur in Germany (F), sell-
ing arrangements (F), Edam cheese in
France (F), vinegar in Italy (F), beer
in Germany (F), resale at a loss (F),
advertising restrictions (F), dis-
tribution restrictions (F), measures
having equivalent effect (L), Article
34 TFEU, Article 28 EC, Article 30
ECT, Article 30 EECT
etc.
```

At present, we are in the process of finishing the first prototype of this representation of European and Austrian law. The next step would be a verification of the conceptual structure using the knowledge acquisition and text analysis tools. Such a process is very time-consuming and requires financial resources not available so far. However, the existing ontology provides already a very helpful tool for legal work as it represents legal and fact concepts and its links.

Dynamic Electronic Legal Commentary (DynELCom) on State Aid Law

The DynELCom on State aid law represents a practical decision for a manageable text collection. Building the ontology, the refinement process but also testing of the various functions is much easier.

Due to my personal familiarity with EU state aid law[2] this domain was chosen for first extensive experiments. EU state aid law is characterised by a dominance of so-called "Guidelines" as law-making documents, Commission decisions as the main implementation document and ECJ court decisions only as a relatively small but decisive source of law.

The text corpus consists of EU state aid guidelines, over 1000 decisions of EU state aid law in the field of agriculture and some court decisions. The ontology is composed of about 100 legal (e.g. investment aid in agricultural holdings) and 100 factual (e.g. fallen stock) descriptors. All entries contain a definition, the relevant source (e.g. State aid guidelines for agriculture), relations (broader term, narrower term, synonym, homonym, antonym, etc.), classification, legal conceptual structure, legal inference and other information.

The ontology was checked and refined using methods of corpora-based text analysis. Establishment and verification is still a very time-consuming process; however, such tools are an indispensable support for a legal analysis that is presently used for our work on a legal commentary in state aid law in agriculture.

We have chosen the example of fallen stock for description of the ontology:

Fact concept:

```
Header: fallen stock (F)
Definition: Animals have died or
killed by euthanasia on a farm or
during transport but not slaughtered
for human consumption.
Source: Agriculture Block Exemption
Regulation, Article 2 (14)
Relations: fallen stock (A)
Classification: 03.50.30
Legal conceptual structure: animal
health, state aid
Other information: none
```

Legal concept:
```
Header: fallen stock - state aid(L)
Definition: A Member State may grant
aid IF fallen stock exists AND only
100% of costs of removal of fallen
stock are supported AND only 75% of
the costs of destruction of such car-
casses are supported AND have notified
a block exempted aid scheme OR got
approved an aid scheme under point
134 of the AGL.
Source: Agriculture Block Exemption
Regulation, Article 2 (14), Article
16 (d), Agriculture State Aid Guide-
lines, point 134
Classification: 08.60
Legal conceptual structure: state
aid, agriculture
Relations: fallen stock (A), BT ag-
riculture, BT State aid.
Other information: none
```

Legal concept:
```
Header: fallen stock - animal health
(L)
Definition: Fallen stock has to be
removed and destructed according to
EU law and national law.
Relations: BT animal health, fallen
stock (A)
etc.
```

Anchor (link):
```
Header: fallen stock (A)
Links: fallen stock (F), fallen stock
- state aid (L), fallen stock - ani-
mal health (L), BT animal health, BT
state aid, BT agriculture; Agriculture
Block Exemption Regulation, Article
2 (14), Article 16 (d), Agriculture
State Aid Guidelines, point 134,
etc.
```

The example of fallen stock may explain the structure of the ontology: Fallen stock is a factual concept but also twice a legal concept. Rules of animal heath oblige farmers to collect and dispose animal carcasses. Rules of state aid allow support for concerned farmers.

A search in this ontology will show the two main legal concepts and thus also the legal consequences of a fact concept fallen stock. The result is similar that those of a traditional handbook. However, due to corpora-based text analysis and proper formalisation, updating can be done much faster.

FUTURE RESEARCH DIRECTIONS

The future research consists of three main topics: firstly, the extension and verification of our ontology on Austrian and European law; secondly, the building of a dialogue system with a sketchy inference machine, and thirdly the automation of administrative decisions.

The existing first prototype of our ontology on Austrian and European law will be tested with the whole text corpora in Austrian and European law. The testing is important for the improvement of definitions, classifications, relations and text analysis rules. The enhancement consists in better quality of text analysis and categorisation tools, e.g. more rules with higher efficiency. Further, more links between fact and legal concepts will be established. These tests with the Austrian and European legal information systems will be started subject to sufficient resources. At present, we are working on establishing such a team of about 20 experts. The tests will show if such a system provides sufficient quality in indexing and analysis to be of use for lawyers and, in particular, provide an alternative to the traditional commentary.

The first prototype of the dialogue system with a sketchy inference engine will be built in the domain of State aid law. Here, the extended version of the DynELom ontology with rules, concepts, persons, things, actions etc. and the inference engine (a production system) will be implemented. A dialogue system will support the establishment of relevant facts but also completing the required statements and clarifications for a legal analysis. The main part of the dialogue system consists of a well-structured list of questions in order to establish the facts for the case. A sequence of dialogue boxes leads the user through a series of questions.

The automation of administrative decisions is based on a prototype developed already in 2003 (Schweighofer et al., 2003). Existing electronic notifications on state aid will improved in order to allow an automated interpretation of facts with a decision tree. Such a system will provide an automated check of standard submissions. A knowledge base allows cross-checking of the information provided in the notified forms. Thus, standard checks can be handled automatically leaving more time for manual examination of difficult questions.

CONCLUSION

Law will not be excluded from the "industrialisation of the intellect". Even if the automation of contradictory legal reasoning remains out of question due to poor and insufficient understanding of text and visual data by the computer, paralegal work but also standard cases are prone for automation.

The next steps in legal reasoning should be in indexing and not a refinement of the various logic or concept-based knowledge representations. A powerful ontology with social and legal entries provides a strong framework for a conceptual structure of the legal system. Semi-automatic text analysis and text categorisation will support the building of such a broad ontology.

The DynELCom offers the following functions: conceptual information retrieval, dynamic commentary, a "sketchy inference machine" etc., that can be used by legal practitioners but also legal writers.

A first prototype of the ontology has been built in the domains of Austrian and European law. First experiments in the field of State aid law have shown the high potential of this approach.

Future research will concentrate on an improvement of the ontology, a testing of the ontology with text analysis tools and an extension of the knowledge base with the results of this analysis. In State aid law, a question/answering system and automated handling of state aid notifications are envisaged.

Indexing cannot replace legal reasoning in difficult cases but it offers a practical way for orientation of the huge knowledge domain of law. Bringing relevant information together is very often already the most difficult part of legal work. The DynELCom will support this process.

REFERENCES

Alexy, R. (1991). *Theorie der juristischen Argumentation: Die Theorie des rationalen Diskurses als Theorie der juristischen Begründung* (2nd ed.). Frankfurt, Germany: Suhrkamp.

Allen, L. E. (1957). Symbolic logic: A razor-edged tool for drafting and interpreting legal documents. *The Yale Law Journal*, *66*, 833–879. doi:10.2307/794073

Ashley, K. (2009). Ontological requirements for analogical, teleological, and hypothetical legal reasoning. In *Proceedings of the 12th ICAIL International Conference on Artificial Intelligence and Law* (pp. 1-10). New York: ACM Press.

Ashley, K. D. (1990). *Modeling legal argument: Reasoning with cases and hypotheticals*. Cambridge, MA: MIT Press.

Ashley, K. D. (2006). Case-based reasoning . In Lodder, A. R., & Oskamp, A. (Eds.), *Information technology and lawyers: Advanced technology in the legal domain, from challenges to daily routine* (pp. 23–60). Dordrecht, The Netherlands: Springer.

Bench-Capon, T., & Gordon, T. F. (2009). Isomorphism and argumentation. In *Proceedings of the 12th ICAIL International Conference on Artificial Intelligence and Law* (pp. 11-20). New York: ACM Press.

Bench-Capon, T., & Prakken, H. (2006). Argumentation . In Lodder, A. R., & Oskamp, A. (Eds.), *Information technology and lawyers: Advanced technology in the legal domain, from challenges to daily routine* (pp. 61–80). Dordrecht, The Netherlands: Springer.

Bench-Capon, T., & Visser, P. R. S. (1997). Ontologies in legal information systems: The need for explicit specifications of domain conceptualisations. In *Proceedings of the International Conference on Artificial Intelligence and Law* (pp. 132-141). New York: ACM Press.

Berger, A. (Ed.). (2009). *Access to legislation in Europe: Guide to the legal gazettes and other official information sources of the European Union and the European Free Trade Association*. Luxembourg: Publications Office of the European Union.

Berners-Lee, T., Hendler, J., & Lassila, O. (2001). The semantic web. *Scientific American, 284*(5), 34–53. doi:10.1038/scientificamerican0501-34

Bing, J. (1987). Designing Text Retrieval Systems for Conceptual Searching. In *Proceedings of the 1st ICAIL* (pp. 43-51). New York: ACM Press.

Bing, J. (1995). Legal Text Retrieval and Information Services. In: J. Bing & O. Torvund, 25 years Anniversary Anthology in Computers and Law (pp. 525-586). Oslo: Tano.

Bing 1998 Hypertext – the deep structure. [Unpublished invited talk, DEXA 1998. University of Vienna].

Brüninghaus, S., & Ashley, K. D. (2001). Improving the representation of legal case texts with information extraction methods. In *Proceedings of the 8th ICAIL* (pp. 42-51). New York: ACM Press.

Buder, M., Rehfeld, W., Seeger, T., & Strauch, D. (Eds.). (1997). *Grundlagen der praktischen Information und Dokumentation*. Munich, Germany: K.G. Saur.

Bund, E. (1990). *Einführung in die Rechtsinformatik*. Berlin: Springer.

Casanovas, P., Biasiotti, M. A., Francesconi, E., & Sagri, M. T. (Eds.). (2007). *Proceedings of LOAIT 07, II: Workshop on Legal Ontologies and Artificial Intelligence Techniques*. Retrieved from http://www.ittig.cnr.it/loait/LOAIT07-Proceedings.pdf

Casellas, N., Francesconi, E., Hoekstra, R., & Montemagni, S. (Eds.). (2009). *Proceedings of LOAIT 2009: 3rd Workshop on Legal Ontologies and Artificial Intelligence Techniques joint with 2nd Workshop on Semantic Processing of Legal Text*. Barcelona, Spain: IOT Series.

Church, K. W., & Mercer, R. L. (1993). Introduction to the special issue on computational linguistics using large corpora. *Computational Linguistics, 19*, 1–24.

Cross, G. R., & de Bessonet, C. G. (1985). Representation of legal knowledge for conceptual retrieval. *Information Processing & Management, 21*, 35–44. doi:10.1016/0306-4573(85)90126-8

Dickson, J. (2005). Interpretation and coherence in legal reasoning. *Stanford Encyclopedia of Philosophy.* Retrieved September 9, 2009, from http://plato.stanford.edu/entries/legal-reas-interpret/

Dworkin, R. (1986). *Law's empire*. London: Fontana Press.

English Wikipedia. (2010). Dialogue systems. Retrieved March 3, 2010, from http://en.wikipedia.org/wiki/Dialogue_system

Gonçalves, T., & Quaresma, P. (2005). Is linguistic information relevant for the classification of legal texts? In *Proceedings of the 10th ICAIL* (pp. 168-176). New York: ACM Press.

Gordon, T., & Walton, D. (2009). Legal reasoning with argumentation schemes. In *Proceedings of the 12th ICAIL International Conference on Artificial Intelligence and Law* (pp. 137-146). New York: ACM Press.

Gordon, T. F. (1995). *The pleadings game: An artificial intelligence model of procedural justice*. Dordrecht, The Netherlands: Kluwer Academic Publishers.

Grewendorf, G., & Rathert, M. (Eds.). (2009). *Formal linguistics and law*. Berlin: Mouton de Gruyter.

Gruber, T. R. (2003). A translation approach to portable ontology specifications. *Knowledge Acquisition*, *5*(2), 199–220. doi:10.1006/knac.1993.1008

Haft, F. (2009). *Juristische Schreibschule: Anleitung zum strukturierten Schreiben*. Munich, Germany: Normfall.

Haft, F., Jones, R. P., & Wetter, T. (1987). A natural language based legal expert system for consultation and tutoring: The LEX project. In *Proceedings of the 1st International Conference on Artificial Intelligence & Law* (pp. 75-83). New York: ACM Press.

Haft, F., Werner, M., & Gerbliner, M. (1992). *Untersuchung der Möglichkeiten des Einsatzes von Expertensystemen zur Unterstützung von Richtern*. Staatsanwälten und Rechtspflegern. [Unpublished research report, University of Tübingen]

Haley (2009). Oracle and Haley. Retrieved September 17, 2009, from http://www.oracle.com/haley/index.html

Hart, H. L. A. (1994). *The concept of law* (2nd ed.). Oxford: Clarendon Press.

Herrestad, H. (1996). *Formal theories of rights*. Oslo: Juristforbundets Forlag.

Hoekstra, R., Breuker, J., De Bello, M., & Boer, A. (2007). The LKIF core ontology of basic legal concepts. In P. Casanovas, M. A. Biasiotti, E. Francesconi, M. T. Sagri (Eds.), *Proceedings of LOAIT 07, II: Workshop on Legal Ontologies and Artificial Intelligence Techniques* (pp. 43-64). Retrieved from http://www.ittig.cnr.it/loait/LOAIT07-Proceedings.pdf

Kelsen, H. (1960). *Reine Rechtslehre* (2nd ed.). Vienna: Verlag Deuticke.

Kerschner, F. (2007). *Wissenschaftliche Arbeitstechnik und Methodenlehre für Juristen* (5th ed.). Vienna: WUV Universitätsverlag.

Koit, M., Gerassimenko, O., Kasterpalu, R., Rääbis, A., & Strandson, K. (2008). Developing a dialogue system. In P. Sojka, A. Horak, I. Kopecek, & P. Karel (Eds.), *Text, Speech and Dialogue, TSD 2008* (Lecture Notes in Artificial Intelligence 5246, pp. 593–600). Berlin: Springer.

Kowalski, R., & Sergot, M. (1991). The use of logical models in legal problem solving. *Ratio Juris*, *3*, 201–218. doi:10.1111/j.1467-9337.1990.tb00058.x

Lauritson, M., & Gordon, T. F. (2009). Towards a General Theory of Document Modeling. In *Proceedings of the 12th ICAIL* (pp. 202-211). New York: ACM Press.

Leith, P. (1986). Fundamental errors in legal logic programming. *The Computer Journal*, *29*(6), 545–552. doi:10.1093/comjnl/29.6.545

Lodder, A. R., & Oskamp, A. (Eds.). (2006). *Information technology and lawyers: Advanced technology in the legal domain, from challenges to daily routine*. Dordrecht, The Netherlands: Springer.

Luhmann, N. (1968). *Vertrauen: Ein Mechanismus der Reduktion sozialer Komplexität* [Republished 2000]. Stuttgart, Germany: Lucius & Lucius Verlagsgesellschaft.

Luhmann, N., & Baecker, D. (Eds.). (2002). *Einführung in die Systemtheorie*. Heidelberg, Germany: Carl-Auer-Systeme-Verlag.

Manning, C. D., Raghavan, P., & Schütze, H. (Eds.). (2008). *Introduction to information retrieval*. Cambridge: Cambridge University Press.

McCarty, L. T. (1977). Reflections on TAXMAN: An experiment in artificial intelligence and legal reasoning. *Harvard Law Review, 90*, 89–116. doi:10.2307/1340132

Merkl, D., & Schweighofer, E. (1997). The exploration of legal text corpora with hierarchical neural networks: A guided tour in public international law. In *Proceedings of the 6th ICAIL* (pp. 98-105). New York: ACM Press.

Mital, V., & Johnson, L. (1992). *Advanced information systems for lawyers*. London: Chapman & Hall.

Moens, M. F., Uyttendaele, C., & Dumortier, J. (1997). Abstracting of legal cases: The SALOMON experience. In *Proceedings of the 6th ICAIL* (pp. 114-122). New York: ACM Press.

Oskamp, A., & Lodder, A. R. (2006). Introduction: Law, information technology and artificial intelligence. A. R. Lodder & A. Oskamp (Eds.), *Information technology and lawyers: Advanced technology in the legal domain, from challenges to daily routine* (pp. 1-22). Dordrecht, The Netherlands: Springer.

Prakken, H. (1997). *Logical tools for modelling legal argument: A study in defeasible reasoning in law*. Dordrecht, The Netherlands: Kluwer Academic Publishers.

Reisinger, L. (1987). *Strukturwissenschaftliche Grundlagen der Rechtsinformatik*. Graz, Austria: Leykam-Verlag.

Rose, D. E. (1994). *A Symbolic and Connectionist Approach to Legal Information Retrieval*. Hillsdale, NJ: Lawrence Erlbaum Associates Publishers.

Salton, G. (1989). *Automatic Text Processing: The Transformation, Analysis, and Retrieval of Information by Computer*. Reading, MA: Addison-Wesley.

Sator, G., Rudnianski, M., Rotolo, A., Riveret, R., & Mayor, E. (2009). Why lawyers are nice (or nasty): A game-theoretical argumentation exercise. In *Proceeding of the 12th ICAIL International Conference on Artificial Intelligence and Law* (pp. 108-117). New York: ACM Press.

Schweighofer, E. (1999). *Legal knowledge representation, automatic text analysis in public international and European law*. The Hague: Kluwer Law International.

Schweighofer, E. (2008). Reduktion von Komplexität: Durch Recht und IKT . In Schweighofer, E., Geist, A., Heindl, G., & Szücs, C. (Eds.), *Komplexitätsgrenzen der Rechtsinformatik: Tagungsband des 11. Internationalen Rechtsinformatik Symposions IRIS 2008* (pp. 42–48). Stuttgart, Germany: Boorberg.

Schweighofer, E., Haneder, G., Rauber, A., & Dittenbach, M. (2002). Improvement of vector representation of legal documents with legal ontologies. In *Proceedings of the 5th BIS.* [CD-ROM]. Poznan: Poznan University of Economics Press.

Schweighofer, E., Liebwald, D., & Drachsler, M. (2003). Electronic notification of block exempted state aid. In Proceedings of the 16th Annual Conference on Legal Knowledge and Information Systems (pp. 147-148). Amsterdam: IOS Press

Schweighofer, E., Rauber, A., & Dittenbach, M. (2001). Automatic text representation, classification and labeling in European law. In *Proceeding of the 8th ICAIL International Conference on Artificial Intelligence and Law* (pp. 78-87). New York: ACM Press.

Searle, J. R. (1990). Ist der menschliche Geist ein Computerprogramm? *Spektrum der Wissenschaft, 3*, 40–47.

Simitis, S. (1970). *Informationskrise des Rechts und Datenverarbeitung*. Karlsruhe, Germany: C. F. Müller.

Smith, J. C., Gelbart, D., Maccrimmon, K., Atherton, B., Mcclean, J., Shinehoft, M., & Quintana, L. (1995). Artificial intelligence and legal discourse: The Flexlaw legal text management system. *Artificial Intelligence and Law, 3*(1-2), 55–95. doi:10.1007/BF00877695

Steinmüller, W. (1993). *Informationstechnologie und Gesellschaft: Einführung in die Angewandte Informatik*. Darmstadt, Germany: Wissenschaftliche Buchgesellschaft.

Stranieri, A., & Zeleznikow, J. (2006). Knowledge discovery from legal databases: Using neural networks and data mining to build legal decision support systems . In Lodder, A. R., & Oskamp, A. (Eds.), *Information technology and lawyers: Advanced technology in the legal domain, from challenges to daily routine* (pp. 81–118). Dordrecht, The Netherlands: Springer.

Susskind, R. (1987). *Expert systems in law*. Oxford: Oxford University Press.

The 12th International Conference on Artificial Intelligence and Law. (2009). [Proceedings]. New York: ACM.

Turtle, H. (1995). Text retrieval in the legal world. *Artificial Intelligence and Law, 3*(1-2), 5–54. doi:10.1007/BF00877694

Wittgenstein, L. (1984). *Tractatus logico-philosophicus, Werkausgabe* (*Vol. 1*). Frankfurt am Main, DE: Suhrkamp.

Zeleznikow, J., & Hunter, D. (1994). *Building intelligent legal information systems: Representation and reasoning in law. Computer Law Series, 13*. Deventer, The Netherlands: Kluwer.

Chapter 13
A Case for the Re-Use of Community Reasoning

Andrew Stranieri
University of Ballarat, Australia

John Yearwood
University of Ballarat, Australia

ABSTRACT

In software engineering, the re-use concept is a design principle that improves efficiency, quality and maintainability by ensuring that software artifacts are developed once and re-used many times. In an analogous way, a group's reasoning can be imagined to be re-used by that or another group to enhance efficiency, transparency and consistency in decision-making. However, the re-use of reasoning is difficult to achieve because group reasoning cannot easily be captured and the way in which a group reasoning artifact is subsequently used is not obvious. This chapter explores the case for the re-use of community reasoning and concludes that individuals can benefit from a representation of a previous group's coalesced reasoning if the reasoning to be modeled and the scheme to represent the reasoning have been selected to suit the task. The authors contend that specifying the future community likely to re-use the reasoning, called the intended audience, informs a decision regarding whether an exercise aimed at coalescing a group's reasoning is best performed verbally, in writing or with the use of more structured schemes such as Argument visualization.

INTRODUCTION

In practice, reasoning is rarely performed by an individual in isolation from others but involves the exchange of assertions between participants in a group that Yearwood and Stranieri (2009) call a reasoning community. A reasoning community is identified as a group of individuals connected by their desire to apply reasoning to reach a decision. Members of a reasoning community may or may not belong to the same organisation, community of practice (Wenger 1998) or commitment (Kofman and Senge 1993), hold the same values, aim for the same outcomes or share much else in common except the need to reason toward a solution to a problem.

DOI: 10.4018/978-1-60960-091-4.ch013

Each time participants in a community engage in reasoning that leads to a decision, they draw on relevant information and knowledge and engage in communication with others, in a process McMahon (2001) calls cognitive cooperation until, ultimately a decision is reached. The next time a community commences reasoning about a similar problem, background information, reasons and perhaps even the decision adopted by the first community is typically not readily available as a starting point for deliberation. Minutes typically record decisions and do not register the reasons, evidence or dialectic flow advanced to arrive at the decision. As a consequence, the reasoning of one group is not readily re-used by another.

The re-use concept is prevalent in software engineering as a design principle that improves efficiency, quality and maintainability of computer programs by ensuring software artifacts are designed once and re-used many times. In that field, attention is typically focused on the re-use of software, however the collection and organisation of reasons underpinning decisions, known as rationale management, has attracted considerable attention in recent years (Dutoit et al. 2006). Despite its well accepted utility, rationale management is rarely applied in practice. Schneider (2006) calls this the Rationale Paradox and attributes it, in part to difficulties associated with the collection of reasons during decision dialogues.

Reasons are typically generated to support assertions during group decision-making dialogues when the flow of consciousness is focused on the creation and analysis of solutions and not on documenting or recording reasons(Csikszentmihalyi 1990). Further, reasons are difficult to collect because their elicitation often involves the translation of tacit knowledge into explicit knowledge, a process known to be difficult, time consuming and, according to (Nonaka 1991), achieved in practice only when there is a clear social incentive for doing so.

Another explanation for the reluctance a group has to record reasons derives from difficulties in the use of representational formats. Simple transcripts of dialogue are relatively easy to collect but too cumbersome to subsequently to re-use. The many approaches for representing and expressing reasoning can be categorised for the purpose of exploring the case for re-use as follows:

- Transient
  - Narrative expressed verbally
- Permanent
  - Representations of a single stream of reasoning
    - Classical syllogisms
    - First order logics
    - Concept maps
  - Representations of multiple streams, ie group reasoning
    - Narrative expressed in writing
    - Non monotonic logics
    - Argument visualization based on instance argumentation
    - Argument visualization based on group argumentation

Individuals have been reasoning to assimilate and infer new knowledge for thousands of years without any representation or embodiment outside the thinker. New knowledge was expressed verbally and a permanent representation only emerged relatively recently with the advent of writing. Verbal reasoning can be performed in an explanatory cause-effect, quasi-logical form or with the use of a story-telling form. However, regardless of the form, verbal reasoning soon reaches its limit as the concentration and memory of the listener is exhausted. Also, a complex verbal argument can easily involve circular reasoning or fallacies, which may be difficult to detect.

Schmandt-Besserat (1996) argues that writing was invented in Mesopotamia in the fourth millennium BC to create records that enabled traders and taxation officers to deal with arguments from traders that were too complex for verbal recollections. As an external and permanent representation of reasoning, writing is a significant advance over verbal processing because it brought reliable memory and permitted greater examination and rigor to be applied to a piece of reasoning. In a fundamental sense, writing enables the reasoning to be re-used.

However, a written argument is linear and abstract and this generates a considerable cognitive load in terms of understanding, structuring and reproducing what the piece of prose says. For instance, in a community involved in discussing the benefits of nuclear power generation, written submissions from each participant represent arguments related to aspects of nuclear power generation including the disposal of waste, the risk of radioactive pollution and the low cost of production. The submissions contain considerable overlap in concepts and draw on conflicting supporting evidence. The task of contrasting and comparing arguments against each other is made difficult because of the linearity inherent in written narrative.

A more structured representation of reasoning than narrative emerged during the rise of democracy in Ancient Greece. The classical syllogism advanced by Aristotle (1984) led to numerous logics over ensuing centuries. Despite their substantial differences, the majority represented reasoning along a single stream to a true or preferred outcome by an individual reasoner. In recent decades the limitations of representations that were restricted to a single chain of reasoning became apparent and a suite of non-monotonic logics emerged. Some of the non-monotonic logics, including Zeno advanced by Gordon and Karacapilidis (1997) drew directly on concepts from argumentation.

In parallel, a number of schemes that captured arguments without the overhead of logical formalism emerged. Argument visualisation methods including Toulmin's (1958) layout of arguments, Rittel and Weber's (1973) IBIS, gIBIS (Conklin and Begeman 1988), Belvedere(Suthers et al 1997), GoReason by van Gelder (2003), ArguMed (Verhaij 2003), the generic actual argument model (GAAM) by Yearwood and Stranieri (2006). Perhaps the most advanced visualization approach to date is the Compendium system described by Shum et al (2006).

Many Argument visualization schemes have been advanced with the implicit assumption that their main benefit lies in a capacity to enhance deliberation within a group leading to better decisions. Although designing studies to demonstrate enhanced deliberation or improved decision-making is methodologically challenging, according to van den Braak et al(2006) the few studies performed reveals little evidence that these approaches enhance the quality of deliberation, learning or outcomes. This suggests that a main benefit for Argument visualization schemes lies in their capacity to be used to organise large and complex reasoning that involves data from diverse media, into a single coherent representation suited for re-use.

The need for explorations into the re-use of reasoning has emerged in recent years primarily due to increasing decision complexity, associated with an information society and increasing expectations for consultative dialogue by groups. Dutoit et al (2006) identifies the pressing need for the management of reasons underpinning software engineering decisions, particularly as implementations become incredibly complex. Outside of software engineering, many decisions are increasingly complex and underpinned by a great deal of often conflicting and highly technical information yet are decisions are required to be made following widespread consultation. An anecdotal example of this is a 3000 page draft report into the dredging of Port Philip Bay in Australia recently released for public consultation. The length and complexity of this document make it very difficult for most commentators to read and understand all issues and their inter-relatedness, yet norms for community consultation require deliberation on it.

The issues explored in this chapter relate to identifying the reasoning that a group performs that can or ought to be captured for re-use. The claim is made that not all reasoning a group performs is appropriate for re-use but that a core can be identified. An additional question involves the representational scheme best suited for modeling reasoning for re-use. The claim made here is that for some communities, re-use is best achieved with permanent representations of a single stream, whereas others require representations that permits multiple paths and multiple truths. A final issue raised in this chapter relates to how a representation of a community's reasoning can be re-used by a future community.

Questions related to what reasoning is best captured, how it is ideally represented and how it should be used by a future community are discussed in the context of reasoning community phases introduced by Yearwood and Stranieri (2009). This is discussed in the next section dealing with what reasoning to capture. Following that sections on how best reasoning should be captured and re-used, is presented.

WHAT REASONING CAN BE CAPTURED?

The reasoning community phases described by Yearwood and Stranieri (2009) does not model fine grained assertions and counter-assertions within groups as non-monotonic logics do, nor on longer dialogues comprising sequences of exchanges as the Walton and Krabbe (1995) typology of dialogue does, but advances four phases to describe the process of group reasoning. The four phases, engagement, individual reasoning, group coalescing and decision-making are described as follows:

- Engagement. This phase involves the the formation and focus of the community. Tasks to be performed include:
  - the selection of participants
  - agreement on a protocol for communication that governs exchanges between participants. This may include a specification of the role of a facilitator, procedural rules such as those advanced by Roberts (2000), rules for the kinds of assertions and kinds of media permissible
  - agreement on a protocol for how the group will ultimately make a decision; by voting, consensus or other mechanisms.
  - The specification of the problem to be solved. The issue or problem specification may change during discussion
  - The specification of the intended audience. The intended audience in some communities extends to all communities dealing with similar problems in the future, whereas in others no future community is expected to use the reasoning. In software engineering design rationale management, the intended audience is typically identified as a community of future engineers charged with modifying the implemented system. This audience will require quite a detailed explanation of the constraints, reasons and assumptions that were deployed in the system's initial implementation. The intended audience for a reasoning community involving a marital dispute that introduces no novel factor, is limited. However, the intended audience of the highest appeal court reasoning community includes a far wider audience that includes every future marital dispute. The intended

audience does not refer to the group interested in the reasoning but is restricted to those that will, in their own deliberations, re-use the reasoning. Much of the public may be interested in every aspect of a celebrity couple's divorce but the couple would not expect anyone to re-use their reasoning.

- Individual reasoning. Each participant ascertains their own facts and beliefs, using them to draw inferences that lead to assertions. Tasks to be performed by each participant includes:
    - An individual's coalescing of reasoning that involves identifying and making sense of reasons in order to assert their own claims.
    - An individual's judgment to arrive at a position on the specified issue
- Group coalescing. The coalescing of reasoning for the entire community involves organising the assertions and reasons advanced by participants into an explicit coherent representation. A summary document, white paper or argument map exemplifies group coalescing outputs. Group coalescing leaves a permanent record of the reasoning and provides a starting point for engagement and individual reasoning phases of future communities.
- Decision-making. This phase involves reaching a decision about the performance of an action or the solution to a problem. The resolution of the problem requires the implementation of the decision-making protocols in order to reach a final decision.

There is little reasoning in the engagement phase that warrants collection for re-use. The selection of participants and the identification of communication and decision-making protocols that occurs in the engagement phase are often constrained by social, legal, cultural or organisational norms which are adopted without a great deal of reasoned discussion. As little reasoning is involved, there is little scope for the re-use of reasoning regarding the selection of participants, and communication and decision-making protocol. For instance, participation in a doctor/patient community is typically restricted to the patient, guardians and a practitioner by cultural and professional norms. Participation in many workplace meetings is prescribed by organizational policy and similarly, often not the subject of reasoned discussion. For communities in which there are minimal norms governing participation, reasoning leading to the selection of participants is typically made using a small set of criteria such as a participant's knowledge or status as a stakeholder and does not involve reasoning that is particularly deep or complex.

Reasoning toward the identification of a communication protocol and decision-making protocol is similarly often governed by norms that are simply adopted without reasoning. In communities where no norms for communication or decision-making exist, the possibility for reasoned selection of protocols is limited because little is known about optimal communication or decision protocols for particular groups or issues. Consequently, re-use of reasoning about communication and decision protocols is only marginally useful for a future community.

The issue to be addressed is also specified and perhaps modified during the engagement phase. Again, this typically does not arise from chains of reasoning. Socio-political or organizational reasons account for the topicality of issues to a greater extent than inference. Once formulated, an issue may change from its initial conception and similarly, the reasons for the change are relatively straight forward and ought to be recorded for re-use but do not require the overhead of sophisticated representations.

Although participants to a group dialogue, actively perform reasoning, a single individual's reasoning is typically not ideally represented for re-use unless it is placed in the context of other's views. The individual reasoning phase represents the work done by a participant to understand the issue at hand, other participants claims and ultimately, to arrive at, and assert to others, his or her own claims. Individual reasoning involves two main elements:

- Individual coalescing. Assembling background knowledge, facts, claims and reasons asserted by others, so that the individual understands the context.
- Individual judgment. Determination of which facts, claims and reasons the individual will hold, and the assertion of these to other participants

Participants rarely reason from a blank starting point but invariably draw on knowledge that has emerged in the past. Each participant coalesces background knowledge with assertions and reasons from other participants to yield an understanding of the argument sufficiently deep so as to infer facts, claims and reasons for assertion to others. The wife in a family law dispute reads widely and discovers that health is a relevant consideration because this factor was central to deliberations in a past appeal court reasoning community. Once health is adopted as relevant, individual judgment occurs by assembling evidence to support the claim that her own health is poor. Inference is deployed to derive the conclusion that she should receive a greater share of assets given the fact that her health is poor. This is communicated to other participants and their feedback is again coalesced for the assertion of new or revised facts, claims and reasons.

The exercise of modeling an individual's reasoning is useful for the automation of reasoning for the development of intelligent systems. It may also be beneficial to organize and clarify and record the reasoning that a individual performs for him or herself however, the contention raised here is that it is less useful for the re-use of reasoning. Rather, the representation of reasoning performed by an entire group where individual and often conflicting assertions are coalesced into a representation of the group's reasoning that includes all participant's reasoning, is far more useful for re-use. In this way, an individual's claims are juxtaposed against those of others; the reasons and backings are placed in the context of competing claims.

The generation of an explicit and structured representation of an individual's reasoning using, for instance, an Argument visualization tool is not advanced here as a desirable exercise for the re-use of reasoning. Rather, a structure already coalesced from a previous groups reasoning, is more aptly re-used by an individual; as a starting point for their own coalescing. The structured argument enables each participant to succinctly view the relevant factors and the inter-relatedness that a previous group arrived at. This is akin to recording Niemeyer and Dryzak's (2007) meta-consensus in a sufficiently structured format that a future community may assimilate.

The group coalesced product of a previous group's reasoning is used by each individual in a future community. McMahon (2001) notes individuals assert claims to form a pool of reasons and then elect to embark on collective judgment to deem some reasons stronger than others and ultimately determine an outcome. In a community of reasoning, the assembly of a pool of reasons occurs as formulated by McMahon however an additional step involving the generation of a group coalesced product involves agreement that all reasons advanced are acceptable and plausible, even though some participants will not value all of them highly, if at all during a collective judgment exercise.

The coalescing of group reasoning involves the construction of an explicit representation of the reasoning performed by all participants. The explicit representation is an organization of claims and reasons into a structure that serves three purposes:

- Group coalescing enables a permanent record of the community's reasoning to remain for future reasoning communities to re-use.
- Group coalescing facilitates individual reasoning
- Group coalescing serves to make reasoning democratically transparent to all

The group coalescing process does not necessarily result in agreement or dispute facilitation. The main benefit of a group coalescing process is the clarity in communication and the potential for re-use of the reasoning. A new divorce dispute may take as a starting point, the group coalesced reasoning from a similar community in the past. The positions of the husband and wife may not be the same, however the hierarchy of concepts will be relevant. In common law countries, the group coalesced reasoning of the highest court is typically used as the starting point for deliberations by a new community.

The group coalescing process involves the generation of a representation of the participants reasoning at a level of abstraction sufficient for the perceived needs of an intended audience, a future reasoning community. As Dutoit et al (2006) notes, this is often difficult to do by the group themselves, though Argument visualization tools can facilitate the task. Currently, whether an independent scribe can perform the coalescing better than the group is an open question.

The decision-making phase involves the determination of an outcome or solution to the problem at hand. Strategies for the final judgment include voting, referring to a sole arbiter, deferring to an algorithm or deploying various forms of consensus. The field of decision sciences has focused in depth on the discovery of models for decision-making. There is an implicit assumption that the existence of a decision making model that has been mathematically or empirically validated as arriving at an optimal decision will naturally be adopted by all reasoning communities making the same decision. The assumption is perhaps too strong in reality because most decision models are based on a theoretical perspective such as utility theory (Fishburn 1970) or Pareto-optimality(Feldman 1980) that some participants may question. Ultimately, agreement on the decision making approach to be adopted by a group is itself, a decision made by the group, in the engagement phase.

Reasoning communities involving a legal case at court, are typically obliged to accept that the final decision will be made by one participant, the judge. In a conventional patient doctor reasoning community comprising a doctor and a patient,the assumption made by both participants is that the doctor will ultimately determine a diagnosis and treatment plan. In many committee meetings, consensus loosely defined as agreement by all, is the norm whereas in more formal meetings, all decisions are made by a majority vote.

Group decision-making approaches and the reasons for their adoption are appropriately captured for re-use. Although initiatives in decision sciences aim to advance new and better models for the decision making phase, in many reasoning communities, the decision-making protocol is determined by social norms and little reasoning is applied, so the scope for re-use is limited. A future community, for instance may benefit to know the voting algorithm used by a previous community, precisely how it had been implemented and reasons for its use, in order to re-use the protocol. Similarly, if more complex decision-making models are used such as those based on multi-objective criteria, then the precise nature of the models need to be captured in detail in order for another community to re-use them in their own decision-making phase.

To summarize what kind of reasoning is appropriately captured for re-use we claim that essentially group coalesced reasoning and reasoning about the decision-making protocol is most beneficially captured. Reasoning used in the engagement phase is appropriate for re-use if inference rather than recourse to social norms was used to select participants, refine the issue or determine the communication and decision-making protocol the community adopted. The question of how best to represent group coalesced reasoning is discussed in the next section.

HOW TO REPRESENT GROUP REASONING

A verbal account of deliberations, though transient, is appropriate for situations where the reasoning is unlikely to be re-used; where an intended audience cannot readily be imagined. In most jurisdictions, lower courts hand down decisions and reasons verbally. The verbal account delivered by the magistrate coalesces the reasoning on a case by summarizing the charge, the evidence presented and his/her decision-making process. The intended audience for a lower court does not extend beyond the parties present at court so there is little need to record the reasoning in any more permanently accessible way. Many others may be interested in the decision, however no other judge is likely to re-use the reasoning because the lower court is not established to set precedents but typically follows precedents set by higher courts.

A lower court judge could conceivably represent his or her reasoning using a scheme such as first order logic, a series of classical syllogism or concept maps. This entails making a permanent representation of a single stream of reasoning. A permanent record may be useful for the parties and for posterity but is not useful for re-use because, as described above, an intended audience to re-use the reasoning cannot be imagined. However, formalisms that are restricted to single streams of reasoning cannot represent contradictions. Even in lower courts, judges make findings of fact because the facts presented by plaintiff and defendant are contradictory. Indeed as Waissmann (1951) notes law is inherently open textured because a concept cannot be defined prior to its use. Narrative expressed in writing is a natural and convenient format for the representation of group coalesced reasoning where permanency is required and the conflicting assertions or evidence is to be modelled.

Narrative expressed in writing is appropriate where reasoning is likely to be re-used by an intended audience. A higher court judgment is intended to be followed by every lower court and every legal practitioner in the jurisdiction. Consequently, narrative expressed in writing is appropriate. However, narrative expressed in writing quickly gets too difficult to re-use if there are many factors, inter-related in complex ways and also if the evidential base crosses media. As Shum et al (2006) note inputs into a group reasoning process in the form of interactive 3D simulations, video, or audio are difficult to weave into a written narrative. In software engineering design rationale, diagrams such as Use Case or Swimlane diagrams, two or three dimensional simulations, schematic diagrams are important to describe an implementation, yet cannot readily be expressed in written narrative.

Numerous Argument visualization approaches have been advanced in recent years and many have the potential to represent coalesced group reasoning that includes cross media data for use by a wide intended audience more effectively than written narrative. According to Freeman (1991), argumentation involves a family of concepts that includes concepts related to the processes and protocols that occur during an argument and concepts related to the product of an argument process. Stranieri, Yearwood and Zeleznikow (2001) discern argument models that emphasize the exchanges within a community of social agents as dialectical models and those that focus on the structure of arguments as non-dialectical models.

Approaches that emphasize dialectical exchanges include numerous non-monotonic logics (NML). NML extensions to first order logic (See Carbogim, Robertson and Lee (2000) for a survey) typically include operators such as attack, rebut or accept to represent elements that occur during dialectical exchanges. The Walton and Krabbe (1995) typology of dialogue that includes persuasion, inquiry, negotiation and deliberation dialogues represents dialectical exchanges at the level of entire sequences of utterances. In persuasion dialogues, for instance, the dialogue involves the advancement of those reasons that help a proponent advance an argument whereas deliberation dialogues involve reasons that appeal to a higher rationale.

Toulmin (1958) presented an influential approach that emphasized non-dialectical elements of argumentation. He concluded that most arguments have a structure which consists of six basic invariants: claim, data, modality, rebuttal, warrant and backing. The argument in Figure 1 is drawn from reasoning regarding a family law distribution of property following divorce. The claim of the argument in Figure 1 is the statement that the property split between husband and wife should be equal. This is made on the basis of two data items, that the Wife has contributed more to the marriage and the Husband has greater future needs. A mechanism is required to act as a justification for why the claim follows from data. Known as the warrant which is, in Figure 1, the statement that 'Contributions counter balance future needs'. The backing provides authority for the warrant and in a legal argument is typically a reference to a statute or a precedent case. The rebuttal component specifies an exception or condition that obviates the claim and in this case is omitted. The data items in Figure 1 can be claims for arguments in their own right as illustrated in Figure 2. In this way a tree of arguments, each with a warrant and backing supporting the claim can be specified.

The Toulmin formulation is identified as an instance level argument representation because the structure represents one position. An entirely independent diagram is required to represent the claim that the split should not be equal. That argument may be based on quite different data items than those in Figure 1, which are themselves claims of different data arguments to those in Figure 2. There is therefore no direct way to in-

Figure 1. Toulmin argument for equal property split

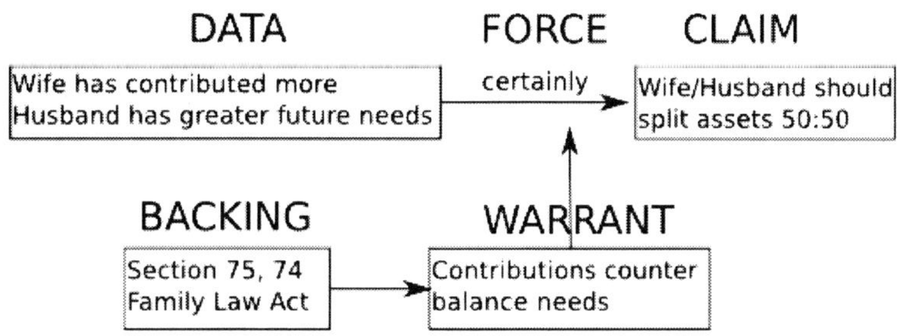

Figure 2. Toulmin argument for Wife contributed more

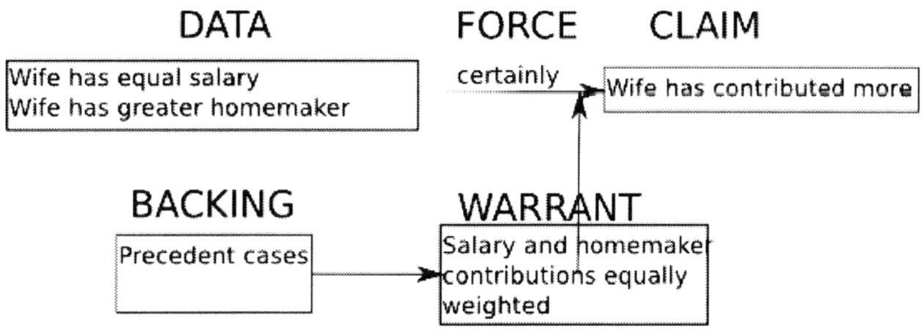

tegrate diverse positions into a single Toulmin layout.

Consequently, the Toulmin formulation is limited in representing the coalescing of a group's reasoning. Stranieri, Zeleznikow and Yearwood (2001) illustrate that almost all applications of the Toulmin layout varied the original structure in one or another. Many of the variations can be seen to enable the integration of competing arguments into the same diagram to facilitate comparisons. The representation of a group of related arguments into a coherent whole requires more generic argument schemes discussed next.

The Issue Based Information System advanced by Rittel and Webber (1973) is an argument map that makes pro and con arguments for diverse positions explicit. All viewpoints are accommodated in the one map representing the coalescing of the community's reasoning. Intelligent computer systems based on IBIS by Gordon and Karacapilidis (1997) attest to the compactness and expressiveness of this approach. Figure 3 illustrates an IBIS map for a marriage split. A position that the split be equal is juxtaposed directly with a competing argument reflecting the position that the split should favour the husband and another that the split should favor of the wife. For each, arguments for and against are listed.

Some of the limitations of the IBIS approach include the contention that the representation does not capture the interactions between arguments for or against a position. For instance, the arguments for an equal split are simply listed as if each contributes independently to the position whereas some arguments may be dependent on others. Further, some arguments may be so contentious that they become issues in their own right for a separate dialogue leading to a range of issues that were often related in ways that are difficult to record.

A graphical version of IBIS, gIBIS advanced by Conklin and Begeman (1988) enabled maps to be saved, modified and recalled more readily than hardcopy ones but recording the rationale underpinning a decision remains cumbersome. Dutoit et al (2006) survey other argument visualisation approaches and tools that aimed to make the capture of reasoning as effortless as possible. The Compendium system described by Shum et al (2006) represents the state of the art in Argument visualization systems. This system enables data from diverse sources on a very large number of issues to be structured for access with a highly useable software.

The Generic Actual Argument Model (GAAM) advanced by Yearwood and Stranieri (2006) represents diverse arguments on an issue in a single scheme comprising two levels. Figure 4 illustrates a structure central to that model, called a Generic Argument Structure (GAS) in which a variable/value format is adopted whereby each IBIS position on an issue is assigned within the domain

Figure 3. IBIS map for marriage split

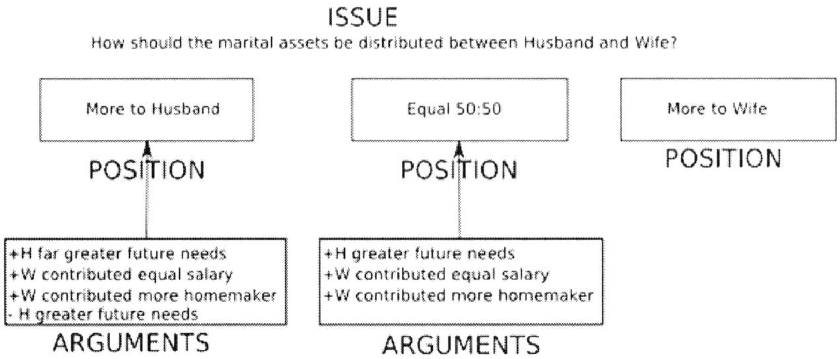

Figure 4. Generic actual structure for marriage split

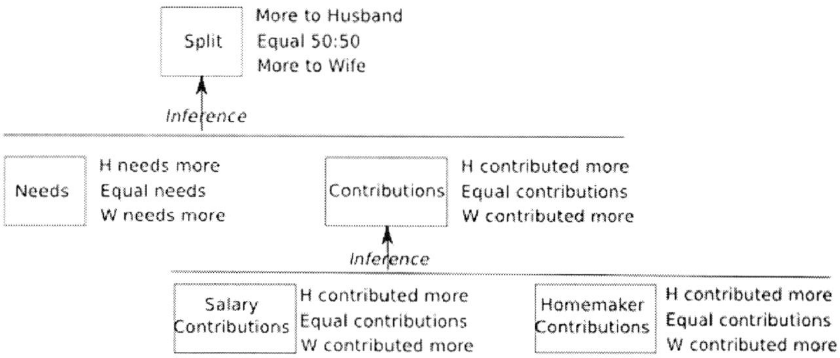

of values for a variable. The Actual level is an instantiation of that node performed by assigning a value to the variable.

Subordinate nodes represent relevant variables for the parent node linked by a slot known as an inference. An inference slot represents a mapping between subordinate variable values and parent values. This allowed a machine implementable inference such as a set of rules or a neural network to be specified in place of a human inference at any branch of the tree.

The production of a group coalescing artifact need not be performed by the group itself but can be generated by observers. In modelling the decision-making process that judges used when making sentencing decisions, Hall et al (2005) generated a generic argument structure as independent observers to a series of discussions by a group of experienced lawyers. Polletta and Stark (2005) describe how a panel merged and grouped views from hundreds of participants in small groups in town hall discussions related to rebuilding Manhatten following the 9/11 attacks. The panel can be seen to be performing group coalescing; the artifact generated was a suggestions with associated reasons that was fed back to the groups for further deliberation.

In summary, the appropriate representational format for modeling group coalescing depends upon the conceptualization of an intended audience. Where there is no intended audience, verbal reasoning would seem to suffice. Where the intended audience is broad and the issue is not overly complex or involve diverse media, narrative expressed in writing suffices. In other contexts, a more structured representation will enable individual's in future communities to adopt the coalesced product at least as a starting point. Instance based argumentation is more structured however, is limited because it does not juxtapose diverse views.

HOW TO RE-USE A GROUP COALESCED PRODUCT

Initially, in the individual reasoning phase, individuals can re-use a group coalesced product to enable individual coalescing to occur more easily. Individual coalescing is made difficult if there is no artifact at all from previous communities or if an artifact is in narrative form and is not very general. Reading single cases in family law is a tedious way to understand how judicial reasoning occurs in that domain. However, an argument map that is an abstraction of relevant factors and inferences has a wider scope and is potentially more useful for individual coalescing.

A group coalescing artifact can be re-used by individuals in the the group during deliberations, particularly if all participants use the same map. For instance, the identification of points of difference and agreement is facilitated if all individuals in a reasoning community coalesce their knowledge using the same artifact. This has been illustrated by Yearwood and Stranieri (2009) using their generic argument structure in ethics debates. Stranieri, Yearwood and Mays (2008) demonstrate that groups of secondary school children from year 8 (approximately aged 13) to year 10 readily adopted the Generic Argument Structure, IBIS and Toulmin's argument structure with very brief training sessions. Further, the use of a structured representation of factors proved intuitive for parties to an online dispute resolution system described by Muecke and Stranieri (2007).

CONCLUSION

The concept of a reasoning community is not advanced as a new utopian way for groups to rationally co-operate. The central claim is that humans often reason collectively and that a recognition of this enables the extrication of phases within the collective reasoning process that can better inform the technologies developed to support the reasoning. Further, as the reasoning is exalted, demonstrably sub-optimal decision-making on the basis of power plays and information hiding is diminished. The end product is to enable groups of people to reach decisions by supporting and elevating reasoning.

Adopting the view that reasoning often occurs in groups and identifying the phases of a reasoning community's processes facilitates the identification of schemes and technologies that could enhance the individual or group reasoning by re-useing the reasoning of other groups. Re-use of reasoning is seen to be a vehicle for enhancing the quality and efficiency of reasoning but, in elevating the importance of the creation of a reasoning artifact, re-use is central to the shift in perspective on reasoning advanced in this chapter. The re-use of reasoning under the conceptualization of group reasoning advanced here involves the generation and retrieval of an artifact produced in the group coalescing phase. This artifact is an explicit representation of all participants assertions, beliefs and reasons in one coherent whole that is sufficiently abstract to encapsulate all points of view and sufficiently detailed to link to knowledge that is highly specific to the problem.

REFERENCES

Aristotle, . (1984). *The complete works of Aristotle: The revised Oxford translation (Chapter on logic)*. Princeton, NJ: Princeton University Press.

Carbogim, D., Robertson, D., & Lee, J. (2000). Argument-based applications to knowledge engineering. *The Knowledge Engineering Review*, *15*(2), 119–149. doi:10.1017/S0269888900002058

Conklin, J., & Begeman, M. L. (1988). gIBIS: A hypertext tool for exploratory policy discussion. *ACM Transactions on Office Information Systems*, *6*(4), 303–331. doi:10.1145/58566.59297

Csikszentmihalyi, M. (1990). *Flow: The psychology of optimal experience*. New York: Harper Perennial.

Dutoit, A., McCall, R., Mistrík, I., & Paech, B. (2006). Rationale management in software engineering: Concepts and techniques . In Dutoit, A., McCall, R., Mistrík, I., & Paech, B. (Eds.), *Rationale management in software engineering* (pp. 1–48). Berlin: Springer. doi:10.1007/978-3-540-30998-7_1

Feldman, A. M. (1980). *Welfare economics and social choice theory*. Boston: Kluwer.

Fishburn, P. C. (1970). *Utility theory for decision-making*. New York: Operations Research Society of America.

Freeman, J. B. (1991). *Dialectics and the macro-structure of arguments*. Berlin: Floris Publications.

Gordon, T. F., & Karacapilidis, N. I. (1997). The zeno argumentation framework. In *Proceedings of the Sixth International Conference on Artificial Intelligence and Law* (p. 10-18). New York: ACM Press.

Hall, M. J., Calabro, D., Sourdin, T., Stranieri, A., & Zeleznikow, J. (2005). Supporting discretionary decision-making with information technology: A case study in the criminal sentencing jurisdiction. *University of Ottawa Law & Technology Journal*, *2*(1), 1–36.

Kofman, F., & Senge, P. M. (1993). Communities of commitment: The heart of learning organizations. *Organizational Dynamics*, *22*(2), 5. doi:10.1016/0090-2616(93)90050-B

McMahon, C. (2001). *Collective rationality and collective reasoning*. Cambridge: Cambridge University Press.

Muecke, N., & Stranieri, A. (2007). An argument structure abstraction for bayesian belief networks: Just outcomes in on-line dispute resolution. In J. F. Roddick & A. Hinze (Eds.), Fourth Asia-Pacific Conference on Conceptual Modelling (APCCM 2007), Ballarat, Victoria, Australia, January 2007. *Conferences in Research and Practice in Information Technology, 67*, 35-40.

Niemeyer, S. J., & Dryzek, J. S. (2007). The ends of deliberation: Metaconsensus and intersubjective rationality as deliberative ideals. *Swiss Political Science Review*, *13*(4), 497–526.

Nonaka, I. (1991). The knowledge creating company. *Harvard Business Review*, *69*(6), 96–104.

Polletta, F., & Stark, D. (2005). NSF project highlights: Policy made public: Technologies of Deliberation and Representation in Rebuilding Lower Manhattan. In Proceedings of the 2005 National Conference on Digital Government Research. *ACM International Conference Proceedings, 89*, 153-154.

Rinner, C. (2001). Argumentation maps: Gis-based discussion support for on-line planning. *Environment and Planning. B, Planning & Design*, *28*, 847–863. doi:10.1068/b2748t

Rittel, H., & Webber, M. (1973). Dilemmas in a general theory of planning. *Policy Sciences*, *4*, 155–169. doi:10.1007/BF01405730

Robert, H. M. I. (2000). *Robert's rules of order newly revised.* Cambridge, MA: Perseus Publishing.

Schmandt-Besserat, D. (1996). *How writing came about.* Austin, TX: University of Texas Press.

Schneider, K. (2006). Rationale as a by-product . In Dutoit, A., McCall, R., Mistrík, I., & Paech, B. (Eds.), *Rationale management in software engineering* (pp. 91–109). Berlin: Springer. doi:10.1007/978-3-540-30998-7_4

Shum, S., Selvin, A., Sierhuis, M., Conklin, J., Haley, C., & Nuseibeh, B. (2006). Hypermedia support for argumentation-based rationale . In Dutoit, A., McCall, R., Mistrík, I., & Paech, B. (Eds.), *Rationale management in software engineering* (pp. 111–132). Berlin: Springer. doi:10.1007/978-3-540-30998-7_5

Stranieri, A., Yearwood, J., & Mays, H. (2008). Explicit representations of reasoning to support deliberation within groups. In Burstein, F., Johanson, G & Thompson, H (Eds), Harnessing Knowledge Management to Build Communities. Proceedings of the 11th Australian Conference on Knowledge Management and Intelligent Decision Support, 2008. Monash University, Australia, pp 1-14.

Stranieri, A., Zeleznikow, J., & Yearwood, J. (2001). Argumentation structures that integrate dialectical and non-dialectical reasoning. *The Knowledge Engineering Review*, *16*(4), 331–358. doi:10.1017/S0269888901000248

Suthers, D., Lesgold, A., Katz, S., Weiner, A., Toth, E., & Harrigal, K. …DeLeon, Y (1997). An interface for collaborative and coached approaches to learning critical inquiry. IUI '97, Proceedings of the 1997 International Conference on Intelligent User Interfaces, January 6-9, 1997, Orlando, Florida, USA. *ACM*, 217-220.

Toulmin, S. (1958). *The uses of argument*. Cambridge: Cambridge University Press.

Van den Braak, S. W., Van Oostendorp, H., Prakken, H., & Vreeswijk, G. (2006, August 28-29). A critical review of argument visualization tools: Do users become better reasoners? Workshop Notes of the ECAI-06 Workshop on Computational Models of Natural Argument Riva del Garda, Italy.

van Gelder, T. (2003). Enhancing deliberation through computer supported argument mapping . [Chapter] In *Visualizing argumentation: Software tools for collaborative and educational sense-making*. London: Springer Verlag.

Waissmann, F. (1955). Verifiability . In Flew, A. (Ed.), *Logic and language*. Oxford: Blackwell.

Walton, D., & Krabbe, E. (1995). *Commitment in dialogue: Basic concepts of interpersonal reasoning*. Albany, NY: State University of New York Press.

Wenger, E. (1998). *Communities of practice: Learning, meaning and identity*. New York: Cambridge University Press.

Yearwood, J. L., & Stranieri, A. (2006). The generic/actual argument model of practical reasoning. *Decision Support Systems*, *41*, 358–379. doi:10.1016/j.dss.2004.07.004

Yearwood, J. L., & Stranieri, A. (2009). Deliberative discourse and reasoning from generic argument structures. *AI & Society*, *23*(3), 353–377. doi:10.1007/s00146-006-0069-y

Section 3
Collaborative Reasoning in the Real World

Many initiatives toward enhancing group reasoning have been performed to date in diverse application areas. Insights from application areas can be categorised into two groups; those that seek to enhance or understand group reasoning in real world communities, and those that seek to enhance or understand group reasoning in virtual communities. Significant insight into collaborative reasoning can be gleaned from the experiences in understanding how collaborative reasoning actually occurs in real world groups, and how challenges in developing technologies to support the reasoning have been tackled. This group of chapters presents contributions from a variety of real world applications in diverse industries.

Chapter 14
Collaborative Transportation Sharing:
From Theory to Practice via a Case Study from France

Jesus Gonzalez-Feliu
Université Lumière Lyon 2, France

Joëlle Morana
Université Lumière Lyon 2, France

ABSTRACT

This chapter explores the role that information sharing has on the collective decisions made in supply chain management. The authors identify information sharing as the critical factor in reasoning that occurs as stakeholders along the supply chain collaboratively make decisions. However, a shared conceptual model is required for determining what information must be shared. Their model identifies five elementsin information system management related to shared logistics projects: the enterprise's solutions (for each stakeholder and for the entire reasoning community), their deals, sharing management, organizational features and information and communication technologies related to the management of shared information. Moreover, the main accelerating and limiting factors are overviewed.The chapter illustrates the applicability of the model with a case study on the distribution of newspapers in France.

INTRODUCTION

The freight transportation industry is a major source of employment and supports the economic development of the country. However, freight transportation has many negative aspects including congestion and environmental disturbance, which negatively affect quality of life, particularly in urban areas. Both the new trends in retail and commerce organization and the technological innovation in supply chain management and distribution planning have led decision makers to consider collaborative strategies to reduce overall cost and pollution emissions, and improve social management of the supply process (see papers concerned with sustainable development and transport and logistics management). In freight distribution, the most popular collaborative strategy is that of logistics sharing. This can take place

DOI: 10.4018/978-1-60960-091-4.ch014

at the transport level, but also in warehousing, inventory and other operations. These strategies are based on collaborative decision making and information sharing. They usually take the form of agreements and partnerships. The main aspects of collaborative logistics in production and supply management have been recently reviewed, however logistics sharing in freight distribution remains a less explored subject in the literature, but commonly observed in several real-life cases.

The aim of this chapter is to define the main concepts related to logistics sharing agreements and to present a conceptual schema representing the most important organisational aspects. We will focus on socio-economic and legislative aspects by making an analysis of feedback of several freight distribution experiences and introducing the main concepts of collaboration applied to this field. First of all we will present the main concepts of logistics sharing, based on the main definitions of collaborative logistics. After a brief presentation of the main background issues, the most important organisational aspects of this type of approach will be presented. Then, the socio-economic and legislative aspects of collaborative logistics sharing will be described. To illustrate the presented concepts and schemas, we propose an analysis of several experimental cases from the literature, and a case study from the French press distribution sector; a distribution company that started a collaborative logistics sharing project at the end of 2008.

BACKGROUND

In the last years, several strategies and logistics models have been developed in order to increase the supply chain effectiveness. Collaboration is one of the most promising areas of study in supply chain management (Lambert et al., 1996; Barrat, 2004; Min et al., 2005; Simatupang & Sridharan, 2005; Lambert, 2008). In supply chain management, collaboration can take place at several stages of the chain and with different levels of interaction. These different levels of interaction can be classified as follows, by Crainic and Laporte (1997) for freight transportation and by Baglin et al. (2009) for collaborative logistics:

- *Transactional collaboration*: Logistics and transportation need consistent administrative practices and document exchanges. The first stage of collaboration consists of the common coordination and standardization of administrative practices and exchange techniques, requiring information and communication systems.
- *Informational collaboration*: This level of collaboration concerns mutual exchange of information such as sales forecasts, stock levels and delivery dates. It is important to note, that confidentiality and the process of competition can hinder collaboration.
- *Decisional collaboration*: This category concerns the different collaboration possibilities in planning and management decisions within logistics and transportation. These decisions can belong to different planning stages:
    - *Operational planning*: This planning stage is related to daily operations that can be coordinated or shared, like freight transportation or cross-docking.
    - *Tactical planning*: The middle-term planning stage involves several tactical decisions, like sales forecasts, shipping operational decisions, stock and production management and quality control. At this level, sharing decisions can establish a relation of trust between the collaborators.
    - *Strategic planning*: The highest collaboration stage is related to long term planning decisions such as network design, facility location, finance and production planning.

Although in a narrow sense the word "sharing" refers to the joint or the alternating use of inherently finite resources, both material and immaterial, it can also refer to the process of dividing and distributing. In logistics, the main shared resources are information, infrastructure, management/planning tools, vehicles and human resources. In transportation and logistics sharing, operational decisions are in general individually made. Tactical and strategic decisions can be made by different actors or groups. We can define three types of transportation and logistics sharing approaches:

- *Non-collaborative sharing*: the shared resources are managed by their users independently. There is no collaboration between these users.
- *Collaborative sharing with hierarchical decision making*: the shared resources are commonly managed by their users but the main decision processes are hierarchic.
- *Collaborative sharing with non-hierarchical decision making*: this type of sharing differs from the above in the fact that the different users take part into the decision processes.

In the first and the second types of sharing, strategic decisions are taken by a single decision maker. In the third type, these decisions are made by the members of a collaborative group of actors under a partnership contract or other type of agreement. In this chapter, we focus on the third type of sharing approach. Collaborative sharing can be defined as the joint use of material and immaterial logistics resources, and a good management of the collaborative actions is necessary to ensure the continuity of this type of sharing approach. When two or more actors decide to collaborate in a sharing approach, we can call them sharing partners. Although sharing can be formalized by agreements that are not formal written partnership contracts, we observe several analogies between logistics partnerships and collaborative logistics sharing approaches (Lambert, 2008). The relations between partners are complex, and several papers have defined the main principles of collaboration in partnerships (Newbourne, 1997; Lambert et al., 1996; Cooper et al., 1997). Lambert et al. (1996), define the main principles a logistics partnership and the differences between other forms of inter-enterprise relationships, as well as a first model to analyze their feasibility from a management point of view. Other works define the factors that accelerate collaboration, most of them related to production and warehousing (Lambert et al., 2004; Lambert, 2008), and involve in general multiple participants.

In collaborative logistics sharing, the involved actors in the approach can be considered as a reasoning community (Evangelou & Karacapilidis, 2007; Yearwood & Stranieri, 2009), in other words, a group or community of individuals that engage in dialogue with each other in order to reason toward action. As such the term is broader than communities of practice, communities of action or communities of purpose. Members of a reasoning community may or may not belong to the same organisation, hold the same values, aim for the same outcomes or have anything in common, except the need to reason toward a solution of the same or similar problem. From these statements, we can define a Logistics Sharing Community (LSC), as the group of stakeholders involved in collaborative transportation sharing with non-hierarchical decision making and as the extension of the logistics partnership principles to a larger group of reasoning. An LSC can also be considered as a small reasoning community (Yearwood & Stranieri, 2006), where their members are well identified and defined, and have different objectives. They can also have different points of view and decision values at different planning stages. A reasoning community engages in a process that involves three main components (Yearwood & Stranieri, 2009):

- In the *individual reasoning* phase each individual seeks evidence, organises it and ultimately forms claims that represent his or her preferred position or beliefs.
- The *communication of reasoning* phase describes the transmission of all aspects of individual and coalesced reasoning to others.
- Finally, in the *coalescing of reasoning* phase a form that represents the reasoning processes acceptable to the entire community is obtained. A coalescing of reasoning does not mean that an agreement about a solution is reached. Rather, coalescing of reasoning reflects the state where each individual's reasoning is understood and accepted as valid by the community even if there is such a divergence of views that agreement is impossible.

In classical logistics research, most of the literature deals with individual reasoning, even when dealing with collaborative approaches. We will propose in the following sections two models, taking into account the logistics principles and applying them to a reasoning community in order to define the main organisational and socio-economic aspects of a logistics sharing community.

ORGANISATIONAL ASPECTS

Collaboration happens through information and knowledge management (Laudon & Laudon, 2007; Evangelou & Karacapilidis, 2007; Lambert, 2008; Baglin et al., 2009). For this reason, we can assert that collaborative transportation sharing needs an efficient shared information system to ensure a high standard of performance. We propose an organisational model to represent the organisational aspects of logistics sharing, based on that of Laudon and Laudon (2007) for the conception of a shared information system. These ideas are summarised in the chart of Figure 1.

This conceptual model, initially proposed for individual decision making situations, is composed of five modules. The sharing management module contains all the elements of the management of sharing services and collaboration. In this case, a focus will be made on the elaboration of new business models based on the collaboration; as well as on the location of new customers. The information technologies module contains the Information and Communication Technologies (ICT) that are used in the proposed sharing solution. Here, we find all the internal and external tools. For instance internal tools include personal computers, but also professional softwares and Intranets. External tools include traceability tools like bar codes or Radio Frequency Identification. The organisation module lists all of the actors involved in the sharing solution, in both

Figure 1. Flowchart for the conception, design and management of an information system for logistics sharing (adapted from Laudon & Laudon, 2007)

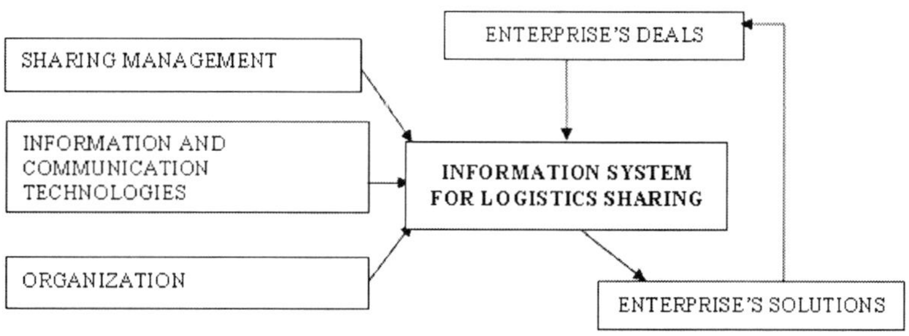

internal and external contexts. These three modules are related to tactical and operational decisions. The enterprise's deals and solutions are related to strategic planning. The deals of each enterprise concern the decreasing of costs and value creation for all stakeholders. The solutions are connected with the reduction of cycle time or the improvement of customer service, for example. All the modules are more precisely described above, starting by those utilising strategic planning then focusing on the tactical ones.

Enterprise's Deals Module

The enterprise's deals module presents both the project's expectations and the risks that are studied in that project's preliminary developments. Considering the technologies, tools and their usage levels, several choices must be made in order to set up the best solution of logistic sharing services. In order to make these choices, it is important to formulate questions related to the goals and the risks of the project, and to find the appropriate answers. Accordingly, it is important to make a deep analysis of the possible risks that the project may encounter. From the work of Seiersen (2006), we can list the main categories of risks related to a collaborative transportation sharing project:

- The risks related to the project accounting itself, more precisely to the different type of resources that can be affected by the project, in financial, economical, technical, technological or human terms. In general, each member of a LSC knows its implications in financial and accounting terms, and these decisions are individually taken.
- The risks related to the organization of the project and its continuity. Two types of risks can be identified in this category: the risks related to the logistics organisation; for operational decisions, which are in general individually made, and those of the LSC tactical and strategic actions, where misunderstandings and other obstacles to collaboration have to be seen as potential risks. It is important to note that the reorganisation of a project (including changes on the LSC status) can be considered only when the project is operative and stable.
- The technological risks; in general, the technologies present problems related to functionality, robustness and compatibility, among others. Before choosing a technology, it is important to think about these questions.
- The risks related to policies, processes and current practices. The development and usage of new logistics solutions can need an important change in the way people think and act to make them operative. Continuous social analysis during all conception and development phases are crucial to the stability and success of very innovative solutions. These social analyses need to be made on both the LSC members and the external actors involved in the different stages of the project.
- The risks related to the impact of the systems in the current and future operations, at both human and technical levels. In this category, the collaboration rules and the respect of them have to be considered.
- The dependence risks; If the information system is based on several technologies, the risks related to the dysfunction of these technologies have to be considered. When a technological tool presents a dysfunction, the system can be less efficient, or can stop because of it. These risks have to be studied in a preliminary phase of a project.

Enterprise's Solutions Module

The enterprise's solutions are the main objectives of the project and the evaluation of its performance (Laudon & Laudon, 2007). Although at the end of the 20th century the notion of performance has been basically related to economic indicators, the notion of sustainability is nowadays a central element in transportation and logistics planning and management (Morana & Gonzalez-Feliu, 2009). Sustainable development can be represented as the junction of three spheres: the first one deals with the economic aspects, the second one contains the social and the societal elements, and the third is related to the environment. The economic analysis is different, since it is necessary to reach each member to reach economic sustainability to ensure that of the entire LSC (i.e., if one member does not reach the economic balance in order to reach that of the group, it will quit the community).

Some recent French research (Marais & Reynaud, 2007) proposes a ranking table for the different components of Sustainable Development actions, based on an exhaustive literature review analysis. The authors propose 5 central subjects: Strategy, Enterprise's policy, organisation, systems and key competencies. We can also refer to the main environmental and social norms and recommendations, as for example the AFNOR ISO 14000 environmental norms, the SA 8000, OHSAS 18001 and AA 1000 social norms or the SD21000 norm for sustainable development (Morana & Gonzalez-Feliu, 2009). Sustainability Balanced Scorecards (Hockerts et al., 2002) can be considered as sustainability performance indicators. However, in order to take into account the group reasoning component of an LSC, collaborative decision making (Raghu et al., 2001; Beroggi, 2005) has to be valued more highly than individual decision making while analysing the social sphere of sustainable development. For example, collaborative multicriteria analysis (Evangelou & Karacapilidis, 2007) can be used to develop decision support systems that model the different expectations and key objectives of the LSC, considering those of each member of the community. Moreover, the durability of collaborative transportation sharing approaches can be modelled using group reasoning decision support systems.

Organisation Module: Types of Actors in Supply Chains Related to Sharing

Reasoning occurs in a social context and usually involves numerous participants with differing views (Yearwood & Stranieri, 2009). Over the length of a supply chain, several actors interact in order to complete all the tasks necessary to produce and distribute a product to a retailer. In this section, we will introduce the main categories of actors in logistics and their potential domains of sharing, focusing on the logistics of distribution. We will outline the categories of actors in the freight transportation field that can be directly concerned to LSC so to group reasoning.

First of all, we can describe the "loaders" (Ambrosini & Routhier, 2004), who are the actors that send or receive the freight. We can find in this category the producers of the various raw products and components as well as the final product manufacturers, the logistics providers, the distribution and gross commerce enterprises, then the retailers. These actors can be considered as "senders", if they act at the origin of the transport, and "receivers" if they are at its destination. Another important category is the "transporters". The transporters can be the "loaders" that make self-transport operations, or the third-party transportation companies (Ambrosini & Routhier, 2004). These companies can be workers that have only one vehicle, small and medium enterprises or big companies and multinational groups, as well as postal and courier operators and logistics providers. A third category is that of the platform management companies (Patier et al., 2007). The heterogeneity of the LSC and the different goals and the degrees of implication for each member have to be considered while modelling the community decision process.

Yet other actors participating indirectly to transportation sharing, like public administrations, highway companies, customs operators, are not considered in this classification for two reasons:

the first is that their possible implication as logistics sharing partners are much less important with respect to the three main categories and the second is that they are external actors to the LSC, although they can have influence to several decisions related to this community's strategic and tactical actions.

Sharing Management Module: Sharing Approaches in Freight Distribution

In supply chain management, we observe sharing approaches in different processes, involving both production and distribution sub-chains. More precisely, focusing on distribution, we can distinguish two main domains of application: the complementary activities to transportation, as for example warehousing or supplying, and the transportation itself. We will describe briefly the organisational models of sharing in both fields.

The Efficient Consumer's Response (ECR) is a concept developed in the grocery distribution context. It is defined as a cooperative approach whose goal is the total satisfaction of the consumer by an improvement on the economic performance of the different actors within the supply chain. The ECR optimises the retailer's supply and improves the promotional actions and the freight availability by the use of ICT and logistics information systems, as well as the usage of activity based management tools. The Vendor Management Inventory (VMI) can be considered the next step in respect the ECR. In this collaborative approach, the supplier is co-responsible for the warehouses' re-supplying using the sales forecasts. This involves using collaborative actions. This approach implies a commitment of the distribution company to give real time information to the producer, who will be able to make a re-supply proposal and then make his demand predictions in order to adapt his production phases and his resources to these demand predictions. A new form of VMI, which can be called "shared VMI", has been developed in the UK and France, and involves several producers, who agree to work with the same distribution company and share information with this company (Simonot & Roure, 2007). At the infrastructure level, shared platforms and infrastructures are very common. However, most of them are only physically shared, and the actors that operate in these platforms do not collaborate. Another model of shared platform is the "collaborative warehouse", where several producers and distribution companies share a physical space and logistics information to improve the global performance of the overall distribution processes (GCI and Capgemini, 2008). This idea can also be found in consolidation platforms, like classical cross-docks, regional platforms, urban consolidation centres or urban logistics spaces.

In freight transportation, collaboration between two operators is an action that is usually informal and not documented. These actions are taken to increase the loading rate of a vehicle, or to make a "friend" company deliver to a customer that the contracted operator is not able to get to (Patier, 2004; Morana & Gonzalez-Feliu, 2009). In frequent collaboration cases, the approaches can be formalized by agreements. Another form of collaboration is the networks of transportation companies. Most of these networks involve small and medium companies. A network is presented as an association, although some of them assume the form of a cooperative company (Simonot & Roure, 2007). A more collaborative sharing approach is the open e-marketplace. This approach is based on an electronic information exchange system, where the transportation offer actors meet the transportation demand ones. The offer comes from transport companies, and the demand can come from "loaders" or from transporters that do not have enough quantity of goods to transport in a particular area.

Information Technologies' Module

Information is a central key of sharing. Without information sharing, the other levels of sharing cannot take place. In transport management, the role of ICT has been recently overviewed (Fabbe-Costes, 2007). Two types of information technologies are identified by the author: the transportation management modules, related to transportation planning, and the information exchange tools, that allow transportation to be integrated into the supply chain.

In logistics planning, decisions on the transportation network settings have a direct impact on the service quality but also on their costs. It is then important to adapt the transportation network to the economic, geographic, organisational and quality constraints (Crainic & Laporte, 1997; Wieberneit, 2008). In tactical and operational freight distribution planning, these questions are related to warehousing, transportation management, vehicle assignment and crew scheduling. The two last points derive from the two first, and take place after them. In research, both inventory and vehicle routing and scheduling problems are very popular, and several algorithms for multistage transportation systems have been proposed in recent surveys (Dullaert et al., 2007, Gonzalez-Feliu, 2009). Moreover, a periodic survey on operative software for vehicle routing management can be found in (Hall & Partyka, 2008).

In transportation planning and management, ICT play a crucial role, and are usually combined with optimisation modules in order to improve the performance of the different operations. A special attention has to be given to the main technologies which allow the freight transport operations to be included in the global supply chain of a product. Fabbe-Costes (2007) individuates three categories of IS, i.e. document exchange systems, communication systems and traceability systems. The document exchange systems ensure data exchange amo\ng actors and record several transactions, following several mediums, like the written transportation documents, the Electronic Data Interchange (EDI), or the web-EDI. Then, the communication systems ensure the enterprise flow's work. Finally, the traceability systems are developed to find and follow freight movement. In this field, several standards are used, in general related to Radio Frequency Identification (RFID) or to classical reading systems (bar codes).

COLLABORATIVE SHARING APPROACHES FOCUS

We have outlined the main organisational aspects in collaborative transportation sharing. These approaches need the collaboration of different actors, each of them having their own strategies, targets and processes. It should be noted that the works of Laudon and Laudon (2001) follow an enterprise's perspective, not those of a reasoning community. To introduce the different socio-economic and legislative factors that are related to collaborative sharing, we propose a conceptual model for sharing analysis based on those of Lambert et al. (2004) and Lambert (2008), including several elements of group reasoning to logistics partnership models.

Table 1. Information systems and technologies in supply chain (Fabbe-Costes, 2007)

Information Systems	Technologies and Tools
Document exchange systems	Fax; Electronic Data Interchange tools (web-based or intranet-based); Internet
Communication systems	Onboard radio; onboard/portable terminals; Fixed phone; Mobile phone; Internet; Multifunction portable terminals
Traceability Systems	Identification/codification (bar codes, RFID); Electronic readings; Waymarks; Vocal systems; Recorders; Memory systems

Solutions and Recommendations

In order to illustrate the model, we have done a comparative analysis of 18 logistics sharing communities, at different levels. The data used for this analysis has been obtained from various sources. Most of the feedback experiences have been identified in research reports and other scientific documents (Patier, 2004; Simonot & Roure, 2007; TL&Associés and LET, 2009; Semet et al., 2009). We have extracted the information about sharing approaches filling in the missing elements with information found in technical documents and professional logistics articles. Four cases are obtained from only technical and professional information, with no direct relation to a scientific study, and nine have been done by face-to-face interviews with logistics and sustainable development managers involved in the project. This has provided information about the capacities and the adaptability competencies of each member of the corresponding LSC, and the motivations and decision processes in the analysis and development of logistics sharing solutions.

In terms of sharing management (see details in above paragraph), we have considered four VMI cases (E4, E5, E6, E18), three different transportation sharing agreements (E1, E8, E9), five transportation networks and consortiums (E10, E11, E12, E13, E16), four cases of shared logistics platforms with mutual usage of distribution systems to retailers (E2, E3, E7, E15), one open e-marketplace (E17) and two non-transportation based collaboration agreements (E14, E15). We present first the main categories of motivators, then, we study both the facilitators and limitations. Finally, the economic and environmental performance feedbacks are presented.

Motivators

The motivators are the factors that incite the development of a sharing approach. These factors are defined from the socio-economic and legislative context of the practices. More exactly, we can distinguish the following groups of motivators:

1. *Economic, environmental and value motivators.* Although the economic efficiency is the strongest motivators found in the experiences reported, other elements are also considered, usually related to the prestige of the partners. Therefore, collaboration among transportation operators is chosen as a cost optimisation approach when the favourable conditions are met. The two main

Figure 2. Conceptual model for sharing analysis in freight distribution planning and system design

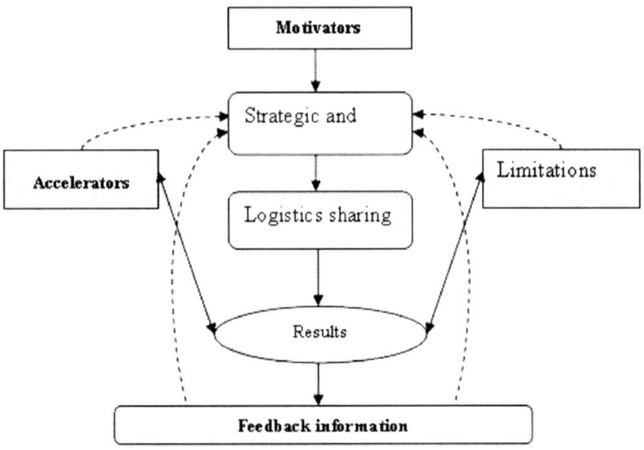

Table 2. Description of the 18 cases

Num.	Type	Description	State
E1	Transportation agreement	Two concurrent logistics services companies made a transportation sharing agreement to avoid empty returns	Demand changed
E2	Production platform sharing	Three concurrent automotive industry groups made almost identical models with the same patterns in the same factory	Strategies changed
E3	Shared warehouse and distribution	Two concurrent tyre brands built a particular shared warehouse to optimise the distribution operations by using the same transport operator	Operative
E4	Shared VMI	Shared VMI between a cleaning products international group and a healthcare industry international group	Operative
E5	Shared VMI	Shared VMI among non competing food industry companies	Operative
E6	Classical VMI	A grocery distribution group developed a VMI approach with several furnishers and collaborators	Operative
E7	Collaborative warehouse	First case of collaborative warehouses in the food industry (two non competing companies)	Simulation
E8	Transportation agreement	Two non competing fresh food industry small enterprises developed a mutual transportation and representation strategy, sharing vehicles and salespeople	Operative
E9	Transportation agreement	Three competing fresh food transportation companies collaborated to increase their vehicle's loading rates, developing several sharing approaches	Changed. Partners' fusion
E10	Network	Network of small and medium French transportation enterprises	Operative
E11	Network	Network of LTL transportation enterprises in France	Operative
E12	Consortium	Last mile distribution system based on a urban consolidation centre and a consortium of transportation companies	Operative
E13	Consortium	Collaborative last-mile transportation sharing approach for small and medium transportation companies	Partner's search
E14	Agreement	Printing and logistic chain solutions for two competing newspaper publishing companies in The Netherlands	Operative
E15	Agreement	A newspaper distribution company made collaboration agreements with book editors to increase their loading rates	Operative
E16	Network	Closed e-marketplace (network with membership fees and entry conditions) for overseas transportation	Operative
E17	Open marketplace	Open e-marketplace for national road transportation	Non efficient

reasons of vehicle sharing are related to the increase of the loading rates (E1 to E11, E14 to E18). Urban logistics measures are related to environmental aspects, from which CO_2 reduction is more important than the other environmental aims (E6, E7, E8, E12, E13, E15). Also quality is considered as important, but is in general seen as a feedback indicator that motivates new developments for logistics sharing services. For example, the main results of exploratory collaborations (E5 and E9) led to the research of new partners to start bigger partnership-based sharing services. In the same way, the transportation networks and consortiums (E10 to E13) have followed evolutions inciting collaboration among their partners.

2. *Legislative motivators*. Normative and jurisprudence aspects of sharing are related to public administrations. Nowadays, the most important facilitators in this category are the different local laws that help the development of sharing approaches in urban and regional freight transportation (Dablanc, 2008). We observe two types of policies: restrictions to non-sharing approaches and

incentives to sharing approaches. We can find in these categories the access permits of Genova (Italy) and Amsterdam (The Netherlands), restrictions based on minimum loading factors of the vehicles or their CO_2 emissions in Padova and Bologna (Italy) or the reduction of restrictions to sharing-promoting transportation systems in Ferrara (Italy). More restrictive policies are those of the Principality of Monaco or Vicenza (Italy), where only one operator is allowed to access the limited traffic zone, imposing transportation sharing to the other actors in urban logistics. These questions are observed on four cases (E8, E12, E13 and E15).

3. *Relation motivators*. When actors have already collaborated, because of common interests (this is the case of the small transportation operators that led to E10 and E11, as well as some of the participating operators of E12 and E15) or because they belong to the same supply chain, so they are complementary (E4 to E6, E8, E15 and E18), transportation sharing is more naturally taken into account than in cases where those conditions are not met. Moreover, non-concurrent or complementary enterprises are more concerned to these types of approaches in absence of legislative or financial motivators.
4. *Financial motivators*. They are related to subventions and financial help that can come from public, private or mixed entities. Several approaches have emerged from research and innovation projects financed (totally or partially) by public entities (E4, E7, E12, E13).

Facilitators

The facilitators are the conditions and situations that have a positive impact on the daily operations of a sharing approach. They are similar to those of collaboration and logistics partnerships (Lambert et al., 2004; Lambert, 2008). These factors are not only related to the logistics organisation but also to the evolution of the strategic planning relations between sharing partners. A history of relations between two actors can facilitate a durable partnership. For example, most transportation networks are born from a small group of "friendly" companies, who have collaborated or started cordial relations (E10, E11, E12, E16). Shared VMI approaches are related to grocery distribution, and have a central distribution company that organises the supply system for their collaborators. In general, a project where partners already had good relations has positive results in terms of collaboration management.

The boundary between the motivators and the facilitators is not always clear. However, we can distinguish these two categories by the fact that the motivators have an impact on the strategic decisions before the project is operative, and the facilitators have impacts which are observed at the operational level for a given sharing approach.

Limitations and Obstacles

Closely related to the facilitators, the limitations and obstacles are those factors that can become an impediment to the successful development of a logistics sharing approach and in preceding studies they have not been defined explicitly. We can find several types of limitations and obstacles, which can be grouped into the following categories:

1. *Legislation*. Restrictive legislation to sharing approaches is legislation that is related to freight compatibility, i.e. the norms and laws that forbid the loading of a vehicle with products of different sorts (for example dangerous goods, fresh food, waste, raw materials, etc.) or to competition laws that can limit the development of sharing approaches (E8, E9, E15).
2. *Organisation*. In this category we find the physical and organisational conditions for freight compatibility, like dimensions,

freight, type of packaging, loading unit and loading operations main characteristics. Another organisational factor is the acceptability of the sharing approach, which also has to be taken into account when defining the main characteristics of the collaboration for logistics sharing (E3, E7, E8, E15).

3. *Confidentiality*. The main issues related to confidentiality that can become an obstacle to logistics sharing are when two competing actors decide to collaborate in a distribution system that implies sharing some of their logistics resources. Since information is the base of good collaboration, if one or more partners manage confidential information that they don't want to share for competition reasons, the efficiency of the sharing approach can decrease considerably. These issues are seen in most of the initiatives involving competing enterprises that do no have the support of public entities (E1, E3).
4. *Responsibility*. The factors related to the transportation operation's responsibility are strictly derived from the contract between the different actors of these operations. If the collaborations for logistics sharing follow a contract or a chart where the questions of responsibility are well defined, these questions will not constitute an obstacle to sharing. On the other hand, if these questions are not clearly specified in a contractual document, legal disputes related to responsibility can easily occur or the quality of service decreases because of these questions (E17).

Case Study: The Press Distribution in France

From the detailed experiences, we observe that sharing approaches are mainly developed as an answer to economic performance questions, and are related to motivating contexts. Only a few of them show that environmental performance is an important evaluation factor. Open marketplaces do not present very positive results, because of the responsibility transfer aspects. We observe that most of the detailed experiences present an important information sharing system, but only the transportation networks and some agreements involve more than one transportation operator. Only four of them present an important platform sharing approach based on collaboration. These results are related to facilitators and limitations. Sharing approaches among complementary actors seem to be easier to implement than those that involve competing companies. However, these approaches come about due to economic factors that incite their development and effectiveness.

Another important question is the evolution of the sharing approach. In networks and agreements involving two or more transportation operators, if sharing is giving very good results, we observe two types of evolutions. One is the reinforcement of the network or the creation of strong partnerships. The second possible evolution is the union of the sharing partners into a group to optimize their overall resources and give a better service to their customers. We present the main results of a case study research for the biggest press distribution company in France. This is a qualitative study. We will present the main results of the organisational model, whose methodology and extended results can be found in a recent study (Morana & Gonzalez-Feliu, 2009), and complete the study by an analysis of the logistics sharing approach using the conceptual model presented above.

The Press Distribution in France: General Context

The press distribution sector is experiencing decreasing trends in sales and distribution flows, because of internet information sources and the free press. These factors have a repercussion in the distribution system. In France, there are only two distribution companies for classical press, i.e., written press excluding free journals, which are not real competitors (the competition exists at the editor level, but not in terms of distribution costs). Moreover, they are already collaborating in non-Parisian areas. The specifics of the metropolitan region of Paris, which represents approximately 1/4 of the total population of France (overseas territories not included), justify the usage of separate distribution systems for each of the two distribution companies. The distribution system is based on intermediary platforms for storage and consolidation. Moreover, two different types of products are managed: daily press, which has very tight time constraints, and magazines, which can be managed with less restrictive time limitations. Another important characteristic of press distribution is the variability of magazines' quantities to distribute, which make the transportation demand difficult to estimate. Therefore, the last mile distribution system is rigid and based on small transportation operators, to ensure a personal contact with selling points, that can need to give a key to the vehicle driver because of the distribution period (from 2:00 a.m. to 6:00 a.m.) and the risks of having the freight stolen.

COMPANY'S OVERVIEW

The case study deals with the Presstalis group, an action-based company that delivers 80% of the traditional press distribution in France. The main activities of the company deal with the logistics planning and organisation for all the distribution and reverse supply chains of the written press, excluding the free press distribution, whose supply chains differs from those of the traditional written press. The company's key information is summarised in the following table:

In 1994, because of a constant decrease of the traditional press in France, Presstalis started to consider a set of strategies projecting in a ten year horizon to modernise their company and increase the efficiency of its distribution system. From this perspective, Presstalis invested in a new collaborative transportation sharing solution project, constituting a mixed group including logistics operators, software developing companies, research centres and consultants, with a double aim (1) to optimise their ' land logistics' (press depots and warehouses) and (2) to find new products to distribute, in a transportation sharing perspective. The case study presents the preliminary phase of the project, as well as its current state.

Organisational Model Results: Qualitative Analyses on 9 Months (January-September 2009)

We will present the most significant results of the qualitative analysis made on the basis of a directive interview plus two site visits, and concluded by a small interview at the end of the period (September 2009) with those responsible for the project in order to have feedback on the recent developments, like the partner research phase and the last organisational advances in the collaborative logistics sharing platform.

The logistics chain of Presstalis (*sharing management module*) is similar to that of the other French press distribution groups, and can be divided into two macro-echelons: the inbound distribution (from the publishers' printing platforms to the press depots), and the outbound distribution (from the depots to the retailing activities selling the distributed products). Presstalis had in 2009 about 172 press depots and 30000 selling points. The inbound distribution presents a more flex-

Table 3. Summary of Presstalis' key information

Key information
2008 Turnover 2560 M € 2007 Turnover 2677 M € 3000 employees in 2008 2.5 million customers per day
Presstalis distributed products sells (end 2008)
- Global sales: 1235 million units (2560 million euros) - Unsold rate/ (49.9% value) - Sales in overseas France = 1.1% of the global sales - Export sales (100 countries) = 8.9% of the global sales

ible transportation demand than the outbound distribution; inbound flows are contracted each day and well optimised, because consolidation and rationalisation is possible. However, this is not possible for the outbound flows, because of the retailer's constraints and Presstalis quality control settings.

One of the goals of the shared platform is to reduce the number of storage and consolidation platforms that will in turn reduce the number of echelons in each category of distribution flows; by including other categories of products in the outbound distribution. Doing this, Presstalis can find two questions to answer: the first is in respect of the constraints specific to the written press distribution, and the second is the security of the products that can cohabit with the written press. These two questions are crucial to find the categories of products that can be compatible with written press distribution. A first study identified several categories, like books, catalogues, stationery products, collectibles, CDs, and DVDs, and a deeper compatibility study started in September 2009. The chosen system is based on bilateral partnerships first, with the aim of constituting a small transportation consortium and finding regular customers that ensure the minimum freight volumes to justify a sharing approach like this. More specifically, the chosen system is a hybrid model that mixes a VMI solution with a consortium of transporters. The VMI solution has already been defined, and the consortium of transporters is in a development phase. Presstalis can be compared to a logistics operator without vehicles under direct ownership. For this reason, the initial consortium will include the current Presstalis transportation operators and two other parcel operators that are specialised in urban distribution.

Currently, Presstalis does not have very expansive technological information systems. This choice derives from the outbound distribution system's rigidity as well as from the low value of the written press products. Indeed, the network design and transportation planning tools for the last mile are commercial software programs. The transportation plans are established for a given period (in general, 6 months), and do not change unless under unexpected or urgent constraints, as for example the presence of special events or the addition of a new publishing company. The inbound distribution system is more flexible, and a personalised software, developed by an internal service, was chosen. These transportation plans are established every day and can be easily modified. The company is equipped with standard communication technologies (mobile phone for the terminal-vehicle communication, e-mail and fax for the transportation plans exchanges), because no more complex systems are required for non high-value products, which is the case of the written press. Moreover, the standard barcode system is used for freight traceability, as well as for managing the pickup and delivery transactions. The current technologies used by Presstalis can become an obstacle to the introduction of other products; if more strict follow-up constrained freight is transported, barcodes are not enough to ensure a correct follow-up, and RFID solutions will have to replace the current traceability systems, leading to changes that need a large monetary contribution. Depots are managed manually and few automation systems are used. Moreover, the EDI standard is not considered, since the transportation operators for the inbound distribution accept the commands made by fax and the outbound distribution system is based on medium and long term agreements with a rigid routing plan for each transportation operator. However, a SAP technology is in a pre-implementation phase to improve the warehousing performance, and is partnered with Presstalis' warehouses and their own and associated press depots.

The organisation module defines the different actors in Presstalis' LSC. We observe two types of members: Presstalis' internal actors, who are those related to the echelons of both distribution systems and their sub-contractors, i.e., their contracted transporters, the logistics platforms managers and employees, the press depots operators, Presstalis' partners, and their current collaborators. These actors are not enough to maintain an efficient sharing system (the loading rates are still too low to reach Presstalis' performance targets), so new collaborators will have to be found. Four transportation operators have given a tentative agreement to implement a sharing system based on a consortium, and new potential customers are being looked at. These operators are environmentally-friendly last mile transporters, who deal with potentially compatible products with respect to Presstalis' distribution system. Since their goals are similar, agreements can be stated, but not all the freight will be shared. In the partner research phase, also the juridical and regulation aspects of sharing are studied. The public administration systems are also an important part because lawful standards depend on these entities, and their implication can have non negligible impacts to the acceptability of potential LSC members to enter the reasoning community. Finally, the potential customers, i.e., the "loaders" and logistics operators that can contribute to the community by bringing transportation demand, will be sought after the basic version of the sharing virtual platform is tested.

Sharing is considered a good solution because of the system rigidity during the press distribution period, and the possibility to deliver other products after the last selling point has been visited. The risks of this approach have to be considered, and a proposed sharing distribution system has to be developed. To do this, the main tactical decisions have to be spelled out, then, using the analysis model, the main strategic decisions can be defined. Moreover, considering that the project is in a development phase (the completion is less than 1/3 of the total respectful of the initial planning), the performance evaluation indicators are starting to be defined. Two main groups of indicators will be taken into account: first, the economic sustainability of the proposed solution will be tested, then, in order to evaluate its environmental performance, a comparison between different situations will lead to an analysis on the effects of the proposed solution on CO_2 reduction. The other pollutants and the noise nuisances will not be initially estimated.

Logistics Sharing Analysis

After presenting the main organisational aspects, a first analysis of the logistics sharing is proposed to link the logistics organisational model to group reasoning theory. Based on this analysis, a conceptual model is deduced. If we focus on the collaborative logistics sharing motivators, we observe that the context of the press distribution had an important influence on the decision to consider collaborative sharing. The rigidity of the current system and the economic and environmental factors motivated the development of a sharing approach (the economic efficiency has to be increased to maintain the current distribution costs or reduce them, and the environmental efficiency is important to improve the company's popularity and quality), and to find the potential collaborators for a collaborative sharing platform. A high-quality internal organisation and a good relation with partners is necessary to ensure system operability. For these reasons, a preliminary study of the partners has been made. The chosen partners are enterprises that were already important collaborators with the distribution company or had similar and non contradictory aims in a potential sharing approach. These collaborators have in general similar interests, but a dialogue and debating phase is necessary before a "principle of agreement" can be established. Moreover, these actors must be able to collaborate in the different planning horizons. The sharing approach design has been developed by the press distribution company and the partners have been chosen on the basis of their adaptability and compatibility with respect to this system. In this way, the main limitations and obstacles are able to be avoided. However, some legislation questions, specific to competition rights and to responsibility transfers, are under deliberation and analysis because they are considered the main obstacles to the system. These limitations are also considered in the risk analysis.

The first results are good but not enough to ensure a better control over transport demands. However, they show that the developed approach can be efficient and could be extended to a bigger consortium. The main results are used to redefine some details of the main strategies and to focus on questions not initially taken into account, such as specific legislative questions, and the potential of other freight categories such as packaged non-fresh food, small electronic products, software and tobacco derivates, among others, to be considered.

A crucial question has however, to be solved: the management of the logistics sharing virtual platform, and in consequence that of the LSC. The community manager needs to mediate between the parties, without having a predominant or a discriminative attitude toward one or more members. This figure, that is starting to be defined, can have a significant influence on the other members, and constitute both a facilitator and an obstacle.

In order to summarise the proposed results, we propose a conceptual model that summarises and connects the organisational model and the sharing analysis factors in the context of a sharing community. It can be illustrated in the following chart as shown in Figure 3.

The relations are similar to those of the organisational model based on the work of Laudon and Laudon (2007), but have taken into account the differences in the decision processes between a reasoning community and an individual decision maker. In addition to the former organisational model, this one introduces the socio-economic elements and proposes a first relation schema between the logistics concepts and the collaborative decision making notions.

Strategic decisions are influenced by the motivators as we have seen above. The main difference between this model and that of Laudon and Laudon (2007) is that strategic decisions follow three phases: first, individual strategic decisions are made, followed by a contribution of each member to the community strategic decision

Figure 3. Logistics sharing community decision processes. The shadowed blocs indicate that they can affect the LSC as a facilitator or an obstacle

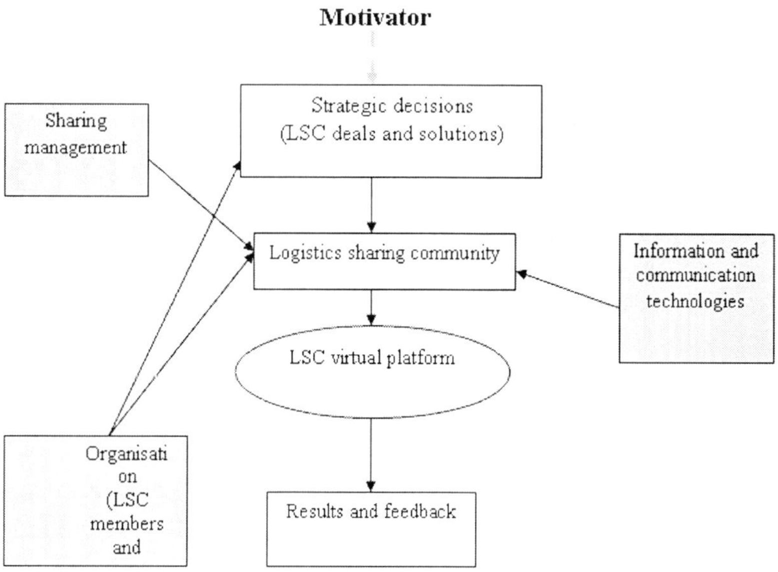

process; finally, after a collaborative discussion process, group decisions can be stated.

The three shadowed blocs (organisation, sharing management and information and communication technologies) constitute both the tactical decisions and the main facilitators and obstacles to collaborative transportation sharing. The two last modules are quite similar to those explained in the organisational aspects' conceptual analysis. From the case study (Eisenhardt, 1989), we can reformulate the organisation module as follows. First, the different actors of the sharing approach have to be identified, and their relation to the LSC defined (members, collaborators, regulators, etc.) in order to define the reasoning community dealing with sharing management and planning and the extended reasoning group, including the involved authorities (legislation and regulation). These actors have already been defined as well as their contributions by either facilitating or slowing down the collaboration among members of the community. Each actor's individual decisions will have an impact on the argumentation phase of the group decision stage. From these decisions and the derived actions, the main facilitators and obstacles to group reasoning can be identified. The impacts of these factors have a repercussion on the logistics sharing community decisions and on the ability to reach agreements and equilibriums. Then, indirectly, they have an impact on the performance of the sharing approach as well as on tactical and strategic decisions.

As we have seen, collaborative transportation sharing involves different actors and cannot be analysed using only the enterprise's perspective, like in classical supply chain management. The introduction of group reasoning and collaborative decision making can constitute an important research direction in logistics, more precisely by developing collaborative decision support systems for logistics sharing communities.

CONCLUSION

Logistics sharing is becoming a popular approach to reduce the distribution costs of a product. However, the subject has not been deeply studied in the literature. In this chapter, we presented the main concepts of logistics sharing in the freight distribution sector, focusing on collaborating transportation-based sharing approaches. We presented both an organisational model for the development of a sharing-based information systems and an analysis model for its strategic decisions, defining the categories of factors that can have an impact on them.

A collaborative sharing approach must pass through an efficient information sharing system. In this perspective, several aspects have to be considered in the different planning horizons. The tactical decisions deal with technologies and planning tools, with sharing management and with the possible actors of the sharing-based supply chain, both at internal and external levels. The strategic decisions consider the possible tactical choices to define the main objectives and define the sharing solutions to develop, considering also the risks of these choices.

The factors that influence strategic decisions can be grouped into three categories. First of all, the motivators, i.e. the reasons that incite the actors to collaborate in logistics sharing solutions. Then, the facilitators and the limitations and obstacles, which have an impact on the evolution of a conceived sharing service.

Several requirements can be deduced from the presented models and examples of logistics sharing approaches. The first is to consider the transportation field explicitly in the supply chain management methods, providing multidisciplinary research that includes system engineering, information science, economy, management, sociology and decision sciences, among others. In this sense, group decision theory becomes a field of investigation that can produce interesting results for logistics sharing management decision support. The second requirement, for the logistics strategic decision makers of enterprises, is to make a preliminary analysis of the factors having an impact on the main decisions in order to choose the best approach to meet their goals. Third, for managers, to identify the external factors facilitating and limiting the logistics sharing solution in order to better assure its performance. Finally, the public decision makers have to consider the legislation that can have a positive or negative impact on the development of collaborative sharing services for freight distribution in respect of the market and the competition rules of the current macroeconomic context.

However, this field is yet to be studied thoroughly and this work remains exploratory, aiming to establish standards and patterns to support logistics sharing decision making. New variables will appear with the development of sharing management and planning. Moreover, a measurement scale will be implemented to facilitate the evaluation and the strategic decision support. An extrapolation of the proposed models from (and to) other fields has to be considered to generalise and enhance them to other applications of sharing and collaboration. Finally, collaborative decision making methods, like collaborative multicriteria analysis, and group reasoning approaches, like argumentation models, can be applied to logistics sharing communities to build specific decision support systems for these collaborative approaches in logistics and transportation.

ACKNOWLEDGMENT

We should like to acknowledge Mr. Paul Leperck and Mr. Dennis Scardoni, from Presstalis, for their crucial help in participating in the main interviews and organising the various visits in order to complete our case study. We would also like to acknowledge the anonymous referees for their valuable suggestions that allowed us improve to the present chapter.

REFERENCES

Ambrosini, C., & Routhier, J. L. (2004). Objectives, methods and results of surveys carried out in the field of urban freight transport: An international comparison. *Transport Reviews*, *24*(1), 57–77. doi:10.1080/0144164032000122343

Barrat, M. (2004). Understanding the meaning of collaboration in the supply chain. *Supply Chain Management: An International Journal*, *9*(1), 30–42. doi:10.1108/13598540410517566

Beroggi, G. E. G. (2005). *Designing and evaluating e-management decision tools*. New York: Springer.

Cooper, M., Lambert, D. M., & Pagh, J. (1997). Supply chain management: More than a new name for logistics. *The International Journal of Logistics Management*, *8*(1), 1–13. doi:10.1108/09574099710805556

Crainic, T. G., & Laporte, G. (1997). Planning models for freight transportation. *European Journal of Operational Research*, *97*, 409–438. doi:10.1016/S0377-2217(96)00298-6

Dablanc, L. (2008). Urban goods movement and air quality policy and regulation issues in European cities. *Journal of Environmental Law*, *20*(2), 245–266. doi:10.1093/jel/eqn005

Dullaert, W., Bräysy, O., Goetschalckx, M., & Raa, B. (2007). Supply chain (re)design: Support for managerial and policy decisions. *European Journal of Transport and Infrastructure Research*, *7*(2), 73–92.

Eisenhardt, K. (1989). Building theories from case study research. *Academy of Management Review*, *14*(4), 532–550. doi:10.2307/258557

Evangelou, C., & Karacapilidis, N. (2006). A multidisciplinary approach for supporting knowledge-based decision making in collaborative settings. *International Journal of Artificial Intelligence Tools, 16*(6), 1069–1092. doi:10.1142/S0218213007003631

Fabbe-Costes, N. (2007). Systèmes d'information logistique et transport. *Techniques de l'Ingénieur, AGL2*.

GCI & Capgemini. (2008). *Future supply chain 2016: Serving consumers in a sustainable way*. Germany: Global Commerce Initiative.

Golden, B. L., Raghavan, S., & Wasil, E. A. (2008). *Vehicle routing: Latest advances and challenges*. Boston: Kluwer. doi:10.1007/978-0-387-77778-8

Hall, R., & Partyka, J. (2008). On the road to mobility. *OR/MS Today, 35*(1).

Hockerts, K., O'Rourke, A., & Zingales, F. (2002). Balanced scorecard and sustainability: State of art review. *CMER* [working paper series], *2002/65*. Fontainebleau, France: INSEAD.

Lambert, D. M. (Ed.). (2008). *Supply chain management: Processes, partnerships, performance* (3rd ed.). Sarasota, FL: Supply Chain Management Institute.

Lambert, D. M., Emmelhainz, M. A., & Gardner, J. T. (1996). Developing and implementing supply chain partnerships. *The International Journal of Logistics Management, 7*(2), 1–17. doi:10.1108/09574099610805485

Laudon, J. P., & Laudon, K. C. (2007). *Management information systems*. London: Pearson Prentice Hall.

Marais, M., & Reynaud, E. (2007). Comparaison entre les entreprises françaises publiques et privées face aux exigences du développement durable. *Centre d'Etudes et de Recherche en Gestion d'Aix-Marseille (CERGAM) Working Paper Series, 800*.

Min, S., Roath, A. S., Daugherty, P. J., Genchev, S. E., Chen, H., Arndt, A. D., & Richey, R. G. (2005). Supply chain collaboration: What's happening. *International Journal of Logistics Management, 16*(2), 237–256. doi:10.1108/09574090510634539

Morana, J., & Gonzalez-Feliu, J. (2009). *Les décisions tactiques et opérationnelles d'une mutualisation d'un système de distribution: Le cas du groupe NMPP*. Paris: Research Workshop in Sustainable Transportation and Logistics.

Newbourne, P. T. (1997). The role of partnerships in strategic account management. *The International Journal of Logistics Management, 8*(1), 67–74. doi:10.1108/09574099710805592

Patier, D. (2004). *La place du transport de marchandises en compte propre*. La Direction de La Recherche et des Affaires Scientifiques et Techniques & Centre pour la Communication Scientifique, Hyper Article en Ligne.

Patier, D., Dufour, J. G., & Routhier, J. L. (2007). *Du transport de marchandises en ville à la logistique urbaine*. Techniques de l'Ingénieur.

Raghu, T. S., Ramesh, R., Chang, A. M., & Whinston, A. B. (2001). Collaborative decision making: A connectionist paradigm for dialectical support. *Information Systems Research, 12*(4), 363–383. doi:10.1287/isre.12.4.363.9705

Seiersen, N. (2006). Systèmes d'information logistique. *Techniques de l'ingénieur*.

Semet, F., Routhier, J. L., Meunier, F., & Stumm, M. (Eds.). (2009). *Logistique Urbaine Mutualisée Durable: Etat de l'art scientifique*. Report LUMD – Advancity.

Simatupang, T. M., & Sridharan, R. (2005). An integrative framework for supply chain collaboration. *International Journal of Logistics Management, 16*(2), 257–274. doi:10.1108/09574090510634548

Simonot, P. Y., & Roure, J. (2007). *Logistique collaborative. Une question d'avenir. Paris: Economica. TL&Associés & LET (2009). Flexibilité et Impacts de la Demande de transport des différents secteurs Economiques, et simulation de Scénarios d'évolution*. Paris: PREDIT.

Wieberneit, N. (2008). Service network design for freight transportation: A review. *OR-Spektrum, 30*, 77–112. doi:10.1007/s00291-007-0079-2

Yearwood, J., & Stranieri, A. (2006). The generic/actual argument model of practical reasoning . *Decision Support Systems*, *41*, 358–379. doi:10.1016/j.dss.2004.07.004

Yearwood, J., & Stranieri, A. (2009). Group structured reasoning for coalescing group decisions. *Group Decision and Negotiation*. doi:. doi:10.1007/s10726-009-9162-1

Chapter 15
Spatial Ontologies in Multi-Agent Environmental Planning

Dino Borri
Polytechnic University of Bari, Italy

Domenico Camarda
Polytechnic University of Bari, Italy

ABSTRACT

Landscapes and townscapes have been studied by many disciplinary areas over time. This study addresses the cognitive and perceptual dimensions of environmental spacescapes in planning by human agents. In fact, because of their dynamic complexity, environmental spacescapes create challengesfor the typical spatial behaviour of an agent perceiving and navigating in it. Therefore, environmental planning activities need to identify and manage the 'fundamentals' of spacescapes from the viewpoints of living single agents or multi-agent organizations, those to whom the planning effort is addressed. In this framework, the chapter deals with spatial ontologies in multi-agent systems. Some recent experiments are described and discussed here, highlighting spatial features of navigated environments from an environmental planning perspective.

INTRODUCTION

Landscapes and townscapes have been studied by many scholars over many past decades, from the milestone essay by Lynch (1960) to the more recent spatial-cognition literature (e.g., Golledge, 1998; Amedeo et al., 2008; Lloyd, 2009). The present study aims at contributing, by addressing a specific issue, to a broader discussion. The starting point is the increasing awareness that today spacescapes (i.e., townscapes, cityscapes and the like) represent not merely backgrounds, but rather the main, knowledge-intensive, meaningful spaces and entities that the humans adapt for their lives. They show fair dynamic complexity, where the spatial behaviour of the agents proves to be atypical, ill-structured, particularly (but not only) when needing to be simulated in AI environments, such as cybernetics or robotics. In this picture, an understanding of spatial-cognition issues can be attempted by looking at the 'fundamentals' of

DOI: 10.4018/978-1-60960-091-4.ch015

spacescapes, located in the resilient needs and ways of living and reasoning of multiple agents.

Indeed, the impact of spaces on human agents' behaviours is traditionally steeped in prejudicial common sense. Accordingly, well designed space architectures or artworks seem to be more able to give significance to the space for humans to live in, than formless, unstructured spaces. Also, conversely, urban social marginality is seen as connected to the abundance of landmarks and symbols in city spaces, supposedly able to deprive or alienate poor people. However, common sense and prejudices catch on easily, since no practical and theoretical distinction is clearly put down between content and form, in real world. Yet space is often claimed as being characterized by diverse attributes and features where primary, essential qualities (spacescape 'fundamentals') are distinct from additional, secondary qualities (spacescape 'ornaments') (see, e.g., Goodman, 1951, albeit with some sceptical remarks).

Human agents think and operate in living spaces, both as single and as multiple agents. Their perception and representation of space determine space ontologies, whose importance is commonly essential for AI robotics. As there is circularity between AI and cognitive science, it is evident that developing robotic devices may in turn increase knowledge on human behaviours in space. Therefore, an exercise of identification and comprehension of space features in human agents can be critical in multi-agent social and environmental reasoning, particularly when dealing with strategic participatory planning. In fact, in these cases, an increasing knowledge on space fundamentals may induce the shared identification of structures, pillars, invariant, resilient characters of the environment, on which to build/plan the development of cities and towns.

In this context, decision-support systems (DSS) would largely benefit from an involvement of multiple agents that is able to be non-reticent on the intrinsic complexity in which agents themselves live, perceive and reason.

Basing on the above points, the present chapter looks at space ontology in multi-agent systems, by presenting and discussing some experimentations carried out by our research group in recent years. [1]In particular, the next section 2 starts from the discussion of the spatial-cognition outcomes of a decade of group reasoning for scenario-building activities in local planning. Then it ends up by drawing out results from a multi-agent experimentation on spatial knowledge in navigating a built environment. In section 3, brief remarks try to account on achievements and future perspectives of multi-agent spatial cognition in planning.

NOTES ON SPATIAL COGNITION AND REPRESENTATION IN MULTI-AGENT REASONING AND PLANNING

Multi-Agent Spatial Representation in the Mediterranean

From their onset as a proper research-in-action branch, multi-agent-system researches have been intrinsically drawn upon space perception and cognition (Ferber, 1999; Wooldridge, 2002; Weyns et al., 2007a). Ranging from interactive planning processes to the building up of shared visions and scenarios, interactive system architectures have been built trying to involve structurally human and/or artificial agents in the spatial domain and toward spatial themes. Particularly urban and regional environments have been dealt with, in different contexts (Arentze & Timmermans, 2006; Barkowsky et al, 2007; Khakee et al., 2008). Spatial ontologies prove to be decisive in such contexts, intensely space-based, particularly in their crucial role for the building up –and management- of systems relying on ICT architectures, with peculiar languages and rules.

The present chapter examines some results of space-based multi-agent processes, particularly in terms of spatial cognition and representation ontologies. Most of such experiences have been developed in real policymaking processes, with strong objective-oriented approach: therefore, spatial-cognition results are basically by-products of the main research context. Nonetheless, interesting insights can be drawn out from those experiences.

In particular, some of them have been carried out by the Planning Research Group of the Department of Architecture and Urban Planning (DAU) of the Technical University of Bari, Italy. A first group of research activities (1999-2003) was financed by the European Union, aiming at building up sustainable development scenarios in some Mediterranean Countries to support planning decisions. A second group was less homogeneous, concerning case-studies developed in southern Italian cities (2004-2008), aimed at setting up hybrid multi-agent architectures to allow human/artificial cognition exchanges.

In the first set of research initiatives—on the interchanges between agriculture and urbanization in Tunis, sustainable coastal management in Izmir, Turkey, and the impact of market globalization on local resources in Rabat, Morocco—interactive and community-based processes to build future (shared) scenarios were set up, dealing with socio-economic and environmental domains. The scenario-building activity was based on a variant to the *strategic choice* approach by Friend and Hickling (1997), as developed by subsequent evolutions of futures studies (Jungk & Mullert, 1996). Group working and reasoning was structurally embedded in the process methodology, allowing *au-paire*, democratic interactions among participants. This objective was pursued by setting up virtual 'rooms', which participants could access without the hierarchical or power constraints that are usually present in real meetings (Khakee *et al.*, 2002a). An ICT-based architecture was built with this purpose, so as to support reasoning interactions among diffused agents, both indoor and remote, using open internet platforms (Khakee *et al.*, 2002b; Barbanente *et al.*, 2007). In the second set of initiatives, interactive community-based sessions were set up in various towns in Apulia region, Southern Italy. Implementation fields and aims were different, but the vision-building process was basically the same, i.e., an ICT-based architecture aimed at supporting the cognitive exchanges of agents, so as to allow more aware decisions by public policymakers. Group reasoning was embedded in the process also in these cases, but interactions did not occur in virtual 'rooms' but in real face-to-face meetings, so being more influenced by social and other ties (Borri & Camarda, 2006; Camarda, 2010).

In this section, the attempt is to analyze the interactions occurred in those multi-agent group-reasoning sessions, in order to explore the mode of conceptualizing and representing the space by human agents, when exchanging spatial-environmental contents within a real planning process. In particular, some features of space-environment, as put down by branch literature (Goodman, 1951; Kelly & Bischof 2008; Twersky & Hard, 2009), were referred to, so as to infer possible spatial ontologies. Among the protocols analyzed, about 60 showed spatial cognition references, similar to the excerpt reported in Figure 1.

Within such protocol analysis, comparisons were carried out against couples of antonymic features that characterize different cognitions of the space-environment dealt with, namely substance/ornament (Goodman, 1951), beacon/landmark (Kelly & Bischof, 2008), geometry/function and sensitive/emotional (Lourenco & Huttenlocher, 2007; Shettleworth, 2009), Euclidean/non-Euclidean dimension (Hartley *et al.* 2004). As an outcome of such comparison, an ontological representation of spatial features was drawn out, first focusing on each protocol statement and then clustering features on similar spatial contexts. Examples of such process are respectively shown in Figures 2 and 3.

Figure 1. Excerpt from the interaction protocols

Izmir (Turkey) 2002					
The urban space in which I live does not show character of amenity, salubrity and livability because of the building up of waterfront, that does not let wind blow and refresh the area.	Because of serious domestic illegal waste water discharges pollution the urban space is perceived as a sewage dump and does not allow livability	When some friends come and visit me, they realize to be near to my urban living space when they see the chimneys and smell the industry, because of wind	In my city sometimes it is not clearly possible to distinguish if a space is urban, agricultural or natural	If the space is made up of few built volumes, not more than 5%, then the space can be still perceived as an agricultural environment. If not, it becomes unrecognizable or urban environment	The perception and the awareness of environmental spaces should be accepted mentally and felt by heart.

Tunis 2001	Rabat (Morocco) 2003		South Italian cities (2003-06)		
the urban environment must be a comfortable habitat, spacious and convenient	The presence of modern architectures is not necessarily in contrast with the urban spaces of the Moroccan tradition	Today urban space can is recognized through air pollution, waste and litter on the soil, car klaxons, vehicular congestion, absence of green areas.	The spaces of city often create alienation because they are felt as hostile by poor people, especially when full of landmarks	The football playground alters the perception of a unitarian urban space	Open space is perceived as such only by bounding and bordering urban areas

Figure 2. Features of space representation from the analysis of protocols (excerpt)

statement	type of feature													
	substance	ornament		beacon	landmark		geometry	function		euclidean	non-eucl.		sensitive	emotional
The urban space in which I live does not show character of amenity, salubrity and livability because of the building up of waterfront, that does not let wind blow and refresh the area.		•						•						•
The identification of spaces deserving environmental protection is not always univocal by external persons. Only people who live the space is able to perceive its structural significance					•		•				•		•	
My son does not recognize this area as an agricultural area. Because of rare but evident houses, he perceives a potentially urban environment, rather then an agricultural one					•									
Today urban space can is recognized through air pollution, waste and litter on the soil, car klaxons, vehicular congestion, absence of green areas.		•			•						•		•	
the urban environment must be a comfortable habitat, spacious and convenient		•			•			•			•			•
The football playground alters the perception of a unitarian urban space							•							

Figure 3. Features of space representation for a urban space (excerpt)

	TYPE OF FEATURE					
	Emotional	Landmark	Sensitive	Functional	Geometrical	Non-Euclidean
URBAN SPACE	poor people's sense of alienation	litter on the ground	vehicular congestion	presence of infrastructure	existence of complementary spaces, such as agriculture and industry	feeling of comfortableness
	low livability	absence of green areas	sounds of klaxons	services	variable configuration	existence of spaciousness
		historical and cultural aspects	air renewal (*pleasant u.s.*)	presence of new technologies		adequacy to people needs
		modern architectures	pollution (*degraded u.s*)	social services		
			low pollution (*future u.s.*)	presence of industry		
				absence of agriculture		
				abundance of water resources (*future u.s.*)		

In particular, protocol statements allow possible clustering of at least three spatial contexts: urban space, natural space and agricultural space. Figure 3 shows the example of urban space.

The picture coming from the above figures is of course incomplete and perhaps misleading in some cases. This is particularly due to its deductive origin from a process explicitly oriented to other aims, carried out as an ex-post activity by an external knowledge engineer. Nonetheless, some interesting considerations can be drawn out, which can be helpful in better understanding the building up of space perceptions by real agents involved in a planning process.

1. The groups of features are not mutually exclusive, but intersecting in many cases. This suggests that the process of construction of each space conceptualization starting from features is not always easy. Examples are the absence of agriculture, characterizing not only urban space but also industrial ones.
2. A prevalence of functional features seems evident, confirming the traditional concept of function-based spaces in spatial planning (eg., Rodwin & Hollister, 1984). However, the origin of such associations may be highly context depending. In our case studies, many agents are public officers whose daily activity regularly draws upon technical-functional semantics.
3. There is a diffused presence of features concerning the environmental pollution, which is certainly a sign of increased environmental sensibility. However, this occurrence is may be context dependent, too, because of the particular research field of action, explicitly pivoted on sustainable development.
4. Emotional and non-Euclidean features seem to be important keys for spatial perception. This seems to confirm interesting recent research activities (Damasio, 2005; Shettleworth, 2009; Tucker et al., 1999) and brings about the legitimate importance of non-measurable attributes in the representation of space.
5. The eccentric, not obvious use of some perception attributes of a given space-environment is very interesting. Examples are such images as 'existence of spaciousness' or 'litter on the ground' to represent a urban space. Further example is the allegedly inescapable need of using only emotional landmarks to perceive a natural environment –in that recalling some positions of deep environmentalism (Fischer, 2000; Golledge, 1998).
6. A final interesting aspect is the time dimension in perceiving spaces, and the dependence on different agent types, particularly in some kinds of features. For example, protocol often reports that young agents do not perceive agricultural landmarks when time has determined some residential transformations in the area. The agent/time viewpoint may open suggestive perspectives, especially when dealing with the dynamic representation of space in agent navigation (Denis & Loomis, 2007).

We must remember here that the collected set of spatial-cognition outcomes is an ex-post exercise of the analysis of the interaction protocols of all scenario-building experiences carried out. Spatial features resulted entirely from a multi-agent approach, in which arguments occur in reasoning processes carried out within (virtual or real) group contexts of discussion. There is no evidence about a possible similarity of outcomes in case of a single-agent approach, where each agent declines her conception of space features autonomously. Admittedly, although increasing literature is interested today in single-agent approaches about the theme of spatial cognition, its major focus is on memory and the role of creativity in organizing spatial activities and decisions, rather that spatial ontologies. An interesting field in this context is architecture, with significant and promising research perspectives at hand (Kluwe et al., 2003; Mallgrave, 2009).

In sum, outcomes seem to envisage interesting research perspectives in the field of the cognition of space, particularly under the viewpoint of possible knowledge-engineering applications.

That is why some aspects of spatial cognition have been explored experimentally, focusing on spatial architecture creativity and spatial navigation support. The next chapters will be devoted to such subjects.

Spatial Features and Primitives of a Navigated Environment

The Essential Background

In AI robotics, where spatial features have been long studied for operational planning purposes, there is a fundamental distinction between structured and unstructured (or ill-structured) spaces. Basically, it mirrors the distinction between spaces with simple geometries (characterized by elementary profiles, rare unexpected events, rare secondary components, rare decision needs) and spaces with complex geometries (characterized by composite profiles, frequent unexpected events, frequent secondary components and decision-demanding needs) (Georgiev & Allen, 2004; Kelly & Bischof, 2008; Dazinger & Rafal, 2009). It is quite understandable that a robot may carry out movements and the related learning process more easily in simple than in complex geometries, so implying more recognizable cognitive configurations. This occurrence seems rather evident, even if robots are able to typify a major part of the real world, and basically lack the difficulty of moving from structured to unstructured world, which is typical of human agents.

When shifting to humans, the situation changes substantially. In order to define a reasonably elementary space, we can consider a geometrically simple, empty, perhaps monodimensional space. This is, for example, an indoor, long and desert corridor, with a sequence of doors and/or windows, an origin and an end without lateral connections. This space possibly resembles the minimum arc of a graph, it is simple and can be navigated by the human agent without any particular attention.

A complex space for humans, on the other side, can be represented by a highly crowded context, with hardly recognizable multidimensional geometry. For example, this is an outdoor space, such as a community fair, without any clear form, origin, end, and requiring particular attention. A human agent moves randomly through it, risking being overwhelmed by the unclear paths of the crowd, or being afraid of losing herself, her friend or a child perhaps not yet (self-)trained to move in such complex space-environment. Such complex spaces may suggest agents to previously identify and memorize characteristic signs –such as a 'beacon' or a 'monument'-, aiming at replacing a 'structure' and/or a 'geometry' that are absent or unintelligible (de Hevia & Spelke, 2009; Gero & Tversky, 2009; Hirtle, 2003).

In a corridor of a school or university building a human agent-navigator moves routinely. If she has been already there, at least once (hopefully more than once, better if periodically), the significant space-environment and some 'objectives' are already in her memory, reinforced by iterated experiences. The experiment developed here does rely on a large class of engineering students at the Technical University of Bari, in Italy. Their daily routine consists in using a long and apparently amorphous corridor to reach their lecturers' offices for information about exams and lecture clarifications.

As Goodman (1951) puts down –with sceptical yet explorative reasoning- there is a possible distinction between 'essence' and 'ornament', when dealing with individual and/or social (multi-agent) human perceptions of the space-environment. In order to investigate on such distinction, an interesting experiment is to analyze the behaviour of a human agent-navigator who had never scanned that space-environment before. If she had been scanning that space before, then she could have easily memorized the execution of the task and the problem would be solved. In this case, the problems generated, concerning for example the management of the spatial memory connected to that task, would be different although similarly interesting.

Rather clearly, there are two diverse perspectives at hand. A first one is an individual perspective, in which the navigation task is carried out independently and autonomously by an isolated agent. An additional perspective is a social, multi-agent one, in which the navigation is carried out by a non-isolated agent, who is (directly or indirectly) assisted by an adviser-agent, so resembling the setting up of a dynamic organization of spatial reasoning and learning.

The experiment focuses on a human agent-navigator looking for and scanning a corridor serving the professors' rooms in a university department. The corridor is considered a well structured space-environment here, and the agent's activity can be instantaneous or can require some time. It can be an instantaneous activity because the navigator faces a structure of rooms served by a corridor, with no intermediate environments between the agent's starting point (the external space of origin) and endpoint (the internal space of destination). Conversely, some time can be needed to understand the building layout and the mutual relationships among its different parts (such as the professors' rooms, entrance etc.). In particular, if the localization of the corridor is not easily perceivable by incoming agents, or it is not helped by a human agent-adviser at the entrance, the human agent-navigator will execute the task with a *trial-and-error* approach (Whitehead & Ballard, 1991). As a matter of fact, the agent will perform an 'active wandering' in the space-environment in order to carry out the navigation task, possibly facilitated by a map in the department entrance.

Now we need to generalize a position problem. In fact, a big building could have a multiplicity of corridors, each of them serving a number of professors' rooms, and in that case the agent should previously find out the right corridor among that multiplicity. However, once the right corridor has been entered, and if it is simply shaped (for example, the corridor is not crossed in its mid-way by another transversal corridor with other rooms), the agent can perform simple actions. At the beginning of the corridor, she can ask an occasional agent-adviser the location of the desired room, and then she can walk through the entire corridor in order to find the room and the professor. Alternatively, if the human agent-navigator prefers to avoid asking an adviser, she can look at the usual series of nameplates with the name of each professor (particularly if doors are closed) or follow a previous direction by friends, or use a map available *in loco*, or find the professor with an active navigation—perhaps by glimpsing him through an ajar door or by hearing his voice—or ask occasional professors or persons in the corridor. Obviously, a normal context rationality will prevent the navigator from using the mere intuition, dream, fantasy, or a dice throwing after numbering all rooms.

In investigating the distinction between 'essence' and 'ornament' when navigating intelligently in a space-environment, sequences of actions can be identified from the above framework, able to envisage possible agent behaviours to be checked out in the experiment. They can be listed as follows:

i. the agent-navigator drives herself by using active navigation (i.e., meeting the professor);
ii. the agent-navigator drives herself by using localized signals (i.e., nameplates on doors);
iii. the agent-navigator drives herself by using geometrical signs (i.e., a departmental map with professors' names and rooms);
iv. the agent-navigator drives herself with the help of an accompanying or occasional human agent-adviser.

Action type (ii) suggests that an ornament (the nameplate) allows the retrieval of an essence (the room) in a continuum (the corridor). This situation is important, because the continuum would not be usefully qualified *per se* for the task execution. Should the nameplate be missing, the action type (iv) suggests that the agent-adviser at the corridor starting point could advise the navigator to number the rooms in her searching, being the desired room number ten on the left side. Also, the adviser could qualify the professor's room through the ornaments on the door, which denote that room within all rooms. With reference to the inherent particular task, the ornament is a non-secondary feature in this specific case -although being perhaps secondary with reference to the general task conceptualization.

Description of the Experiment

The questionnaire layout. As said before, a questionnaire survey was carried out to start the experimentation, and the layout of questions was organized as shown in Figure 4 (Borri & Camarda, 2009, p. 201). Although the questions lied all in the domain of interest of the study, some of them were put down with the peculiar aim of raising the agents' attention and comments on the main themes under the specific scope of analysis.

Figure 4. Excerpt from the survey framework

QUESTIONNAIRE

Situation 'A': You are at DAU (Town Planning Dept) and decide to visit a professor in that department, crossing the space in between.

Question A1: Find out the professor you want to know and declare the reasons for your visit.

Question A2: Describe the actions you carry out to reach the professor of your interest

Question A3: Describe in detail the "substantial elements" of the space in which you move, being of help or obstacle for your reaching the professor of your interest *("substantial elements" are intended as spatial elements and their physical qualities, or substances such as materials, dimensions, physical barriers/helpers etc...)*

Question A4: Describe in detail the "ornamental elements" of the space in which you move, conditioning the actions in your reaching the professor of your interest *("ornamental elements" are intended as objects, shapes, colours, lights, aesthetics etc...)*

Question A5: Describe in detail your general sensations and preferences concerning the "substantial elements" and the "ornamental elements" of the space in which you move.

Situation 'B': You are at DAU and decide to visit professor Selicato in his room, crossing the space in between.

Question B1: Describe the actions you carry out to reach the professor of your interest

Question B2: Describe in detail the "substantial elements" of the space in which you move, being of help or obstacle for your reaching professor Selicato.

Question B3: Describe in detail the "ornamental elements" of the space in which you move, conditioning the actions in your reaching professor Selicato.

Question B4: Describe in detail your general sensations and preferences concerning the "substantial elements" and the "ornamental elements" of the space in which you move.

The questionnaire was delivered to 260 students of a Town planning course program in the Technical University of Bari, from October 2008 to February 2009, mostly aging about 20. The survey took place through an Internet homepage linked to the institutional didactical portal of the Town Planning Department, and the related answers were collected by email.

In the end, the answers to the questionnaires are 180 (96 women and 84 men), that is about 70% of the entire student population of the course.

Some aspects of the resulted knowledge base. Following the literature on spatial representation and ontology (Egenhofer & Golledge, 1998; Kikiras et al., 2006; Amedeo et al., 2008), a major aim of this exercise was to find out information on some important issues needed to navigate in such a simple spatial environment as a university corridor. In this context, the answer protocols were analyzed with a double approach, aiming at achieving a reliable robustness of results, given the large quantity of data gathered. First, a statistical analysis was carried out, concerning the answer texts, the categories of questions and the gender profiles of the participants. 'Essences' and 'ornaments' were identified within A-type answer (known target perspective) and B-type answer (unknown target perspective). They were further cross-analyzed with the categories of 'landmark' and 'beacon' (Kelly & Bischof; 2008), for significant clues. Ad-hoc text-mining software (particularly Concordance 3.2 and WordStat 5.0) was used for statistical analysis, basically focusing on text keywords, i.e., their frequencies, deviations, grouping and clustering. Therefore, following literature warnings about the biases involved by the excerpting of words from organic discourses, a subsequent analysis aimed at contextualizing those findings in the text of the answer protocols (Warren & Gibson, 2002). This second analysis was developed without software support, by reading and rescuing answers page by page.

Due to the large quantity of texts and keywords, only the statistical analysis was developed on the entire population of the database elements. Text fine-reading was carried out as an iterated random check on population samples.

Within this general approach, an initial analysis aimed at understanding the agents' interpretations of spatial elements in navigating the environment. In particular, the 'essence' or 'ornament' nature of the elements was first investigated (Goodman 1951). In this concern, responses to questions A3 and B2 (essence), as well as A4 and B3 (ornament) were analysed and statistically described. A preliminary check revealed that separate A3/B2 and A4/B3 comparisons were not significant in this investigation. Therefore, essences and ornaments were analyzed as keywords occurrences in two aggregate groups, and subsequently verified in the actual response context.

A number of spatial elements occur in both groups with similar frequency. However, a check with the response context reveals that some words are used by agents just as a general reference to better explain/locate listed essences or ornaments. Examples of such cases are sentences like "...the entire *wall* of the *corridor* is an essence", or "... *pictures* on the *wall* are ornaments". Admittedly, this situation is intriguing per se, particularly highlighting the risks of acritically identifying lexical/logical ontologies with spatial ontologies.

The above situation, together with a search for similar words and lemmas, induced a simplification in the list of keywords, which is reported in Figure 5.

The table illustrates the results of an analysis clustered on the gender profiles of respondents. Such descriptors as standard deviation show that the two groups generally disagree on the nature of spatial elements, as statistically expected. This seems to be basically independent from the nature itself, except a slightly lower disagreement on ornaments, perhaps due to a misleading text in the questionnaire (Figure 6).

By looking at the two different profiles, their preferences do not substantially differ from each other, especially as far as ornaments are concerned.

Figure 5. Some statistical descriptors of essences (l) and ornaments (r)

	men	womer	sum	st.dev
WOOD	822	1825	2647	709,23
DOORS	1182	1079	2261	72,83
GLASS	591	1233	1824	453,96
FRAMES	539	1131	1670	418,61
CORRIDORS	796	694	1490	72,12
STEEL	462	1028	1490	400,22
HANDLES	334	1130	1464	562,86
SEGRETERIA	1002	462	1464	381,84
SPACE	797	539	1336	182,43
ROOMS	693	411	1104	199,40
LIBRARY	514	514	1028	0,00
WALLS	385	616	1001	163,34
OFFICES	796	205	1001	417,90
HALL	411	436	847	17,68
ROOMS	539	231	770	217,79
FURNITURE	359	308	667	36,06
ENTRANCE	436	128	564	217,79
STUDENTS	487	77	564	289,91
WC	154	308	462	108,89
WINDOWS	154	282	436	90,51
FLOOR	282	154	436	90,51
ZONE	308	102	410	145,66
INLET	180	180	360	0,00
STAIRS	256	77	333	126,57
NAMEPLATES	231	102	333	91,22
WATER	102	205	307	72,83
LABORATORY	102	205	307	72,83
CHAIRS	256	51	307	144,96
CORNER	154	102	256	36,77

	men	womei	sum	st.dev
DOORS	1440	720	2160	509,12
MODELS	976	771	1747	144,96
WALLS	848	565	1413	200,11
NOTICE BOARD	745	616	1361	91,22
CHAIRS	822	488	1310	236,17
FURNITURE	591	385	976	145,66
STUDY ROOMS	359	514	873	109,60
LIGHT	565	282	847	200,11
WOOD	513	308	821	144,96
FLOOR	565	205	770	254,56
TABLES	436	334	770	72,12
WINDOWS	411	308	719	72,83
PICTURES	411	308	719	72,83
PLANTS	385	308	693	54,45
GLASS	308	231	539	54,45
ENQUIRY DESK	385	128	513	181,73
NAMEPLATES	179	334	513	109,60
IMAGES	359	102	461	181,73
POSTERS	154	308	462	108,89
MARBLE	334	102	436	164,05
FURNITURE DOORS	179	257	436	55,15
PLASTIC	231	205	436	18,38
CORRIDOR	308	102	410	145,66
ENTRANCE	308	102	410	145,66
LIBRARY	282	102	384	127,28
ROOMS	102	257	359	109,60
DESKS	205	154	359	36,06
PHOTOS	102	231	333	91,22
BOOKS	154	179	333	17,68
OFFICES	282	25	307	181,73
SPACE	179	128	307	36,06
HALL	256	25	281	163,34
PATROL	51	205	256	108,89

Figure 6. Histograms of essences and ornaments

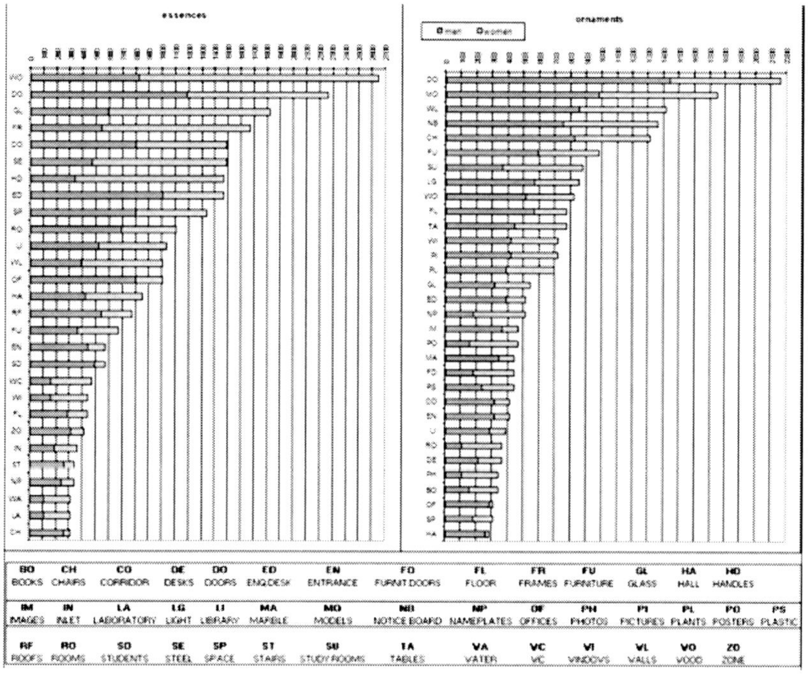

Figure 7. Most cited essences and ornaments in the survey

- Ornaments: **doors**, models, **walls**, notice board, chairs, **furniture**, **rooms**, light, **wood**, floor, tables, windows, pictures, plants, **glass**.
- Essences: **wood**, **doors**, **glass**, frames, corridor, steel, handles, enquiry room, space, **rooms**, library, **walls**, offices, hall, **furniture**.

Just a slight presence of perhaps more 'traditional' essences can be found in women (wood, frames, glasses) rather than in men, for mere accuracy sake. Overall, Figure 7 synthesizes a top-down ranked list of spatial elements (Borri & Camarda, 2009, p. 202).

It is evident that a number of elements are present both as ornaments and as essences (in bold in Figure 7), although the lists have been previously adjusted to minimize biases. This can be explained by the fact that each of those elements is recurrent and plays a major role in characterizing built spaces, both as primary structures and as ornaments.

An additional analysis on functional and geometrical features has been carried out, in order to work out the extent to which they have been used to describe the elements in the protocols, among a number of argumentative and/or qualitative features (Figure 8).

Overall, geometrical features are preferred to functional ones, with a significant prevalence of geometrical essences on geometrical ornaments. This might be due to a more established use of geometry to describe structures or structural parts of buildings, traditionally occurring in engineering schools. Also, it is interesting that women using geometric feature to describe ornaments are about 1/2 of men, whereas a similar number of women and men uses geometries in dealing with essences. A sample analysis of answer protocols confirms that qualitative features are used more often by women to identify and discuss ornamental elements of space.

A final analysis in the study has dealt with the landmark/beacon (L/B) features of spatial elements (Kelly & Bischof, 2008). This features had not been intentionally considered when preparing the questionnaire text. However, after looking at the responses, some important connections emerged with a number of experimentations recently carried out on similar subjects or situations (Gillner *et al.*, 2008). Therefore, it seems interesting to highlight some aspects at least

Figure 8. Functional and geometrical features

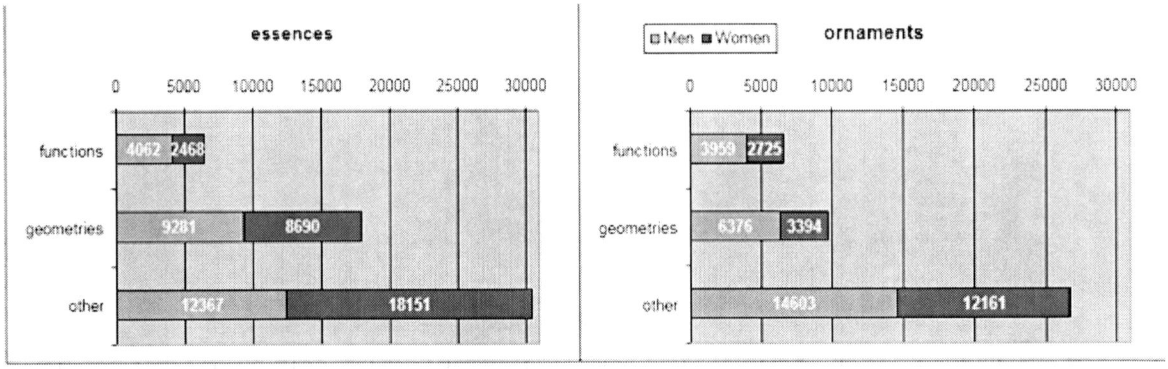

linkable to that dichotomic issue, particularly concerning the role of elements perceived by agents in navigating spatial environments. The fundamental issue to be verified is then the possibility that (perceived) spatial elements can be somehow considered either landmark or beacon features within a navigation activity.

As the questionnaire survey did not contain explicit questions addressing L/B characters, therefore indirect proxies were needed to investigate such issues. A fundamental difference between A and B questions was useful in this concern. In fact, in situation 'A' the navigation objective (a professor in her room) was surely known to respondents, who had selected that objective in answer A2. In the corresponding B1 question, the target professor is fixed and even unknown to many survey participants. This circumstance is shown in the elements cited by the agents in the different situations, described in Figure 9 (Borri & Camarda, 2009, p. 202).

In particular, situation 'A' seems then more suitable when looking for possible beacons for the navigation target. Specifically, occasional students (35%) and nameplates (15%) can reasonably act as beacons for the navigating agents, although addressing different features (in fact, nameplates are ornaments for few participants, whereas students encountered in the corridor are mainly considered as essences, particularly by men. See Figure 5). On the contrary, situation 'B' seems more suitable for the identification of navigation landmarks, being the target mostly unknown to agents and being therefore signs and indications critical for its reaching. Specifically, the enquiry desk (20%) and occasional students (45%) are mostly cited, followed by nameplates (10%) and departmental map (10%). Occasional students and enquiry desk are considered as essences in Figure 5 (the latter with greater emphasis), whereas the departmental map is cited only *en passant* in the answer protocols, just to stigmatize its ineffectiveness, as professors are not indicated.

Features form a basic flowchart for space navigation. The results of the above analysis induce further complex insights when compared to a simple summarizing model to support the target-oriented navigation of students, based on *IF-THEN-ELSE* logical rules. The previous discussions suggest that the agents can recognize some features of the simple space-environment investigated for navigation. These features are considered as 'essences' structuring the space, as well as 'ornaments' characterizing the same space. Therefore, such indications are sufficient to organize a simple expert system based on logical trees of rules, whose general layout is shown in Figure 10 (Borri & Camarda, 2009, p. 203).

The model is really simplified, in order to enhance its readability. It is divided into two parts, addressing the main activity, i.e., navigation support, and the needed environment maintenance. We are interested in the main activity, here, therefore we assume the user as an agent needing navigation support. Then the system asks the agent's gender, which is significant given the differences found in women/men protocols, for

Figure 9. Proxy features addressing 'landmarks' and 'beacons'

1. Situation 'A', question 2		2. Situation 'B', question 1	
• Nameplates	15%	• Nameplates	10%
• Direct knowledge	35%	• Direct knowledge	15%
• Enquiry desk	10%	• Enquiry desk	20%
• Students	35%	• Students	45%
• Map	5%	• Map	10%

Figure 10. A qualitative model of the navigation support system

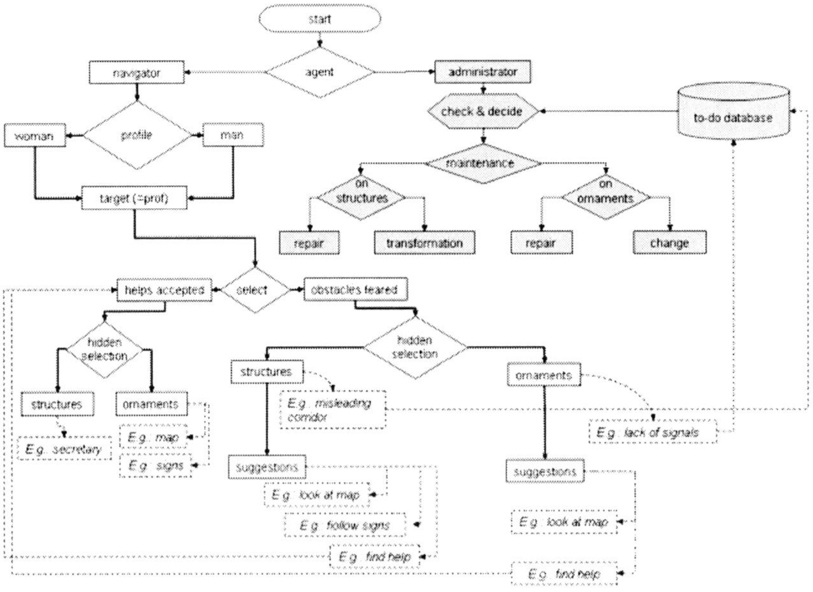

example in singling out essences and ornaments. Yet, gender differences are not considered in this paper, because of the basic nature of the study, and a unique gender profile has been instead considered.

Therefore, the agent selects her target (a professor in her room), and accesses the knowledge base previously built on the survey results. In order to ease navigation, the system asks if the agent prefers to learn about obstacles toward her target, or to accept hints and signs to reach it. A plausible condition of the first type is an agent who already has information on the target, and wants to avoid wasting time (beacon-like condition). A condition of the second type, instead, is characterized by less information and then a closer need for support (landmarks-like condition). We will now briefly describe the beacon-like condition, by looking at the actual web-browser page of the interaction module (Figure 11) (Borri & Camarda, 2009, p. 204).

The page shows a number of conditions of obstacles possibly occurring along the navigation. Each statement is essentially connected to the lists of space features –i.e., as 'essence' or 'ornament' nature-, previously delivered in the knowledge base. Of course, in order to avoid possible influences and misinterpretations, the nature of the features is not revealed to agents –although remaining associated to keywords as hidden characters in the knowledge base, in order to be used in subsequent steps. If the selection reveals the need of improving or repair the space-environment, then the system stores the selected condition in a ad-hoc database, for the administrators to carry out consequent activities.

Once feared obstacles have been singled out, the system compares them with the contexts where they had been conceptualized in the survey protocols, by drawing upon the knowledge base. This action may rely on either automated or semi-automated routines, by ad-hoc software for text mining (e.g., Concordance, Wordstat, TLab etc.). The reason for that comparison is to allow the retrieval of suggestions to overcome the obstacles, hidden in the protocols themselves (see Figure 10). If no suggestion can be drawn out, then a lack of information is recognized by

Figure 11. Browsing and selecting the different obstacles

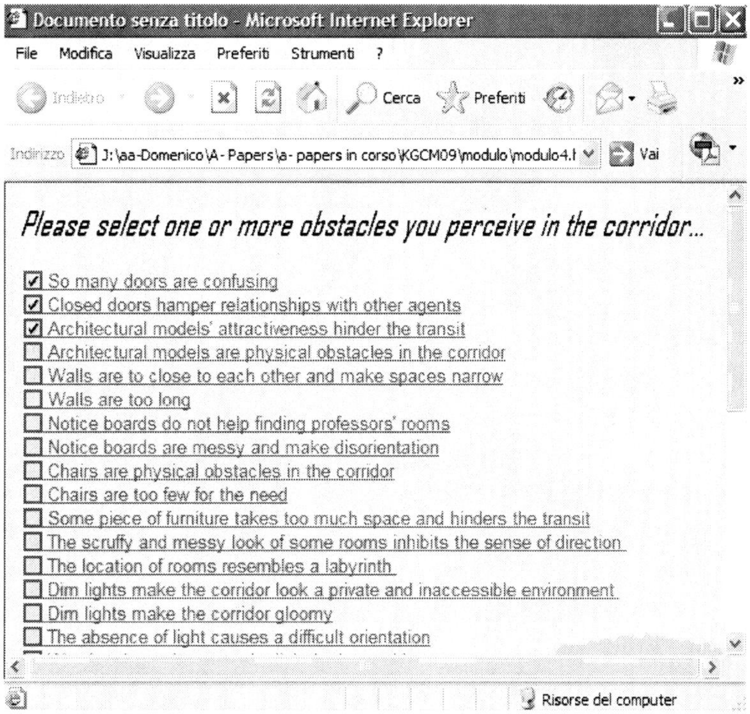

the system, which drives the agent to the second support condition, i.e., help needed (Figure 12) (Borri & Camarda, 2009, p. 205).

The landmarks-like condition support process is rather similar to the beacon-like one. In this case, the agent selects the "helps accepted" con-

Figure 12. Selecting the different help conditions

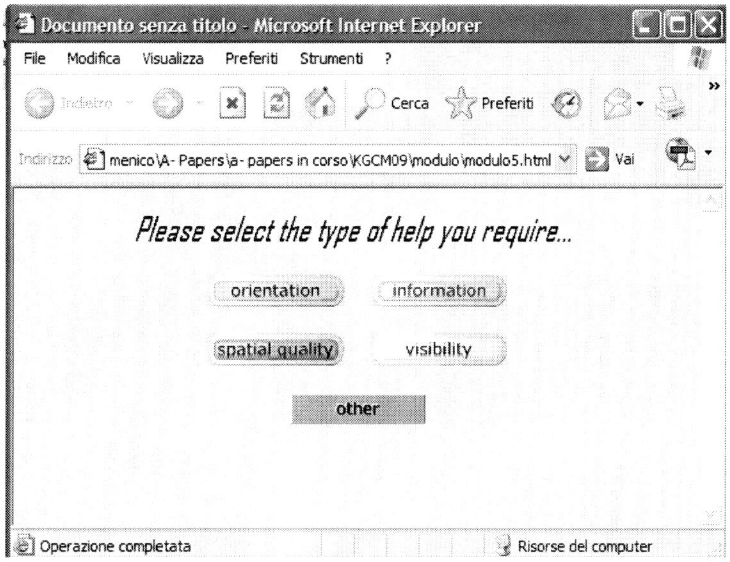

dition and moves forward in the process. Should she find no help from the automatic scanning of protocols, then a number of agents is singled out for possible support (enquiry desk, students etc.).

Both in landmarks-like and beacon-like conditions, the main database of clues and suggestions subsequent to the "obstacle feared" or "helps accepted" selection relies on the results of protocol scanning, but also on other sources. For example, external databases can be drawn upon, or experts in diverse knowledge domains (sociology, psychology, engineering, architecture, etc.) previously involved to fill up the main database. In both conditions, the process layout can be iterated progressively for different conditions and elements, in a Delphi-like fashion, until the agent's satisfaction (Khakee et al., 2002b).

The description above concerns the example of an agent-navigator: an agent-administrator instead, would follow the process developing toward maintenance-oriented paths.

As far as the system platform is concerned, the formal layout is today based on an architecture built on Corvid Exsys 5.0, integrated with text-mining software (Simstat 2.06 with Wordstat 5.0) and fine-tuned with Dreamweaver 8.0 in its html interface.

After carrying out the experiment and its subsequent analysis and modelization, the conceptualization of the corridor space seems more clearly connected to the perception and representation of separate features, whose presence, scarcity or lack is able to make more or less executable the navigation task. In particular, ornament and essence elements emerge from the analysis as important features of the space-environment considered, frequently using similar words for different features (Figure 13).

However, from a logical point of view, their sets do not seem completely disjointed from each other, being rather filled up with a number of common members. It is not easy to characterize each set logically, because this non-empty intersection is paralleled by exclusive members that are themselves not antonymic nor mutually exclusive. Hence, Goodman's (1951) scepticism about the mere transposition of the robotics distinction between essences and ornaments seems to be somehow confirmed. Yet, a fuzzy distinction still remains clear, emphasizing the degree of membership of terms to each set. Like in many

Figure 13. Feature sets and members

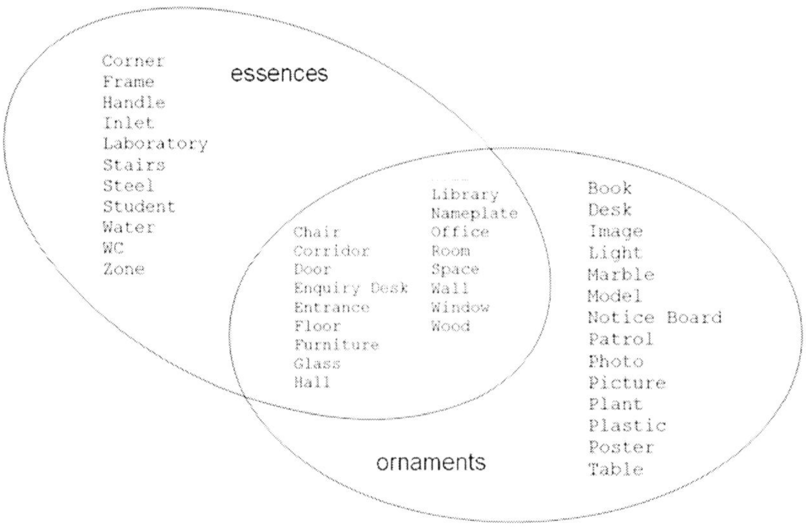

fuzzy logic applications, it is basically built around complementary concepts, such as physical core vs. decorative device, or structural vs. typifying attribute etc., which are still able to denote each set even in a complex way (Sanchez, 2006). This confirms the difficulty of inferring logical rules and constructs from robotics to human-based domains, but enhances the potentials of embedding complex reasoning in multi-agent DSSs (Aboutalib & Veloso, 2007; Weyns et al., 2007b).

Within a non-deterministic approach to handling complex spatial conceptualizations, space-environment features can be valuably dealt with, with useful perspectives. As a matter of fact, in our experiment, ornaments allowed the agent's finding out of essences (e.g., the professor's room) and enhanced the structuring degree of the corridor (therefore, its usability: Goodman, 1951) for the task execution. In turn, this basically confirms our introductory suggestion about the characterization of a continuum (the corridor), which would not be usefully qualified *per se* without feature search and retrieval.

In this concern, questionnaire responses allow some considerations about the question of structuring the space-environment. As mentioned earlier, robotics studies typically consider the environment in which a robot can perform its activities as a structured space, as opposed to unstructured spaces in which robots are disoriented (Aboutalib & Veloso, 2007). From this point of view, the present research has dealt with high degrees of space structuring, since the navigated space is a long simple corridor, with no intermediate environments between the agent's starting and end points. However, some elements mentioned by the questionnaire participants can decrease the degree of structuring significantly. The most evident element is certainly a geometrical one, representative of physical obstacles (doors, furniture, chairs) or physical shapes (recesses, narrowings, deviations). But some intrinsic features of the environment are evident as well. For example, agents realize modifications in the structuring degree of the corridor also due to non-geometrical features changes, such as floor material ("this section is not slippery because of the linoleum"), or wall colour ("this warm colour allows a better orientation"), or light ("where doors are not vitreous, the corridor is dark and disorienting"). This shows a comparable importance –then- in both geometrical and non-geometrical, qualitative characters for the real structuring degree of a space-environment. In the general process of spatial perception-cognition-decision, this would confirm the commonsense difference between a robot and a human agent navigating in an indoor space-environment, for the execution (perception, cognition, decision) of a specific task. Furthermore, this would confirm the importance of emphasizing (instead of reducing) the complexity of different environmental elements and features in spatial decisions –a rather important achievement in the domain of spatial planning.

Unlike the previous experiences carried out in urban or regional environments, this experimentation was addressed to single students, by using questionnaires. Therefore, outcomes in terms of spatial cognition features should not be considered as the result of a group-based interaction and reasoning. However, there was no restriction put on questionnaire filling up, concerning a single vs. cooperative analysis of the corridor in order to execute the requested navigation task. Admittedly, some students strolled around the corridor in groups, and many of them delivered pretty similar responses to the survey. Therefore, even if the corridor experimentation should not be considered as a perfect single-agent navigation approach, the related interaction process can be represented as a hybrid group reasoning approach. Yet, the database built on responses can be well considered a multi-agent-generated knowledge base, because of its multiple-source acquisition approach (Wooldridge, 2002).

At the present stage, the aim of this experimentation has not been to build up a decision-support system (DSS) for the navigation toward reaching the room of a professor. However, the study allowed the setting up of a navigation-support layout based on simple logical if-then rules, aimed at helping reaching a simple spatial objective. Therefore, the actual building up a real system architecture to support spatial navigation in simple environments could be a fair perspective of future research.

DISCUSSION AND FINAL NOTES

The representation of the space of landscapes and townscapes is nowadays considered as a multifaceted issue. Its complexity is increasingly recognized to be an intrinsic and non-reducible character of its very core. Therefore, being perceived under several viewpoints, space is not identifiable as a single, static and a-priori snapshot for a whole community, as traditionally intended. In fact, it does vary greatly among community members, either expert or commonsense agents, and throughout a number of situational occurrences and viewpoints. It also varies with time, but with features that are not always so obvious as traditionally expected (Pouget *et al.*, 2002; Day, 2008).

The research shown in this paper is carried out in different spatial contexts. The set of experimentations in the Mediterranean deals with urban regions or macro-regional contexts. In this case, spaces conceptualized by agents are made up of different entities, diverse and multidirectional trajectories, depicting arcs that are mutually interconnected geometrically, as well as functionally and perceptively. According to logical and topological literature (Steen & Seebach, 1995; Simon, 1996; Shirali & Vasudeva, 2005), such space is an ill-structured entity overall, whose structuring degree, for representation or navigation aims, strongly depends on a number of quali-quantitative relevant features. The research tried to single out some of them, drawing on group-reasoning-based and multi-agent perceptions and drawing out cooperative conceptualizations of space. On the other hand, the scanning of a departmental corridor allowed a sort of fine-tuning of some features, introducing a drastic decrease in the navigation scale. In it, the space-environment analysed is clearly a highly structured (although not completely) (Shirali & Vasudeva, 2005) space, especially when compared to an urban region. In this framework, the results of both experimentations can be integrated toward ontological representations of different spacescapes, perceived and conceptualized by using a multi-agent, navigation-based approach.

Figure 14. Map of the space-environment ontology (SpaceElement class is partial: all elements are reported in Figure 5)

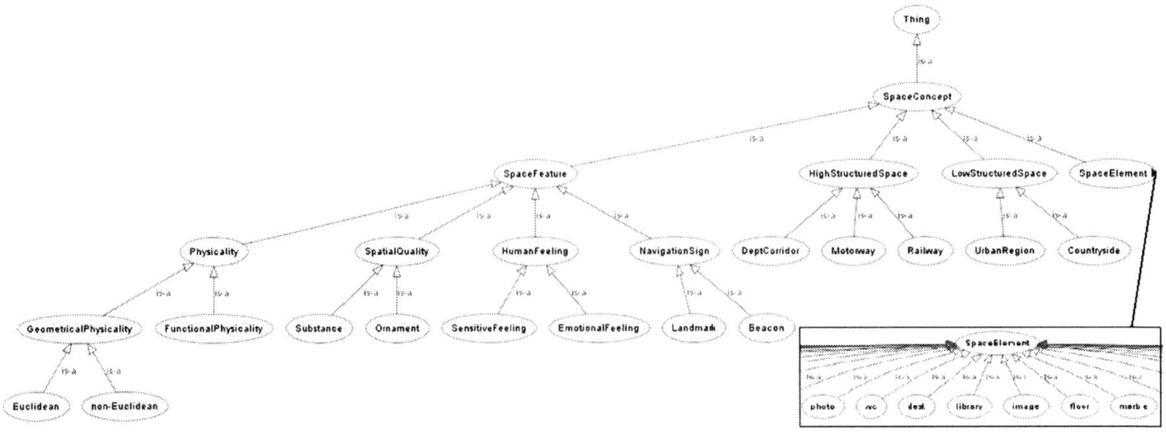

An attempt was carried out in this sense, at the end of the experimentation sessions, by using the Protégé 4.0 software by Stanford University. The ontology starts from a simple concept tree, where the ontologies induced by the two experimentations are integrated ("merged") and represented in classes and subclasses, which are linked by *IS-A* logical properties (Figure 14).

The ontology is written in web ontology language (OWL 2.1), which is a markup language used to represent meanings and semantics through shared vocabularies and relationships between verbal terms. The explicit objective of OWL is to allow the processing of human-generated information through software agents rather than merely by human agents (Lacy, 2005). The Space-environment ontology is then suitable to be processed by reasoning and query engines for decision support about space navigation or maintenance, as represented in previous sections. A snapshot of the Protégé framework built for that aim is reported in Figure 15.

Of course, the above ontological description is a simplified proxy of the highly complex cognitive conceptualizations involved when dealing with space perception and representation in real spacescapes.

Fortunately, under a planning or decisionmaking perspective, this complex structure of space perception is increasingly considered by DSSs as a richer framework to allow more aware decisions. Therefore, DSS architectures have started embedding approaches and methodologies, in order to grasp such multifaceted representation of spatial perceptions. Integrated attempts have been difficult and with mixed results, to date, in that they still fail to provide a rich but manageable synthesis to decisionmakers in need of support (Borri & Camarda, 2006).

Under an environmental planning perspective, however, the involvement of a plurality of agents with multiform conceptions about space still remains essential. The knowledge raised and exchanged in real group-reasoning processes represents a critical issue to be managed in order to draw and attain realistic development scenarios for communities. The re-composition of the knowledge–action dichotomy in spatial planning, long

Figure 15. The space-environment ontological representation with Protégé 4.0

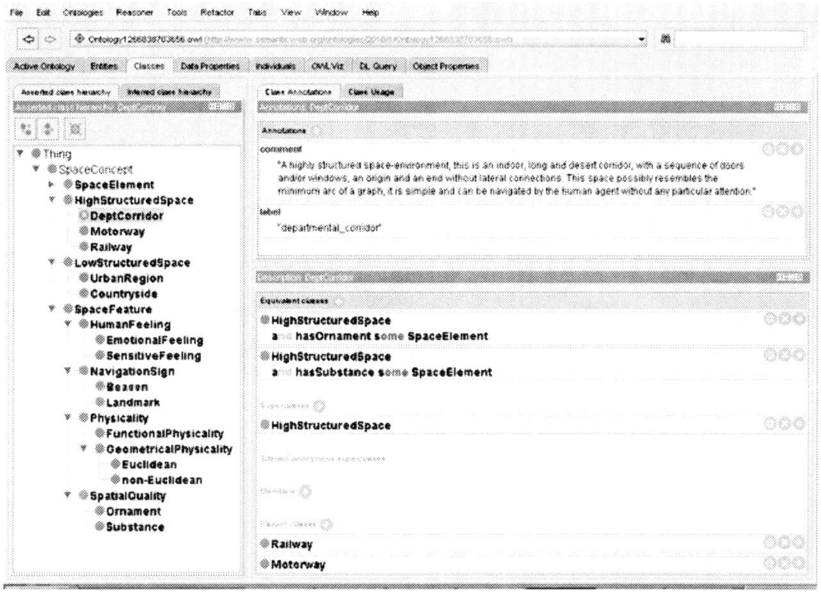

discussed by literature (Friedmann, 1987; Schon, 1983; Forester, 1999), can find a catalytic layout in multi-agent DSS, where interactions enhance knowledge-intensive and action-oriented planning processes. In this framework, research aiming at involving and improving spatial cognition by group reasoning represents a critical effort toward effective management, besides its contingent and still partly unresolved difficulties.

The experience in the Mediterranean context is very explicit in this sense, sketching out multiform spatial perceptions and consequent categorizations, in turn involving multiform possibilities for DSS in enhancing development scenarios. Needless to say, the aim of this study is not the building up of a decision-support system for navigation. There is, of course, an increasingly enormous number of alternative ways to construct an expert system to aid navigation in the way proposed, nowadays. Instead, the chapter has reflected on space ontologies as elements and features able to strongly influence and even shape (at least) some spatial initiatives of the agents. The possibility of singling out logical if-then rules has proved useful for the drawing out of navigation-support layouts to help the reaching of multi-agent spatial objectives.

However, the cognitive interactions of the agents in reflexive-creative tasks, as for example depicted in specific planning literature (Schon, 1983; Argyris, 1982), would benefit from system architectures based on spatial ontologies. Far from being depressed in narrow patterns, professional creativity would in fact focus its creative effort on organizing originally and effectively spatial concepts and primitives from ontologic libraries and database previously collected. In this complexity-intensive context, building up an actual system architecture based on spatial ontologies could have a fair motivation in environmental planning. Therefore, it could be a fair perspective for future research.

NOTE

The present chapter has been conceived jointly by the authors. Nevertheless, D.Borri wrote sections 1 and 2.2.1, D.Camarda wrote the remaining sections. Authors are grateful to A.Barbanente, M.Cera, A.DeLiddo, L.Grassini, A.Khakee, M.Patano and M.Puglisi for their cooperation in the research activity that made the present chapter possible.

REFERENCES

Aboutalib, S., & Veloso, M. (2007). Towards using multiple cues for robust object recognition. In *Proceedings of AAMAS'07*. Honolulu, Hawaii. May 14–18.

Amedeo, D., Golledge, R. G., & Stimson, R. J. (2008). *Person-environment-behavior research: Investigating activities and experiences in spaces and environments*. New York: Guilford Press.

Arentze, T., & Timmermans, H. (2006). Multi-agent models of spatial cognition: Learning and complex choice behavior in urban environments . In Portugali, J. (Ed.), *Complex artificial environments*. Berlin: Springer. doi:10.1007/3-540-29710-3_12

Argyris, C. (1982). *Reasoning, learning, and action*. San Francisco, CA: Jossey-Bass.

Barbanente, A., Camarda, D., Grassini, L., & Khakee, A. (2007). Visioning the regional future: Globalisation and regional transformation of Rabat/Casablanca. *Technological Forecasting and Social Change*, *74*, 763–778. doi:10.1016/j.techfore.2006.05.019

Barkowsky, T., Knauff, M., Ligozat, G., & Montello, D. R. (Eds.). (2007). *Spatial cognition: Reasoning, action, interaction*. Berlin: Springer.

Borri, D., & Camarda, D. (2006). Visualizing space-based interactions among distributed agents: Environmental planning at the inner-city scale. *Lecture Notes in Computer Science*, *4101*, 182–191. doi:10.1007/11863649_23

Borri, D., & Camarda, D. (2009). The cooperative conceptualization of urban spaces in AI assisted environmental planning. *Lecture Notes in Computer Science*, *5738*, 197–207. doi:10.1007/978-3-642-04265-2_29

Camarda, D. (2010). Beyond citizen participation in planning: Multi-agent systems for complex decision making . In Nunes Silva, C. (Ed.), *Handbook of Research on E-planning*. Hershey: IGI Global. doi:10.4018/978-1-61520-929-3.ch010

Courrieu, P. (2005). Function approximation on non-Euclidean spaces. *Neural Networks*, *18*, 91–102. doi:10.1016/j.neunet.2004.09.003

Damasio, A. (2005). *Descartes' error: Emotion, reason, and the human brain*. London: Penguin.

Danziger, D., & Rafal, R. (2009). The effect of visual signals on spatial decision making. *Cognition*, *110*, 182–197. doi:10.1016/j.cognition.2008.11.005

Day, S. B., & Bartels, D. M. (2008). Representation over time: The effects of temporal distance on similarity. *Cognition*, *106*, 1504–1513. doi:10.1016/j.cognition.2007.05.013

de Hevia, M. D., & Spelke, E. S. (2009). Spontaneous mapping of number and space in adults and young children. *Cognition*, *110*, 198–207. doi:10.1016/j.cognition.2008.11.003

Denis, M., & Loomis, J. M. (2007). Perspectives on human spatial cognition: Memory, navigation, and environmental learning. *Psychological Research*, *71*, 235–239. doi:10.1007/s00426-006-0079-x

Denis, M., Michon, P. E., & Tom, A. (2006). Assisting pedestrian wayfinding in urban settings: Why references to landmarks are crucial in direction-giving . In Allen, G. L. (Ed.), *Applied spatial cognition: From research to cognitive technology* (pp. 25–51). Mahwah, NJ: Erlbaum.

Egenhofer, M. J., & Golledge, R. G. (1998). *Spatial and temporal reasoning in geographic information systems*. Oxford: Oxford University Press.

Ferber, J. (1999). *Multi-agent systems: An introduction to distributed artificial intelligence*. London: Addison-Wesley.

Fischer, J. (2000). *Citizens, experts, and the environment: The politics of local knowledge*. Durham, NC: Duke University Press.

Forester, J. (1999). *The deliberative practitioner: Encouraging participatory planning processes*. Cambridge, MA: MIT Press.

Friedmann, J. (1987). *Planning in the public domain: From knowledge to action*. Princeton, NJ: Princeton University Press.

Garling, T., & Evans, G. W. (Eds.). (1991). *Environment, cognition, and action: An integrated approach*. New York: Oxford University Press.

Georgiev, A., & Allen, P. K. (2004). Localization methods for a mobile robot in urban environments. [TRO]. *IEEE Transactions on Robotics and Automation*, *20*, 851–864.

Gero, J. S., & Tversky, B. (Eds.). (1999). *Visual and spatial reasoning in design*. Sydney, Australia: University of Sydney, Key Centre of Design Computing and Cognition.

Gillner, S., Weiss, A. M., & Mallot, H. A. (2008). Visual homing in the absence of feature-based landmark information. *Cognition*, *109*, 89–104. doi:10.1016/j.cognition.2008.07.018

Golledge, R. G. (Ed.). (1998). *Wayfinding: Cognitive mapping and other spatial processes*. Baltimore, MD: Johns Hopkins.

Goodman, N. (1951). *The structure of appearance*. Cambridge, MA: Harvard.

Hartley, T., Trinkler, I., & Burgess, N. (2004). Geometric determinants of human spatial memory. *Cognition*, *94*, 39–75.

Hirtle, S. C. (2003). Neighborhoods and landmarks . In Duckham, M., Goodchild, M. F., & Worboys, M. F. (Eds.), *Foundations of geographic information science* (pp. 191–230). London: Taylor & Francis. doi:10.1201/9780203009543.ch10

Jungk, R., & Mullert, N. (1996). *Future workshop: How to create desirable futures*. London: Institute for Social Inventions.

Kelly, D. M., & Bischof, W. F. (2008). Orienting in virtual environments: How are surface features and environmental geometry weighted in an orientation task? *Cognition*, *109*, 89–104. doi:10.1016/j.cognition.2008.07.012

Khakee, A., Barbanente, A., Camarda, D., & Puglisi, M. (2002b). With or without?: Comparative study of preparing participatory scenarios using computer-aided and traditional brainstorming. *Journal of Future Research*, *6*, 45–64.

Khakee, A., Barbanente, A., & Puglisi, M. (2002a). Scenario building for Metropolitan Tunis. *Futures*, *34*, 583–596. doi:10.1016/S0016-3287(02)00002-2

Khakee, A., Hull, A., Miller, D., & Woltjer, J. (Eds.). (2008). *New principles in planning evaluation: Urban planning and environment*. Aldershot, UK: Ashgate.

Kikiras, P., Tsetsos, V., & Hadjiefthymiades, S. (2006). Ontology-based user modeling for pedestrian navigation systems. *ECAI 2006 Workshop on Ubiquitous User Modeling (UbiqUM)*, Riva del Garda, Italy.

Kluwe, R. H., Lüer, G., & Rösler, F. (Eds.). (2003). *Principles of learning and memory*. Basel, Switzerland: Birkhäuser.

Lacy, L. W. (2005). *OWL: Representing information using the Web ontology language*. Victoria, Canada: Trafford Publishing.

Lew, A. R., Foster, K., Bremner, J. G., Green, M., & Slavin, S. (2005). Detection of geometric but not topological transformations in 6-12-month-old infants in a visual exploration paradigm. *Developmental Psychobiology*, *47*, 31–42. doi:10.1002/dev.20075

Lloyd, R. (2009). *Spatial cognition: Geographic environments*. Amsterdam: Springer-Verlag.

Lourenco, S. F., & Huttenlocher, J. (2007). Using geometry to specify location: Implications for spatial coding in children and nonhuman animals. *Psychological Research*, *71*, 252–264. doi:10.1007/s00426-006-0081-3

Lynch, K. (1960). *The image of the city*. Cambridge, MA: The MIT Press.

Mallgrave, H. F. (2009). *Architect's brain: Neuroscience, creativity, and architecture*. London: Wiley-Blackwell.

Mockler, R. J. (1989). *Knowledge-based systems for management decisions*. Upper Saddle River, NJ: Prentice-Hall.

Pouget, A., Ducom, J. C., Torri, J., & Bavelier, D. (2002). Multisensory spatial representations in eye-centered coordinates for reaching. *Cognition*, *83*, B1–B11. doi:10.1016/S0010-0277(01)00163-9

Rodwin, L., & Hollister, R. M. (1984). *Cities of the mind: Images and themes of the city in the social sciences*. Berlin: Springer.

Sanchez, E. (Ed.). (2006). *Fuzzy logic and the semantic Web*. Amsterdam: Elsevier.

Schön, D. A. (1983). *The reflexive practitioner*. New York: Basic Books.

Shettleworth, S. J. (2009). *Cognition, evolution, and behavior*. New York: Oxford University Press.

Shirali, S., & Vasudeva, H. (2005). *Metric spaces*. London: Springer-Verlag.

Simon, H. (1996). *The sciences of the artificial*. Cambridge, MA: The MIT Press.

Steen, L. A., & Seebach, J. A. (1995). *Counterexamples in topology*. New York: Dover Publications.

Tucker, D. M., Hartry-Speiser, A., McDougal, L., Phan, L., & Degrandpre, D. (1999). Mood and spatial memory: Emotion and right hemisphere contribution to spatial cognition. *Biological Psychology*, *50*, 103–125. doi:10.1016/S0301-0511(99)00005-8

Tversky, B., & Hard, B. M. (2009). Embodied and disembodied cognition: Spatial perspective-taking. *Cognition*, *110*, 124–129. doi:10.1016/j.cognition.2008.10.008

Warren, T., & Gibson, E. (2002). The influence of referential processing on sentence complexity. *Cognition*, *85*, 79–112. doi:10.1016/S0010-0277(02)00087-2

Weyns, D., Omicini, A., & Odell, J. (2007a). Environment as a first class abstraction in multiagent systems. *Autonomous Agents and Multi-Agent Systems*, *14*, 5–30. doi:10.1007/s10458-006-0012-0

Weyns, D., Parunak, H. V. D., & Michel, F. (Eds.). (2007b). *Environments for multi-agent systems*. Berlin: Springer. doi:10.1007/978-3-540-71103-2

Whitehead, S., & Ballard, D. H. (2004). Learning to perceive and act by trial and error. *Machine Learning*, *7*, 45–83. doi:10.1007/BF00058926

Wooldridge, M. (2002). *An introduction to multiagent systems*. London: Wiley.

ADDITIONAL READING

Allen, G. L. (2007). *Applied spatial cognition: From research to cognitive technology*. New York: Erlbaum.

Casasanto, D., & Dijkstra, K. (2010). Motor action and emotional memory. *Cognition*, *115*(1), 179–185. doi:10.1016/j.cognition.2009.11.002

Damasio, A. R. (1999). *The feeling of what happens*. New York: Harvest Books.

Damasio, A. R. (2003). *Looking for Spinoza: Joy, sorrow, and the feeling brain*. New York: Harcourt.

Dolins, F. L., & Mitchell, R. W. (Eds.). (2010). *Spatial perception, spatial cognition*. New York: Cambridge University Press.

Dove, G. (2009). Beyond perceptual symbols: A call for representational pluralism. *Cognition*, *110*(3), 412–431. doi:10.1016/j.cognition.2008.11.016

Gardner, H. (1999). *Intelligence reframed: Multiple intelligences for the 21st century*. New York: Basic Books.

Hammond, K. (1989). *Case-based planning: Viewing planning as a memory task*. Boston: Academic Press.

Houde, O. (2003). *Dictionary of cognitive science; Neuroscience, psychology, artificial intelligence, linguistics, and philosophy*. New York: Psychology Press.

Jia, L., Hirt, E. R., & Karpen, S. C. (2009). Lessons from a faraway land: The effect of spatial distance on creative cognition. *Journal of Experimental Social Psychology*, *45*(5), 1127–1131. doi:10.1016/j.jesp.2009.05.015

Johnson-Laird, P. N. (1993). *Human and machine thinking*. Hillsdale, NJ: Lawrence Erlbaum.

Lei, P. L., Kao, G. Y. M., Lin, S. S. J., & Sun, C. T. (2009). Impacts of geographical knowledge, spatial ability and environmental cognition on image searches supported by GIS software. *Computers in Human Behavior*, *25*(6), 1270–1279. doi:10.1016/j.chb.2009.05.003

Lipinski, J., Simmering, V. R., Johnson, J. S., & Spencer, J. P. (2010). The role of experience in location estimation: Target distributions shift location memory biases. *Cognition*, *115*(1), 147–153. doi:10.1016/j.cognition.2009.12.008

Mallot, H. A., & Basten, K. (2009). Embodied spatial cognition: Biological and artificial systems. *Image and Vision Computing*, *27*(11), 1658–1670. doi:10.1016/j.imavis.2008.09.001

McCullough, S., & Emmorey, K. (2009). Categorical perception of affective and linguistic facial expressions. *Cognition*, *110*(2), 208–221. doi:10.1016/j.cognition.2008.11.007

Meusburger, P., Funke, J., & Wunder, E. (Eds.). (2009). *Milieus of creativity: An interdisciplinary approach to spatiality of creativity*. Heidelberg, Germany: Springer.

Minsky, M. L. (2006). *The emotion machine: Commonsense thinking, artificial intelligence, and the future of the human mind*. New York: Simon and Schuster.

Petrini, K., Russell, M., & Pollick, F. (2009). When knowing can replace seeing in audiovisual integration of actions. *Cognition*, *110*(3), 432–439. doi:10.1016/j.cognition.2008.11.015

Pluhar, Z. F., Piko, B. F., Uzzoli, A., Page, R. M., & Dull, A. (2010). Representations of the relationship among physical activity, health and perceived living environment in Hungarian urban children's images. *Landscape and Urban Planning*, *95*(4), 151–160. doi:10.1016/j.landurbplan.2009.12.012

Pollak, S. D., Messner, M., Kistler, D. J., & Cohn, J. F. (2009). Development of perceptual expertise in emotion recognition. *Cognition*, *110*(2), 242–247. doi:10.1016/j.cognition.2008.10.010

Richardson, D. C., Spivey, M. J., Barsalou, L. W., & McRae, K. (2003). Spatial representations activated during real-time comprehension of verbs. *Cognitive Science*, *27*(5), 767–780. doi:10.1207/s15516709cog2705_4

Ron, S. (2005). *Cognition and multi-agent interaction: From cognitive modeling to social simulation*. New York: Cambridge University Press.

Schank, R. C., & Childers, P. G. (1984). *The cognitive computer: On language, learning, and artificial intelligence*. London: Addison-Wesley Publishing.

Shanahan, M. (1997). *Solving the frame problem*. Cambridge, MA: MIT Press.

Shemyakina, N. V., & Dan'ko, S. G. (2004). Influence of the emotional perception of a signal on the electroencephalographic correlates of creative activity. *Human Physiology*, *30*(2), 145–151. doi:10.1023/B:HUMP.0000021641.41105.86

Uttal, D. H. (2001). Making sense of the development of spatial cognition. *Trends in Cognitive Sciences*, *5*(7), 316–317. doi:10.1016/S1364-6613(00)01663-6

Zeelenberg, R., & Bocanegra, B. R. (2010). Auditory emotional cues enhance visual perception. *Cognition*, *115*(1), 202–206. doi:10.1016/j.cognition.2009.12.004

Zetterberg, A., Mörtberg, U. M., & Balfors, B. (2010). Making graph theory operational for landscape ecological assessments, planning, and design. *Landscape and Urban Planning*, *95*(4), 181–191. doi:10.1016/j.landurbplan.2010.01.002

KEY TERMS AND DEFINITIONS

Agent: Human or artificial (or hybrid) element involved and acting in planning, decision-making or routinely activities.

Environmental Planning: The set up and organization of strategies and actions to achieve mainly social and economic objectives, with particular attention to environmental impacts and issues.

Multi-Agent Reasoning: An organization of agents in which exchanges of information and/or general cognitive contents occur.

Ornamental Attributes: Decorative, non-structural elements of a space, able to guide, hinder or to simply decorate a navigation task. Examples are objects, shapes, colours, lights, aesthetics.

Scenario Building: Setting up an interacting organization and environment aimed at cooperatively working out future scenarios for given areas and/or situations, particularly in social communities.

Spatial Cognition: The abilities of achieving, organizing, using, and amending knowledge contents dealing with spatial environments, enabling human agents to handle cognitive tasks in daily-life activities.

Spatial Ontology: A formal representation of spatial knowledge by using spatial concepts and mutual relationships in order to create a language-free ground to discuss about a spatial domain.

Substantial Attributes: Structural elements and physical qualities of a space, able to guide, hinder or to simply decorate a navigation task.

Chapter 16
Chewing the Communal Cud: Community Deliberation in Broadsheet Letters and Political Blogs

Jane Mummery
University of Ballarat, Australia

Debbie Rodan
Edith Cowan University, Australia

ABSTRACT

Contending that media users are more than consumers and that the mass media are able to achieve more in the public sphere than simply meet market demand, Mummery and Rodan argue in this chapter that some types of mass media may in fact fulfil public sphere responsibilities. The authors demonstrate how forums such as broadsheet letters to the editor and online political blogs—despite their commonly recognised limitations due to influence by private/commercial ownership, editorship, and the requirements of authorship—may exemplify, enable and support community deliberation over issues of public concern. More specifically, via engaging with Jürgen Habermas' conceptions of the necessary conditions for rational and communal deliberation, and critically examining recent debates in these forums, the authors argue both that these mediated forums can enable and exemplify community deliberation and, more generally, that community deliberation itself does not need to be strictly consensus-oriented to be productive.

INTRODUCTION

Communities need to chew their cud; they need their members to not only publicly and critically deliberate and debate what matters to them – whether via gossip, public meetings, social activism or democratic deliberation – but resolve disputes and come to rational and defensible agreed-upon decisions, decisions that in turn impact upon and establish a sense of collective agreement and, indeed, community. These kinds of communal and dialogic involvements of individuals in discussion and decision-making have also been recognized as key to practices of deliberative democracy (see, e.g., Bohman & Rehg, 1997; Chambers, 2003; Fishkin, 1991). Dewey (1927), for instance, has contended that inquiry and communication are the basis for democratic society and has stressed the

DOI: 10.4018/978-1-60960-091-4.ch016

merits of group deliberation over the decisions of any single individual. Such a view is also held by Jürgen Habermas, who additionally contends that the inquiry and communication making up group deliberation should be carried out in accordance with "norms of reasonable public dialogue" (Dzur, 2002, p. 323).

The question however, is how these "norms of reasonable public dialogue" should be understood, and how this process can best be supported, given it requires specific attributes from not only individual participants but social, political and technological systems and institutions. This chapter cannot hope to tackle all of these issues and their various nuances; instead it will concentrate on the role of one particular institution – that of the mass media – and how it can be understood as supportive of these norms. It will, in other words, consider how the mass media may exemplify, enable and support community deliberation. Just what we understand by 'community deliberation' here needs clarification. There are several alternative proposals according to which these terms can be understood, ranging from McMahon's (2001) conception of 'collective reasoning' to Yearwood and Stranieri's (2009) conception of 'reasoning community'. In this instance, however, whilst it seems that our discussion is indeed of a reasoning community, to use Yearwood and Stranieri's term, we will be drawing on the broad conceptions of deliberation and the public sphere, or *Öffentlichkeit*, developed and revised by Habermas (1989; 1998; 2006). Although these conceptions have often been critiqued for their utopian nature – critics seeing them as setting impossibly high and/or exclusionary standards for deliberation (e.g. Fraser, 1992; Schudson, 1997) – we contend Habermas provides a useful outline of the required procedures and standards for both a well-functioning public sphere and rational, communal deliberation. We do however accept that these procedures and standards operate in the way of ideals, a point highlighted by what we see as the most useful aspect of Habermas' delineation: that, being the statement of procedural conditions, it remains formal and empty of content.[1] On this basis, then, we will be drawing on Habermas' conceptions of the necessary conditions for rational and communal deliberation to evaluate two of the sites offered by the mass media for public deliberation: letters to the editor and political blogs.[2] First, however, we need to outline the relations envisaged by Habermas to hold between his conceptions of the mass media, the public sphere and public deliberation.

MASS MEDIA, THE PUBLIC SPHERE AND THE IDEAL OF DELIBERATION

Envisaged as "the space between civil society and the state in which citizens can debate issues of common concern" (Haas, 2004, p. 179), the Habermasian public sphere, as initially outlined in *The Structural Transformation of the Public Sphere*, is not only required to be a space in which all citizens can participate freely and equally, but free from explicit governmental or commercial interference. The Habermasian conception further rests on two key points. First, he assumes that the public should actively deliberate and not merely be "informed" (Haas, 2004, p. 179). Second, and relatedly, he assumes that such deliberation is necessary for quality popular participation in liberal-democratic society (Calhoun, 1992, pp. 1-4), and indeed that the outcomes of such deliberation can "in part hold governments to account" (Garnham, 2007, p. 206). The corollary of this is his assumption that decisions made on behalf of a democratic society should be neither simply handed down by some higher state authorities nor the result of "rubber-stamp[ing]" the needs of a capitalist economic system (Garnham, 2007, p. 206). Instead the society as a whole needs some way of collectively making decisions that affect the society as a whole. Citizens, in other words, need to make and debate "genuine choices" and nut-out genuine "dilemmas" (Garnham, 2007, p. 206), and this decision making process is an important part of the way in which liberal democracies function.

Given these points, Habermas contends that the aim of debate in the public sphere is to reach collective agreement on issues of public concern, with the corollary of this – common to all theorists of deliberative democracy – being the use of communication strategies that contribute to greater understanding and the chance for genuine collective agreement. In his view, then, communication strategies that attempt to manipulate and persuade for strategic aims, use propaganda or blackmail in attempting to coerce individuals can never be considered genuine attempts at public reasoning (Habermas, 1984; Dahlberg, 2005, p. 120). Appropriate communication strategies are rather those that prioritise and promote reasonableness and dialogue, with agreement developing "only as the result of more or less exhaustive controversy in which proposals, information and reasons can be more or less rationally dealt with" (Habermas, 1998, p. 362). Participants would thus ideally take part in public debate in rational, logical ways and would be not only upfront about their interests, bias, and prejudices (practices that may be quite different to those typically utilized in everyday debates and discussions), but committed to reflexivity, discursive equality and inclusion and what Dahlberg has called "ideal role-taking", by which he means a commitment to trying to see arguments from the perspectives of others (Dahlberg, 2001). These, then, are the norms in Habermas' eyes of any kind of reasonable public dialogue. More specifically, Habermas (1987) contends that in the public deliberation of reasoning communities, agreement on issues develops only after proposals are subjected to what he calls "*discursive verification*", a process whereby a proposed "truth claim is first rendered problematic and then redeemed" (pp. 361, 364)

As is noted above, this process requires truth claims to be "justified" not simply by participants on a personal level (where one would draw on one's personal experience, for instance, as many callers to talkback radio programs in Australia do), but also through a broader community-level process of dialogue and "argumentative reasoning" (Habermas, 1987, p. 364), according to which all participants would be "free to raise and challenge claims without fear of coercion, intimidation, deceit, or the like and … have equal chances to speak, to make assertions, self-presentations, and normative claims and to challenge others" (Warnke, 1995, p. 126). Under this process, participants are expected to a) recognize, raise problems about, and justify all proposed claims to truth, and b) accept that their individual preferences may be "mistaken: inaccurate, incoherent, or incomplete" (Dzur, 2002, p. 325). Indeed, for Habermas, truth claims can only be said to be redeemed at the public level if all "disputed claims" are so debated and redeemed (Warnke, 1995, p.126). Meaning first that, in properly deliberative communities, it is no more and no less the capacity of the better argument to emerge through reasoned dialogue that enables the possibility of consensus on specific truth claims. And second, that deliberation can best be understood as that "debate and discussion aimed at producing reasonable, well-informed opinions in which participants are willing to revise preferences in light of discussion, new information, and claims made by fellow participants" (Chambers, 2003, p. 309).

What becomes clear here is that Habermas' conception of deliberation can be read as stressing the procedural over any actual or envisaged content or indeed outcomes. As McCarthy (1994) puts it, he "leaves all substantive questions to the public use of reason itself" (p. 49). This is worth mention as it means that while particular outcomes are desirable – specifically communally developed and justified consensus – deliberation itself is not invalidated by the lack of any tangible outcomes. At the very least, such deliberation helps the community determine what issues are of communal concern and what the relevant factors are for constructive discussion. It also helps community members to express the public good they support (Turner, Tomlinson & Pearce, 2006, p. 117). Habermas' conception also reminds us that all outcomes are in fact partial and contingent, open to further debate and revision. That is, while they are useful for the community that discursively constructed and verified them, they cannot claim universality in any absolute sense (Knops, 2007). This, in turn, makes clear that deliberative quality rests on the meeting of procedural conditions, specifically "the presumption that reasonable or fair results are obtained insofar as the flow of relevant information and its proper handling have not been obstructed" (Habermas, 1998, p. 296).

Finally, it is worth noting that despite the claims of a number of critics (e.g. Buckingham, 1997; Peters, 1993; Phillips, 1996) that he over-prioritises the rational, creating "untenable distinctions between rationality and irrationality, reason and emotion" (Haas, 2004, p.179), Habermas' proposed process for public reasoning does make room for the "aesthetic-affective aspects of discourse" (Dahlberg, 2005, p. 116). For instance, public reasoning can absolutely involve "intuition and imagination" and draw on "feeling" (see Lash cited in Dahlberg, 2005, p. 117). It can enable the emergence of an ethic of "fairness" (see Benhabib cited in Dahlberg, 2005, p. 117), and the contesting of claims can certainly "be articulated and reasoned with a degree of passion and commitment" (Dahlberg, 2005, p. 117). In other words, public reasoning – under Habermas' model – can include embodied and empathetic perspectives; and this capacity we consider to be instrumental to understanding not only community deliberation but the way the mass media can exemplify, enable and support such deliberation.

These points, of course, mean that reasoning communities that want to achieve consensus on issues of public concern have to create the conditions necessary for productive structures of communication. And, obviously, for this process to function well, one key condition is that participants have access to relevant and accurate information. After all, it is only when citizens have this information that they can so "engage in rational-critical deliberation" (Haas, 2004, p. 179). But how do citizens get the pertinent information they need to deliberate? Habermas and others claim that the public communications enabled by the mass media are one of the key contributors to the discursive makeup of the public sphere (Habermas, 2006, p. 9). (A claim supported by the fact that the typical remit of the public broadcasting media, for instance, is to educate and inform the populace.) He explains their role:

A dispersed public interconnected almost exclusively through the electronic media can keep up to date on all kinds of issues and contributions in the mass media with a minimum attention, even in fleeting moments during the day, in small, private circles. People can take affirmative or negative positions on issues, and they do this implicitly all the time. In this way, they contribute to evaluating competing public opinions, if not to their articulation. Public communication operates as a hinge between informal opinion-formation and the institutionalized processes of will-formation – a general election or a cabinet meeting, for example. For this reason, the discursive constitution of the public sphere is important (Habermas, 2006, p. 9)

So, for Habermas, the mass media can positively contribute to the open critical debate and rational deliberations required in the public sphere, both through providing full and accessible reportage and by remaining independent of crude economic interests and state control. However many writers over the last twenty years – including Habermas himself in *The Structural Transformation of the Public Sphere* – have questioned whether the mass media can in fact achieve this goal given its weakening by a public relations and advertising ethic, a weakening Habermas (1989) describes as the abandonment of consensus negotiated via the "criteria of rationality" for the "consensus created by sophisticated opinion-molding" (p. 195). Nonetheless, despite this concern, many media theorists, analysts and politicians do see the role and responsibility of the media as much more than simply meeting market demand and, thereby, media users as more than consumers (e.g. Schultz, 1994; Jacka, 2000; Raboy, Abramson, Proulx, & Welters, 2001). The media is rather portrayed as playing a social role – as having a responsibility to audience members as national citizens, voters, members of a public sphere etc.– and it is this which gives rise to the still broader question that underpins this chapter: whether the media can play a more active role in the public sphere than simply the provision and evaluation of information.

One possibility is that the mass media could and should provide a venue for public deliberation itself. Here theorists and practitioners of what has been commonly called 'public journalism'[3] argue that the "promotion of reasonable and informed dialogue among citizens should be part of the role journalists play in a democracy" (Dzur, 2002, p. 313). This, they suggest, could be achieved in a number of ways. It might entail the sponsoring and convening real and virtual public meetings and community discussion groups, carrying out polls and surveys of the community in order to connect editorial and reportage decisions with community interests, and even the partnering of citizens with journalists in gathering the news. Less controversially,[4] it might be achieved simply via the provision of such forums for public deliberation as letters to the editor and, more recently, online blogs (other examples would include talkback radio and edemocracy sites, etc.). Indeed it is these kinds of forums that, we suggest, despite their commonly recognized limitations, may represent effective instances in the media of the kinds of public deliberation of issues that Habermas sees as necessary to both a community and deliberative democracy. It is to consideration of letters to the editor and online political blogs as sites of public deliberation that we now turn.

LETTERS TO THE EDITOR AND PUBLIC DELIBERATION

As a space for genuine public deliberation, the letters forum is not an uncontroversial proposal. Despite the fact that letters to the editor do appear to be "widely read and highly valued" (Raeymaeckers, 2005, p. 200), and can certainly be envisaged as part of a well functioning public sphere, they are typically seen as a highly constructed and mediated sphere due to four overlapping points:

1) Media ownership is often seen as impacting upon editorial decisions as to the selection and publication of letters (Bromley, 1998, p. 150). Added to this point is the issue of a highly concentrated newspaper media ownership. In Australia, for example, 90% of the print media is owned by News Ltd., a subsidiary of News Corporation, and Fairfax. Such a concentrated media environment suggests two possible outcomes: "alternative coverage" may become more restricted (Hessing, 2003, p. 27); or there will arise a "close interaction between media and political interests" (Ward, 1995, p. 131).
2) The second factor is that the media is a business concerned with competition and profit (Schultz, 1994, p.18). This can have two effects. First, the selection of letters for publication may be based on what the newspaper thinks best represents the interests and concerns of their readership (e.g. Sotillo & Starace-Nastasi, 1999; Wahl-Jorgensen, 2002b). Second, letters may well be seen as needing to "satisfy the news values of conflict and negativity" (Wahl-Jorgensen, 2004, p. 98; Raeymaeckers, 2005, p. 218; Hessing, 2003, p. 36; Richards & King, 2000, p. 483), and are therefore selected accordingly. This also means on the flip side, of course, that letters discussing issues "that are not currently being discussed" in the community

are unlikely to be published (*The Media Report*, 2001).

3) The third factor is again an issue to do with editorial selection and intervention, this time to do with form and style. More specifically, letters are typically selected by letters' editors on the basis of four rules: they need to be well written, grammatically correct, concise and persuasive (Wahl-Jorgensen, 2001, 2002b). Further, letters may be edited to remove overly emotional or insulting phrases, or shortened to make them "punchy and more radical" (Raeymaeckers, 2005, p. 219).
4) The fourth concern is that research has found that most letter writers are Anglo white males, with many letters pages showing little inclusion of women or those from non-Anglo ethnic and race backgrounds. Further, letter writers tend to be "firmly situated in a community and have the excess time and energy required for a commitment to political activism" (Wahl-Jorgensen, 2004, p. 101; 2002b, p. 77; cf. Hart, 2001; Sparks & Perez, 1991).

Given these various factors, the question needs to be asked again: to what extent can letters to the editor really exemplify community deliberation? This is a vexed question. Certainly, given the above issues, it is clear that we cannot accept the view (apparently typical to many letters' editors) that letters to the editor instantiate a "wide open public forum" (Wahl-Jorgensen, 2002a, p. 130). At the same time, however, to dismiss the letters forum as too mediated to allow or represent 'real' deliberation seems just as problematic. Individual letters, after all, do reflect real concerns and the letters forum does represent instances of public debate however limited (Lynn & Lea, 2003). Active citizens do use this forum to express their opinions on public issues of common concern; politicians and readers do read the letters page. Letters in other words do act as a "barometer of the [everyday] concerns of readers" (Morrison & Love, 1996, p. 46), and they can play a decisive role in the formation and legitimation of public opinion insofar as they can identify and evaluate "interests and actors, positions and arguments that configure the policy agenda" (Hessing, 2003, p. 41; Sotillo & Starace-Nastasi, 1999, p. 273). Further, the very tendency of letters' editors to publish letters that are already seen as representative of issues of public concern specifically works to legitimate those issues as proper for community deliberation.

One way to demonstrate this is by considering some of the debates carried out in letters to the editor pages in Australian broadsheets over the past few years. For instance, since the events of late 2001, public attention has often focused on questions of national identity and values, on who should be counted as Australian and why. And this attention and the correlative community deliberation is manifest in the letters forum. To illustrate how public deliberation is manifest within this forum, we can consider the letters to the editor devoted to the above issues published in Australia's national broadsheet *The Australian*[5] over the period of September 2001 to February 2002. Reported events of the time (the Tampa crisis, 9/11 and the beginning of the War on Terror, the Bali Bombings, Australia's 2001 federal election and then Australian Prime Minister Howard's 'fear' campaign, the mandatory detention of asylum-seekers, and the Woomera Detention Centre hunger strikes) meant that public attention was directed specifically to questions concerning Australianness and un-Australianness, and to how and when these descriptors should be used.

Debating Australianness in Letters to the Editor in *The Australian* 2001-2002

In letters to the editor published in the 2001 to 2002 time period, two different answers to the question of the demarcation of Australianness from un-Australianness were presented and debated (Mummery & Rodan, 2003). First there was the common view that being un-Australian or Australian came down to the matter of seemingly incompatible values, with many letter writers of this period suggesting that certain groups of people held values that were intrinsically un-Australian. As such, a relatively common argument presented in this time was that Islamic asylum seekers were constitutively un-Australian and thereby a "danger to Australia" (Christy, 2002, January 24). They were perceived as being both too other and unwilling to be successfully assimilated into the Australian way of life (e.g. Schmulow, 2001; Christy, 2002; Hallandal, 2002). As one letter writer wrote in November 2001:

Perhaps many Australians supported a tough stand by the Government because they saw the latest boatpeople as different from all others that have come. They saw people who were aggressive in their demands and who belonged to a religious culture that in various parts of the world, rightly or wrongly, seemed to be in conflict with the mainstream host nations. They saw a religious culture that in Western democracies seemed to put its identity and concerns ahead of those of its adopted country ... They instinctively and genuinely feared for social cohesion and the impact of a culture that seemed to be making its own rules, unlike any previous migrant group ... They saw it as being careful (James, 2001).

Given these beginning assumptions, this view, as expressed in letters, effectively took the form of a discourse of 'protectivism' (Mummery & Rodan, 2003) which argued that we – those who are unambiguously Australian – should protect our way of life from those who hold to different values and practices (Schmulow, 2001).

This same discourse was also used to critique those who objected to the sorts of protectivist strategies and policies that were seen as making up the Australian government's 'tough stand'. (Strategies ranged from the ramping up of policies and practices of mandatory detention to Australia's involvement in the ongoing War on Terror.) For example, objectors to this 'tough stand' were described by letter writers as "simple-minded" (Bird, 2001), "the chattering classes" (Bretz, 2001), "bleeding hearts" (Crichlow, 2002), a "soft-touch" (Cohen, 2002), "sideline commentators" who have had the "great luxury of never actually having to make the hard decisions" (Cameron, 2002), and as unpatriotic. As one letter writer puts it:

A WHITE feather to all those people who seek to denigrate Australia's commitment to the action against terrorism. If our nation were threatened I guess you would invoke the Brisbane line, and allow us patriots to man the barricades, while you sipped your pinots and sent appeasing emails to the invaders (Lovelock, 2001).

Further, in their suggestion of an unwillingness to protect and defend the Australian way of life, these descriptors were also used to suggest the holding of anti-western and un-Australian attitudes. As one such objector described the situation, proponents of the government's 'tough stand' typically justify it as "patriotic defence of our boundary", and accuse critics of an "un-Australian lack of concern for our sovereignty and security" (Stretton, 2002). In other words, to be Australian is to take the 'tough stand', to be prepared to put your life on the line, and to accept the need for both of the above. More specifically, being Australian means reserving all criticism "for bin Laden and the like" (Saunders, 2001). That is, simply to express criticism of the 'tough stand' is to be un-Australian (Stretton, 2002).

Such views, however, were strongly critiqued by proponents of a second discourse, according to which to be un-Australian is to be inhumane and to support inhumane practices (e.g. Bruhns, 2001; Groessler, 2002; Farran & Curnow, 2002; Supple, 2002). In the eyes of these letter writers, those who are really un-Australian are those who do not decry the government's tough stands, and those who do not accept that Australia has certain obligations towards all asylum-seekers under international law (e.g. Gray, 2001; Reynolds & McDonald-Toone, 2001; Hoy, 2001). Furthermore, the descriptor of un-Australian is also applied to those who do not recognise that the sorts of practices inherent in the government's tough stands are actually irreconcilable with the attitudes and values traditionally lauded as Australian: tolerance and compassion, the beliefs in a 'fair go' and in equality, not to mention in inclusive democratic practices. As one letter writer laments:

This Australia Day, it has become apparent that Australians have undergone a fundamental change. At every nationalisation ceremony and at every backyard Barbie, we should realise our national character has been altered ... On Australia Day, so much for the compassion and tolerance to which Australians once proudly laid claim (Wurth, 2002).

Finally, these writers critiqued the attempted silencing of criticism of Australian government policy, seeing it as a form of censorship and as undemocratic. These writers thus described proponents of the 'tough stand' as prejudiced, racist and xenophobic (e.g. MacKenzie, 2001), not to mention as promoting a deplorable stifling of reasoned views and debate. In a word: un-Australian.

Now, while letter writers in this time clearly do not reach a consensus as to who or what precisely should be considered Australian or un-Australian, the give and take of their debate does demonstrate one key point on which they have reached an accord: that Australianness and un-Australianness should be understood in terms of behaviour and values, particularly with regards to the perceived desirability of said behaviours and values. That is, the key issue for all letter writers of this period is the protection of Australianness from un-Australian influences, whether this takes the form of excluding groups and critiquing behaviours that are seen as too other, or criticising those who do not seem to care about the need to protect Australia from such groups. The second point that became clear through this debate was the recognition by all letter writers – regardless of which views they espoused – that the values of Australianness could and should be vigorously debated, and that this debate could and should be carried out by more than just policy-makers or presumed experts.

What seems clear, then, is that while the letters forum may be as limited in scope as initially discussed, letters can certainly be seen as one of the Australian public's attempts to 'talk through events' and can thereby be conceived as the work of engaged citizens concerned to nut-out genuine dilemmas. Certainly it is fair to say that a series of theme-related letters published over several weeks or months – selected according to the vagaries of an editor's decision-making – does not equate to a robust example of deliberative exchange leading to consensus. Nonetheless, in the period we surveyed, letter writers did explicitly and typically pick up and comment on each others' points, albeit often acrimoniously (to the point that the letters' editor Judith Elen of *The Australian* intervenes (2001), stating that no more 'name-calling' letters will be published). Exchanges, then, might not display much coming to agreement via discursive verification, but they certainly display an ongoing communal contesting of truth claims, an ongoing commitment to chewing the cud of public views regarding Australianness and un-Australianness.

Despite these points, however, perhaps a question arises from the purported limitations of the print media itself. This is the argument that the rapid development and utilization of new communications technologies spells the demise of print media. That is, that the print media will be unable to compete with the (presumed) increased interactivity – important for effective popular participation in deliberative democracy – offered by the new communications technologies, and the (presumed) decreased costs and ecological footprints of these same technologies. This argument suggests not only that the demographic for print media such as broadsheets may be dwindling, but that letter writers, their views and their debates are perhaps even less representative of their community than initially thought. Perhaps, then, an exemplification of public deliberation in the mass media needs to be sought in another domain.

POLITICAL BLOGS AND PUBLIC DELIBERATION

A second, perhaps less problematic, site for public deliberation provided by the mass media could, we suggest, be found in political blog sites. After all, blogging "in its most basic form is a type of online publishing best characterized as an interactive diary" (Maynor, 2007, p. 5), and the following of such sites as political blogs – as opposed to purely social blogs – can allow participants to "check out information found in other media", as well as providing "convenience", a venue for "information seeking" and a sense of "community" (Johnson, Kaye, Bichard, & Wong, 2008, p. 104). Indeed a number of researchers argue that many political blogs "function much as 'reporters'" in that bloggers:

select, aggregate, interpret, and sometimes independently produce information. They gather news from the mainstream media and highlight key stories and events; they provide commentary, often ideological or partisan, that helps readers understand these stories and events; and they themselves generate stories, much as mainstream news outlets do, by emphasizing events ignored by those outlets and even by understanding their own reporting... (Lawrence, Sides, & Farrell, 2009, p. 6; Johnson et al., 2008, p. 115; Kenix, 2009, p. 812).

Additionally, however, unlike the majority of mainstream journalism, political blogs are constitutively "interactive and allow viewers to post comments", and this, added to the possibilities intrinsic in hyperlinking and the fact that political bloggers – readers and writers – tend to be active seekers of information, effectively exposes participants to potentially ever increasingly "complex networks of ideas" (Lawrence, et al., 2009, p. 2).

In general, then, the main activity of political blogging is to comment on original sources, seek information, and provide in-depth discussion regarding specific topics of interest to participants (Johnson et al., 2008, p. 104). Such discussion furthermore has – via its interactivity – the potential to be "self-correcting through the process of contestation and deliberation" (Maynor, 2007, p. 18), even though bloggers themselves – like letter writers – may not be overly concerned with objectivity or fairness. Given, then, its focus on "carrying out virtual conversations and debates about public policies in the hope of positively shaping their resolution", Maynor (2007) takes the view that blogging may enable the public presentation and consideration of "reasons and arguments acceptable in deliberative forums" (pp. 12, 18). All this means that political blogs may be seen as "credible sources for information inquiry" (Johnson et al., 2008, p. 115) and may attract a significant readership. There are, however, three main limitations with seeing political blogs as sites for public deliberation:

1. The demographic of those who use and write politically interested blogs closely matches that of letter to the editor writers – political bloggers "tend to be well-educated, white middle-class", male with an average age of "43.3 years" (Johnson et al., 2008, p. 108; Lawrence, et al., 2009, p. 10).
2. Second, Lawrence, et al., have found that about "94% of political blog readers consume only blogs from one side of the ideological spectrum" (2009, p. 11). Because political blog networks tend to seek and reinforce existing political opinions they tend to be "homophilous", falling "well short of the deliberative ideal" (Lawrence, et al., 2009,

p. 6-7). A number of researchers further argue the "balkanization" of the blogosphere, meaning a) "the isolation of individuals within groups that share their values and beliefs" (Maynor, 2007, p. 13; Xenos 2008, p. 501), entailing b) their limited "opportunity for revising their opinions" (Lawrence, et al., 2009, p. 17).

3. Thirdly bloggers may be limited in their ability to enable public debate because they serve what Habermas has called a "'parasitical' function of criticizing and correcting the mainstream press" (cited in Lawrence, et al., 2009, p. 4). Indeed Kenix (2009, p. 814) states that that "rather than 'answering' to public concerns about content", the political blogs she sampled "were driven by individual interests created within the culture of mainstream media" (also see Xenos, 2008, p. 495).

Nonetheless, despite such limitations, we would still contend that this blogosphere can and does exemplify processes important for public deliberation. To begin with, political bloggers are "likely to nurture blog readers' intrinsic interest in politics and suggest avenues into which blog readers can direct their activism" (Lawrence, et al., 2009, p. 16). Because of the "prevalence of non-experts" and the conversational tone in much political blogging, as well as the "information-driven character of blog coverage and discussion", the practice could further the potential of "mediated deliberation" (Xenos, 2008, p. 485). Additionally, these virtual communities exemplify a "self-directing networking" – that is, "users can find their own destination on the internet or make one themselves easily should they so desire" (Maynor, 2007, p. 19). Furthermore, following and participating in political blog sites could, to some degree, provide 'training' for participants to develop the skills to participate in other public forums (see Pateman (1970) on this point of skill development for citizens in deliberative democracy). These points can best be demonstrated through a brief critical consideration of some of the discussion generated in one well-known and respected political blog sited in Australia: Larvatus Prodeo founded by sociologist Mark Bahnisch.

Debating the 2007 Australian Federal Election in Larvatus Prodeo

Described by founder Mark Bahnisch (2009) as "an Australian group blog which discusses politics, sociology, culture, life, religion and science from a left of centre perspective", Larvatus Prodeo (LP) is the joint work of Bahnisch and a "diverse range of contributors" ("About LP Bloggers", 2009). These comprise about fifteen guest bloggers and of course the other followers and commentators of the site. Like letters to the editor, there are guidelines to the writing of LP blogs: topics in this instance are decided by the founder and his guest bloggers, blog responses are typified by extended pieces of analysis that add to Mark's initial blog, and although posts are not edited as such, the site is moderated ("Comments Policy", 2009). As Bahnisch states: "Our blog, our discretion" and "No public discussion will be entered into regarding moderation decisions" ("Comments Policy", 2009).

Some comments about the structural way in which discussion works on the site are necessary at this point. One of the LP bloggers (often the founder Mark) starts a discussion thread, usually in response to an item in the independent online media such as *Crikey*, a political report, a new political biography, an item in the mainstream media such as News Limited, or an online article on a political website, etc. This thread is then given a tag name and is available for public discussion. For instance, with regard to the LP tag 'Federal Election 2007', we found that bloggers sourced material from a variety of places: namely, the Australian Election Study from the 2007 election; Fairfax media (in particular *The Sydney Morning Herald* and *The Age* newspaper); *Crikey* (the independent online newspaper which explicitly describes its role as "part of the so-called fourth estate" ("About Crikey", n.d.)); News Limited papers (in particular *The Courier-Mail* and *The Australian*); the New Republic website (at http://www.tnr.com/); Australian parliamentary library papers on the 2007 federal election[6]; *On Line Opinion* (Australia's e-journal of social and political debate at http://www.onlineopinion.com.au/); the Possum political blog (sourced for graphs at http://possumcomitatus.wordpress.com/[7]); The Poll Bludger (sourced for trends at http://blogs.crikey.com.au/pollbludger/); FiveThirtyEight Politics Done Right (at http://www.fivethirtyeight.com/). Of these, the most referenced source was *Crikey*, with the next being News Limited print, in particular *The Australian* and the *Courier-Mail*. Most references, however, are to online websites and print newspapers.[8]

There are several additional points worth noting with regards to the discussion carried out on these threads. First, bloggers can and do respond to each other and not just the moderator; second, bloggers can and do explain things to each other as requested and required; third, discussion is carried out by both experts and non-experts; and finally posts are left on site and categorized, thus establishing an archive of all of the deliberation regarding a particular tag. Pragmatically, such categorizing allows people to join in discussions easily and hone in on what interests them. Furthermore, similar to Xenos's (2008, p. 485) findings on other blogging sites, in LP bloggers deliberate with each other – indeed most bloggers on this tag are repeat bloggers who post in a number of the threads [9] – and do not necessarily defer to, or agree with, the moderator of the thread. Participants other than the guest bloggers finally have the option to start a discussion thread themselves on a topic that interests them on the Saturday Salon thread.

Turning then to the kinds of deliberation carried out on the site, LP has been recognized within Australia as facilitating public deliberation, winning three categories in the 2006 Australian Blog Awards – Best Political Blog, Best Collaborative Blog and Best Queensland Blog – and being named as joint winner of *Crikey*'s 2006 Blog of the Year award ("Accolades for LP", 2009). More specifically, it has been recognized as explicitly facilitating public deliberation concerning the 2007 federal elections. The site, according to Bruns, Wilson, Saunders, Kirchhoff, Nicolai (2008), "provided a highly critical corrective to mainstream journalism"; "seemed to influence public opinion on key election themes"; and "offered a coverage of political events" which was a diversion from the "customary focus of political leaders". For instance, the tag archive of 'Federal election 2007' has 319 posts spread across fifteen discussion threads ranging from 'Rudd and Gillard attack News Limited, Hartigan punches back' to 'The middle-classing of Labor' to 'Inside Kevin07'. Comments for each thread ranging from sixty to two.

With regards to the discussion within the tag 'Federal election 2007' itself, it is primarily formal in tone.[10] While there are some colloquialisms and/or slang words used throughout the posts, the majority of the language is for an educated/professional audience; bloggers and readers would need to have some understanding of the political process and institutions. For instance, in the thread 'The great pretender', reader knowledge is assumed in the posts regarding a journalist called Milne (Ambigulous, 2008), while others contain many references to political books and the institution of journalism. Additionally, while there are many instances of bloggers disagreeing with Mark (the founder and moderator), or each other, the language remains respectful with little sledging of other bloggers. Indeed, most negative views are directed at News Limited and the mainstream media (see the thread 'Rudd and Gillard attack News Limited, Hartigan punches back'). Even when a poster commented on someone else's blog (for instance, Hannah's Dad (2008), in response to Merkel & wpd, in the thread 'Issues and the

2007 election'), he moderated his response by saying: "Not having a go at you two, just riffing [sic] on the theme". In the thread, 'John Howard: media tart?', there is no direct sledging of other bloggers although there are impassioned remarks made about John Howard (the previous Prime Minister of Australia). For instance one blogger says: "If you want to make yourself truly sick listen to this clip of Howard discussing the moral superiority of the anglosphere to some American shock jock" (cosmicjester, 2008).

Although most bloggers deliberate over issues raised in the blogs, deliberation does occur in varying levels throughout the threads. For instance in the thread, 'The great pretender', the initial comments on the blog can be described as deliberative in that they build constructively on Mark's commentary. For example, the first commenter Ambigulous (2008) extends Mark's argument in providing the case of the Gippsland By-election and the ensuing results as deduced by the Nationals in a private poll. Paul Burns (2008) then builds on Ambigulous' comment by suggesting that the real results and analysis will come out in 30 years time. In the thread, 'Rudd and Gillard attack News Limited, Hartigan punches back', there is one theme that is deliberated throughout the thread. The theme is initiated by Jack, who suggests that the newspapers are not as one-sided as asserted by Mark. Jack (2009a) says: "Actually the Courier Mail had loved to rubbish the LNP/Coalition for years before the election... Oh, and it's interesting how you just mention *The Australian* as the only News Limited paper supporting Howard, because I remember an awful lot of the company's newspapers coming out in support of Rudd". The dialogue goes back-and-forth as bloggers deliberate on Mark's position:

Comment 13: "The Courier Mail was so anti the Bligh government it was embarrassing... The Australian is no better really" (Wombat, 2009).

Comment 15: "Wombat most of the Courier's criticism of QLD Labor was due to the poor services it was providing... The criticisms for both parties were true, the problem is the party faithful have a habit of only seeing the negative publicity they get and then all of a sudden the press is against them!" (Jack, 2009b).

Comment 17: "Jack, whilst I think the Courier Mail is more anti-ALP than anti-LNP, that's not the biggest issue here... It doesn't do much analysis, or big ideas." (Down and Out of Sài Gòn, 2009).

Another example, 'The state of Rudd nation', starts with Mark's response to the quarterly Newspoll geographical and demographic analysis published one year after the election. Specifically, Mark (2008c) directs other bloggers to Possum's "spiffy graphs" and The Poll Bludger's notes which outline "interesting trends in the aggregate poll", and explains how reform works and why some policies will work and others will not. He points to two complex issues that the Labor party is going to find difficult to solve: the "Murray-Darling" and "student performance in schools" (2008c), both of which have also been discussed in Australian newspapers and television programs such as *Insight* on SBS. Other bloggers on this thread also clearly have some in-depth knowledge about politics and different electorates. For instance, jinmaro's response (2008) to Mark's second post points to another complex issue (maintenance of infrastructure) about which the "parliamentary ALP and the citizenry at large" would also be "deeply unhappy" about the options presented. Further responses include Matt C (2008) responding to Jinmaro with a clarification as to what is and is not a "broken promise" in politics and Professor rat (2008) commenting that if Kevin Rudd's reforms and policies "don't measure up on some of the big challenges then its golden parachute time baby".

In his original post, Mark (2008c) also said he would be "interested to hear from South Australian LP-ers about what they think is going on". A number of LP bloggers responded to his request for more information: for example SofaMan (2008) lists some of the factors he/she considers to "play in SA right now". Paulus (2008a) also provides more information about the state of South Australia and disagrees with SofaMan's "insufficiently Left analysis", going on to say that SofaMan's "reasons, whatever their validity, are hardly going to drive people to vote Lib" (2008a). Mark (2008d) then clarifies a point about "federal voting intention disaggregated by state" with Paulus and SofaMan. Paulus (2008b) responds to Mark and then Paul Walter (2008) joins the discussion and Bingo Bango Boingo (2008) queries a point made by Paulus. PinkyOz (2008) queries Mark's use of the word "'honeymoon'"; saying the word is "deceptive [in that] it makes it seem like a temporary effect rather than a trend (short/long, yet to be seen)". Mark concedes PinkOz's point. And so on.

Overall, of the fifteen discussion threads in the tag 'Federal election 2007' about 60% of the bloggers clarified and/or explained the points they made. In the thread, 'Slow progress in health funding reform', the blogger Francis Xavier Holden (2008a) provides an explanation on government processes and the reasons why progress appears slow. He identifies himself as part of a submission and suggests that one of the reasons "these things proceed slowly is that after the 20 second news grabs and simple rhetoric from politicians someone actually has to do the hard work". In response to Merkel (2008), who asks, "Do you know if they've formally asked for submissions anywhere?", Holden (2008b) responds "I'm not sure – I can find out...". In the thread, 'Marketised celebrity politics, Australian style' (Mark, 2008a), where bloggers disagree with Mark (the founder), he responds by clarifying his position in an exchange with Bismarck. Bismarck (2008) suggests (in regards to Iguanagate staffer Batten) that "the issue is an allegation that she was leant on to change her version of events". Mark (2008b) rejoins: "And the most appropriate forum to make such an allegation, Bismark, is *A Current Affair*, rather than making a statement to police?"

What becomes very clear with these examples, then, is that bloggers on these and other threads do not necessarily accept the truth claims made by others – they often question and contest these claims. Certainly some bloggers in this site do not, through discussion, appear to change their views (see, e.g., OldSkeptic, 2008 on the thread 'The state of Rudd nation'), but most are clearly engaged in a public debate to nut-out issues and dilemmas of public concern. Furthermore, bloggers are providing, clarifying and debating information sourced from the full range of institutions they participate in, deliberations that are in turn shared in an open public forum with other interested citizens who themselves can ask for further information and clarification.

CONCLUSION

What then can be concluded from this discussion? Two key points. First, that such forums do seem to enable, support and exemplify community deliberation over issues of public concern. They not only provide a space in and by means of which participants can debate issues of concern to themselves and their community, but they are open to all to participate in, albeit with some normative constraints: both kinds of forum are moderated in various ways. Certainly writers, letters and blogs are not reflective of the full gamut of community concerns – and nor are letter writers and bloggers representative of the full community – but then again it is also clear that not everyone participates in public deliberation within a community. In fact, it seems clear that a deliberative democracy could not operate functionally if everyone did (cf. Goodin, 2000). Nonetheless, we can say that participation in such forums demonstrates the point that all of the reasoning that takes place within the public sphere is a form of life that is only instantiated by the doing of it. Hence we contend that letter writers and bloggers are, by definition, interested, involved and active citizens, and that writers and bloggers, along with other active participants in the public sphere – those, for instance, who participate in civic forums such as GetUp! (see Rodan & Balnaves, 2009), edemocracy sites such as E-Democracy.org[11], or even talkback radio – comprise a key part of a community's reasoning community.

Our second conclusion responds to a possible objection to the above point. More specifically, our conclusion that such mediated forums as letters to the editor and blogs do represent sites of public deliberation may be challenged on the grounds that while participants might debate in these forums, they do not formally meet the criteria for deliberation in the public sphere insofar as participants do not come to rational and consensual agreements on issues. That is, whilst they certainly render various truth claims problematic, they do not redeem them in the way Habermas calls for. Truth claims are rather left open and contested. We would contend, however, that this criticism is based on a more outcomes oriented stronger conception of deliberation than we are drawing on here. In our conception, deliberation is not invalidated by either the lack of any tangible outcome, or the fact that any outcomes achieved can only be partial and contingent, open to revision. In this account, what is important for a reasoning community is not the actual reaching of consensus but the practice and procedure of a genuine deliberation performed without strict adherence to or regulation by any pre-set and idealised telos or endpoint other than that of public deliberation itself. Such a procedure could only take the form of continuous and transformative negotiations giving rise to temporary conclusions and hegemonic articulations always open to further negotiation and transformation.

It is thus keeping in mind both the idealist and normative aspects of Habermas' conceptions of the public sphere and public deliberation that we contend that the mass media forums of letters to the editor and political blogs do indeed exemplify instances of public deliberation. Not only that, but we agree with Lawrence et al. (2009) that deliberation in such public mediated forums is fundamental to any reasoning community because it helps "individuals refine their own opinions, develop greater tolerance for different opinions, and perhaps identify common ends and means" (p. 2). We also agree with Meraz (2007) that as a "result" of participating in public deliberation, citizens may "increase political participation, autonomy and individual self-efficacy" (pp. 60-61). This would in fact be the best outcome mass media forums could offer for public deliberation, but this can be enough to better enable citizens to challenge any symbolic manipulation (see Dryzek & Niemeyer, 2006, p. 646) by politicians and those with vested interests, including particular mainstream presses. For this reason, even non-consensus oriented public deliberation is highly valuable and the mass media clearly provide sites in which such deliberation can be carried out.

ACKNOWLEDGMENT

Our thanks to Judith Elund who has assisted us in the content analysis of the Larvatus Prodeo blog, and to the two anonymous reviewers of the chapter. You have helped us enormously in clarifiying our discussion in this chapter.

REFERENCES

Ambigulous. (2008, July 21). [Comment 1 to blog post: The great pretender]. *Larvatus Prodeo* [Blog], Federal election 2007. Retrieved October 1, 2009, from http://larvatusprodeo.net/2008/07/21/the-great-pretender/

Bingo Bango Boingo. (2008, October 02). [Comment 22 to blog post: The state of Rudd nation]. *Larvatus Prodeo* [Blog], Federal election 2007. Retrieved October 1, 2009, from http://larvatusprodeo.net/2008/10/02/the-state-of-rudd-nation/

Bird, L. (2001, October 02). Single mind could be a liability [Letter to the editor]. *The Australian*, p. 12.

Bismarck. (2008, June 24). [Comment 2 to blog post: Marketised celebrity politics, Australian style]. *Larvatus Prodeo* [Blog], Federal election 2007. Retrieved 01 October, 2009, from http://larvatusprodeo.net/2008/06/24/marketised-celebrity-politics-australian-style/

Bohman, J., & Rehg, W. (Eds.). (1997). *Deliberative democracy: Essays on reason and politics*. Cambridge, MA: MIT Press.

Bretz, L. (2001, September 27). Court's role is to see that we act legally [Letter to the editor]. *The Australian*, p. 12.

Bromley, M. (1998). Watching the watchdogs?: The role of reader's letters in calling the press to account . In Bromley, M., & Stephenson, H. (Eds.), *Sex, lies and democracy* (pp. 147–162). London: Longman.

Bruhns, C. (2001, October 10). Minds, pockets agents for peace [Letter to the editor]. *The Australian*, p. 14.

Bruns, A., Wilson, J. A., Saunders, B. J., Kirchhoff, L., & Nicolai, T. (2008, October 15-18). Australia's political blogosphere in the aftermath of 2007 federal election. Paper presented at the Association of Internet Researchers 2008: Internet Research 9.0: Rethinking Community, Rethinking Place. Copenhagen, Denmark. Retrieved September, 10, 2009, from http://eprints.qut.edu.au/15116/1/15116.pdf

Buckingham, D. (1997). News media, political socialization and popular citizenship: Towards a new agenda. *Critical Studies in Mass Communication*, *14*, 344–366. doi:10.1080/15295039709367023

Burns, P. (2008, July 21). [Comment 2 to blog post: The great pretender]. *Larvatus Prodeo* [Blog], Federal election 2007. Retrieved October 1, 2009, from http://larvatusprodeo.net/2008/07/21/the-great-pretender/

Calhoun, C. (1992). Introduction: Habermas and the public sphere . In Calhoun, C. (Ed.), *Habermas and the public sphere* (pp. 1–50). Cambridge, MA: MIT Press.

Cameron, R. (2002, January 26-27). Caring, tolerant Aussie is a thing of the past [Letter to the editor]. *The Australian*, p. 18.

Chambers, S. (2003). Deliberative democratic theory. *Annual Review of Political Science*, *6*(1), 307–326. doi:10.1146/annurev.polisci.6.121901.085538

Christy, W. (2002, January 24). Gaps in arrival process must be queried [Letter to the editor]. *The Australian*, p. 10.

Cohen, B. (2002, January 29). Refugees evince hope and mental health [Letter to the editor]. *The Australian*, p. 12.

Coleman, R. (1997). The intellectual antecedents of public journalism. *The Journal of Communication Inquiry*, *21*(1), 60–76. doi:10.1177/019685999702100103

cosmicjester. (2008, September 29). [Comment 16 to blog post: John Howard: media tart?]. *Larvatus Prodeo* [Blog], Federal election 2007. Retrieved October 1, 2009, from http://larvatusprodeo.net/2008/09/28/john-howard-media-tart/

Crichlow, F. (2002, February 02-03). Settler families sold out and browned off [Letter to the editor]. *The Weekend Australian*, p. 16.

Crikey. (n.d.). About. Retrieved December 10, 2009, from http://www.crikey.com.au/

Dahlberg, L. (2001). Computer-mediated communication and the public sphere: A critical analysis. *Journal of Computer-Mediated Communication, 7*(1). Retrieved November 30, 2009, from http://jcmc.indiana.edu/vol7/issue1/dahlberg.html

Dahlberg, L. (2005). The Habermasian public sphere: Taking difference seriously? *Theory and Society*, *34*, 111–136. doi:10.1007/s11186-005-0155-z

Dewey, J. (1927). *The public and its problems*. New York: Holt.

Down and Out of Sài Gòn. (2009, July 01). [Comment 17 to blog post: Rudd and Gillard attack News Limited, Hartigan punches back]. *Larvatus Prodeo* [Blog], Federal election 2007. Retrieved October 1, 2009, from http://larvatusprodeo.net/2009/07/01/rudd-and-gillard-attack-news-limited-hartigan-punches-back/

Dryzek, J., & Niemeyer, S. (2006). Reconciling pluralism and consensus as political ideals. *American Journal of Political Science*, *50*(3), 634–649. doi:10.1111/j.1540-5907.2006.00206.x

Dzur, A. W. (2002). Public journalism and deliberative democracy. *Polity*, *34*(3), 313–336.

E-democracy.org. (n.d). About. Retrieved December 10, 2009, from http://forums.e-democracy.org/

Elen, J. (2001, November 30). Agenda set by losers? [Letter to the editor]. *The Australian*, p. 10.

Farran, B. & Curnow, A. (2002, January 29). Refugees evince hope and mental health [Letter to the editor]. *The Australian*, p. 12.

Fishkin, J. S. (1991). *Democracy and deliberation: New directions for democratic reform*. New Haven, CT: Yale University Press.

Francis Xavier Holden. (2008a, July 29). [Comment 1 to blog post: Slow progress in health funding reform]. *Larvatus Prodeo* [Blog], Federal election 2007. Retrieved October 1, 2009, from http://larvatusprodeo.net/2008/07/29/slow-progress-in-health-funding-reform/

Francis Xavier Holden. (2008b, July 29). [Comment 5 to blog post: Slow progress in health funding reform]. *Larvatus Prodeo* [Blog], Federal election 2007. Retrieved October 1, 2009, from http://larvatusprodeo.net/2008/07/29/slow-progress-in-health-funding-reform/

Fraser, N. (1992). Rethinking the public sphere: A contribution to the critique of actually existing democracy . In Calhoun, C. (Ed.), *Habermas and the public sphere* (pp. 109–142). Cambridge, MA: MIT Press.

Garnham, N. (2007). Habermas and the public sphere. *Global Media and Communication*, *3*(2), 201–214. doi:10.1177/1742766507078417

Goodin, R. E. (2000). Democratic deliberation within. *Philosophy & Public Affairs*, *29*(1), 81–109. doi:10.1111/j.1088-4963.2000.00081.x

Gray, R. (2001, September 25). Standing force a UN imperative [Letter to the editor]. *The Australian*, p. 12.

Griffen-Foley, B. (2009a, December 2). Dial-in democracy? Forty years of talkback, from Andrea to Kyle [Audio podcast]. Sydney, Australia: The Sydney Institute. Retrieved December 10, 2009, from http://www.thesydneyinstitutepodcast.com/2009/12/02/BridgetGriffenFoleyDialinDemocracyFortyYearsOfTalkbackFromAndreaToKyle.aspx

Griffen-Foley, B. (2009b). *Changing stations: The story of Australian commercial radio*. Sydney: UNSW Press.

Groessler, R. (2002, January 29). Refugees evince hope and mental health [Letter to the editor]. *The Australian*, p. 12.

Haas, T. (2004). The public sphere as a sphere of publics: Rethinking Habermas's theory of the public sphere. *The Journal of Communication*, *54*(1), 178–184. doi:10.1111/j.1460-2466.2004.tb02621.x

Habermas, J. (1984). The theory of communicative action: Reason and the rationalization of society.: *Vol. 1. T. McCarthy, Trans*. Boston, MA: Beacon Press.

Habermas, J. (1987). *Knowledge and human interests* (Shapiro, J. J., Trans.). Cambridge, UK: Polity Press.

Habermas, J. (1989). *The structural transformation of the public sphere* (Burger, T., Trans.). Cambridge, UK: Polity Press.

Habermas, J. (1998). *Between facts and norms: Contributions to a discourse theory of law and democracy* (Rehg, W., Trans.). Cambridge, MA: MIT Press.

Habermas, J. (2006). *Time of transitions*. (C. Cronin, Trans., & M. Pensky, Ed.). Cambridge, UK: Polity Press.

Hallandal, G. (2002, January 26-27). Caring, tolerant Aussie is a thing of the past [Letter to the editor]. *The Australian*, p. 18.

hannah's dad. (2008, June 23). [Comment 3 to blog post: Issues and the 2007 election]. *Larvatus Prodeo* [Blog], Federal election 2007... Retrieved October 1, 2009, from http://larvatusprodeo.net/2008/06/23/issues-and-the-2007-election/

Hart, R. (2001). Citizen discourse and political participation: A survey . In Bennett, L., & Entman, R. (Eds.), *Mediated politics and the future of democracy* (pp. 407–432). New York: Cambridge University Press.

Hessing, M. (2003). Green mail: The social construction of environmental issues through letters to the editor. *Canadian Journal of Communication*, *28*, 25–42.

Hoy, N. (2001, October 18). Public debate: voters want more [Letter to the editor]. *The Australian*, p. 12.

Jack. (2009a, July 01). [Comment 10 to blog post: Rudd and Gillard attack News Limited, Hartigan punches back]. *Larvatus Prodeo* [Blog], Federal election 2007. Retrieved October 1, 2009, from http://larvatusprodeo.net/2009/07/01/rudd-and-gillard-attack-news-limited-hartigan-punches-back/

Jack. (2009b, July 01). [Comment 15 to blog post: Rudd and Gillard attack News Limited, Hartigan punches back]. *Larvatus Prodeo* [Blog], Federal election 2007. Retrieved October 1, 2009, from http://larvatusprodeo.net/2009/07/01/rudd-and-gillard-attack-news-limited-hartigan-punches-back/

Jacka, E. (2000). Public service TV: An endangered species? In Turner, G., & Cunningham, S. (Eds.), *The Australian TV book* (pp. 52–68). St. Leonards, Australia: Allen & Unwin.

James, D. (2001, November 21). Careful vote, not racist. [Letter to the editor]. *The Australian*, p. 12.

Jinmaro. (2008, October 02). [Comment 4 to blog post: The state of Rudd nation]. *Larvatus Prodeo* [Blog], Federal election 2007. Retrieved October 1, 2009, from http://larvatusprodeo.net/2008/10/02/the-state-of-rudd-nation/

Johnson, T., Kaye, B., Bichard, S., & Wong, W. (2008). Every blog has its day: Politically-interested Internet users' perceptions of blog credibility. *Journal of Computer-Mediated Communication*, *13*, 100–122. doi:10.1111/j.1083-6101.2007.00388.x

Kenix, L. J. (2009). Blogs as alternative. *Journal of Computer-Mediated Communication, 14*, 790–822. doi:10.1111/j.1083-6101.2009.01471.x

Knops, A. (2007). Debate: Agonism as deliberation – On Mouffe's theory of democracy. *Journal of Political Philosophy, 15*(1), 115–126. doi:10.1111/j.1467-9760.2007.00267.x

Larvatus Prodeo. (n.d.). About LP Bloggers. Retrieved September 15, 2009, from http://larvatusprodeo.net/

Larvatus Prodeo. (n.d.). Accolades for LP. Retrieved September 15, 2009, from http://larvatusprodeo.net/about-larvatus-prodeo/accolades-for-lp/

Larvatus Prodeo. (n.d.). Comments Policy. Retrieved September 15, 2009, from http://larvatusprodeo.net/

Lawrence, E., Sides, J., & Farrell, H. (2009, March). *Self-segregation or deliberation? Blog readership, participation, and polarization in American politics*. Unpublished paper. Retrieved September 10, 2009, from http://home.gwu.edu/~jsides/blogs.pdf

Lovelock, J. (2001, October 19). Choice of war should be subject to vote [Letter to the editor]. *The Australian*, p. 12.

Lynn, N., & Lea, S. (2003). A phantom menace and the new Apartheid: The social construction of asylum-seekers in the United Kingdom. *Discourse & Society, 14*(4), 425–452. doi:10.1177/0957926503014004002

MacKenzie, S. (2001, November 28). Closed culture shuts out complex world [Letter to the editor]. *The Australian*, p. 14.

Mark. (2008a, June 24). [Blog post: Marketised celebrity politics, Australian style]. *Larvatus Prodeo* [Blog], Federal election 2007. Retrieved October 1, 2009, from http://larvatusprodeo.net/2008/06/24/marketised-celebrity-politics-australian-style/

Mark. (2008b, June 24). [Comment 3 to blog post: Marketised celebrity politics, Australian style]. *Larvatus Prodeo* [Blog], Federal election 2007. Retrieved October 1, 2009, from http://larvatusprodeo.net/2008/06/24/marketised-celebrity-politics-australian-style/

Mark. (2008c, October 02). [Blog post: The state of Rudd nation]. *Larvatus Prodeo* [Blog], Federal election 2007. Retrieved October 1, 2009, from http://larvatusprodeo.net/2008/10/02/the-state-of-rudd-nation/

Mark. (2008d, October 02). [Comment 19 to blog post: The state of Rudd nation]. *Larvatus Prodeo* [Blog], Federal election 2007. Retrieved October 1, 2009, from http://larvatusprodeo.net/2008/10/02/the-state-of-rudd-nation/

Matt, C. (2008, October 02). [Comment 5 to blog post: The state of Rudd nation]. *Larvatus Prodeo* [Blog], Federal election 2007. Retrieved October 1, 2009, from http://larvatusprodeo.net/2008/10/02/the-state-of-rudd-nation/

Maynor, J. W. (2007, April). *Blogging for democracy: Deliberation, autonomy and reasonableness in the blogosphere*. Paper presented at the Midwest Political Science Association, Chicago, Illinois.

McCarthy, T. (1994). Kantian constructivism and reconstructivism: Rawls and Habermas in dialogue. *Ethics, 105*(1), 44–63. doi:10.1086/293678

McMahon, C. (2001). *Collective rationality and collective reasoning*. Cambridge, UK: Cambridge University Press.

Meikle, G. (2002). *Future active: Media activism and the internet.* New York: Routledge.

Meraz, S. (2007). Analyzing political conversation on the Howard Dean candidate blog. In Tremayne, M. (Ed.), *Blogging, citizenship and the future of media* (pp. 59–81). New York, London: Routledge.

Merkel, R. (2008, July 29). [Comment 2 in blog post: Slow progress in health funding reform]. *Larvatus Prodeo* [Blog], Federal election 2007. Retrieved 01 October, 2009, from http://larvatusprodeo.net/2008/07/29/slow-progress-in-health-funding-reform/

Mickler, S. (2004). Talkback radio, anti-elitism and moral decline: A fatal paradox? In Nile, R. (Ed.), *Us and them: Anti-elitism in Australia* (pp. 97–116). Perth, Australia: API Network, Curtin University of Technology.

Morrison, A., & Love, A. (1996). A discourse of disillusionment: Letters to the editor in two Zimbabwean magazines 10 years after independence. *Discourse & Society*, *7*(1), 39–75. doi:10.1177/0957926596007001003

Mummery, J., & Rodan, D. (2003). Discourses of democracy in the aftermath of 9/11 and other events: Protectivism versus humanitarianism. *Continuum: Journal of Media & Cultural Studies*, *17*(4), 433–443. doi:10.1080/1030431032000152014

OldSkeptic. (2008, October 06). [Comment 29 to blog post: The state of Rudd nation]. *Larvatus Prodeo* [Blog], Federal election 2007. Retrieved 01 October, 2009, from http://larvatusprodeo.net/2008/10/02/the-state-of-rudd-nation/

Pateman, C. (1970). *Participation and democratic theory*. Cambridge, UK: University Press.

Paul Burns. (2008, July 21). [Comment 2 to blog post: The great pretender]. *Larvatus Prodeo* [Blog], Federal election 2007. Retrieved 01 October, 2009, from http://larvatusprodeo.net/2008/07/21/the-great-pretender/

Paul Walter. (2008, October 02). [Comment 21 to blog post: The state of Rudd nation]. *Larvatus Prodeo* [Blog], Federal election 2007. Retrieved 01 October, 2009, from http://larvatusprodeo.net/2008/10/02/the-state-of-rudd-nation/

Paulus. (2008a, October 02). [Comment 18 to blog post: The state of Rudd nation]. *Larvatus Prodeo* [Blog], Federal election 2007. Retrieved 01 October, 2009, from http://larvatusprodeo.net/2008/10/02/the-state-of-rudd-nation/

Paulus. (2008b, October 02). [Comment 20 to blog post: The state of Rudd nation]. *Larvatus Prodeo* [Blog], Federal election 2007. Retrieved 01 October, 2009, from http://larvatusprodeo.net/2008/10/02/the-state-of-rudd-nation/

Peters, J. (1993). Distrust of representation: Habermas on the public sphere. *Media Culture & Society*, *15*, 541–571. doi:10.1177/016344393015004003

Phillips, K. (1996). The spaces of public dissension: Reconsidering the public sphere. *Communication Monographs*, *63*, 231–248. doi:10.1080/03637759609376391

PinkyOz. (2008, October 02). [Comment 23 to blog post: The state of Rudd nation]. *Larvatus Prodeo* [Blog], Federal election 2007. Retrieved 01 October, 2009, from http://larvatusprodeo.net/2008/10/02/the-state-of-rudd-nation/

Possum. (n.d.). About. Retrieved December 10, 2009, from http://possumcomitatus.wordpress.com/about-possum/

Professor rat. (2008, October 02). [Comment 7 to blog post: The state of Rudd nation]. *Larvatus Prodeo* [Blog], Federal election 2007. Retrieved 01 October, 2009, from http://larvatusprodeo.net/2008/10/02/the-state-of-rudd-nation/

Raboy, M., Abramson, B. D., Proulx, S., & Welters, R. (2001). Media policy, audiences and social demand. *Television & New Media*, *2*(2), 95–115. doi:10.1177/152747640100200202

Raeymaeckers, K. (2005). Letters to the editor: A feedback opportunity turned into a marketing tool: An account of selection and editing practices in the Flemish daily press. *European Journal of Communication, 20*(2), 199-221.

Reynolds, M. & McDonald-Toone, T. (2001, October 16). Prized for peace [Letter to the editor]. *The Australian*, p. 12.

Richards, T., & King, B. (2000). An alternative to the fighting frame in news reporting. *Canadian Journal of Communication, 25*, 479–496.

Rodan, D., & Balnaves, M. (2009). Democracy to come: Active forums as indicator suites for e-participation and e-governance. In A. Macintosh & E. Tambouris (Eds.), *Proceedings of First International Conference, ePart 2009,* (pp. 175–185). Berlin: Springer-Verlag

Sadler, R., Hayllar, T., & Powell, C. (1985). *Senior language*. South Melbourne, Australia: The MacMillan Company.

Saunders, H. (2001, October 11). PM acted well in affairs not of his making [Letter to the editor]. *The Australian*, p. 12.

Schmulow, A. (2001, October 04). Stand and fight for free society values [Letter to the editor]. *The Australian*, p. 10.

Schudson, M. (1997). Why conversation is not the soul of democracy. *Critical Studies in Mass Communication, 14*(4), 1–13. doi:10.1080/15295039709367020

Schultz, J. (1994). Media convergence and the fourth estate. In Schultz, J. (Ed.), *Not just another business: Journalists, citizens and the media* (pp. 15–33). Leichhardt, Australia: Pluto Press.

SofaMan. (2008, October 02). [Comment 11 to blog post: The state of Rudd nation]. *Larvatus Prodeo* [Blog], Federal election 2007. Retrieved 01 October, 2009, from http://larvatusprodeo.net/2008/10/02/the-state-of-rudd-nation/

Sotillo, S. M., & Starace-Nastasi, D. (1999). Political discourse of a working-class town. *Discourse & Society, 10*(2), 249–276. doi:10.1177/0957926599010002006

Sparks, M. K., & Perez, E. (1991). Women not heard in letters columns. *The Masthead, 43*(2), 14–15.

Start an Issues Forum. (n.d.). Retrieved December 10, 2009, from http://forums.e-democracy.org/

Stretton, H. (2002, February 09-10). Detainees must have legal refuge [Letter to the editor]. *The Weekend Australian*, p. 18.

Supple, A. & Supple C. (2002, February 11). Refugee stories reach happy ending [Letter to the editor]. *The Australian*, p. 12.

The Media Report [Television show]. (2001, July 12). Radio National. Sydney, Australia: ABC.

Travers Scott, D. (2007). Pundits in muckrakers' clothing: Political blogs and the 2004 US presidential election. In Tremayne, M. (Ed.), *Blogging, citizenship and the future of media* (pp. 39–57). New York, London: Routledge.

Turner, G., Tomlinson, E. & Pearce, S. (2006). Talkback Radio: Some notes on format, politics and influence. *Media International Australia incorporating Culture & Policy, 118*, 107-119.

Wahl Jorgensen, K. (2001). Letters to the editor as a forum for public deliberation: Modes of publicity and democratic debate. *Critical Studies in Media Communication, 18*(3), 303–320. doi:10.1080/07393180128085

Wahl-Jorgensen, K. (2002a). The normative-economic justification for public discourse: Letters to the editor as a "wide open" forum. *Journalism & Mass Communication Quarterly, 79*(1), 121–133.

Wahl-Jorgensen, K. (2002b). Understanding the conditions for public discourse: Four rules for selecting letters to the editor. *Journalism Studies*, *3*(1), 69–81. doi:10.1080/14616700120107347

Wahl-Jorgensen, K. (2004). A "legitimate beef" or "raw meat"? Civility, multiculturalism, and letters to the editor. *Communication Review*, *7*, 89–105.

Ward, I. (1995). *Politics of the media.* South Melbourne, Australia: Macmillan.

Warnke, G. (1995). Communicative rationality and cultural values . In White, S. K. (Ed.), *The Cambridge companion to Habermas* (pp. 120–142). Cambridge, UK: Cambridge University Press.

Wombat. (2009, July 01). [Comment 13 to blog post: Rudd and Gillard attack News Limited, Hartigan punches back]. *Larvatus Prodeo* [Blog], Federal election 2007. Retrieved October 1, 2009, from http://larvatusprodeo.net/2009/07/01/rudd-and-gillard-attack-news-limited-hartigan-punches-back/

Wurth, B. (2002, January 26-27). Caring, tolerant Aussie is a thing of the past [Letter to the editor]. *The Australian*, p. 18.

Xenos, M. (2008). New mediated deliberation: Blog and press coverage of the Alito nomination. *Journal of Computer-Mediated Communication*, *13*, 485–503. doi:10.1111/j.1083-6101.2008.00406.x

Yearwood, J., & Stranieri, A. (2009). Deliberative discourse and reasoning from generic argument structures. *AI & Society: Knowledge . Culture & Communication*, *23*(3), 353–377.

ENDNOTES

1 Dryzek and Niemeyer (2006, p. 647) in their discussion of pluralism and consensus come to similar conclusions: "What this might seem to suggest is that we need to introduce some procedural norms for the evaluation of particular instances of meta-consensus. One candidate would be Habermasian standards of communicative rationality".

2 Talkback radio is another instance of a media forum which can facilitate public deliberation. However, like all media forums it has limitations for public deliberation and we have not included it in the above discussion for a number of reasons. One main reason is that many studies of the medium have revealed that callers can tend to present their personal experience and/or mainly praise and/or agree with the presenter's views (a specific example here being practices surveyed in the John Laws program) (Turner, Tomlinson & Pearce, 2006, pp. 111-113, 117). While participants in talkback radio are *reasonably* diverse – especially if we consider the different kinds of talkback programs in Australia – generally most audiences "skew old", are male and Anglo (Turner et al., 2006, p. 117; Griffen-Foley, 2009a; 2009b). Additionally, as in letters to the editor and political blogs, the presenters and producers of talkback radio do have some editorial control. For example Mickler (2004) found that there are "basic technological capacities" that enable presenters and producers to vet the content and the callers. These include: "the seven-second delay between words spoken and their transmission" which enables presenters to delete repetition and/or *crazy* callers; "'voice channel override'", enabling the presenter "to cut in on a caller, simply by talking, at any time"; callers being "stacked by a producer, who puts up a list on the presenter's monitor screen of incoming callers with their name, locality and the gist of the point they wish to make" (also see Griffen-Foley, 2009a; 2009b); and the "'ring-back' or 'call-back'" where callers with predictable opinions are phoned by the production staff so they can be "plugged

into the program at any time" (pp. 103-104). Taking all these techniques together – including the "theatrical and rhetorical skills of the presenter in setting up and framing topics" – the presenter has remarkable command "over the tone, direction and biases of talkback discussion segments". Mickler further points out that these techniques used by presenters are disguised so what listeners hear are seemingly "free individuals speaking their minds in spontaneous, one-to-one conversation and debate, on perfectly equal terms with absolutely minimal mediation and second-order representation" (2004, p. 104). Having made the above points, clearly many talkback radio callers are drawing on their experience to contribute to the conversation about "the public good" they support (Turner et al., 2006, p. 117). Callers do comment on previous callers and expert opinions offered in the program and in this sense talkback radio offers a limited form of dialogue. But the nature of talk also often means that the conversation is ephemeral. We take the view that the process of writing letters and blogs enables a different kind of deliberation to that of talkback radio: in particular, through the writing process citizens can reflect on the issues that concern them; and responses in writing enable citizens to deliberate more on the issue because of the enduring nature of writing.

3 As Coleman (1997, p. 62) points out, there have been a number of different names for this movement: public journalism, civic journalism, new community journalism, public service journalism, communitarian journalism. All, however, would subscribe to the view "Journalism can and should play a part in strengthening citizenship, improving public debate and reviving public life" (Rosen cited in Coleman, 1997, p. 62).

4 Some theorists, such as Dzur, argue that media outlets are in fact mistaken when they attempt to represent themselves as "a public forum" as it is not possible for one institution to meet all the requirements of a deliberative democracy (Dzur, 2002, p. 334). He says that "public journalists underestimate how difficult it is to foster the rigorous and representative deliberation that would offer a clear view of community interests – both practical and ethical" (p. 334). One of the major problems is that journalists can present and give undue weight to views and interests that are not actually representative of "a deliberative majority" because they have not given systematic attention to the inequalities that already exist within a society (p. 334). Another is the fine line between activism and paternalism, with journalists perhaps coming to overly drive deliberation in the public sphere.

5 *The Australian* is distributed nationally six days a week, each day devoting nearly a full page to letters. This page also typically includes the editorial and a cartoon, both of which typically address current issues. This format changed in July 2005 to include a section of comments from industry/semi-industry people, for example, editors, Australian book review writers etc.

6 These papers can be accessed at http://www.aph.gov.au/library/pubs/RP/2007-08/08rp30.htm

7 Possum is an "accidental poll analyst" (see "About Possum", n.d.). Possum Comitatus is the "pseudonym of a single writer, Scott Steel", "a 34 year old from Northern Bayside Brisbane [with] a keen interest in politics and econometrics".

8 In contrast, Travers Scott (2007, p. 46) found that political bloggers in his US study tended to source mainstream news about "49.5%" and these were dominated by specific print sources whereas "alternative news outlets appeared in only 19%" of posts.

9 Apart from the founder Mark, who starts a number of threads, bloggers Paul Burns, Adrien, and Ambigulous posted in at least a half dozen threads.

10 Here we are making a distinction from other levels of language by drawing on Sadler, Hayllar and Powell's definition in *Senior Language* (1981, p. 82): "Formal usage is serious in purpose. It is polished and thoughtful and has a dignity that separates it from other levels of usage. Very often it is directed at an educated audience".

11 E-Democracy.org operates as a mutual host, facilitating public forums through the provision of technology such as the GroupServer (See "Start an Issues Forum"), and claims to build "online public space in the heart of real democracy and community" ("About – E-Democracy.Org"). Forums are organised so that dialogue is not monopolised by participants with specific interests or agendas ("Start an Issues Forum", "E-Democracy. Org"). More specifically, a process has been established by E-Democracy.org for initiating issues forums: the process includes drafting a discussion charter and rules; designating a forum manager; and a citizen-based steering committee. It is recommended that around 100 subscribers should be on each forum. E-Democracy claims civil dialogue can occur because participants are required to use real names when they participate and can only participate twice in a twenty-four hour period. These rules are also in place to enable a diverse range of opinions and participants. Nonetheless, E-Democracy forums do not appear to be conversational or intercreative in the way Meikle uses these terms in relation to the possibilities for new *forms* of debate and action (2002, p. 31). As a consequence, E-Democracy forums could be described as transmissional in nature – that is, participants have some choice and/or control within a generally "chronologically fixed or programmed information flow" (Meikle, 2002, p. 30). Despite these limitations, the relationship between form and content in E-Democracy.Org's forums is worth considering because the procedures draw attention to the conditions of possibility for public deliberation.

Chapter 17
Decision Support in Interest Based Negotiation Support Systems:
The AssetDivider System

Emilia Bellucci
Victoria University, Australia

John Zeleznikow
Victoria University, Australia

ABSTRACT

Negotiation Support Systems (NSS) model the process of negotiation from basic template support to more sophisticated decision making support. The authors attempt to develop systems capable of decision support by suggesting possible solutions for the given dispute. Current Negotiation Support Systems primarily rely upon mathematical optimisation techniques and often ignore heuristics and other methods derived from practice. This chapter discusses the technology of several negotiation support systems in family law developed in their laboratory based on data collected and methods derived from practise. The chapter explores similarities and differences between systems the authors have created and demonstrates their latest development, AssetDivider.

INTRODUCTION

Alternative dispute resolution (ADR) is becoming a popular way of resolving disputes quickly, amicably and economically. We define ADR as any process considered an alternative to public dispute resolution, for example as an alternative to court battles. The different forms of ADR include Negotiation, Mediation and Arbitration. Negotiation is often between the disputants themselves; there is no third party whose role is to act as facilitator or umpire in the communications between the parties as they attempt to resolve their dispute (Astor & Chinkin, 2002). Mediation involves a third party neutral, known as the mediator, to assist parties in formulating their own resolution of the dispute. Arbitration is an adversarial process whereby an independent third party, after hearing submissions from the disputants, makes an award binding upon the parties. The arbitration process

DOI: 10.4018/978-1-60960-091-4.ch017

could be as close to judicial determination as one can get (Charlton, 2000).

Negotiation can be described as an avenue by which social conflict is resolved. Negotiation is often the first method of dispute resolution that is called upon to resolve social conflicts. The non-violent nature of negotiation (if conducted properly) explains why negotiation is often preferred to harsher substitutes such as court trials or violent war.

There are a number of theories to describe how people should negotiate; the two major being Positional and Principled negotiation. Positional Negotiation is based on the premise that one takes a position in a dispute and argues it. Occasionally concessions will be made in order to avoid a stalemate and ultimately any solution from the negotiation will reflect a win-lose (one disputant will win, while the other loses). Principled Negotiation, developed by the Harvard Negotiation Project, advocates proposing an argument based on the disputant's interests which support their position. It also promotes cooperation between disputants by advocating a joint search for options and objective criteria. Principled Negotiation is the most widely used negotiation theory in practise and as such we have used it as a basis to our research.

Cooperative Negotiation describes the communication of parties when the outcomes are the result of coordinated behaviour of both participants (Robertson et al., 1990). Parties are more likely to be satisfied with (and most importantly adhere to) suggested solutions if they participated in reaching the solution. Whilst reaching a solution can indicate a successful negotiation, more important is how well the parties adhere to the solution. Our research is based on the assumption that disputing parties co-operate in negotiation, hence increasing the likelihood that solutions arrived by negotiation are successfully implemented.

Our work has focused on supporting the negotiation process through technology. Otherwise known as Negotiation Support Systems (NSS), we have developed a number of computer systems to support the decision making aspect of negotiation.

An emerging area within the realm of computerised alternative dispute resolution is On-line Dispute Resolution (ODR). ODR systems are web-enabled, primarily to provide parties who for various reasons cannot or should not meet face-to-face, with the opportunity to conduct negotiations and to facilitate easier and faster communication. We see Negotiation Support Systems as a subset of On-line Dispute Resolution systems, since negotiation is viewed as a type of dispute resolution.

This chapter will also classify NSS by the type of decision making it offers and how it models negotiation support on computer. The following section outlines the systems in each of these classifications, and not surprisingly, we will find many will overlap across classifications.

In line with the theme of this book, we will focus on NSS that specifically support decision making, whether that be by simply enabling disputants to communicate online to systems capable of proposing a solution to a dispute. Our review will not be limited to those providing support in Family Law (as there appears limited research outside our laboratory).

We detail Principled Negotiation as our primary negotiation theory and link this with a thorough investigation of NSS developed in our laboratory. Following this, we discuss the development of our latest computer system, AssetDivider and conclude with a brief description of future work currently being explored in our laboratory.

NEGOTIATION SUPPORT SYSTEMS

On-Line Negotiation Support Systems

According to some early pioneers of Electronic NSS, conflict in the new internet environment seemed inevitable: "Cyberspace seemed to us to be too active, too entrepreneurial, too competitive, and too lucrative a place for it not to have many conflicts" (Katsh & Rifkin, 2001). In particular, the continual growth in the development of the Internet has lead to increased numbers of Electronic NSS. (Bichler et al., 2003) describes electronic negotiations as 'processes that involve computer and communication technologies in one or more negotiation activities'. These technologies include the use of e-mail and multimedia, databases, decision support systems and knowledge-based systems. On-line NSS can be classified into the following categories: Web-based NSS, On-line Auctions, Automated Negotiation and Automated Agent-based Negotiations.

Web-based NSS refer to systems implementing the use of email and visual aids such as multimedia objects to facilitate effective communication between disputants. Negotiation support packages assist parties to overcome the challenges of conventional negotiation through a range of analytical tools to clarify interests, identify tradeoffs, recognise party satisfaction and generate optimal solutions (Thiessen & McMahon, 2000). Their aim was to better prepare parties for negotiation or to support them during the negotiation process. A primary player in this area is SmartSettle (www.Smartsettle.com), which uses graphs to illustrate the satisfaction ratings of disputants towards packages. INSPIRE (Kersten, 1997) was among the first electronic negotiation support systems developed. INSPIRE enabled disputants to negotiate through the Internet, making extensive use of email and web browser facilities. Another example of a text-based electronic negotiation support system is CBSS (Yuan et al., 1998). The system enables 'full process support' by enabling communication in real-time through hot-line coordination, message exchange and the editing of common documents. Both INSPIRE and CBSS are examples of On-line Systems that fully support the standard processes of negotiation.

Automated negotiation involves a process of 'blind bidding', where parties submit settlement offers and a computer program automatically notifies them when a settlement is reached (Schultz et al., 2001). A major provider of automated negotiation is Cybersettle (www.cybersettle.com). It is an example of on-line NSS in the area of legal claim settlement, that allows disputants to settle claim disputes by lawyers to choose the best option from three confidential offers provided by the claims professional.

There are many advantages and disadvantages of using On-line NSS. One advantage includes the seemingly private submission of offers. Most systems allow party details, offers and demands to be kept confidential, so as to protect a parties' interest should negotiations fail. There is also a considerable reduction in time attending meetings, and settlements are often achieved faster as on-line facilities operate continuously. Since the use of Internet technology tends to lower costs, it is also likely there will be a greater pool of assets to distribute. Personality conflicts or human bias can be minimised using computer systems to facilitate negotiation due to the distance a computer screen provides between disputing parties (Bellucci, 2004).

Disadvantages include the necessary use of text-based communication methods, which may reduce important cues that can lead to misinterpretations, negative interpersonal behaviour and frustration. Online communication necessarily advantages those who are familiar and comfortable with online technology and tools. It is expected, though, that the continual and widespread use of the Internet throughout society will ensure fewer people continue to feel intimidated by on-line tools.

Template vs. Decision (Making) Support Systems

In classifying NSS (online or otherwise) with respect to how they model decision support, it is useful to categorise them into template and decision (making) support systems. The earliest Negotiation Support Systems were primarily responsible for tracking past preferences and informing disputants about progress being made towards a solution to a conflict. We refer to these systems as template systems. DEUS (Zeleznikow et al., 1995), INTERNEG (Kersten, 1997), CBSS (Yuan et al., 1998), Negotiator Pro and The Art of Negotiating (Eidelman, 1993) are all template based systems.

DEUS (Zeleznikow et al., 1995) represented our earliest attempt at building negotiation support systems in Australian Family Law. It is a template-based system that displays the level of disagreement between disputants. The model underpinning the program calculates the level of agreement and disagreement between the litigants' goals at any given time. The disputants reached negotiated settlement when the difference between the goals was reduced to nil. DEUS is useful to gain an understanding of what issues are in dispute and the extent of the dispute over these issues.

Template systems are useful in the way they support human negotiation. It is the disputants who will be trying to resolve the dispute – not the computer system. DEUS for example calculated how much progress had been made in the negotiation as a way to motivate parties towards reaching a resolution.

Early decision (making) support negotiation systems primarily used Artificial Intelligence techniques to model negotiation. LDS (Peterson & Waterman, 1985) used rule-based reasoning to assist legal experts in settling product liability cases. SAL (Waterman et al., 1986) also used rule-based reasoning to help insurance claim adjusters evaluate claims related to asbestos exposure. NEGOPLAN (Matwin et al., 1989) is a rule based system written in PROLOG which advised upon industrial disputes in the Canadian paper industry. Mediator (Kolodner & Simpson, 1989) used case retrieval and adaptation to propose solutions to international disputes, while PERSUADER (Sycara, 1993) integrated case based reasoning and decision-theoretic techniques to provide decision support to United States' industrial disputes.

Mediator, Persuader, NEGOPLAN and Family_Negotiator are considered to be intelligent systems since they can generate solutions using the system's internal knowledge as well as users input. All incorporate some level of negotiation support, together with the ability to provide users with a resolution to the current problem. Solutions from these systems can be used to compare against other proposals including those arrived at by human negotiation. This will give disputants more options when deciding on which solution is best for them.

Artificial Intelligence techniques such as case-based, rule-based and hybrid reasoning have had mixed degrees of success in providing negotiation support. The Mediator proved quite successful in its retrieval and adaptation of previous cases. NEGOPLAN used rule-based reasoning to successfully model Canadian industrial disputes, while PERSUADER successfully modelled US industrial disputes through the use of a hybrid case and rule-based methodology. Family_Negotiator however, did not perform to its initial expectations, primarily due to its relatively simple modelling of the domain.

Negotiation environments, in which a number of negotiation support services are available, are becoming increasingly popular. We have developed two major negotiation support systems IMODRE, an integrated multi-agent online dispute resolution environment (Abrahams et al., 2010), and Online Dispute Resolution Environment (Lodder & Zeleznikow, 2005). Lodder and Zeleznikow's On-line dispute environment uses a dialog system and negotiation support system (based on our research) to support the resolution of conflict. Both support dispute resolution by their ability to employ various services to resolve disputes amicably.

Modelling Negotiation on Computer

Yet another classification describes how the technology is modelled in NSS; whether we model according to rational decision making (science) or human behaviour (social science). Numerous models have been developed from detailed studies of how people negotiate. Formal models, such as Game theory, rely upon a mathematical concept of optimal convergence. But do such models realistically simulate human behaviour? (Kalai & Stanford, 1988) notes 'humans are more correctly modelled as having bounded rationality, that is choosing strategies from less-than-complete considerations and striving for satisfactory rather than optimal levels of utility'. Game theory, for example, seems to ignore satisfaction as an underlying requirement of ensuring a mutually acceptable outcome. It supports a win-lose approach contrary to promoting cooperation among the parties.

Similarly, (Raiffa, 2002) describes two major analysis techniques which could be applied to how negotiation is modelled on computers.

1. Using normative analysis. This type of modelling uses rational decision making to model decisions. Researchers in normative analysis use economic theories and game theory to suggest solutions based on mathematical modelling.
2. Using descriptive analysis. Descriptive analysis focuses on how decisions are made in reality. Researchers in descriptive analysis use models derived from heuristics and actual data to mimic the negotiation process.

Examples of systems based on predominately normative advice include INSPIRE (Kersten, 1997) and SmartSettle (Thiessen & McMahon, 2000), which use Pareto Optimisation techniques to suggest optimal solutions. These contrast sharply with our work in negotiation support, namely Family_Winner (Bellucci, 2004) and Asset Divider – our systems provide advice based on the way people negotiate or to help them make more informed decisions based on simple mathematics. Our goal is to provide feasible suggested solutions that are acceptable to the user, rather than searching for purely optimal solutions.

Negotiation Support Systems Developed in our Laboratory

The underlying negotiation principle we have chosen to model is Principled Negotiation, developed by the Harvard Negotiation Project[1]. Part of our data collection phase (detailed in (Bellucci, 2004)) included interviews with a number of mediators. One of the questions asked was on the principle of negotiation support they advocate and promote with their clients. All mediators emphasised the importance of collaboration and the search for common interests. As such, and in keeping with our goal to develop systems from actual experience, we have employed Principled Negotiation as the underlying theory to our Negotiation Support Systems.

Principled Negotiation essentially emphases parties look for mutual gains. When interests conflict, Principled Negotiation advocates parties arrive at a ruling that is independent of the beliefs of either side. The essential features of Principled Negotiation as a problem-solving task are as follows:

Separate the people from the problem. This is to ensure that persons with stronger personalities cannot influence others into a decision that is biased towards a party or group of parties. This aspect is perhaps most relevant in disputes between people who are involved in an on-going relationship, for example in family law disputes that concern the welfare of children. In modelling this aspect, we use the seemingly un-biased and removed medium of a computer screen to promote a sense of separation.

Focus on Interests, not on positions. Participants must distinguish and make known their underlying values in order to justify their position. In most negotiations, each party will have interests they would like satisfied by settlement, and it is important these be understood as separate from their positions. By isolating the reasons why a position is most appealing, participants in a negotiation will increase the chance of achieving agreement. We have chosen to use ratings to model interests (as is the case in Family_Winner and Asset Divider).

Invent options for mutual gain. Even if the parties' interests differ, there may be bargaining outcomes that will advance the interests of both parties. Once interests have been ranked to determine the relative importance of each, a range of options is discussed before deciding on an outcome. Next, the negotiators need to invent options for mutual gain. This is what constitutes the decision-making aspect of the strategy. (Wertheim et al., 1992) insists brainstorming is one way of encouraging cooperative decision-making. Other approaches include Expanding the pie, awarding Compensation and Log-rolling.

Compensation and Log-rolling are similar in that both seek to resolve differences between disputants in their interests and preferences. An interest is defined as what a person truly desires from a situation, consisting of a person's wants, needs, concerns and fears. An agreement is far more likely if at least some of these interests are satisfied in the final agreement. Compensation allows for parties to be rewarded as a method to promote fairness in the final settlement. Log-rolling does not assume compensation, entirely resting on considering priorities (and the differences between them) to form an agreement. We have attempted to model decision making by using a combination of log-rolling and compensation strategies to support trade-off strategies, such as those employed in Family_Winner and Asset Divider.

Insist on objective criteria. Some negotiations are not susceptible to a win—win situation. The most obvious of these is haggling over the price of an item: since the more money one side negotiates, the less their opponent receives. In these cases, unbiased independent evaluations of an item will guide a price for the item that both parties will agree on. Wherever appropriate, our research assumes the participants agree on objective criteria prior to the commencement of a negotiation. For example, in Asset Divider, parties are expected to enter the value of an item - it is assumed negotiators have sought advice on the price of an item, and this figure has been agreed to by both parties.

Our expertise is in the domain of Australian Family Law, and in particular divorce negotiation. As mentioned above, our work has concentrated on modelling how negotiations proceed in practise, using heuristics and other methods derived from professionals in the area. Systems such as Family_Negotiator accept single offers around priorities in its attempt to arrive at a solution that is likely to be accepted by both parties. Our later work, namely Asset Divider and Family_Winner, arrive at decisions by conducting a thorough analysis of user input, including priorities, likely trade-offs and rules. We use heuristics and other methods derived from practice to arrive at a solution both parties may find acceptable. In addition, our research has concentrated on the extension of template systems to systems capable of providing offers that are acceptable to both parties.

Our laboratory has resulted in a number of NSS capable of supporting negotiation at varying degrees. Our earliest Negotiation Support System was Family_Negotiator. It utilises a hybrid rule-based and case-based system to provide disputants with advice on how to best resolve issues in an Australian Family Law dispute. Whilst evaluating the Family_Negotiator system, we discovered Family Law did not lend itself to the use of either Case-based or Rule-based Reasoning. Nor did the overall framework of Family_Negotiator provide in-depth solutions expected from real-life negotiations.

AdjustWinner (Bellucci & Zeleznikow, 1998), uses a utility function to achieve equal distribution of the common pool. The algorithm used in the system was the Adjusted Winner procedure (Brams & Taylor, 1996). AdjustWinner resolves a dispute by dividing issues and items among disputants, through a mathematical manipulation of numeric preferences. Although not classed as a NSS, AdjustWinner provided the framework for decision-making support that was later incorporated into a NSS to form Family_Winner (Bellucci & Zeleznikow, 2006).

Family_Winner is an interest based Negotiation Support System that takes a common pool of items and distributes them between two parties based on the value of associated ratings. The basic premise of the system is that it allocates items based on whoever values them more. Once an item has been allocated to a party, the ratings of the remaining items are modified (according to the actions of trade-offs) to ensure the items (and their associated ratings) are ready for the next round of allocation. The system implements a well known phenomenon in mediation, that disputants do often change their initial ratings in light of the allocation of a related issue. The system develops this aspect of face-to-face negotiation through trade-off manipulation and compensation strategies, developed from heuristics embedded in data. Family_Winner is synonymous with the practises of family mediators. For example, Family_Winner can only operate on the assignment of ratings by parties (which indicate the relative importance of each issue to a disputant – a priority) and supports the sequential resolution of issues. Mediators involved in our data collection indicated they collected priority settings from their clients, and often resolved disputes by considering each issue separately.

AssetDivider, developed as an extension of Family_Winner, provides support to asset division. Family_Winner takes ratings assigned to items by the parties involved and develops a list of allocations to each party; based on trade-offs inherently present in the dispute. AssetDivider similarly develops a list of allocations based on tradeoffs, except it also takes into account the monetary value of assets and accepts a percentage split (provided by the mediator or other user) to guide the development of an allocation list for parties.

The following sections will detail the theory of negotiation support in AssetDivider (based predominately on Family_Winner) and will demonstrate its actual use through a case study.

AssetDivider

AssetDivider uses the major principles underlying Family_Winner – namely in their use of interests. The theory which best supports our definition of negotiation support is Principled Negotiation (Fisher & Ury, 1991), developed under the Harvard Negotiation Project. It emphasizes parties look for mutual gains and focuses on the underlying values (or interests) that justify a disputant's position, as opposed to attempting negotiation solely from their positions.

Family_Winner takes a common pool of items and distributes them between two parties based on the value of associated ratings. Each item is listed with two ratings (a rating is posted by each party), which signify the item's importance to the party. A rating in Family_Winner is a number in value from 0 - 100 (0 being of no importance; 100 to signify absolute importance). The algorithm to determine which items are allocated to whom works on the premise that each parties' ratings sum to 100; thereby forcing parties to set priorities. The program always checks this is the case, and if not, it realigns ratings to ensure all sum to 100. The system's basic rule of thumb is that it allocates items based on whoever values them more. Once an item has been allocated to a party, the ratings of the remaining items are modified (according to the actions of trade-offs) to ensure the items (and their associated ratings) are ready for the next round of allocation (Bellucci, 2004).

Family_Winner was evaluated by a number of family solicitors at Victoria Legal Aid (VLA). Whilst the solicitors were very impressed with the way Family_Winner suggested trade-offs and compromises, they had one major concern

– that in focusing upon negotiation, the system had ignored the issues of justice (Zeleznikow & Bellucci, 2006). For example, Family_Winner simply allocates property to parties based on their interest in the item. It does not allow for monetary values to influence the allocation process. The dollar value of items is important to the dispute because each party wants to be allocated the right or 'just' amount of money. This concept contrasts with linking an interest value to an item, which is intrinsically different. An interest is an evaluation based on the significance of the item to a person. For example, party A may be very fond of a lamp that has been passed down throughout the generations, and consequently they give it a rating of 50. The remaining items are not as important to party A, and so are given much lower ratings. Whilst using interests to negotiate is a very interesting exercise, it does not in any way reflect the dollar value of the item. This is where Family_Winner fails to support the mediation process effectively. Whilst our industry partners, Relationships Australia Queensland (RAQ) consider the way Family_Winner supports interest-based negotiation by setting priorities as useful; they were also concerned with the missing impact of monetary values. Hence, our new theory of negotiation support (implemented in AssetDivider) incorporates the basis of Family_Winner's allocation and trade-off strategy by utilizing both interests and an item's monetary value.

THEORY IMPLEMENTED IN ASSETDIVIDER

Whilst AssetDivider is a standalone system, many of the ideas used to form the system were a developed in Family_Winner. For this reason, we will be drawing attention to AssetDivider's differences and similarities in relation to Family_Winner. We will be assuming the reader has no prior knowledge of AssetDivider's predecessor.

Family_Winner and AssetDivider's Input and Output

Family_Winner takes a list of issues (usually items for distribution between two parties) and allocates them based on a rating given by the parties in dispute. Two sets of ratings are provide, one for each party in dispute. This rating (a numerical value between 0 and 100) does not represent the monetary value of the item, instead it symbolises how important the item is to the party. We assume a party wants to keep an item they feel is important to them.

Similarly, AssetDivider accepts a list of items together with ratings (two per item) to indicate the item's importance to a party. In addition it also accepts the current monetary value of each item in dispute. We assume this dollar value has been negotiated (if necessary) before AssetDivider is used[2]. Hence, only one dollar value is entered per item. The proposed percentage split is also entered; this reflects what percentage of the common pool each party is likely to receive in the settlement. The system is not capable of determining the percentage split; this figure has to be derived from the mediator's knowledge in past cases or from computer systems such as SplitUp (Stranieri et al, 1999), which can provide a percentage split given certain characteristics and features of divorce cases.

AssetDivider's output consists of a list of items allocated to each party. All of the items (except one) on the allocation lists were provided by disputants in the intake screen. The additional item is a "payout" item, which reflects the amount of money a disputant would need to pay the other party for the items they have been allocated and collectively are valued greater than the percentage split offers them. For example, party A have been allocated a total value of $100,000 in assets, and party B $115, 000. Under a 50/50% split, party B will need to pay $15,000 to party A to satisfy the percentage split. Family_Winner in contrast does not accommodate payouts since its focus was to distribute items determinate on only interest ratings.

Family_Winner and AssetDivider's Allocation Strategy

The order by which issues are allocated is of paramount importance in a negotiation. Professional mediators have indicated issues attracting little disputation should be presented foremost for allocation, so as to help foster a positive environment in which to negotiate. By summing the ratings of issues to 100, the level of discourse surrounding an issue can be measured by calculating the numerical distance between the ratings of an issue assigned by each of the parties. For example, if two parties assign the same high rating to an item, then it is expected the level of disputation surrounding the issue to be substantial (because both parties want the item), whereas large differences between the ratings of parties indicate the issue will be resolved much more quickly. Both Family_Winner and AssetDivider use this strategy in deciding the order by which items are presented for allocation.

Family_Winner allocates items to parties according to whoever values them the most. Once an item has been allocated to a party, the remaining ratings (of items still in dispute) are changed by trade-off equations. These modifications try to mimic the effect losing or gaining an item will have on the rest of the items still in dispute. The equations directly modify ratings by comparing each rating against that of the item recently lost or won (each party's set of ratings are modified as a result of an allocation). The equations update ratings based on a number of variables - whether the item allocated was lost or gained, the value of the allocated item in relation to items still in dispute and the value of the item whose rating will change as a result. In Family_Winner, the extent to which ratings were modified was determined through an analysis of data we collected from mediation cases provided by the Australian Institute of Family Studies. These are detailed in (Bellucci, 2004).

AssetDivider accepts items, a rating per issue and the monetary value of an item (unlike Family_Winner, which does not consider the monetary value of items at all). The allocation strategy as described above is similar in Family_Winner, except that the equations have been modified to reflect greater fairness by considering the price of an item. AssetDivider's allocation strategy works by provisionally allocating an item to the party whose rating is the highest. It then checks the dollar value of items it has been allocated previously (that is, their current list of items), the dollar value of the item presently allocated and the dollar amount permitted under the percentage split given by mediators. If by allocating the item in question the party exceeds its permitted amount, the item is removed from its allocation list and placed back into negotiation. In this case, the item has not been allocated to a party. If the dollar value of the item was within the limits of the amount permitted under the percentage split rule, then the allocation proceeds. Once an allocation has occurred the 'losing party' is compensated by the trade-off equations modifying ratings (whereas in Family_Winner both winning and losing parties were affected). Analysis from the calculations in Family_Winner revealed it was not always fair to the losing side if the winning side received a higher rating.

Family_Winner had also attracted some criticism concerning the scaling of ratings to sum to 100 only once (at the initial intake). After the system removed an item from the negotiation (upon allocation); it was argued that the remaining ratings in dispute should be scaled to 100 again. The reasoning here is to ensure that every item has been allocated with the same rules in place (that is all ratings add to 100) as was the case in the first item's allocation. Whilst theoretically this reasoning is quite sound; there was a problem with implementing this in practise. Once we implemented this new methodology, we found that as the number of issues in dispute reduced, so did the difference between the ratings of the same item. This result defeated the reason why we introduced trade-offs in the first place, ie trade-offs should cause some effect. It is also a problem when ratings are dissimilar because we do not know to which party we should allocate the item. The reason why AssetDivider does not scale all ratings to 100 following allocation is to ensure the ratings still reflect disputant priorities set in the first instance. The trade-off equations allow for small changes to apply to ratings, and it is these changes that decide to who the item will be allocated.

User Interface Issues

In comparing user interface issues between the two systems, significant improvements have been made to AssetDivider. There is more space on the screen for users (we presume will be Mediators or lawyers) to enter additional information about the case. We have added reporting services, which will print case details such as case identifiers (case number), initial ratings given by users, ratings upon allocation and a final summary of the solutions arrived at by the system. This summary will include the list of allocated items and their associated monetary value of each party.

In Family_Winner, diagrams were shown on screen to describe the current 'state of play', that is the items in dispute, their values (ratings), and Relationship Ratings (RR) between items. Relationship ratings are used to reflect the importance a party places on one item in relation to another. Mathematically, the RR is the absolute difference between the ratings of two items. We named these diagrams Trade-off Maps, which are based on the structure of Constraint Diagrams. They were shown on screen just before an allocation occurs to help users understand how Family_Winner allocates items. In developing AssetDivider, we decided not to include these Trade-off Maps, as informal discussions with users revealed they simply helped to confuse users – and contrary to the reason why they were developed – did not aid user understanding of how the system arrived at its solution. In conjunction with displaying Trade-off Maps, Family_Winner would display new ratings as they change; that is every allocation was displayed sequentially on screen. This made using Family_Winner quite tedious; as the user had to clear each screen for every allocation that occurred. AssetDivider displays the solutions it has arrived at after all items have been allocated; at which point the user can choose to print or save the allocation list.

In addition, the system has been designed so users can print a number of percentage split scenarios per case very easily. Once the information pertaining to a case has been entered, the user can press the back button on the screen to arrive at the screen where the user can change the percentage split, and then press the 'allocate' button on the next screen to see the results. As a mediator commented to us; it is a very useful feature if it allows clients to view allocation lists given different percentage split scenarios.

Demonstration of AssetDivider

AssetDivider is a decision support tool to help mediators develop scenarios very quickly for the resolution of property mediations in family law. The program accepts as input data pertaining to a single case, and with only a couple of screens, is able to generate a list of property allocations to each party. The user will need to enter the following information:

1. A list of items in dispute together with ratings (two per item), which indicate the importance or priority of an item to a party. There is a slider bar which can be used to enter a number, or one can enter the number directly in the grid. The rating can be any number (except 0 if both parties enter 0 for an item). Users should be aware the program

will take the ratings of a party and scale them to 100 for the purposes of the calculations made by Asset Divider. It might be easier to interpret the results if users enter the ratings so that a party's rating list adds to (or near to) 100.
2. The current monetary value of each item in dispute. We assume the dollar value has been negotiated prior to Asset Divider's use. Hence, only one dollar value is entered per item.
3. A percentage split is also entered; this reflects what percentage of the common pool each party is to receive in this scenario.

The program also allows for the entry of identifying information, such as names, comments and descriptions of the case. This data is optional, as the program only requires data 1- 3 stated above to calculate allocations.

Asset Divider's output consists of a list of items allocated to each party. All of the items (except one) on the allocation lists were provided in the intake screen by the disputants. The additional item is a "payout" item, which reflects the amount of money a disputant would need to pay the other party for the items they have been allocated and collectively are valued greater than the percentage split offers them.

AN EXAMPLE USING ASSETDIVIDER

The case description of this real-life divorce scenario and the relative point allocations have been extracted from (Brams & Taylor, 1996) page 105. The case Jolis v Jolis, began on December 5th, 1980, and concluded on October 30th, 1981. The case was heard in New York City, at a time when a new law subjecting all martial property to a 50 –50 split was being introduced. The couple had been married for 41 years, of which 33 they spent together. The Wife had given up her early and successful career to care for the couple's four sons. The couple had lived together in substantial wealth, primarily due to the expansion of the Husband's diamond business.

There are both real estate and liquid assets to be divided. The Husband's diamond business is not treated as marital property as its growth was primarily due to market forces, especially the diamond boom of the 1970's. The children's welfare is not included as an issue as they are no longer considered minors at the time of separation.

The following screens details the presentation of the above mentioned case to AssetDivider. The relevant case information is entered in the intake screen (Figure 1).

The next screen (Figure 2) appears once the user clicks on Next>>. It is also the final screen of the program. In this screen, users are able to enter their ratings. Once the user clicks "Allocate Items", the allocation summary is filled appropriately. In the Allocation Summary table, ratings for Husband (party A) and Wife (party B) are scaled to add to 100 in columns "Party A Rating" and "Party B Rating" respectively. It is then these ratings that are used to drive the allocation. In this case, the ratings initially entered add to 100.

Table 1. Point allocations and dollar valuations (Brams & Taylor, 1996), page 105

Issues	H's ratings	W's ratings	Dollar value of asset
Paris Apartment	35	55	$642,856
Paris Studio	6	1	$42,850
New York Coop	8	1	$103,079
Farm	8	1	$119,200
Cash And Receivables	5	6	$42,972
Securities	18	17	$176,705
Profit Sharing Plan	15	15	$120,940
Life Insurance Policy	5	4	$24,500
Total	100	100	$1,273,102

Figure 1. Intake screen for negotiation

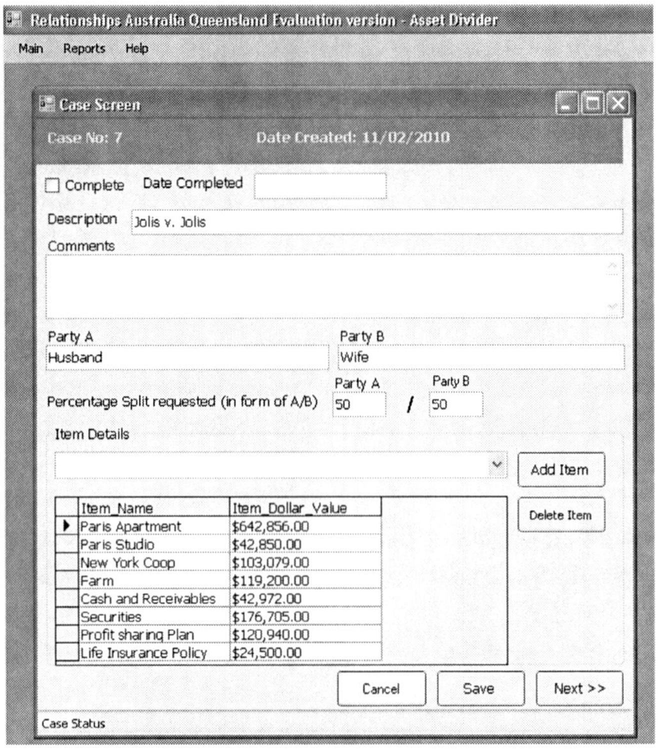

Figure 2. Final screen of AssetDivider. It gives the user the allocation list for each party; which includes a payout figure allocated accordingly

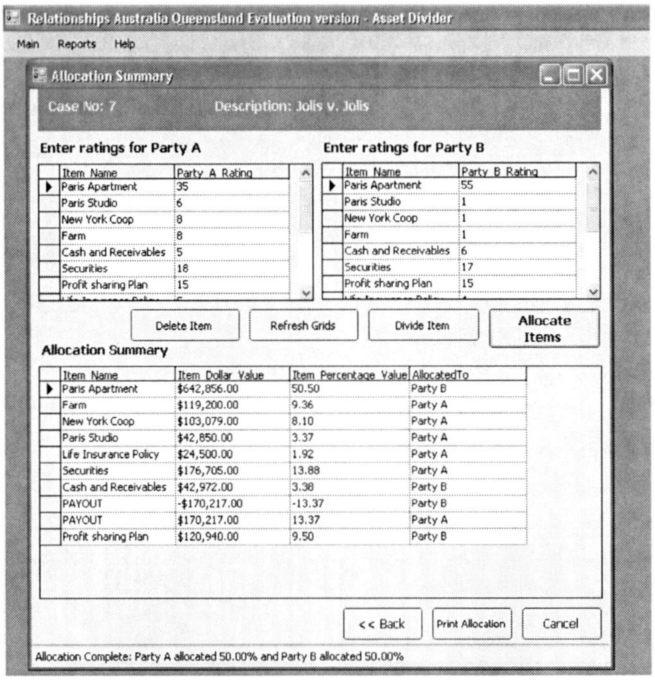

The Allocation Summary lists another issue, which is not initially part of the negotiation. It is the payout figure, which is used to ensure the parties received their portion of the percentage split, entered in the first screen.

Once the user presses the button "Print Allocation", the following report is produced.

According to AssetDivider, the preferred outcome, taking into account each party's' priorities (ratings) and percentage split indicated:

In analysing the case, we can see that both parties wanted the Paris Apartment above all else; though Wife (party B) valued it more than the husband (Party A). As a consequence, both parties gave the rest of the items relatively low values. On the whole, both parties received the items they valued considerably (except for Party A's loss of Paris Apartment to Party B – since she valued it much greater). The only item valued equally by the parties was profit-sharing plan (15). It was given to Party B. Party B also need to pay out Party A the amount of 170,217 to ensure the split is exactly 50%.

AssetDivider allows users to quickly generate different allocation lists for the same case, by allowing the percentage split in screen 1 to be modified as appropriate. Mediators at RAQ commented it is often useful to present clients with a number of scenarios, which detail exactly what items (or lack of) form part of their percentage split. It may be the case clients take a lower percentage if it means they do not have a debt to repay, for example.

To generate different allocation lists for the same user, the user must press on the << Back button to make a change in the percentage split

Figure 3. Allocation report for Jolis vs. Jolis

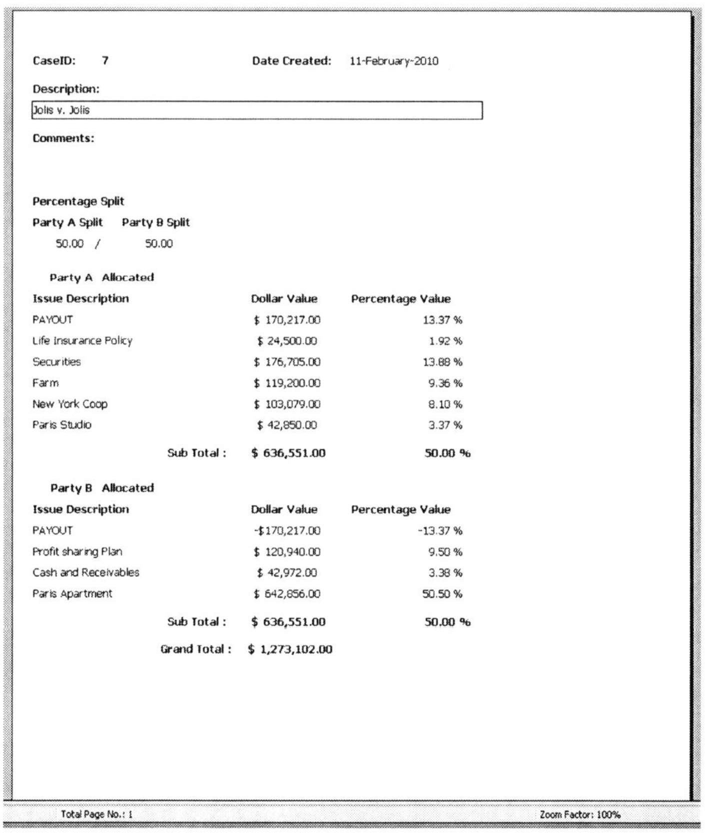
CaseID: 7 Date Created: 11-February-2010

Description:

Jolis v. Jolis

Comments:

Percentage Split

Party A Split Party B Split

50.00 / 50.00

Party A Allocated

Issue Description	Dollar Value	Percentage Value
PAYOUT	$ 170,217.00	13.37 %
Life Insurance Policy	$ 24,500.00	1.92 %
Securities	$ 176,705.00	13.88 %
Farm	$ 119,200.00	9.36 %
New York Coop	$ 103,079.00	8.10 %
Paris Studio	$ 42,850.00	3.37 %
Sub Total :	$ 636,551.00	50.00 %

Party B Allocated

Issue Description	Dollar Value	Percentage Value
PAYOUT	-$170,217.00	-13.37 %
Profit sharing Plan	$ 120,940.00	9.50 %
Cash and Receivables	$ 42,972.00	3.38 %
Paris Apartment	$ 642,856.00	50.50 %
Sub Total :	$ 636,551.00	50.00 %
Grand Total :	$ 1,273,102.00	

Total Page No.: 1 Zoom Factor: 100%

Table 2. Allocation list for husband (party A) and wife (party B) using AssetDivider

Husband (Party A)	Value of	Wife (Party B)	Value of
Farm	$119,200	Paris Apartment	$642,856
New York Coop	$103,079	Cash and receivables	$42,972
Paris Studio	$42,850	Profit Sharing Plan	$120,940
Life Insurance Policy	$24,500		
Securities	$176,705		
Payout	$170,217		-$170,217
Total:	**$636,551**	**Total:**	**$636551**

window, then move to the next screen by pressing the >> Next button and press the "Allocate items" button to enable the changes to affect the allocation. The table "Allocate Summary" is then populated accordingly. For example, the percentage split box in Figure 1 is changed from 50/50 to 40/60. The following screen (Figure 4) shows the result of a 40/60 percentage split (given all other data remains constant).

The corresponding report follows. (see Figure 8)

In this scenario, it is clear that the amount of money Party B must pay to party A is greatly reduced – from 170,217 to 42,906. Hence, the parties may agree to this settlement if neither wants a substantial amount of debt in the settlement.

Figure 4. Allocation Summary screen for Jolis v.Jolis case; percentage split is 40/60

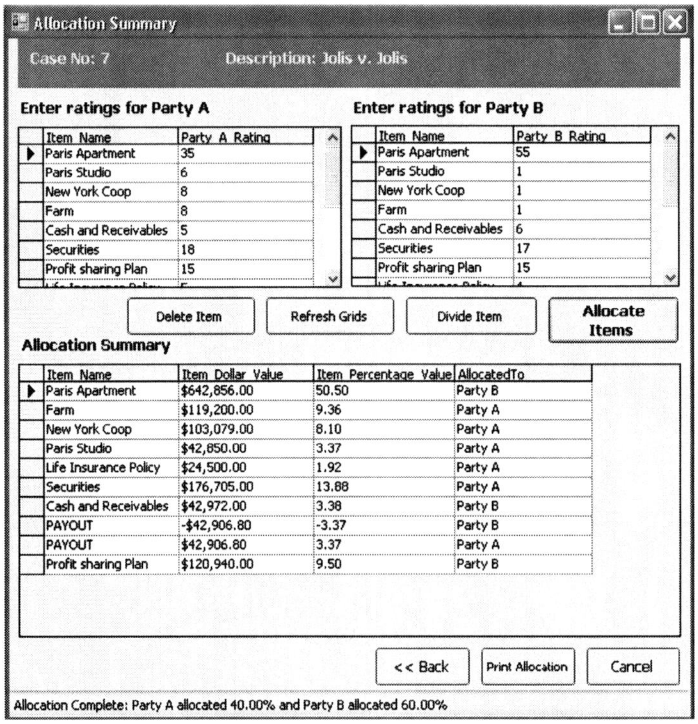

Figure 5. Report for Jolis vs. Jolis case; percentage split is 40/60

CaseID: 7 **Date Created:** 11-February-2010

Description:

Jolis v. Jolis

Comments:

Percentage Split

Party A Split	Party B Split
40.00 /	60.00

Party A Allocated

Issue Description		Dollar Value	Percentage Value
PAYOUT		$ 42,906.80	3.37 %
Life Insurance Policy		$ 24,500.00	1.92 %
Securities		$ 176,705.00	13.88 %
Farm		$ 119,200.00	9.36 %
New York Coop		$ 103,079.00	8.10 %
Paris Studio		$ 42,850.00	3.37 %
	Sub Total :	**$ 509,240.80**	**40.00 %**

Party B Allocated

Issue Description		Dollar Value	Percentage Value
PAYOUT		-$42,906.80	-3.37 %
Profit sharing Plan		$ 120,940.00	9.50 %
Cash and Receivables		$ 42,972.00	3.38 %
Paris Apartment		$ 642,856.00	50.50 %
	Sub Total :	**$ 763,861.20**	**60.00 %**
	Grand Total :	**$ 1,273,102.00**	

Total Page No.: 1 Zoom Factor: 100%

Fairness and Successful Decision Making

Brams and Taylor (Brams & Taylor, 1996) view fairness in a negotiation as ensuring each party receives an equal percentage of their priorities. The Adjusted Winner algorithm (Brams & Taylor, 1996) guarantee fairness and equitability by ensuring an equal number of points (represented by issue ratings) are awarded to each party through a distribution of issues or items.

In an ideal environment, where fairness can be applied with definite certainty, the theories of (Brams & Taylor, 1996) and (Pruitt & Carnevale, 1993) are sustainable. However, our goal of providing negotiation support does not easily lend itself to fairness assessment, due to:

1. The difficulty in assessing fairness to a system whose numerical values fluctuate during the course of negotiation; and
2. A lack of data on which to base comparisons.

Family_Winner nor Asset Divider employ any of the fairness principles mentioned above. It interprets fairness as how satisfied the parties are with the proposed allocation. We argue a disputant's satisfaction is more important than their need for a supposedly fair outcome. The theories promoted in this chapter support satisfaction by allocating issues based on an issue's value to the party. Trade-offs are utilised to enable compensation, satisfying the system's attempt to make the allocation equally satisfactory to both parties.

(Raiffa, 2002) describes prescriptive analysis as a modelling technique in negotiation, which we have chosen to apply to the evaluation of NSS. Prescriptive analysis describes how real people benefit from advice to improve their behaviour. Advice of this kind must be evaluated by its pragmatic value (by its ability to help people make better decisions). Our work evaluates success by judging the usefulness of a NSS to the (human) negotiation.

We have developed two questionnaires (conducted with Mediators) and an interview schedule with managers, based on the CCCF System Operational Context (Hall & Zeleznikow, 2003). These ask users to comment on the operation and use of the system;

1. Mediator Usage census, and
2. Mediator Survey

The mediator's usage census (1) above, is to assess how mediators use the program to assist in mediations. Ideally it should be filled in after a mediation has concluded (whether Asset Divider was used or not), and allows users to enter their thoughts on why Asset Divider was used (or why not), how it was used and their comments on its usefulness.

The survey for mediators (2) should be filled out at the end of the trial (evaluation) period. It allows mediators to reflect on the appropriateness of the program, its user friendliness, quality and overall feel of its usefulness to family mediation.

AssetDivider is yet to undergo an evaluation by our industry partners RAQ. It is expected mediators at RAQ will test and evaluate the system in the near future. We are also in negotiations with other organisations for the purposes of evaluation and commercial use of the software. We expect an extensive evaluation campaign to reveal further improvements to the decision making module and user interface. Our research has revealed a lack of negotiation support systems used in family law. We hope our collaboration with RAQ (or other organisations) will enable AssetDivider to be used in practise, being the first negotiation support systems to do so.

How well will AssetDivider support decision making? This question is difficult to answer without a thorough evaluation of how AssetDivider has been used. Notwithstanding, on a recent visit to RAQ, we were told mediators may use the program to move clients away from trying to attain a particular percentage of the value of the common pool. Often lawyers or family friends may have provided this advice. There may also be issues with a 'loss of face' if they do not fight for a percentage they consider fair. The program used in this way will help clients see what items make up the given percentage split. They may move away from their initial position if they see what items (including the associated payout) they are likely to receive. Mediators believed if the program was used in this manner, it would greatly affect a party's decision making process.

In addition, AssetDivider would provide mediators with confidence to effectively mediate property-related issues. Most family law mediators have degrees in social work or law. Their expertise mainly lies in mediating child-related issues such as visitation schedules, primary care and other associated child related matters. Access to AssetDivider may enable the settlement of both child-related and property issues by mediators; thereby reducing their reliance on lawyers and of course their often exuberant associated costs. AssetDivider's decision making capabilities may enable a mediator to confidently provide advice regarding asset division.

CONCLUSION AND FUTURE WORK

This chapter discusses current research in Negotiation Support Systems particularly in Family Law. Our survey demonstrates the wide number of NSS by classifying existing systems into ADR, on-line, template and decision making support systems. We settle on Principled Negotiation as our underlying theory to negotiation, and then demonstrate how we applied the theory's characteristics to a computer representation. Family_Winner and AssetDivider are examples of our laboratory's attempt to develop solutions which in part mimic and extend the negotiation process between divorcing couples. This chapter devotes most of its space to the development and demonstration of AssetDivider. It is our latest offering which limited informal evaluation has already revealed, will develop as a useful tool to family lawyers and mediators alike.

Future work in negotiation support includes the development of a fully supported integrated negotiation support system, which allows for the negotiation of all issues in family mediation – including asset division and child related issues. The demand for such a service is unquestionable; as is the current demand for family lawyers and mediators. With extensive research, it is believed the development of an all-encompassing negotiation environment will help resolve disputes quickly and cheaply.

Other ideas centre on the use of AssetDivider in different domains. It may be possible to use AssetDivider (or a derivative) to mediate in the dissolution of companies, or in the division of a person's estate once they have passed on – as long as there are two stakeholders involved. This brings us to another point - if we are to move into the development of mediating or division type systems, our research will need to move into multi person dispute resolution. To date we have resisted this direction, simply due to the myriad of complexity it brings to the table.

In March 2008, we commenced a large research grant from the Australian Research Council[3], to work with the Queensland Branch of Relationships Australia, to develop a negotiation decision support system using both interested-based and justice-based negotiation.

Online dispute resolution makes it possible to provide family law services to parties who are geographically dispersed. In such circumstances, the process of separation and divorce could historically feature a large amount of correspondence and expensive litigation. Australia is both a country and a continent, the population is geographically spread and a part of the population lives in remote regions. The provision of government services to such communities is both difficult and expensive.

To meet such needs the Australian Government initiated the Family Relationship Advice Line (FRAL)[4]. The Advice Line is a national telephone service established to assist people affected by relationship or separation issues. The Advice Line provides information on family relationship issues and advice on parenting arrangements after separation. FRAL can also refer callers to local services that can provide assistance.

In a major addition to traditional family dispute resolution services, FRAL provides the organisation of telephone dispute resolution for people assessed as more appropriate for the telephone medium than for face-to-face services. The Telephone Dispute Resolution Service (TDRS) was established through funding from the Federal Attorney General of the Australian Government in 2007 (Thomson, 2009) and is based in Queensland, operated by Relationships Australia (Qld) in partnership with Relationships Australia (NSW). Potential clients cannot directly contact or self refer to the TDRS; rather they need to be referred either through the Family Relationship Advice Line, a Family Relationships Centre, or any other government-funded Family Dispute Resolution provider.

Online Family Dispute Resolution services potentially broaden the options and capacity of existing services by offering another method of supporting families to reach agreements. Given that most people in disputes about the care of children are under 45, arguably, 95% of the disputants will have familiarity with and access to the appropriate technology. For those who do not have access, the technology could conceivably become readily available at locations such as local libraries, Community Legal Centres or

Family Relationship Centres. Furthermore, such an approach could include provision of mobile resources in centres or hire facilities. Compared to the costs of litigation or even prolonged alternative dispute resolution, the investment in online technologies is potentially both value adding and cost saving.

In conjunction with Relationships Australia Queensland, we are developing an Online Family Dispute Resolution Service.

We are investigating principles for the successful negotiation of Information Technology Outsourcing Agreements (Chandar & Zeleznikow, 2009).

Currently we are about to commence two projects that investigate the fairness and consistency of negotiation support systems in the legal domain. In a project with title 'Developing negotiation support systems in law which encourage more consistent and principled outcomes' we argue that unless negotiation support systems are seen to advocate outcomes which arise from consistent and principled advice, disputants will be reluctant to use them. Thus we propose conducting research that will develop measures for assessing the outcomes of online negotiation in the legal domains of sentencing, plea bargaining and family mediation. Such measures will form the basis of a new model for evaluating justice and consistency within online dispute resolution systems. The model will inform the construction of fairer and more consistent systems of IT-based negotiation support in the future.

To meet this goal we will:

1. Develop models of consistency and justice based on two very distinct legal domains: sentencing and family law. Further, the knowledge about these domains will be shared from three distinct Common Law jurisdictions: Australia, Israel and USA.
2. Develop information retrieval techniques to extract knowledge from textual legal and negotiation data.
3. Use KDD techniques (such as association rules, Bayesian belief networks and neural networks) to compare litigated and negotiated family law cases.
4. Develop models of disputation and negotiation in both family law and sentencing. These models will then be tested to examine how closely they align with the notion of Bargaining in the Shadow of the Law (as compared to 'pure' interest-based negotiation).
5. Use Lodder and Zeleznikow's three step model for an Online Dispute Resolution Environment and Toulmin's theory of argumentation to construct a generic online dispute resolution environment. The development of such an environment on which to place various negotiation support systems will increase users' access to justice.
6. Develop and evaluate specific sentencing and negotiation support systems using our newly developed Online Dispute Resolution Environment.

While producing significant information technology and socio-legal advances, the project also develops innovative artificial intelligence techniques to help develop 'consistent' and 'just' negotiation support systems and an environment for online dispute resolution.

(Gray et al., 2007) argue that rather than develop negotiation support systems that are useful in resolving legal disputes, it is more important to develop systems that provide advice about how to avoid conflicts. In domains such as family law and body corporate disputes, it is important for the disputants to have ongoing relationships once the disputes have been resolved. Hence it is vital to develop software that focuses upon developing ongoing relationships, rather than focusing upon interests, justice (using BATNAs) or power. Negotiation support systems that incorporate machine learning and planning can help in this regard.

The next few years will reveal our research directions, and together with the continued support of the NSS community, we hope to continue the all important work in developing research for use in Negotiation Support Systems.

REFERENCES

Abrahams, B., Bellucci, E., & Zeleznikow, J. (2010). Incorporating fairness into development of an integrated multi-agent online dispute resolution environment. *Group Decision and Negotiation*. Retrieved from http://www.springerlink.com/content/2jl7774266830554/

Astor, H., & Chinkin, C. M. (2002). *Dispute resolution in Australia* (2nd ed.). Sydney, Australia: LexisNexis Butterworths.

Bellucci, E. (2004). *Developing compensation strategies for the construction of negotiation decision support systems*. Unpublished doctoral dissertation. La Trobe University, Victoria, Australia.

Bellucci, E. (2008). Development of a mediation tool in family law: AssetDivider . In Francescori, E., Sartor, G., & Tiscornia, D. (Eds.), *Legal Knowledge and Information Systems* (pp. 88–97). Amsterdam: IOS Press.

Bellucci, E., & Zeleznikow, J. (1997). *Family-negotiator: An intelligent decision support system for negotiation in Australian Family Law*. Proceedings of the Fourth Conference of the International Society for Decision Support Systems, Lausanne. *International Society for Decision Support Systems*, 359-373.

Bellucci, E., & Zeleznikow, J. (1998). A comparative study of negotiation decision support systems. In *Thirty-First Hawaii International Conference on System Sciences IEEE Computer Society* (pp. 254 – 262).

Bellucci, E., & Zeleznikow, J. (2006). Developing negotiation decision support systems that support mediators: A case study of the Family_Winner system. *Journal of Artificial Intelligence and Law*, *13*(2), 233–271. doi:10.1007/s10506-006-9013-1

Bichler, M., Kesten, G., & Weinhardt, C. (2003). Electronic negotiations: Foundations, systems, and experiments [Introduction to the Special Issue]. *Group Decision and Negotiation*, *12*, 85–88. doi:10.1023/A:1023064720309

Brams, S. J., & Taylor, A. D. (1996). *Fair division, from cake cutting to dispute resolution*. Cambridge, UK: Cambridge University Press.

Bui, T., & Shakun, M. F. (1997). *Introduction to the negotiation support systems minitrack*. In Thirtieth Annual Hawaii International Conference on Systems Science–Mini track on Negotiation Support Systems, IEEE, Hawaii.

Chandar, S., & Zeleznikow, J. (2009). Risks to consider when negotiating IT outsourcing agreements. Proceedings of GDN2009, the 10th annual meeting INFORMS section on Group Decision and Negotiation, Toronto, Ontario, June 14-17. Retrieved August 26, 2009, from http://info.wlu.ca/~wwwmath/faculty/kilgour/gdn/papers.htm

Charlton, R. (2000). *Dispute resolution guidebook*. Sydney, Australia: LBC Information Service.

Eidelman, J. A. (1993). Software for negotiations. *Law Practice Management*, *19*(7), 50–55.

Fisher, R., & Ury, W. (1991). *Getting to yes: Negotiating agreement without giving in*. Boston: Houghton Mifflin.

Gray, P., Gray, X., & Zeleznikow, J. (2007). Decision negotiating logic: For richer or poorer. In R. Winkels (Ed.), *Proceedings of Eleventh International Conference on Artificial Intelligence and Law, Palo Alto, CA* (pp. 247-251). New York: ACM Press.

Hall, M. J. J., & Zeleznikow, J. (2003). A method for evaluating legal knowledge-based systems based upon the Context Criteria Contingency guidelines framework. In *Proceedings of the Ninth International Conference on Artificial Intelligence and Law, Edinburgh, Scotland* (pp. 274-283). New York: ACM Press.

Kalai, E., & Stanford, W. (1988). Finite rationality and interpersonal complexity in repeated games. *Econometrica*, *56*(2), 397–410. doi:10.2307/1911078

Katsh, E., & Rifkin, J. (2001). *Online dispute resolution: Resolving conflicts in cyberspace*. San Francisco, CA: Jossey-Bass.

Kersten, G. E. (1997). Support for group decisions and negotiations: An overview . In Climaco, J. (Ed.), *Multiple criteria decision making and support*. Heidelberg, Germany: Springer Verlag.

Kolodner, J. L., & Simpson, R. L. (1989). The mediator: Analysis of an early case-based problem solver. *Cognitive Science*, *13*, 507–549. doi:10.1207/s15516709cog1304_2

Lodder, A., & Zeleznikow, J. (2005). Developing an online dispute resolution environment: Dialogue tools and negotiation systems in a three step model. *The Harvard Negotiation Law Review*, *10*, 287–338.

Matwin, S., Szpakowicz, S., Koperczak, Z., Kersten, G. E., & Michalowski, G. (1989). NEGOPLAN: An expert system shell for negotiation support. *IEEE Expert*, *4*, 50–62. doi:10.1109/64.43285

Mnookin, R., Peppet, S. R., & Tulumello, A. S. (2000). *Beyond winning: Negotiating to create value in deals and disputes*. Cambridge, MA: The Belnap Press of Harvard University Press.

Peterson, M., & Waterman, D. (1985). Evaluating civil claims: An expert systems approach to evaluating product liability cases . In Walter, C. (Ed.), *Computer power and legal reasoning* (pp. 627–659). St. Paul, MN: West Publishing Company.

Pruitt, D. G. (1981). *Negotiation behaviour*. New York: Academic Press.

Raiffa, H., Richardson, J., & Metcalfe, D. (2002). *Negotiation analysis: The science and art of collaborative decision making*. Reading, MA: Addison Wesley.

Robertson, S., Zachary, W., & Black, J. (1990). *Cognition, computing and cooperation*. Norwood, NJ: Ablex Publishing Cooperation.

Schultz, T., Kaufmann-Kohler, G., Langer, D., & Bonnet, V. (2001). *Online dispute resolution: State of the art, issues, and perspectives* [E-Com research project]. Geneva, Switzerland: University of Geneva Law School.

Stranieri, A., Zeleznikow, J., Gawler, M., & Lewis, B. (1999). A hybrid—neural approach to the automation of legal reasoning in the discretionary domain of family law in Australia. *Artificial Intelligence and Law*, *7*(2-3), 153–183. doi:10.1023/A:1008325826599

Sycara, K. (1993). Machine learning for intelligent support of conflict resolution. *Decision Support Systems*, *10*, 121–136. doi:10.1016/0167-9236(93)90034-Z

Thiessen, E. M., & McMahon, J. P. (2000). Beyond win-win in cyberspace. *Ohio State Journal on Dispute Resolution*, *15*, 643.

Thomson, M. (2009). *Annual report of the telephone dispute resolution service to the attorney general*. Queensland: Relationships Australia.

Waterman, D. A., Paul, J., & Peterson, M. (1986). Expert systems for legal decision making. *Expert Systems: International Journal of Knowledge Engineering and Neural Networks*, *3*(4), 212–226. doi:10.1111/j.1468-0394.1986.tb00203.x

Wertheim, E., Love, A., Littlefield, L., & Peck, C. (1992). *I win: You win* (pp. 130–135). Ringwood, Australia: Penguin Books.

Yuan, Y., Rose, B. J., Archer, N., & Suarga, H. (1998). A Web-based negotiation support system. *EM - . Electronic Markets*, *8*(3), 13–17. doi:10.1080/10196789800000033

Zeleznikow, J., Meersman, R., Hunter, D., & van Helvoort, E. (1995). *Computer tools for aiding legal negotiation. ACIS95* Sixth Australasian Conference on Information Systems, Curtin University of Technology, Perth, Australia (pp. 231-251).

ENDNOTES

1 See www.pon.harvard.edu

2 Sometimes the parties cannot agree on the monetary value of the item. In this case, mediators would reference standard objective tables and the like to reach a consensus. For example, if parties are arguing over the value of a car, then mediators may access websites that gave independent valuations, such as redbook.com.au.

3 Linkage Project LP0882329 Developing negotiation decision support systems that promote constructive relationships following disputes

4 See http://www.ag.gov.au/www/ agd/ agd.nsf/Page/ Families_FamilyRelationshipServices OverviewofPrograms_ ForFamilyRelationship ServicesPractitioners_ FamilyRelationship AdviceLineResources. Last accessed August 16, 2009.

Chapter 18
A Multiagents System for Coordinating Naval Forces

Larbi Esmahi
Athabasca University, Canada

Elarbi Badidi
United Arab Emirates University, UAE

ABSTRACT

The survivability of a naval surface combatant depends largely on the effective management of combat resources. Within defense models based on platform-centric self-protection, strategies used for situation assessment and engagement policies governing weapon usage influence the effective management of defense units. These strategies enable the surface combatant to adapt to changes in the battle space. In the case of network-centric operations, the task force's ability to adapt to changes in the battle space relies on the information superiority gained through shared awareness. Although shared awareness enables surface combatants to apply situation assessment strategies to self-synchronize to the situation, engagement policies governing weapons usage typically remain platform-centric and rely on centralized command structures to provide overall coordination. This chapter examines the implementation of intelligent agents to create a partially centralized, distributed command structure using Contract Nets to coordinate tactical responses across the task force. In support of this implementation, the NetScheduler application was developed to support Monte Carlo based simulations of engagements involving anti-ship missiles against a naval task force. Simulation results demonstrated improved survivability with increased effectiveness in the management of combat resources.

INTRODUCTION

Information is critical to meaningful interactions between individuals, communities, businesses, and governments. In fact, information is vital to any endeavor where people must collaborate to accomplish specific goals. Information systems aim to facilitate this collaboration by providing capabilities to assemble and act upon information.

Prevalent in business communities, information systems support day-to-day operations while also supporting strategic planning and decision-

DOI: 10.4018/978-1-60960-091-4.ch018

making. In fact, an information system can provide a significant competitive advantage whereby a business can respond more quickly to changes in the marketplace than its competitors. Military organizations have recognized the strategic advantage afforded by information systems and have been moving towards information-enabled, network-centric operations for more than a decade (Cebrowski & Gartska, 1998).

Network-centric operations aim to achieve information superiority and, thus, gain a strategic and tactical advantage. These advantages translate into an increased agility that influences the networked combatant's response to changes in the battlespace that might otherwise jeopardize mission success.

Force protection, the defensive ability of a task force, is a key element to mission success. Current network-centric operations rely on shared awareness to enable individual combatants to self-synchronize (respond to new information in kind rather than through external direction) in a cooperative response to new threats. However, poor quality or conflicting information can hinder the self-synchronization process resulting in a less than optimal response across the networked force.

The project detailed in this chapter examined an approach for the coordination and collaboration of networked combatants that does not rely on self-synchronization. This approach features autonomous agents that, fitted with appropriate goal-oriented behavior, can achieve optimal response coordination through a negotiation mechanism based on Contract Nets.

LITERATURE REVIEW

Command and Control (C2)

The act of command is distinct from the act of command execution thereby allowing for some degree of decentralization. For example, a command center issues orders to units within its command. These units execute the orders thereby decentralizing command execution. Lee and Ghosh (1996) characterize the various C2 organizations noting that the traditional form of centralized command with centralized or decentralized control is inferior to a completely decentralized solution. The attack vulnerability of a single command center feeds this assumption of inferiority. However, the authors note that having a central command provides an effective vehicle for synchronizing combatants where the command center acts as the information and decision gateway.

Dekker (2003) provides guidelines for determining the best combination of centralized and decentralized decision-making within C2. In particular, Dekker explains that although a centralized headquarters typically provides adequate facilities for collecting information and issuing globally optimal solutions, time constraints and communications infrastructure may hinder the ability to develop a globally optimal solution. In fact, Dekker favors decentralization noting that the transmission of orders is typically more compact than the transmission of information required to formulate the orders.

Cebrowski and Gartska (1998) also maintain that "battle time" is a critical factor in the decision-making and execution of tactics and recognize that network-centric operations can provide a significant advantage in terms of reacting to changes in the battlespace. The authors support that networked combatants can synchronize themselves to the mission based on shared awareness. Where centralized command is a top-down method of synchronization and control, Cebrowski and Gartska believe that a bottom-up process, such as self-synchronization through shared awareness, can be just as effective.

Lee and Ghosh (1996) investigate a method of self-synchronization based on the observation that within the chaos of battle, events occur asynchronously with respect to individual combat units. Hence, Lee and Ghosh examine algorithms that promote cooperation between decentralized

C2 centers where mission objectives and common rules of engagement form the basis for cooperation.

In this manner, combatants are free to take independent actions based on beliefs derived from locally available information as well as shared awareness. Decisions formed from these beliefs must adhere to the established mission objectives and rules of engagement. In Lee and Ghosh's approach, a combatant communicates its beliefs and decisions only to those units within its sphere of influence as a means of reducing the risk of information saturation that might otherwise interfere with decision-making.

Through simulation, Lee and Ghosh (1996) compared their self-synchronizing decentralized command structure to that of a fully centralized command. Their comparison linked faster reaction times to better outcomes as evidenced by the decentralized approach. However, as the rules of engagement were common to both command structures, the centralized force did not receive the benefit of a globally optimal solution.

Market-Based Approaches for Coordination

Centralized command organizations are most likely to develop globally optimal solutions. However, time constraints warrant a decentralized approach using some form of coordination to develop an optimal response. When considering this problem as optimizing the allocation of assets to specific tasks, the coordination of combatants becomes a resource allocation problem. Hence, there is merit in considering techniques from other domains, such as market-based approaches.

Market-based approaches deal with the concept of the global good in an environment where self-interested agents apply strategies that would otherwise maximize their own good. The following sections discuss various market-based approaches and their influence on the global good.

Auctions

Auctions are a means for mutually beneficial exchanges ensuring that the seller receives fair market value while bidders obtain items at or below their valuation. Many protocols have been proposed in auction theory (Rasmusen, 2001; Klemperer, 2002; Neumann et al., 2007): the English auction where the bidding process is public, so each bidder has complete information about the auction. At any time, each agent is free to raise his bid. The sealed-bid auction, each bidder submits one bid without knowing the others' bids. The highest bidder wins the item and pays the amount of his bid. The Dutch auction, the seller or the auction manager continuously lowers the price until one of the bidders takes the item at the current price. The Vickrey auction is similar to the sealed-bid auction with some details exception. In fact, each bidder submits one bid without knowing the others' bids and the highest bidder wins, but he pays only the price of the second highest bid.

Typical to most forms of auctions, bidders continue to place ascending bids until no other bids are tendered. The final bid establishes the winner as well as the price paid. The seller can choose the form of auction to influence a better price but it is incumbent on individual bidders to establish a valuation of the auctioned item and implement a bidding strategy in line with that valuation and compliant with the form of auction.

In an English auction each bidder must bid a higher amount than the last accepted bid to remain in the competition. To reduce the risk of having to bid an amount greater than the item's perceived valuation, bidders will implement a dominant strategy whereby bids are intentionally lower than the bidder's valuation of the item and, using small increments, higher than the leading bid. In certain situations, the demand for an item may drive bids above the item's true valuation.

Wellman et al. (2001) examine the English auction for decentralized scheduling of computational resources. Agents bid according to tasks to be completed whereby an agent's bidding strategy aims to maximize the surplus value across all tasks- the difference between the agent's maximum valuation and the actual bid.

Auctioneers may set a reserve price for a time slot based on demand. The reserve price encourages agents to reassess preferences for particular resources to maximize their surplus value. Wellman et al. (2001) notes that this approach is more appropriate for agents competing for a single resource rather than multiple resources. Combinatorial auctions specifically support bidding on a preferred bundle of resources but achieving a globally optimal solution can be problematic (Conen & Sandholm, 2003; Kutanoglu et al., 1999).

Contract Nets

Contract nets were first proposed as a means for distributed problem solving (Smith, 1980; Smith et al., 1981; Sandholm, 2000). Through negotiation, contracts are formed between the contract initiator and selected contractors. Initial contract announcements, issued as a Call for Proposals (CFP), include terms and conditions that contractors must satisfy in order to participate. The initiator evaluates the submitted bids and selects one or more contractors, as necessary. The contract is an explicit agreement between an agent who announces its tasks (the manager) and an agent who proposes to achieve those tasks (the contractor). The manager is responsible for task management and results processing, while the contractor is responsible for tasks processing and results transmission. The protocol proposes a set of performatives that the agents can use either for contract negotiation or contract execution. The contract net has been proposed also for adaptive workflow control (Harker et al., 1996) and for network-centric operations (Beaumont, 2004).

Many extensions have been proposed for the contact net protocol. One of the most important works done in this topic is the extension proposed by Sandholm (1993, 1996, 2000), which is based on marginal costs. In fact, the agents process locally their marginal costs for achieving tasks and use those costs either for announcing a task, bidding, or awarding a contract. Partners' selection is based on profit evaluation using the marginal costs. The integration of utility has then extended the protocol's domain of application to competitive and self-interest agents.

Other Approaches for Agents' Coordination

Other approaches not based on market economies also apply to coordination (Roth, 2000; Jennings et al., 2001; Neumann et al., 2007; Cheung et al, 2003; Kersten & Lai, 2005; Chiu et al., 2005c; Cheng et al., 2007). These range from negotiation or bargaining strategies to cooperative approaches such as blackboard, constraint-based negotiation and voting systems. Although negotiation and bargaining strategies apply to cooperative settings, the participants are typically constrained at two agents. Blackboard or voting systems (Parker, 1995) are not so constrained but, similar to auction, require some form of centralization or mediation.

Token-Based Approach

The token-based approach is inspired by social network models where the team is represented by an acquaintance network.(Xu et al., 2005; Xu et al., 2008; Velagapudi et al., 2009) In the network each node represents an agent and pairs of agents are connected by links and can exchange tokens with each other directly. The tokens can be of three basic types, information tokens, roles tokens and resources tokens. A path and threshold parameters are used to record the route taken by the token and the control information for resources and roles. The coordination is based on individual decisions made by agents of where to pass the tokens they hold. In their individual decisions, agents attempt to maximize the overall team reward based on local information.

Discussion

In general, market-based approaches use bidding strategies to establish globally optimal solutions for resource allocation while the non market-based approaches reviewed tend to focus more on collective problem solving. A common consideration when applying auctions is the role of the auctioneer. The auctioneer acts as a mediator and attempts to resolve conflicts while encouraging a globally optimal solution. Using auctions as a coordination technique in network-centric operations would require a similar form of mediation and mandate a more centralized command organization that could promote the global good. Token coordination is focused on scalability; hence it minimizes communication at the expense of the overall utility. Hence, neither of these approaches lends itself very well to decentralized command and presents levels of communications that might prove unaffordable.

The contract net is one of the most complete models for cooperating agents. The notion of profit and utility recently added to it, resolves the problem of motivation for competitive agents. However, the protocol still has some drawbacks, such as considering the agents as passive actors because no persuasion mechanisms can be used to convince partners for accepting an agreement. Also, dependency relations between agents are not considered. Previous research that introduced Contract nets to network-centric operations applied a central command structure and identified reaction time as a constraining factor.

In our context of force protection, contract net may best minimize the exchange of information and provide roles (i.e., contractor and participant) and messages to support decentralization and coordination. By allowing more than one networked combatant to assume the role of contractor while preventing more than one contractor for any given missile threat, a dynamic, partially centralized organization can be achieved. Similar approaches used to coordinate large groups of agents found a reduction in the overall complexity of coordination when using dynamic, partial centralization (Scerri et al., 2004).

NETSCHEDULER, A SIMULATION TOOL FOR RESPONSE COORDINATION

Simulating Response Coordination

The simulation uses low-fidelity models for both self-directed anti-ship missile threats and for surface combatants. The Java-based platform-centric, Dynamic Engagement Scheduler (DEScheduler) developed at Defense Research & Development Canada (DRDC) provides the necessary low-fidelity models. In this paper we are presenting the resulting extension to the DEScheduler application herein referred to as the NetScheduler.

Extending the platform-centric nature of the simulation to include network-centric coordination involved the introduction of autonomous agents. The agents interact using the Contract Net Interaction Protocol and utilize roles (i.e., Mission Commander and Mission Support) to support a dynamic, partially centralized command structure.

Using Gaia (Woodridge et al., 2000; Zambonelli et al., 2003) an agent-oriented methodology, models of the environment, roles, and interactions facilitate the development of role schemas (see Figure 1).

Schemas depict the protocols and activities supporting the role while also outlining the agent's influence on the environment (permissions). "Liveness" and "Safety" describe a role's responsibility where liveness (the desirable path) expresses the logical flow of protocols and activities, and safety (avoiding the undesirable) denotes the conditions needed to maintain operational integrity.

As a society of agents, rules are used to establish how agents can assume roles within the society's organization. For the purpose of force

Figure 1. Role schema for the mission commander

Role Schema: MISSION COMMANDER (MC)			
Description: This role involves announcing the command of a mission, and soliciting support as required			
Protocols and Activities: IdentifyNewMissions, ExecuteMission, ReviewMissionObjectives, Announce, CallForProposal, FormulatePlan, ReviewProposals, RejectProposal, AcceptProposal, Cancel			
Permissions:	**reads**	*System Tracks*	*// All evaluated and prioritized sensor reports*
	changes	*Mission*	*// All active missions*
	changes	*Call for Proposals*	*// Raises CFPs as required*
	consumes	*Proposals*	*// All proposals received*
	changes	*Contracts*	*// Generates contracts as required*
Responsibilities			
Liveness:	MISSION COMMANDER = (ASSESS || COMMAND)ω		
	ASSESS = (IdentifyNewMissions.Announce)		
	COMMAND = (ExecuteMission. (ReviewProposals.(AcceptProposal || RejectProposal) || Cancel) || ReviewMissionObjectives.CallForProposals.FormulatePlan)		
Safety:	*All threats (system tracks) have a mission commander assigned*		

protection, these organizational rules must specifically address the dynamic nature of the command structure.

The organizational rules are defined as follows where s refers to sensor information (tracks), and i refers to the agent (combatant):

$$\forall s \cdot \forall i, Plays(i, Commander(s)) \wedge \neg Plays(i, Support(s))$$

$$if\ CanEngage(i,s) \wedge \neg\exists i(Commander(s)) \rightarrow Plays(i, Commander(s))$$

$$\forall i, j, Plays(Supercede(i,j), Commander(s))\ iff\ Success(j) > Success(i)$$

The first rule simply states that for all tracks s, agent i cannot play the roles of Mission Commander and Mission Support concurrently. The second organizational rule indicates that an agent can assume the role of Mission Commander if the agent can engage the threat and no other agent is currently in that role. The third rule allows an agent to supersede the Mission Commander if that agent deems its success criteria offer a greater chance of mission success. This is necessary due to delays in communication that would result in more than one agent satisfying the second rule. In this situation, agents in the Mission Support role would defer to the Mission Commander with the higher probability of success. Note, an agent may also choose to assume its own mission and agenda based on its own need for survival.

Implementation of the NetScheduler

The Agent Model (Figure 2) defines the class architecture of the agent system. This model incorporates intelligent agents that, by design, are able to react to the environment, pursue goal-directed behavior, and socially interact (Wooldridge, 2002). The agents consist of the Sensor Manager, Weapons Manager, and Command Link.

The Sensor Manager

The Sensor Manager Agent maintains track (sensor track) information derived from local sensors as well as information received through shared awareness. A track specifically identifies a contact by location and characterizes the contact's behav-

Figure 2. NetScheduler agent model

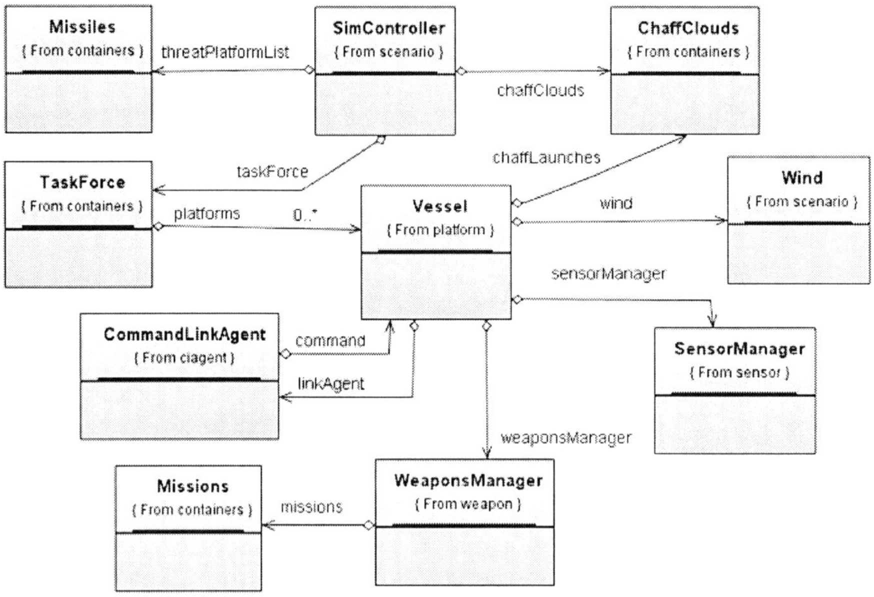

ior in terms of velocity, closest point of approach (CPA), and time-to-go (TTG) – time before impact. The Sensor Manager must prioritize each track based on an assessment of the threat presented by the track information.

For self-protection, the TTG is a major contributor to the threat assessment. However, in the case of force protection, the threat assessment must also consider other units within the task force. A unit vital to mission success receives the designation of High-Value Asset (HVA). Tracks assessed as presenting a higher threat to HVA-designated units are a higher priority than threats to self or other supporting units. The Sensor Manager Agent must maintain track information and provide a prioritized track list to other agents.

The Weapons Manager Agent

The Weapons Manager Agent specifically supports the Mission Commander and Support activities, as per the defined role schemas. In the role of Mission Commander the Weapons Manager Agent reviews the track list provided by the Sensor Manager to determine if a threat exists for which no Mission Commander has been assigned. In these cases, the Weapons Manager determines if it can engage the threat, and if so, announces its intention to assume command of the mission.

The Weapons Manager Agent also reviews missions under its command to determine if there is a sufficient Probability of Success (PS). The Weapons Manager Agent determines the PS based on its situation assessment and stores this value with the track list. If the Weapons Manager Agent's situation assessment deems that the PS is insufficient, it will issue a CFP.

Weapons Manager Agents in the support role will respond to a CFP with a proposal. The proposal reserves the asset and excludes it from further proposals should another CFP be active. Hence, the Weapons Manager Agent relies on the prioritized track list provided by the Sensor Manager Agent to determine the priority of CFP responses.

The Command Link Agent

The Command Link Agent registers the agents within the task force and implements the Agent Communications Language (ACL) used in support of the Foundation for Intelligent Physical Agents (FIPA) Contract Net Interaction Protocol (FIPA, 2002a). The ACL selected implements the message structure specified by FIPA (FIPA, 2002b). A schema was prepared using the Extensible Markup Language (XML) to support the implementation of the ACL message structure (Figure 3) as well as the Contract Net Interaction Protocol captured within the message's content.

An "Announce" protocol was added to allow for Mission Commanders to announce their intention. This message contains information that identifies the Mission Commander, the missions objective (i.e., protect HVA, self, or other unit), and the missions PS. This information allows a Weapons Manager Agent to select an appropriate commander in the case of simultaneous announcements.

EXPERIMENTAL EVALUATION

To determine the viability of using contract nets to establish response coordination for task force protection we have processed some simulations scenarios that allow us to compare results from coordination that relies on a platform-centric response (i.e., common rules of engagement) with those obtained as part of a network-centric (i.e., coordinated task force) response. So for this experimentation we are using two different planners: a reactive planner for non-coordinated defense and a deliberative planner for coordinated defense. The evaluation considers the effective use of weapons in scenarios involving a coordinated response versus scenarios lacking any form of response coordination and relying instead on common rules of engagement. Furthermore, the evaluation must consider if the use of contract nets compromises the ability of the task force to defend itself.

Figure 3. FIPA ACL/XML schema layout

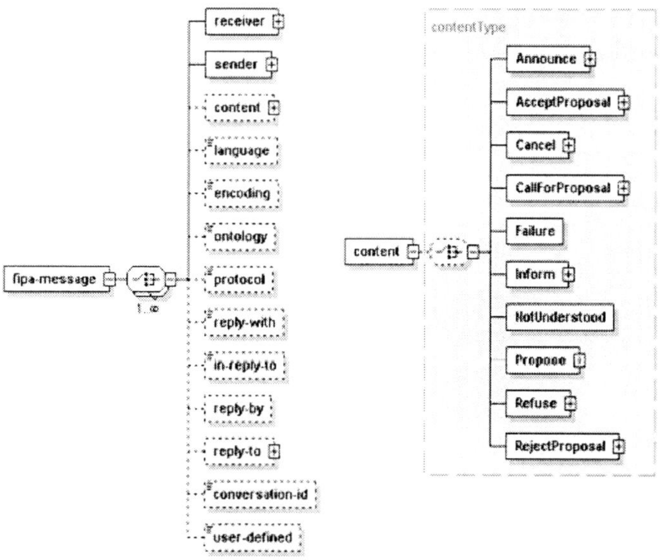

The Engagement Scenario

The NetScheduler application supports Monte Carlo simulation methods that apply user-defined scenarios. For the purpose of this project, the experiments feature an attack scenario containing two groups of subsonic missile threats with the two missile groups appearing within a short period of each other (approximately two seconds) and approaching the task force from orthogonal directions. The engagement scenario attempts to saturate the task force's defenses by introducing multiple threats from two different sectors.

The defense scenario contains a task force consisting of four combatant ships and two HVA (high value assets) ships acting in a non-combatant capacity. The task force's formation involves a defensive screen created by surrounding the HVAs with the four combatants. Approximately two kilometers separates the combatants from the nearest HVA. This formation attempts to represent a situation where a task force might navigate along a prescribed shipping lane in constrained waters (i.e., littoral conditions). All combatants sport the full complement of weapons granted by the simulation, which includes soft-kill and hard-kill weaponry.

The ability of a combatant to defeat a missile threat is largely dependent on the available battle space. In other words, as the missile threat's time-to-go (time before impact) decreases, the combatant has less time to mount a defense and possibly fewer options with which to engage the missile threat. Time-to-go is dependent on the range and speed of the missile threat. Hence, to determine the sensitivity of time-to-go as a factor of mission success, the experiments vary the initial distance of the missile threats from the task force wherein the values vary from beyond the programmed sensor range of 35 km to a distance where reaction time is seriously handicapped (e.g., 15 seconds to impact).

Figure 4 lists the engagement scenarios created for experimentation based on the attack and defense scenarios described. Note, the engagement scenario determines the engagement policies used and determines if the task force operates in non-coordinated mode (no coordinated response) or uses response coordination using the contract net interaction protocol.

The initial range (measures form a common point of origin) given in Figure 4 determines at what range the missile threat first appears in the simulation. Missiles in the simulation will travel along the initial heading defined by the attack

Figure 4. Engagement scenario used for the experimentation

Engagement Scenario	Coordinated Response	Initial Range	Range Variance	Seeker Activation	Activation Variance
Scenario A	No	40 km	1 km	20 km	1.5 km
Scenario B	Yes	40 km	1 km	20 km	1.5 km
Scenario C	No	30 km	1 km	20 km	1.5 km
Scenario D	Yes	30 km	1 km	20 km	1.5 km
Scenario E	No	20 km	1 km	20 km	1 km
Scenario F	Yes	20 km	1 km	20 km	1 km
Scenario G	No	10 km	0.5 km	10 km	0.5 km
Scenario H	Yes	10 km	0.5 km	10 km	0.5 km

scenario until a target has been acquired at which point heading corrections will be introduced into the missile's flight path. Target acquisition does not occur until the missile's seeker has been activated. Figure 4 specifies the activation point, measured as a distance from the point of origin. Note for initial ranges less than 30 km (well inside ship's sensor range) the missile appears with its seeker activated from the onset.

The Engagement Policy Settings

Combatants within a non-coordinated task force utilize a reactive planner that determines the combatant's response to a threat whereas the coordinated response employed by the mission commander within the task force utilizes a deliberative planner. The engagement scenario defines the engagement policy thresholds used by both the reactive and deliberative planners for non-coordinated and coordinated defense, respectively. In particular, policies controlling threat ranking and weapon assignment determine the level at which the planner can respond to a given threat.

Figure 5 denotes constraints applied for threat ranking and response planning based on the planner type employed. The TTG values determine the threat ranking and establish regions wherein ranking increases as the threat missile enters into a particular region. Note the deliberative planner extends the regions to provide coverage for HVAs or other combatants within the deliberative planner's sphere of influence.

In the case of the reactive planner, each combatant is equipped with at most five weapon systems. However, the deliberative planner has access to resources offered by other combatants. So, the deliberative planner can increase the allowable number of engagements without committing all of the combatant's own resources.

In addition to engagement policies that determine the response level, other policies define conditions surrounding the situation assessment (i.e. the planner can consider the missile threat as neutralized by a soft-kill tactic). These conditions assess the lethality of the threat as well as its ability to re-orient onto its intended target. The lethality condition monitors the closest point of approach (CPA). An increasing CPA indicates that the threat is moving away thus reducing its potential lethality.

For platform-centric self-protection, the reactive planner assesses the threat's lethality over a specified period of time and if the threat's CPA increases beyond 500 meters, no further weapon engagements against the threat are initiated. To prevent this condition from hastily reducing the planner's response wherein another combatant in the task force would be under attack, the CPA threshold was increased to 2,000 meters – effectively disabling the condition.

Similarly, the assessment of the missile threat's ability to re-orient onto its intended target was also effectively disabled by increasing the assessment period from three seconds to four seconds, and by increasing the tolerance by a factor of 20%.

Figure 5. Engagement policy settings

Threat Ranking	Time-To-Go (TTG) (sec)		Required Probability of Success ($P_{Success}$)		Maximum Number of Concurrent Engagements	
	Reactive	Deliberative	Reactive	Deliberative	Reactive	Deliberative
Low	$t > 40$	$t > 60$	50%	95%	1	3
Medium	$10 < t \leq 40$	$15 < t \leq 60$	75%	95%	2	5
High	$t \leq 10$	$t \leq 15$	90%	99%	4	7

The Performance Parameters

The measures of merit (MOM) used for evaluation aim to quantify the outcomes in terms of overall effectiveness and resource usage. Overall effectiveness refers to the task force's ability to defeat incoming missile threats while not compromising the task force's ability to defend itself through poor management of the battles pace or over-expenditure of munitions.

Kill Effectiveness

The kill effectiveness measures the total number of kills against the total number of missile threats across all simulation runs. A kill includes the destruction of the missile threat or its seduction away from the task force such that the missile threat cannot engage the task force. Ideally, the kill effectiveness should be 100%, indicating the task force's ability to neutralize all missile threats.

$$E_{Kill} = \frac{\sum_{Runs} \frac{N_{Kills}}{N_{Threats}} \times 100}{Runs}$$

Battle Space Efficiency Ratio

The battle space efficiency ratio (BER) compares the actual amount of time spent by the task force engaging the missile threat versus the window of opportunity or the period commencing at the missile's earliest time of detection by the task force. In typical scenarios, combatants have a short time to assess a threat and initiate a response. Increasing the amount of time spent engaging the threat within the given window of opportunity increases the probability of success and is, hence, a desirable situation.

$$BER = \frac{\sum_{Runs} \frac{\Delta T_{Engaged}(x)}{\Delta T_{Hostile\_Tracked}(x)} \times 100}{Runs}$$

Engagement Effectiveness

The engagement effectiveness measures the total number of successful engagements (i.e., Joys) against the total number of engagements initiated by the task force's combatants. Non-successful engagements are those that fail outright, such as when the counter missile fails to strike the incoming missile threat, or those aborted when another engagement is successful, or when the weapon system can no longer sustain the engagement (i.e., threat enters a blind zone).

$$E_{Engagements} = \frac{\sum_{Runs} \frac{N_{Joys}}{N_{Engagements}} \times 100}{Runs}$$

Hard-Kill Effectiveness

The hard-kill effectiveness measures the number of successful hard-kill engagements against the total number of hard-kill engagements initiated across the task force.

$$E_{Hard-Kill} = \frac{\sum_{Runs} \frac{N_{Successful\_Hard-Kills}}{N_{Hard-Kill\_Engagements}} \times 100}{Runs}$$

Soft-Kill Effectiveness

The soft-kill effectiveness measures the number of successful soft-kill engagements against the total number of soft-kill engagements initiated across the task force. Note for a soft-kill engagement to be successful, the missile threat must not make contact with the task force, nor should the task force destroy it.

$$E_{Soft-Kill} = \frac{\sum_{Runs} \frac{N_{Successful\_Soft-Kills}}{N_{Soft-Kill\_Engagements}} \times 100}{Runs}$$

Long-Range Inventory

The long-range inventory reflects the usage of long-range weaponry (i.e., Surface-to-Air Missile, SAM) by providing a measure of the remaining inventory across the task force.

$$I_{Long-Range} = \frac{\sum_{Runs} \frac{N_{SAM\_Remaining}}{N_{SAM\_Inventory}} \times 100}{Runs}$$

Mid-Range Inventory

The mid-range inventory reflects the usage of mid-range weaponry (i.e., 57 mm gun) by providing a measure of the remaining inventory across the task force.

$$I_{Mid-Range} = \frac{\sum_{Runs} \frac{N_{Gun\_Ammo\_Remaining}}{N_{Gun\_Ammo\_Inventory}} \times 100}{Runs}$$

Short-Range Inventory

The short-range inventory reflects the usage of close-in weapon systems by providing a measure of the remaining inventory across the task force.

$$I_{Short-Range} = \frac{\sum_{Runs} \frac{N_{CIWS\_Ammo\_Remaining}}{N_{CIWS\_Inventory}} \times 100}{Runs}$$

Chaff Inventory

The chaff inventory reflects the usage of off-board counter-measures by providing a measure of the remaining inventory across the task force.

$$I_{Chaff} = \frac{\sum_{Runs} \frac{N_{Chaff\_Remaining}}{N_{Chaff\_Inventory}} \times 100}{Runs}$$

Simulation Results

Given the limited space we have for this paper we present here only the summary of the results in terms of the Measures of Merit (MOM) acquired via Monte Carlo simulation techniques. The MOM presented, summarize the recorded experimentation results denoting the number and duration of engagements, engagement types (i.e., hard-kill, soft-kill), engagement outcomes (i.e., success, fail, or aborted), results (i.e., hit, miss, soft-kill, hard-kill), and weapon inventory usage.

The network-centric response was shown to provide higher effectiveness scores than platform-centric responses, with the noted exception of soft-kill effectiveness (Figure 6). The low score of soft-kill effectiveness is a direct result of the simultaneous deployment of hard-kill tactics.

The destruction of a missile threat by a hard-kill weapon results in the suspension of any soft-kill assessment. The cancellation of the soft-kill assessment results in the soft-kill engagement being flagged as a failure. Hence, the soft-kill effectiveness value does not provide an indication of the efficacy of soft-kill tactics. However when it is viewed in conjunction with the chaff inventory, it is useful to the coordination provided by the network-centric response. In our simulation, the coordination produced fewer soft-kill engagements and ultimately expended fewer munitions.

Although the network-centric response demonstrated better effectiveness, the efficiency

Figure 6. Measures of merit (MOM) summary

Initial Range	Platform-Centric Response	Network-Centric (Coordinated) Response
Kill Effectiveness	96.6%	98.9%
Battlespace Efficiency Ratio (BER)	71.1%	63.1%
Engagement Effectiveness	21.4%	39.3%
Hard-Kill Effectiveness	34.9%	46.6%
Soft-Kill Effectiveness	4.0%	2.9%

score provided by the BER was below that of the platform-centric response. The BER values favor the platform-centric response where BER is a measure of the task force's ability to make use of the window of opportunity to engage a missile threat. The lower BER for network-centric responses is an indication of the time spent on deliberation and coordination of the task force's resources.

While the missile threat is beyond weapon range, the difference in BER values is negligible because deliberation and coordinating communications takes place before the threat has entered weapon ranges. When missile threats appear within weapon range, the time spent on deliberation and coordination occurs within the window of opportunity to engage the threat thereby reducing the actual time spent engaging the threat and, thus, reducing the BER.

Although the BER for the network-centric response is lower than that of the platform-centric response, the time spent on deliberation and communications does not show any adverse effect on the networked combatant's defensive ability. In fact, based on the average number of engagements and remaining inventory, the resource consumption is lower for the network-centric response.

CONCLUSION

In this project we examined the use of Contract Nets for the coordination and collaboration of networked combatants against anti-ship missile threats. This examination involved the development of a simulated environment wherein each surface combatant uses intelligent agents to formulate plans and negotiate a coordinated response. Simulation results demonstrated improved survivability with increased effectiveness in the management of combat resources.

The simulation involved a comparison of surface combatants using platform-centric engagement policies versus those using intelligent agents for network-centric response coordination. Combatants operating in platform-centric mode formed a completely decentralized C2 structure while the combatants using intelligent agents created a dynamic, partially centralized organization. The resulting comparison focused on ship survivability, battle space management, and resource usage. In particular, it was necessary to determine if using Contract Nets for response coordination compromised the task force's defensive ability.

The results of the comparison did not reveal any compromise to force protection. In fact, the results revealed an improvement in survivability in the form of kill effectiveness while also demonstrating a reduction in resource consumption. A reduction in resource consumption increases the task force's ability to react to new situations. In other words, as new threat detections occur, the availability of resources enables the task force to respond with a broader range of options (i.e., long-range or mid-range weapons). This broader range of options increase the task force's agility.

The command structure also influences the task force's agility. In the approach examined in this project, the intelligent agents implemented a dynamic, partially centralized command structure through roles facilitated by the contract net interaction protocol. Beaumont's use of a central coordinator with Contract Nets found good survivability results. However, Beaumont noted that communications and TTG were critical factors (Beaumont, 2004).

This project's results did not reflect this observation. In fact, results based on the dynamic, partially centralized organization showed consistent scores across all TTG values (represented as ranges). In the case of communications, a partial centralization could compensate in the event of failure. This compensation would manifest as multiple Mission Command roles being assumed. In addition, the intelligent agent's goal-directed behavior maintains the surface combatant's ability to take independent action. This independence enables the surface combatant to rely on local information to set priorities, formulate plans, and, when necessary, to give all priority to defending itself.

Although the proposed dynamic, partially centralized command structure was an attempt to evolve organization and doctrine, as suggested by Alberts (Alberts et al., 2000), in line with advances in self-protection, it was necessary to circumvent some of the situation assessment algorithms developed specifically for self-protection. Due to the platform-centric nature of these algorithms (i.e., lethality and soft-kill kill-assessments), a surface combatant could prematurely assign a non-threat status to an anti-ship missile based on its own self-centric evaluation and not consider the threat to other combatants. Hence, further work is required to progress concepts used in self-protection towards more inclusive task force policy.

Coordination of tactics is another area that warrants further work. Investigating how the deliberative planner could decide upon which tactics to initiate from proposed plans was not within the scope of this project. However, based on the feasibility of Contract Nets demonstrated by the results, there would be merit in growing the deliberative planner used in the NetScheduler tool to include tactical coordination.

REFERENCES

Alberts, D., Gartstka, J. J., & Stein, F. P. (2000). Network-centric warfare: Developing and leveraging information superiority (2nd ed.). Washington, DC: DoD C4ISR Cooperative Research Program. Retrieved March 1, 2007, from http://www.dodccrp.org/html3/pubs_download.html

Beaumont, P. (2004). Multi-platform coordination and resource management in command and control. Unpublished masters thesis. Laval University, Quebec, Canada. Retrieved November 29, 2006, from http://damas.ift.ulaval.ca/projets/TeamWork/publications.en.html

Cebrowski, A., & Gartska, J. J. (1998). Network-centric warfare: Its origins and future. *Proceedings of the Naval Institute*, *124*(1).

Cheng, V. S. Y., Hung, P. C. K., & Chiu, D. K. W. (2007). Enabling Web services policy negotiation with privacy preserved using XACML. Proceedings of the 40th Annual Hawaii International Conference on System Science (HICSS) (p. 33a).

Cheung, S. C., Hung, P. C. K., & Chiu, D. K. W. (2003). On the e-negotiation of unmatched logrolling views. Proceedings of the 36th Annual Hawaii International Conference on System Science (HICSS) (p. 29).

Chiu, D. K. W., Cheung, S. C., Hung, P. C. K., Chiu, S. Y. Y., & Chung, A. K. K. (2005). Developing e-negotiation support with a meta-modeling approach in a Web services environment. *Decision Support Systems*, *40*(1), 51–69. doi:10.1016/j.dss.2004.04.004

Conen, W., & Sandholm, T. (2003). Differential-revelation VCG mechanisms for combinatorial auctions. In *Proceedings of the 4th ACM Conference on Electronic Commerce, San Diego, CA* (pp. 196-197). New York: ACM Press.

Dekker, A. H. (2003). Centralisation and decentralisation in network centric warfare. *Journal of Battlefield Technology*, *6*(2), 23–28.

Esmahi, L. (2008). Contract negotiation in E-marketplaces: A model based on dependency relations. *Journal of Electronic Commerce in Organizations*, *6*(3), 74–91.

FIPA. (2002). FIPA contract Net interaction protocol. Retrieved June 14, 2006, from http://www.fipa.org/assets/XC00029G.pdf

FIPA. (2002). FIPA ACL Message Structure Specification. Retrieved on March 22, 2006 from http://www.fipa.org

Harker, T. H., & Ungar, L. H. (1996, May 8-10). A market-based approach to workflow automation. Paper presented at NSF Workshop in Information Systems, Athens, GA. Retrieved October 1, 2006, from http://www.cis.upenn.edu/~ungar/NSF.html

Jennings, N. R., Faratin, P., Lomuscio, A. R., Parsons, S., Wooldridge, M. J., & Sierra, C. (2001). Automated negotiations: Prospects, methods and challenges. *Group Decision and Negotiation*, *10*(2), 199–215. doi:10.1023/A:1008746126376

Kersten, G. E., & Lai, H. (2005). Satisfiability and completeness of protocols for electronic negotiations [INR01/05]. Montreal, Canada: InterNeg Research Center. Retrieved January 4, 2007, from http://interneg.concordia.ca/interneg/research/papers/2005/01.pdf

Klemperer, P. (2002). What really matters in auction design. *The Journal of Economic Perspectives*, *16*(1), 169–189. doi:10.1257/0895330027166

Kutanoglu, E., & Wu, S. D. (1999). On combinatorial auction and LaGrangean relaxation for distributed resource scheduling. *IIE Transactions*, *3*(1), 813–826. doi:10.1080/07408179908969883

Lee, T., & Ghosh, S. (1996). Simulating asynchronous, decentralized military command and control. *IEEE Computational Science & Engineering*, *3*(4), 69–79. doi:10.1109/99.556514

Neumann, D., Schnizler, B., Weber, I., & Weinhardt, C. (2007). Second-best combinatorial auctions: The case of the pricing-per-column mechanism. Proceedings of the 40th Annual Hawaii International Conference on System Science, HICSS (p. 33b).

Neumann, D., & Weinhardt, C. (2002). Domain-independent eNegotiation design: Prospects, methods, and challenges. Proceedings of the 13th International Workshop on Database and Expert Systems Applications, DEXA (pp. 680-686).

Parker, J. R. (1995). Voting methods for multiple autonomous agents. Proceedings of the Third Australian and New Zealand Conference on Intelligent Information Systems, ANZIIS (pp. 128-133).

Rasmusen, E. (2001). Auctions . In *Games and information: An introduction to game theory* (3rd ed., pp. 404–454). Malden, MA: Blackwell Publishers.

Roth, E. (2000). Game theory as a tool for market design . In Patrone, F., Garcia-Jurado, I., & Tijs, S. (Eds.), *Game practice: Contributions from applied game theory* (pp. 7–18). Dordrecht: Kluwer.

Sandholm, T. W. (1993). An implementation of the Contract Net protocol based on marginal cost calculations. In proceedings 11th national conference on artificial intelligence (pp. 256-262).

Sandholm, T. W. (1996). Negotiation among self-interested computationally limited agents. Unpublished doctoral dissertation. Department of Computer Science, University of Massachusetts, USA. Retrieved December 9, 2009, from http://www.cs.cmu.edu/~sandholm/dissertation.ps

Sandholm, T. W. (2000). Agents in electronic commerce: Component technologies for automated negotiation and coalition formation. *Autonomous Agents and Multi-Agent Systems*, *3*(1), 73–96. doi:10.1023/A:1010038012192

Scerri, P., Vincent, R., & Mailler, R. (2004). Comparing three approaches to large scale coordination. In *Proceedings of the First Workshop on the Challenges in the Coordination of Large Scale Multi-agent Systems* (pp. 1-21). Retrieved on August 28, 2009, from http://citeseerx.ist.psu.edu/viewdoc/download?doi=10.1.1.61.1753&rep=rep1&type=pdf

Smith, R. G. (1980). The contract Net protocol: High level communication and control in a distributed problem solver. *IEEE Transactions on Computers*, *29*(12), 1104–1113. doi:10.1109/TC.1980.1675516

Smith, R. G., & Davis, R. (1981). Frameworks for cooperation in distributed problem solving. *IEEE Transactions on Systems, Man, and Cybernetics*, *11*(1). doi:10.1109/TSMC.1981.4308579

Velagapudi, P., Prokopyev, O., Scerri, P., & Sycara, K. (2009). A token-based approach to sharing beliefs in a large multiagent team. In M. J. Hirsch, C. W. Commander, P. M. Pardalos, & R. Murphey (Eds.), Optimization and cooperative control strategies. *Lecture Notes in Control and Information Systems, 381*, 417-429.

Wellman, M., Walsh, W., Wurman, P., & Mackie-Mason, J. (2001). Auction protocols for decentralized scheduling. *Games and Economic Behavior*, *35*, 271–303. doi:10.1006/game.2000.0822

Wooldridge, M. J. (2002). *An introduction to multi agent systems*. West Sussex, UK: Wiley & Sons.

Wooldridge, M. J., Jennings, N. R., & Kinny, K. (2000). The Gaia methodology for agent-oriented analysis and design. *Autonomous Agents and Multi-Agent Systems*, *3*(3), 285–312. doi:10.1023/A:1010071910869

Xu, Y., Scerri, P., Lewis, M., & Sycara, K. (2008). Token-based approach for scalable team coordination . In Pardalos, P. M., Grundel, D., & Murphey, R. A. (Eds.), *Cooperative networks: Control and optimization* (pp. 5–29). Cheltenham, UK: Edward Elgar Publishing.

Xu, Y., Scerri, P., Yu, B., Okamoto, S., Lewis, M., & Sycara, K. (2005). An integrated token-based algorithm for scalable coordination. Proceedings of the fourth international joint conference on Autonomous agents and multiagent systems (pp. 407 – 414).

Zambonelli, F., Jennings, N. R., & Wooldridge, M. J. (2003). Developing multi-agent systems: The Gaia methodology. *ACM Transactions on Software Engineering and Methodology*, *12*(3), 317–369. doi:10.1145/958961.958963

KEY TERMS AND DEFINITIONS

Agent's Mental State: Three main components are in the agent's mental state: beliefs, desires and intentions. However in the market place context we are adding the trust component.

Autonomous Agents: In pure autonomous agents systems, the concern of the designer is with the performance of the individual agent and the system level performance is left to emerge from the agents interactions without taking into consideration the interdependency between the system's components.

Battle Space: Battles pace encompasses all factors and activities within the sphere of operations that influence the ability to engage or protect the force to ensure mission success. These factors include environmental conditions as well as the distribution of friendly and hostile forces and assets. Activities include maneuvers, communications, surveillance, identification, and the application of force by friendly or hostile forces.

Command and Control: Command and Control (C2) encompasses the organizations, resources, and processes necessary for the formulation and execution of military force. As applied to this project, "command" refers to the process of tactical decision-making while "control" refers to the execution of tactics.

Contract-Net Protocol: In the contract-net protocol a contract is an explicit agreement between an agent who announces its tasks (the manager) and an agent who proposes to achieve those tasks (the contractor). The manager is responsible for tasks management and results processing, while the contractor is responsible for tasks processing and results transmission. The protocol proposes a set of performatives that the agents can use either for contract negotiation or contract execution.

Cooperative Agents: In pure distributed problem solvers where agents are supposed to be fully cooperative, the only concern of the designer is with the overall performance of the system and the performance of the individual agent is not important.

Force Protection: Force protection consists of the measures necessary to counter hostile actions taken against personnel, facilities, and assets that might otherwise jeopardize the force's ability to conduct its mission.

High-Value Asset: A high-value asset (HVA) is a resource considered vital to mission success. For example, an aircraft carrier used to provide air support is considered a HVA.

Negotiation Protocol: Set of rules and knowledge structures that provide a means of standardizing the communication between participants in the negotiation process. The negotiation protocol define how the actors can interact with each other and often includes the way in which offers and messages are constructed and send to the opponent.

Negotiation Strategy: Can be defined as the way in which a given party acts within the negotiation protocol rules in an effort to get the best outcome of the negotiation. It is meanly the process by which the agent evaluates offers and generates counter-offers.

Network-Centric Operations: Based on a network of distributed resources, information superiority can be established and exploited into a strategic and tactical advantage.

Section 4
Collaborative Reasoning in the Virtual World

The World Wide Web was once seen as an information repository and resource. However, more recently, it acts as a communication medium and as a forum for the liberal exchange of information and personal expression. As such, it has considerable potential for supporting group reasoning and decision-making. Key considerations needed to move towards implementations of systems to support collaborative reasoning in virtual worlds include a greater understanding of the nature of virtual communities, their differences and similarities with real world communities, the nature of reasoning tasks in both spheres, and the limitations of both. The next group of chapters discusses issues related to collaborative reasoning within or about a virtual world.

Chapter 19
Negotiating Meaning in a Blog-Based Community:
Addressing Unique, Shared, and Community Problems

Vanessa P. Dennen
Florida State University, USA

ABSTRACT

This chapter addresses how members of a blog-based community share problems and support each other in the problem solving process, both sharing knowledge and offering support. Problems are divided into three categories, unique, shared, and community, each having its own particular norms for presentation, knowledge sharing, and resolution. Additionally, processes may differ based on discourse that centers on one individual blog versus discourse that spans multiple blogs. Findings show that intersubjectivity, norms, roles, and individual ownership of virtual space all are important elements contributing to the problem sharing and solving process.

INTRODUCTION

Virtual communities exist for a multitude of reasons. They may support work, learning, and personal development processes; provide a common arena for information sharing; or serve as entertainment and social spaces. In some cases they bring together people already organized as a community, providing additional means of communication. In other cases, the technology provides a platform for uniting people with mutual interests who might otherwise never meet. They may be closed and highly private (e.g. membership only portals and listservs), they may flow and be defined through extended personal networks (e.g. accessible to people noted as friends of friends, such as is done on various social networking sites), or they may live out their existence in plain view (e.g. public web pages such as most blogs and twitter feeds). Regardless of the specific context, interpersonal interaction is the underlying objective of participation in a virtual community.

DOI: 10.4018/978-1-60960-091-4.ch019

Online information seeking is a fairly common activity with ever increasing numbers of people turning to the web for answers to their questions and solutions to their problems ranging from simple and fact-based to complex and subjective. Trustworthiness of information becomes an issue for many of these people. Individuals can readily publish to the web so long as they have a computer and Internet access, with no vetting of content for accuracy or quality. A simple query in a search engine can yield pages of results: some accurate and some not; some empirical and some opinion-based; and some from reviewed or filtered sources and some not. Such is the beauty of the Internet and the simultaneous challenge of using it in an information seeking capacity.

If we marry these two concepts of virtual community and online information seeking, it becomes apparent that in the nexus is a possible solution for narrowing the expanse of the information landscape, increasing the pertinence and (perceived) reliability of one's online knowledge seeking and problem solving experiences, and encouraging a more personalized approach to the whole experience. The result, whether intentional and by design or merely a byproduct of the users' actions, is a knowledge sharing community.

In this chapter, I present a study of how a blogging community serves a knowledge sharing function for its members, with a particular focus on how members share their problems and then help each other by providing support, guidance, and information. The community is one of loosely networked sole-authored blogs written in a diaristic style. The bloggers generally do not know each other off-line, and many use pseudonyms which prevent actions like googling each other, but they are bound by a shared profession: academe. In the blog world, they have sought interactions with like-minded others. Collectively, and in addition to socializing, these bloggers engage in knowledge sharing and problem solving processes on a regular basis. They do so informally, and without labeling their actions as such.

BACKGROUND

Virtual worlds do not exist in isolation from the physical world; what happens in one space readily influences the other. Bloggers, for example, frequently make connections between their online and offline worlds, which may represent actual friendships transcending the different media (Takhteyev & Hall, 2005) or may take the form of shared stories moving from one medium to another. Friendships that form online may not be as strong as those existing in the physical world (Cummings, Butler, & Kraut, 2002), but they can provide social support and contribute to a person's sense of well-being (Baker & Moore, 2008). Further, there are distinct purposes that those online relationships serve, such as providing support in contexts where face-to-face settings might be uncomfortable or unavailable.

The term "community" is one that has been used in an overly general sense, particularly with respect to online communities. Seemingly any collection of people drawn together by common interest or shared membership on a web site has been referred to as a community, a designation which becomes problematic given that community development requires more than just the ability to communicate in the same space (Kling & Courtright, 2003). In this paper, I use the term community to refer to collections of people who interact online on a regular basis, who have established online identities and are known to each other, and who exhibit collective characteristics such as trust, norms, and shared purposes. As such, communities are not mere message boards but are trending toward (if not actually representative of) what Wenger (1998) calls a Community of Practice.

Virtual communities exist based partly on the principle of reciprocity. Both socio-emotional and informational exchanges are likely to occur when a sense of virtual community has formed among participants (Ellonen, Kosonen, & Henttonen, 2007). The rationale for engaging in online communities for knowledge sharing purposes can both be for personal information gain as well as to support one's own social capital within a group (Kosonen, 2009).

Knowledge sharing in online communities is, essentially, a form of problem solving. Indeed, problem solving has been deemed a form of producing assets within a community (Wenger, White, & Smith, 2009). Although it is possible for individuals to simply share their knowledge for general purposes, many people are gravitating toward online communities to seek help in solving their own problems, to find others who share their problems and search for solutions together, and to share their own problem solving experiences in the hopes that others may find them helpful. Josefsson (2005) found that critically ill patients in Sweden used online communities for support purposes and pooled their information from online and offline sources in the interest of helping themselves and each other as they fought their illnesses. Similarly, Fox (2008) found in a survey of American patients that providing help and sharing knowledge about conditions and treatments is a common behavior. Chayko (2008) notes that people frequently cite helping others as a reason for participating in online communities, and telling stories that may help others learn from their experiences is a key motivation for some people to blog (Lee, Im, & Taylor, 2008; Lenhart & Fox, 2006; Ridings & Gefen, 2004). Although many examples of problem sharing in online communities focus on individuals with personal problems, as discussed above, the Internet also has become a platform for supporting larger scale community problem solving initiatives, such as civic engagement activities (Smith, Schlozman, Verba, & Brady, 2009).

Problem solving and knowledge sharing in online communities may be focused internally or externally. Two well known virtual ethnographies provide excellent contrasting examples: Rheingold (2000), in his book about a community called The WELL, discussed how community members came together in each others' times of need to provide support and assist in solving problems existing outside the community. Dibbell (1998), on the other hand, discussed how members of an online community called LambdaMOO approached a problem that existed entirely within the community, namely an offense committed by one community member against various others. Regardless of the locus of the problem – within or outside the community – the process of finding a solution in these as well as in other online communities relies on open communication and knowledge sharing among the membership.

A key element of knowledge sharing in online communities seems to be the attraction of like-minded people. Chayko (2002) notes that online relationships often help people feel less alone and more connected, not just on the level of having someone with whom they might communicate but at the level of being understood and sharing experience. Karlsson (2007) found that diary blogs tend to attract a fairly homogenous reader group who is similar also to the originating blogger. This assertion of homogeneity in blog-based networks is supported by other research (Kumar, Novak, Raghavan, & Tomkins, 2004). Further, bloggers have been found to have a particular audience in mind, even if that audience is comprised of unknown others (Qian & Scott, 2007). This assertion of like-mindedness does not mean that all participants must agree. Differences of opinion can readily be held within an online community, with trust as a critical factor in helping individual members assess the usefulness of information (Greenfield & Campbell, 2006). Still, individuals with differences treat each other respectfully to help maintain the sense of community (Dennen, 2009b). It is thus homogeneity of interest and not always of opinion that brings together a community.

Finally, it should be noted that interaction and community are not inherent to the practice of blogging, and not all bloggers desire feedback from an audience. In a sample of 260 bloggers,

Papacharissi (2006) found that self-expression was the key purpose for many bloggers and that while sharing information was of secondary importance, they were not heavily reliant on or anticipatory of feedback from readers. Further, many blogs exist in isolation, unconnected to others (Herring, et al., 2005). Among such bloggers, the likelihood of a knowledge community developing is slim, although there still might be knowledge to glean from reading these blogs. Thus, knowledge sharing and problem solving among bloggers is a deliberate activity sought out by particular individuals.

Research Context

In this chapter, I describe how problems are presented and discussed within a blogging community, a process that extends from initial sharing of the problem to (potentially) the indication and tale of resolution. In an attempt to organize, I developed a system for classifying problems according to factors that impact their presentation, discussion, and resolution. Specifically, I describe how this classification scheme was developed and then how interactions in a blogging community take place within each category.

When bloggers post about their problems they effectively initiate a knowledge sharing dialogue. That dialogue might include sympathetic or empathetic comments, shared experiences, support, and advice. Further, there are many potential turns and outcomes that can occur via that dialogue. It may present what ends up being the solution, being the catalyst for the blogger finding a solution, or it may lead nowhere at all. When multiple people respond, it may result in divergent or convergent thoughts, and as a result may bring people closer together or further apart, or otherwise alter the nature of their relationship.

We might view problem presentation and resolution within a community of bloggers as existing along two dimensions: number of bloggers posting about the problem and number of bloggers affected by the problem. I use the term problem loosely; it is not necessarily indicative of a negative or conflict situation or experience, but rather of the need to address a situation, perhaps making a decision or taking action. An example of a non-conflict problem situation might be as follows: It's the end of the budget year and a faculty member has been told that she has $2500 in professional development funds to spend or forfeit in the next month. She is initially uncertain how to spend that money and seeks ideas.

For the sake of classification, I differentiate between those problems that are unique (affect one person), shared (affect multiple people in their own ways) or communal (particular to a community). Further, I note whether the problem is raised by one blogger or multiple bloggers. The classification system thus has 5 points:

1. Individual Unique
2. Individual Shared
3. Multiple Shared
4. Individual Community
5. Multiple Community

The reason for the classifications is that the nature of the interaction that surrounds a post about a problem varies based on these dimensions. Note that there is not a classification for Multiple Unique. Inherent in multiple people posting about a problem and there being a collective meaning in that act is some dimension of shared meaning and experience as well as a way to define the bounds of the phenomenon being discussed. Thus, although multiple bloggers can and do post about their unique problems, there is no clear rationale for grouping and discussing these otherwise discrete posts in a collective manner. Further, it is possible that multiple bloggers will post individually about one particular blogger's unique problem, seeking help or support for that problem from as wide of an audience as possible, but these requests tend to be focused more on garnering well wishes (e.g. for someone's dissertation defense) or material support (e.g. for an ill person who lacks health insurance) and occur rather infrequently.

METHODS

Population and Sampling

For this study, I drew upon examples from a blogging community of academics. These bloggers tend to write in a diaristic manner, documenting elements of their personal and professional lives as well as the space where the two intersect. The community is informal and ill-defined, consisting of individual blogs that are interlinked to varying degrees and whose authors read and comment on the other blogs on a regular basis. For further discussion of what makes these blogs a community, see other works by Dennen and colleagues (Dennen, 2006; Dennen, 2009b; Dennen & Pashnyak, 2008).

Examples of each problem situation were selected using a quota sampling procedure. Essentially, at the time I choose to begin sampling, I conducted my regular daily observation of the blogs and each time I encountered an example of one of the problem situations I earmarked it until I had ten in each classification.

Examples to use as case discussions here were selected purposively in an attempt to represent a typical or mainstream case within that category. The determination of a case's suitability was determined not only based on its representativeness of common features in the 10 sampled blogs for each classification but also based on an overall fit noted from my extensive experience within this blogging community. By presenting the cases structurally and with secondary descriptions, each case effectively has been anonymized.

Procedure and Analysis

The procedure for this study involved three phases or levels of observation and analysis. In the first phase, general observations were culled from a year of field notes and archived blog posts. The field notes were related to a larger ethnographic study being done on this community and were not solely focused on documenting instances of problem situations on blogs, but nonetheless captured blogging about problems and their resolutions as a common narrative theme across the blog community. Through these observations, the classification scheme emerged. The classification scheme was then tested or triangulated by looking for evidence of each of these five classes of problem solving interactions in four additional online journaling communities in which I have either participated or researched in the last decade.

In the second phase, I returned to the academic blogging community and identified 10 instances of each problem class. For each of those 10 instances, I looked at trends and differences in how the problem sharing and solving process played out via the blog. The focus at this point was to identify the following elements:

- Nature of the problem type: What types of problems fit this category?
- Typical problem presentation: How did the blogger present the problem and what type of feedback, support, or interaction was being requested? In what way was feedback being requested?
- Response and knowledge negotiation: How did commenters respond to this type of problem? How did they negotiate their understanding of the problem, of what type of help the blogger was seeking, and of what would be a good solution?
- Resolution: Was problem resolution addressed on the blog? Did the commenters play a role in the resolution?

Note that one instance might span multiple blogs and multiple posts; only the simplest of problems were limited to discussion in a single post. From these instances, a description of typical phenomenon (or the range of typical) was developed. These descriptions then were triangulated against my field notes, which covered a far more extensive collection of problem-related posts although in less detail. Further, I returned to two other online communities from phase one to check that my descriptions would apply to problem situations in these communities, too.

In the third phase, I selected a case from each main category or problem type: unique, shared, and communal for further examination. For shared and communal problems I selected cases that demonstrate dialogues across multiple blogs. For the sake of simplicity in diagramming and discussing the cases, I purposely chose cases that, while typical, were relatively small in scope.

After selection, each case was analyzed for structural elements and major narrative events, and then I created a diagram demonstrating the key points of interaction that took place. Note that while these cases are not suitable for generalizing how many comments a problem post will receive or how quickly a problem will be resolved, they are useful for gaining a more vivid sense of how the blog-based problem solving process works.

Limitations

The major limitation of this study is its scope. It is a study of one community whose members are fairly homogenous in terms of career interests and level of education (although homogeneity of membership, as noted above, is not uncommon within virtual communities). It is the very nature of studies such as this one that challenges our definition of generalizability. That said, progression from a single case to a broader application can be part of the inductive theory building process (Reigeluth & Frick, 1999). To ensure rigor, Reigeluth and Frick recommend the consideration of situationality, or the applicability of the theory to varying situations. In this study, the analysis is based on a sample chosen via a quota procedure, and all problem situations encountered during the sampling process were readily assigned to a problem category. Thus, within the community being studied the applicability of findings may confidently be extended beyond the sample.

Still, these findings are not generalizable to other communities and settings in the same way that quantitative findings often are. Stake discusses the concept of naturalistic (1978) or petite (1995) generalizations, in which the reader judges for herself the applicability of research findings to her context based on observed similarities. For this reason, rich description of the research setting and findings is important and has been included in this chapter. I would encourage readers of this chapter to determine the suitability of these findings for use in their own virtual community contexts with these considerations in mind.

Further, the classification system was developed using a process of triangulation (described above) that involved looking for evidence of these types of problems in other online communities with which I am familiar. These other communities represent quite different interests than this academic one (animal breeding, wedding planning, motherhood, and crafting), and yet evidence of all of the problem types and related problem-solving discourse was found in each of these communities. I chose to not include these other communities in the analysis and cases presented in this chapter because I have not studied them as closely and systematically as I have the community of academics.

Presentation of Findings

I discuss each classification of problem situation separately, organized by type of problem: unique, shared, or community. Within each of these areas, I discuss the differences that tend to occur when the problem is presented by individual versus multiple blogs. I wish to note, importantly, that the blogs used in this study are not solely focused on problem solving. They are social spaces, and they are journals. While bloggers may post on problems of various magnitudes, they also may write about topics as diverse as hobbies, likes and dislikes, and current events. At other times, they may merely chronicle the events of or tasks completed on a given day. Some posts may be written for the blogger, as a form of personal documentation, whereas others may be written directly to a reader audience. The posts that focus on problem situations thus represent merely a portion of the discourse that takes place on the blogs, much as problem sharing and solving would be merely one of many activities for a face-to-face community.

UNIQUE PROBLEM

Nature of the Problem

A unique problem is one that affects just one blogger and is so specific in its pertinent details that others are unlikely to have a directly related, shared experience. Thus it is the domain of the individual blogger. Typically these problems are interaction or situation specific and not representative of larger systems (e.g. organizations, institutions, or communities). In other words "I don't have enough time to write" is not a unique problem, but "My department chair is continuously favoring another junior faculty member over me" is. Interpersonal relationships tend to be a frequent topic of unique problems, as do particular life events and opportunities. Not all unique problems are simple and potentially solved in a short time frame. Sometimes the problems presented have been plaguing the blogger for a long period of time – years or decades, even.

Problem Presentation

Individual bloggers tend to present their problems in two main ways: reflective and advice seeking. Those who are reflective appear to be writing out their problem situation primarily for themselves. In contrast, those who are advice seeking tend to ask questions of their readers, such as "What would you do?" "Does anyone have any ideas?" or "Am I overreacting?" Additionally, advice-seeking posts are more likely to include a back-story and are more extensively detailed. In other words, they are written with an audience in mind and the awareness that the audience will need sufficient information to provide useful responses.

Many unique problems unfold over the course of multiple posts; this statement is certainly true of the more complex ones. Follow-up posts that engage in problem elaboration tend to either provide updates since the previous post or additional details that were omitted from the previous post. In the latter case, these posts frequently are prompted by comments that either ask for additional information or make clear that the readers do not fully understand the blogger's problem.

About one-half of the time, the blogger will include potential solutions that she has considered so far. This phenomenon occurs regardless of whether the post seems more focused on personal reflection or is clearly seeking advice.

Response and Knowledge Negotiation

When bloggers post about unique problems, they generally do not anticipate that others will be able to provide a precise solution. Advice seekers are most open to comments and a variety of solution ideas. The elements of personal experience and what is called 'I' language in a therapy context (not to be confused with I-language in a linguistics context) are prevalent in the comments. Although commenters occasionally will trend toward the pedantic in their advice, much more often they will share how they would feel and what they might do, and then offer support. Thus, possible solutions are presented as what I would do and what you might or could do, but not as what you should do. The linguistic distinction there is important. Commenters generally do not assume that they know what will work for the blogger, but rather seem to be engaged in a form of brainstorming for the blogger, using their personal experience, problem solving skills and the Internet (they frequently link to resources). Further, commenters frequently will add lines such as "This worked for me" or "YMMV" (your mileage may vary), indicating their awareness that individual differences and preferences may affect the usefulness of any advice. This type of communication is indicative of the underlying sense of community among the participants.

Reflective bloggers tend to seek support and sympathy more than advice or solutions, and when others offer solutions the blogger often is quick to reply with additional details about how the offered solution will not or will only partially fit their specific problem. Even when commenters relate to the blogger's problem through similar experiences, the blogger typically holds onto a notion of his or her problem being unique. In other words, there may be a bit of resistance to accepting that what seemed like a rather unique problem may, actually, be on a fairly universal theme. Commenters and the blogger must negotiate a comfortable space with plenty of room for individual experience in order to successfully discuss the general problem area and perspectives on both the problem and how it might be addressed.

Occasionally, individual bloggers are simply posting about their problems to vent. In these instances they do not seek a response or reaction from others. If not stated clearly in the original post, upon receiving advice in the comments these bloggers may note in the comments that they are not seeking advice.

Regardless of post type advice-seeking or reflective), most commenters are not only commenting on someone else's problem, but also trying to maintain network relationships. This phenomenon is implicit in many of their actions and word choices. They are sensitive to each other's feelings, particularly those who are in need or experiencing a conflict-oriented problem. Occasionally they even make explicit remarks to this effect. In other words, they do not add to the blogger's stress by introducing additional conflict via comments. That said, at times additional conflict arises inadvertently, as noted when the blogger returns with a post or comment to that effect. In such an instance, the commenter typically takes one of two paths: (a) comments no further; or (b) apologizes and tries to smooth over the situation. The third option, which is to argue one's position, is infrequently taken and generally leads to a breakdown in the relationship between the blogger and commenter (supposing one existed). Occasionally a third party will enter the fray, attempting to make peace between the blogger and commenter who are in conflict, validating both of their feelings and actions.

The above holds true when the commenter is a member of the blogger's community. However, trolls (e.g., people looking solely to cause disruption or conflict) and one-time visitors also are possible commenters. Given their lower level of investment in maintaining a relationship with the blogger and, in the case of trolls, a general desire to create conflict, when these visitors comment they are more prone to violate community norms than others. However, comments that come from community outsiders appear to be much easier for insiders to dismiss as irrelevant. The general wisdom of the community is that one should not engage in discussion with the trolls. Community members may, in comments, call them out as trolls and occasionally the blogger will delete the comment altogether.

Resolution

Typically, if a blogger has either posted extensively or received extensive support or feedback about a problem, she will provide an update as to the status of the problem and any resolution that has occurred. In a resolution post, the blogger typically thanks her readers for their support and ideas. She may explain which ones she tried and which she did not, indicating her rationale for their suitability and ultimate effectiveness.

In eight of the ten instances sampled for this analysis the blogger posted about problem resolution within a month of the original post. In some instances, many follow-up posts were provided, describing the ongoing situation in a blow-by-blow manner. The depth of follow-up was related to the depth or severity of the problem in a parabolic manner. In other words the most minor and most severe of problems tended to have the least amount of follow-up, whereas those problems of medium consequence seemed to have the greatest amount of follow-up. This phenomenon seems likely due to the level of investment and emotional charge. Low levels of investment and emotional charge make an extensive update potentially unnecessary or hardly worth the time it takes to type. High levels of investment and emotionally charge, conversely, may make thorough follow-up overly time-consuming and potentially re-open wounds.

Case Example

In the selected case, the individual blogger composed a post about an interpersonal problem she was having with a colleague. Basically, the colleague had taken an action that made the blogger feel as if she were not respected. The initial post had a clearly emotional tone, and was more reflective than advice seeking. While the blogger included lines like "I'm not really sure what to do now," she did not directly ask for advice. However, that did not prevent her commenters from giving advice. Across 18 comments, 16 people addressed the situation. All offered general sympathy and support, whereas nine shared similar stories of conflicts and resolutions and five offered ideas for solving the problem. These comments spanned three days time, but 15 of them were written within the first 24 hours that the post appeared. This immediacy of response, as depicted in Figure 1, is important because it demonstrates how quickly a problem post is noticed and support is offered by regular readers.

On the second day, the blogger wrote another post about the problem. This follow-up post addressed various issues raised in the comments and added details about the problem. The tone was far less emotional with regards to the problem – presumably the previous post had been written in the heat of the moment whereas this second one was written after a period of cooling off – although somewhat defensive in response to the comments on the previous post, some of which had gently suggested that the blogger was engaging in behavior that might exacerbate the problem. The first few comments on this post were written concurrently with the comments on the first post from the day before; two readers noticed this second post moments after sharing their thoughts on the first and followed up with an immediate

Figure 1. Unique problem posting patterns

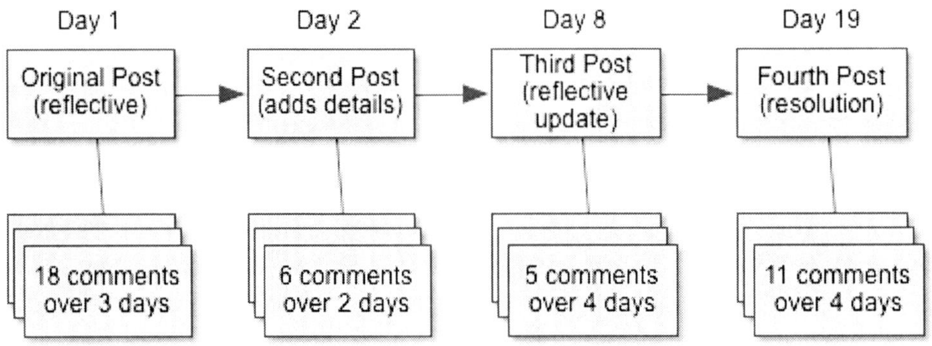

second comment. Each of the six commenters was someone who had posted on the previous post, responding to the additional information posted by the blogger as well as the emotional implications of her tone. Three commenters who perhaps had misinterpreted the first post also made an attempt to smooth over any misunderstandings.

One week after the initial post, the blogger posted an update to the situation. She had both tried a new technique when interacting with her colleague and had spent some time thinking about her initial interpretation of the problem. She had decided that she may have been misreading her colleague's actions. Finally, about 1.5 weeks later, she posted her last post on the problem, providing some resolution. In short, after the third post she decided to open up a dialogue about the conflict with her colleague. It turns out that the colleague was unaware that she had felt there was a conflict between them, and they were able to negotiate an understanding between them. This final post attracted 11 comments, ten of which were written by commenters on the original post.

SHARED PROBLEMS

Nature of the Problem

A shared problem is one that many people have, although the problem may manifest itself in different ways for different people. Within the community of academic bloggers who I have been studying, some shared professional problems revolve around the common work areas of teaching and scholarship as well as issues related to professional development, employment, and sense of self. Some of these problem areas included course design, grading, classroom management, dealing with rejection, job hunting, and imposter syndrome. A variety of personal problems are shared in this community as well, on topics such as personal finance, relationships, and childcare. Note that each of these problems is common to many people, but tends to affect them as individuals. For example, my time management issues and someone else's may be similar but likely are not exactly the same. We know we can anticipate some commonalities as well as some differences in the details. The differences generally are neither so great nor so personal in nature that we feel we cannot relate to each other on the topic.

Problem Presentation

To some degree, the differentiation between unique and shared problem is being made based on the blogger's presentation and perception of the problem. A blogger who is posting about a shared problem will include direct statements about the shared nature of the problem, such as "I know some of you have dealt with this before." The blogger might even indicate or link to specific instances in which others have shared about having the same problem. If the blogger is uncertain as to the degree to which the problem is a shared one, but is inclined to define it as such, she might state something like "Surely I'm not the only person trying to figure this out."

When writing about the problem, the blogger may glide over details with which the intended audience should already be familiar. This phenomenon shows how critical audience is to the problem presentation as well as the definition of the problem as shared. For example, an academic blogger may write about an upcoming dissertation defense without describing committee composition, defense format, or other information that would be known to other academics. That same blogger, when discussing the same problem with non-academic family and friends, would change her problem presentation to either simplify the context, make it analogous to something familiar to the audience, or explain unfamiliar points as asides. The ability to streamline elements of the problem presentation is an appealing one to many bloggers, as is knowing that an audience with shared understanding is likely to provide feedback that fits the context. This phenomenon has been noted directly on blogs, with bloggers stating outright why they at times prefer to blog about certain issues rather than discuss them with family and friends as well as why they get frustrated when explaining elements of academic life to outsiders who have ill-formed ideas about what life as a graduate student or professor is really like.

Response and Knowledge Negotiation

Most of the response to shared problems comes from people who share the problem, which is most likely not a blog's overall audience, but rather a subgroup. That said, people with like backgrounds and thus greater shared experience in the problem area tend to gravitate toward each other in the blogosphere. Commenters on shared problems, then, are most likely to fit one of three categories:

1. People who have solved the problem for themselves
2. People who currently share the problem and commiserate
3. People who anticipate having the problem

Each group of commenters will reply in a different way, based on their own experiences. Further, each seems to serve its own important purpose(s) in the problem response process (see Table 1), ranging from proving the problem can be solved to looking up to the person with the problem.

With a shared problem, knowledge negotiation becomes increasingly complex, but also tends to be less emotionally charged. Mutual problem sufferers willingly share their ideas and resources with the expectation that pooled knowledge and alternate perspectives can only help. There tends to be acceptance that everyone will experience the problem differently. Pressure to accept someone else's solution is lessened (even if only in perception), and attention may not be directly focused on the poster at all times. It becomes increasingly appropriate for others to discuss their problems and seek assistance – although briefly -- in the comment space.

Resolution

Resolution of shared problems typically comes in increments, if it all. Many of these problems represent ongoing issues that face bloggers, problems that are never fully resolved (e.g. time management issues) or that extend for a pre-determined time period and then go away regardless of resolution (e.g. end of semester or tenure review).

However, when there is a resolution post it can be useful to many readers. Whereas the resolution of a unique problem is merely a point of curiosity ("Ah, so that's what happened!") or satisfaction ("She took my advice and it worked!") to most commenters, a resolution statement about a shared problem becomes a post from which others can learn. Given that most blogs are archived and searchable, people who later have that problem (e.g. the initial readers who anticipated facing it at some point) may return to original post when it is more timely for the individual. Further, the resolution can be linked to, and can become part of a knowledge base on the topic.

Table 1. Commenter function in response phase

Category of Respondent	Purpose in Response Phase
Has solved problem	Provide reassurance and support (experienced cheerleader) Share proven solution(s) and resources Set role model (motivational)
Currently shares problem	Remove sense of isolation Share potential solutions and resources Collaborative exploration of solutions
Anticipates having problem	Observe Offer respect and support (inexperienced cheerleader)

Individual vs. Multiple Blogs

The key differences between how shared problems are mentioned on individual versus multiple blogs are related to the dispersion of ideas and the indicators of resolution. Simply put, when a problem is discussed across multiple blogs the dialogue becomes more difficult to follow. Individual readers, by choice or by chance, are likely to read parts of the dialogue on a problem topic, but not all of it. As a researcher diligently and systematically following these blog-based conversations I find it difficult to say with 100% certainty that I have found every last related thread of a cross-blog conversation. By contrast, following a single problem issue on an individual blog is rather simple. If you have the first post and the last post, you merely search all posts in-between. Relevant comments will be attached to the posts

It is useful, at this point, to describe how shared problems are manifest across multiple blogs. The genesis of the problem blogging activity can occur in one of two ways. In the first way, an individual blogger posts about a problem that she has. Her audience reads that post, and some people comment. However, her post also inspires some of her readers to compose their own post on the same topic. Why post to a different blog rather than just as a comment? Typically such actions are due to perceptions of ownership and space; specifically, within this community a blog is like a journal and it is considered a space for one person to write about topics of her choosing. There is a term that is used in the blogosphere to describe the act of writing a lengthy, self-centered comment underneath someone else's blog post: hijacking. Hijacking generally is frowned upon, and people who find themselves inclined to write such comments are encouraged to instead compose a post for publication on their own blog. Additionally, a person may want to attract more attention than would be garnered in the comments or address the specific audience for her own blog, which may differ slightly from the blog where the inspiration was derived.

The rationale of the bloggers is not being intuited here; rather, bloggers outright engage in moments of meta-blogging in which they discuss their intent and actions. For example, one commenter wrote, "Wow. I have a lot to say about this. Rather than hijack your comments, I'm going to post on my own blog." In another case, a blog post begins with an indicator of inspiration and reason for posting on her own blog rather than as a comment elsewhere:

In case you haven't seen (blogger)'s post about (topic), you should. I'm not going to link to her post because I don't want to connect directly to it. I disagree with much of what she says, but I don't want to send people over from here to disagree with her or disrupt the conversation she has going on. I have some very different personal experiences with (topic) and I want to discuss them here, with my peeps.

Alternately, and less common, some bloggers post in comments rather than on their own blogs to achieve a sense of distance between their blogging identity and the comment. As one blogger led in to a comment, "I'm not comfortable saying this on my own blog, and I'm writing anonymously, but you'll know who I am." These forms of meta-blogging are fairly typical among discussions of shared problems and occurred multiple times within the sampled problem cases.

In the second way, individual bloggers post about the same problem or related problems unbeknownst to each other. Eventually one of them sees the other's post or a third party may make the connection and point it out by commenting or composing yet another post that links to the two previous ones and ties them together. In this instance, the conversation has developed in parallel at multiple isolated locations. When the point of connection or synthesis occurs, it is illuminating to see the commonalities that exist (evidence of shared problem). In fact, this process may sway an individual blogger from viewing her problem as unique to viewing it as shared.

Regardless of how and by whom a problem initially is presented within a blogging community, it is typically in other venues and through the actions of other bloggers that synthesis is found. Bloggers use devices such as "round up" posts or "carnivals" that provide links to and sometimes provides synthetic discussion of a group of related and relevant posts. On occasion, a shared problem might result in the formation of a support group, and perhaps even a related group blog. Such was the case when a group of bloggers wanted to share some physical fitness tips and accountability, in one instance, and when a group of graduate students writing their dissertations chose to develop a support group in another.

Finally, in terms of resolution, when a problem is shared across multiple blogs there are multiple locations in which one might look for indicators of resolution. Individual bloggers may or may not follow up with what they did. However, some bloggers may have their initial entry into the fray at the point of resolution. On occasion, a person who did not blog about a problem, but who read about or participated in problem solving comments on other blogs, will write a resolution-style post giving a post-facto account of how she used tips or advice posted elsewhere to improve her own practices.

Case Example

The problem topic of the selected case was about dealing with a disruptive student. Specifically, Blogger 1 (B1) was struggling to maintain control over a class with two outspoken and assertive students who were overrunning the discussion. B1's blog was not heavily trafficked, and the post received two comments during its first two days. However, on the third day two things happened. First, another blogger (B2) also wrote about dealing with a disruptive student and a third blogger (B3) noticed and linked to both of those posts, commenting on how he typically deals with such students. It was after the appearance of B3's post that B1 received an additional 6 comments, two of which explicitly mentioned having come over from B3's blog. Figure 2 depicts this relationship between posts and how they became interlinked by various bloggers.

Figure 2. Shared problem case posting patterns

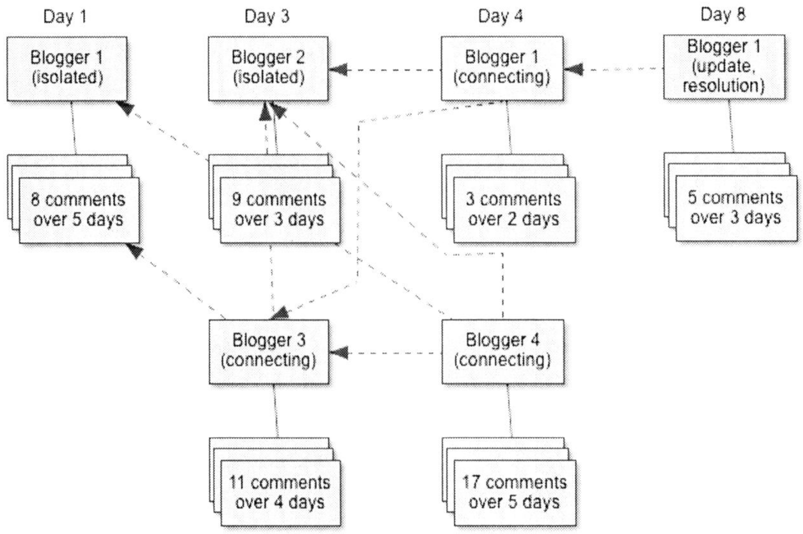

Although the specific details of B1 and B2's problems were different, the more general issues of instructor persona (how to be someone students will respect) and classroom management techniques (what to do in response) were raised. Most of the commenters on their posts as well as on B3's post shared their own stories of disruptive students and how they have either dealt with them or avoided disruption altogether.

On the fourth day, B1 wrote a very brief post, connecting to B2 and B3's post and noting that apparently he was not alone in his problem. He further commented that it was helpful to have so many ideas and so much support from everyone, and directed his interested readers to check out the other related posts. An additional blogger, B4, entered the discourse on that same day and connected to the earlier posts of all three bloggers. B4 had yet to teach her own course but would be a TA during the next semester. She expressed fear of disruptive students and uncertainty about other classroom management issues. Her post generated 17 comments, most of which were encouraging.

Finally, a week later B1 posted an update to the initial post. He had taught the class with the disruptive students again that morning and had tried two of the recommended techniques. One he felt had been successful and the other he was not sure about, but was planning to try it again. Regardless, he had felt much more in control of the class that morning, which was a positive move. The commenters congratulated him on taking control of the situation.

B1 is arguably the center of this shared problem narrative, but it is important to recall that this perspective is representative of a thorough search for and analysis of all interlinked posts. Any individual blogger or commenter's perspective will differ based on where she entered the narrative and how many of the posts and comments she read.

COMMUNITY PROBLEMS

Nature of the Problem

Community problems are, as the title suggests, problems that face a community as a group. Whereas shared problems are common but occur in each individual's own context, community problems impact people within a shared context. Within the community of academics, two types of community problems have been the topic of posts. First, there are problems that impact the blogging community. These tend to be issues related to blog identity (e.g. privacy and pseudonym use), technology (e.g. a bogging, commenting, or photo sharing platform that many of the bloggers use), or community interactions (e.g. civility, norms, and presence of trolls). Second, there are problems that impact a secondary community in which a number of these bloggers participate. In the instance of academic bloggers, sub-groups belong to the same professional organizations (e.g. Modern Language Association, American Historical Association). Members of the American Historical Association found the location of the organization's annual meeting in 2010 problematic because of the hotel owner's political leanings. Blogs provided a forum for discussion of both the organization's response as well as what individual members might do to both attend the conference and remain true to their own beliefs.

(see Jaschik, 2005 for additional information). Regardless of whether the problem is focused on the blogging or an off-blog community, the critical point in this definition is that the problem affects people as a byproduct of their membership in the community.

Problem Presentation

The presentation of community problems is not unlike the presentation of shared problems. It may be initiated on one blog or on multiple blogs with points of connected discourse. Individuals will use their own blogs as the platform for presenting their personal opinions about the problem and, at times, their solutions. Since the problems are things that make it challenging to function within the community, they tend to focus on reasons why someone would not want to remain in the community or changes that will be necessary to be a comfortable or productive participant. These problems generally tend to be less personal in nature (unless, of course, the individual interprets a community problem as a personal attack), and more structural or political.

For problems that are related to the blogging community and how it functions, such as trolls, privacy, or technology failures, the first indication of a problem may occur in the comments, not a main post. For example, a conflict may arise when troll-like comments appear or someone posts a counter-productive statement, or a commenter may alert the blogger to a technology malfunction such as a blogroll (a list of links to other blogs that appears on the sidebar) that is not displaying correctly. Typically these problems then are raised again at the post level.

Posts about external-community problems are likely to link to evidence of the problem (e.g. the other community's web presence or a relevant news story) or quote relevant items within a post. Those presenting the problem demonstrate an awareness of a split in their audience: some who share that external community and some who do not. Some back-story or contextual information might be necessary for that second group of readers and typically is included even though their status typically is that of a distant secondary audience for the post.

Response and Knowledge Negotiation

For within-community problems, there generally is incentive for a strong outpouring of support and free sharing of solutions. Even when one blogger has not been personally afflicted (e.g. a bug has downed comments, but she uses different comment technology, or others have been hit by trolls in the comments but she has not), there is a general sense that any action that weakens the community is worth concern.

A community-level response is evident in many of these cases. The community will respond by pooling its knowledge. Those who have solutions will post them. Early and visible problem solvers will often serve as models for others, and iterations of solutions may appear, each round representing a higher degree of sophistication. The unafflicted will take proactive measures. For example, when one person is affected by a breach of privacy and ratchets up his or her privacy controls and site statistic monitoring, others tend to do so as well in order to avoid the problem. Alternately, a bit of politicking might take place as certain bloggers encourage others to follow their lead (e.g. boycott a particular technology).

It is in response to within-community problems that individual bloggers are most likely to step in and try to directly fix a problem for each other. So, for a technology problem a more tech-savvy blogger may write a piece of code for someone else, call and walk someone else through a fix, or gain access and directly fix the problem themselves.

For a problem like incivility within the comments, which violates a community norm, a commenter from the community may defend the blogger or even attack the negative commenter. Following on to such an event, bloggers may develop or refine their own commenting policies and perhaps even turn on comment moderation. Such actions tend to be noted and discussed when they occur, and often travel through segments of a community like a small wave.

For external-community problems, the within-community response tends to be less great. This phenomenon is not surprising, really; large portions of the community are generally unaffected by the problem and do not anticipate being affected by it, either. Responses from those who share both communities tend to focus on commiseration or presentation of alternate perspectives. Responses from those who do not share the external community tend to focus on support from the outside and examples of how members of other similar communities have addressed like problems. Outsider responses, from people who belong to the external community but not the one where the problem is being discussed, also are likely to appear. One thing is common across all community problem responses: the members flexibly negotiate the allocation of virtual space such that it can at different times be a platform to discuss my individual experience, your individual experience, or our collective experience.

Resolution

Depending on the details of the problem, there may not be a resolution to post. Some of these community problems are ongoing, and some are in areas where there is no individual element of the problem to solve aside from the decision to continue or discontinue affiliation with the community. Else, resolution patterns and norms mimic those for shared problems (see above).

Multiple vs. Individual Blogs

As with resolution, the patterns and norms for shared problems (see above) tend to apply to community problems. It is far more common to see community problems discussed across multiple blogs. Alternately, when a community-level problem is only mentioned on one blog it might be taken as an indicator that few people perceive or agree with the problem within-community problems, in the case of external community problems, few within-community members share that same external community.

Case Example

In the selected case, the discussion topic was the grant application and awarding process within a national funding agency. Again, one blogger's posts rose to form the central part of the narrative, and it is this one blogger (B1) who posted a total of three times compared to one post each for the other seven bloggers who wrote on the topic. Across a four day period of time, eight different bloggers covered the topic, and only two after the first blogger did so without linking to the other posts (see Figure 3). One of those two cases involved Blogger 6, who offered a contrary view and who stated a desire to stay unconnected to the main discourse.

Among the various posts, also, it was B1 who sought a solution to an individual problem, whereas the other bloggers provided commentary about past problems with the agency and offered their opinions about and sharing experiences with the application process. The others perceived the situation as problematic and recognized how it affects them, but did not state a specific current problem of their own from within that context. They called for change writ large, but did not look to take action on their own.

Across the posts, bloggers and commenters alike shared stories and gave advice about dealing with this funding agency. Three commenters represented other disciplines, and commented on how this agency's process differs from the ones with which they are familiar. Collectively, a lot of knowledge was shared across the various posts and comments. In fact, one of the later commenters thanks those who came before him for the wealth of information they have offered which will be useful when he seeks funding in the future.

DISCUSSION

There are some common threads that come out across these categories of problems. First, relationships among the parties involved – specifically the blogger's sense of audience and the commenter's sense of the blogger – permit relatively efficient and effective knowledge exchanges and offer-

Figure 3. Community problem case posting patterns

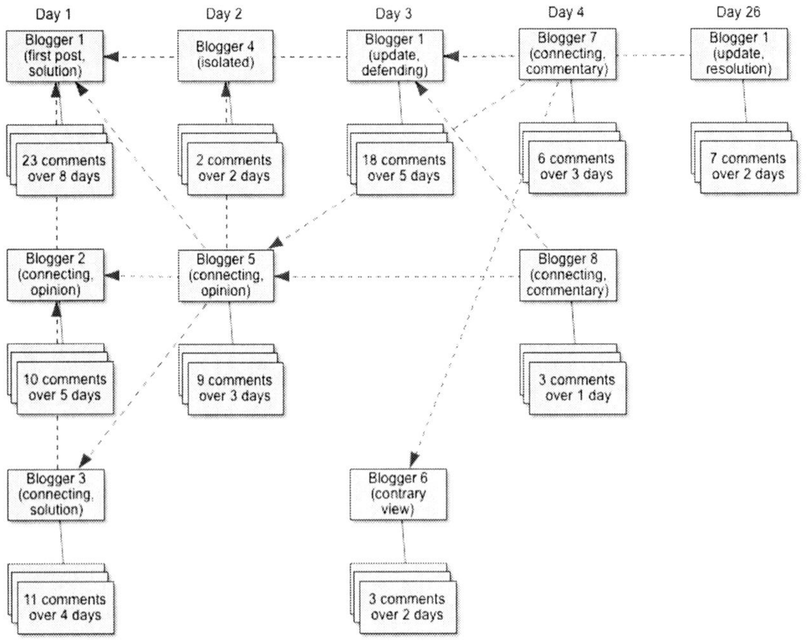

ings of support. These relationships are a factor of prolonged engagement within the blogging community, through which co-bloggers typically develop a sense of intersubjectivity (Vygotsky, 1978; Wertsch, 1985). This sense of mutual understanding is dual, in part existing because of community homogeneity (blogger as academic) and in part because of shared experiences within the community.

Second, and highly related to the first point, community discourse norms help structure the manner in which knowledge is negotiated and problems are solved within the virtual community. These norms dictate when it is appropriate to comment and when it is better to write one's own post, how to productively disagree or offer an alternative perspective, and when and how to link to posts written by others. They help focus or expand topics as is contextually appropriate, and keep the knowledge sharing process productive. In order to achieve this goal, they reflect the characteristics that are the underpinnings of community as defined in this study and by others (e.g., Baym, 1998; Hanisch & Churchman, 2008; Josefsson, 2005; Kling & Courtright, 2003), such as trust, reciprocity, identity development and sense of belonging.

Third, problems are recognized as complex narratives to which there may not be a simple solution. Initial problem posts are, in most cases, open-ended stories. This finding supports the usefulness of sharing multiple perspectives and opinions during the knowledge sharing process over focusing on so-called correct solutions. Commenters enter the narrative and weave in their own threads, which may include opinions, resources, emotional support, or somewhat parallel stories, all with a focus on working toward resolution. Still, the solution to the blogged problem is not certain or predetermined and thus must be shared. The natural human desire for story resolution on the behalf of both blogger and audience, as evidenced via meta-blogging activities, frequently leads to at least some form of update or resolution post. Note that this complexity and the desirability of individual perspectives makes community-based problem solving different from activities such as problem solving on a technical forum in which individuals seek and share concrete solutions more than narratives and emotional support

Fourth, the ownership of virtual space impacts who posts what and where they post it. The discourses that take place on these blogs might not work in another forum. Bloggers might not feel as comfortable posting their problems to a general forum as they do to their own space where they write about various topics related to their lives. To some extent, this assertion involves privacy and individual control. Bloggers enjoy the ability to delete entire blogs as well as specific posts or comments from their space as well as to use password protection and prevent search engine indexing (Dennen, 2009a). Additionally, owning the space permits a blogger to be entirely self-centered when discussing a problem, a luxury not always afforded people posting in shared forums. Similarly, commenters may be situationally more comfortable sharing their experiences on someone else's blog as one comment among many than they might be posting at the top level of their own blog. There is a psychological difference between posting a personal story for the sake of posting it and posting it in response to someone else's need or query. There also is a difference between self-disclosure to a known versus unknown audience, which relates to research on self-disclosure and sense of anonymity (Qian & Scott, 2007).

Finally, there are four common roles that emerge from within the discourse. There can be one or more problem owners, who write their own posts about a problem. There are experts, who have previously handled the problem or worked significantly in that area. There are collectors and archivists who bring together ideas, information, and resources from various sites and synthesize them for others to use in the problem solving process. And finally there are the supporters, who may or may not be offering solutions but who more importantly make the problem owners feel less alone.

There are key differences among the problem types, too. In unique problem situations, the focus is maintained on the original blogger. Stories and experiences can be shared as part of generating knowledge and finding a solution for the blogger, but ultimately the interactions do not deviate from that focus on the blogger. Additionally, unique problems are more likely than shared or community ones to demonstrate emotionally charged writing at an interpersonal level. Shared problem posts are more likely than community ones to be ongoing rather than situational, with bloggers often linking to earlier rounds of discourse, knowledge, and solutions on the topic. Finally, the difference between individual and multiple blog problems, whether shared or community oriented, becomes one of scope and interest. Multiple blog problems begin as individual ones. When individual posts inspire others to write on a topic or someone notices and makes explicit the connections between individual posts, the potential for knowledge sharing increases.

Implications

The value of studying these interactions in a naturalistic setting rests in the insights that arise from how people communicate with each other in a problem-solving context. From these blogs, we can see how particular characteristics of a community poise its members to engage in successful ongoing knowledge sharing processes. In other words, the success that these bloggers have in effectively sharing knowledge and supporting each other is in some ways unique to their community context, but is also indicative of the need for community, norms, individually owned space, and role play among participants. A sense of community helps keep participants focused on a bigger picture and encourages collegiality, trust, and reciprocity – all necessary for creating a safe space for sharing problems and anticipating useful solutions. Norms provide shortcuts for behavior and communication and help prevent misunderstandings; when participants intuitively understand each other's actions, they may more readily focus on the ideas being shared. Individually owned space provides a sense of confidence in terms of both platform (no one else is expected to take center stage) and security or privacy concerns (author controls over editing, deleting, and search engine indexing). Finally, different and dynamic roles allow individuals to contribute to any community conversation in a productive manner regardless of their experience or knowledge (i.e., those who cannot be experts can still be supporters).

These four factors are things that designers of more prescribed online communities might consider in their planning process, and that moderators or facilitators might take into consideration when looking for effective ways of helping people engage in knowledge sharing and problem solving processes in online forums. Although developing a sense of community is heavily reliant on the desire for said community among participants, community builders should be sensitive to the importance of allowing or encouraging that sense of community to form. The other three factors may be more easily supported by deliberate actions. In a highly structured virtual community, desired norms may be set forth as rules, and community facilitators or leaders may readily model both norms and roles. Additionally, tool selection and options may be determined with an eye toward providing users with a sense of both personal space and control over their words once posted.

CONCLUSION

In this chapter I have presented the way in which members of a blogging community present and resolve three classifications of problems – unique, shared, and community – through a knowledge sharing process that is manifest via both blog posts and their resulting comments. Further, I have discussed how problems and knowledge sharing differ based on whether they are centered on one individual blog versus presented across multiple blogs. Regardless of how a problem is classified, this chapter demonstrates that blogs are a useful platform for supporting problem solving and knowledge sharing processes and that both a sense of community and its resulting norms are a key part of making these processes successful.

Classifying problem situations as I have here is important for understanding the relationship between individual bloggers and how support and sharing can occur not despite but rather because of interactions in separate but interlinked virtual spaces. The use of individual blogs as a problem solving and knowledge sharing space is dependent on both blogger and audience, and thus it is the existence of community that helps these virtual spaces thrive. Shared and community knowledge are, in many ways, what bring the community members together. From this sense of belonging, then, and the familiarity and relationships that develop as a result, comes a natural opening for sharing unique problems as well as helping others with unique problems work toward resolution.

The dynamic roles that participants hold in the problem solving process (owner, expert, collector/archivist, and supporter), along with reciprocity (a characteristic of communities) help make these online settings well suited for problem solving and knowledge sharing. Individuals playing the three responsive roles help the problem owner fully develop a problem narrative both implicitly, as intended audience, and explicitly, through their offerings of support and knowledge. The ability to shift in and out of the roles for different problem situations means that one need not be limited to one type of community interaction. Each of these roles is necessary across all three types of problems, and each supports resolution in its own way.

Looking to the future, there is great opportunity for research to be done in this area. The relationship between social elements, individual reflection (e.g. journaling) and knowledge sharing in informal online communities is still not well understood. Many studies – this one included –

look at individual communities or populations. More cross-community research is needed to help increase understanding of what trends and interactions are truly community-specific and which can be generalized to certain types of communities or populations. The challenge to doing such research, however, is the time required to truly come to know an online community. However, an increased focus on research using similar frameworks would, in the long term, facilitate synthesis and triangulation across studies. Similarly, frameworks and classifications themselves might be tested across communities, as was done during the classification development process for this study, to gain confidence in the system's applicability to other people and settings. Finally, related topics such as knowledge management and archiving often are neglected in the context of informal virtual communities, but it seems likely that there is much to be gained from their exploration.

REFERENCES

Baker, J. R., & Moore, S. M. (2008). Blogging as a social tool: A psychosocial examination of the effects of blogging. *Cyberpsychology & Behavior, 11*(6), 747–749. doi:10.1089/cpb.2008.0053

Baym, N. (1998). The emergence of online community . In Jones, S. G. (Ed.), *Cybersociety 2.0* (pp. 35–68). Thousand Oaks, CA: Sage.

Chayko, M. (2002). *Connecting: How we form social bonds and communities in the Internet age.* Albany, NY: State University of New York Press.

Chayko, M. (2008). *Portable communities: The social dynamics of online and mobile connectedness.* Albany, NY: State University of New York Press.

Cummings, J. N., Butler, B., & Kraut, R. (2002). The quality of online social relationships. *Communications of the ACM, 45*(7), 103–108. doi:10.1145/514236.514242

Dennen, V. P. (2006). Blogademe: How a group of academics formed and normed an online community of practice . In Mendez-Vilas, A., Martin, A. S., Mesa Gonzalez, J. A., & Mesa Gonzalez, J. (Eds.), *Current developments in technology-assisted education* (*Vol. 1*, pp. 306–310). Badajoz, Spain: Formatex.

Dennen, V. P. (2009a). *Because I know you: Elements of trust among pseudonymous bloggers.* Paper presented at IADIS International Conference on Web-based Communities 2009, Algarve, Portugal.

Dennen, V. P. (2009b). Constructing academic alter-egos: Identity issues in a blog-based community. *Identity in the Information Society, 2*(1), 23–38. doi:10.1007/s12394-009-0020-8

Dennen, V. P., & Pashnyak, T. G. (2008). Finding community in the comments: The role of reader and blogger responses in a weblog community of practice. *International Journal of Web-based Communities, 4*(3), 272–283. doi:10.1504/IJWBC.2008.019189

Dibbell, J. (1998). *My tiny life: Crime and passion in a virtual world.* New York: Henry Holt and Company.

Ellonen, H. K., Kosonen, M., & Henttonen, K. (2007). The development of a sense of virtual community . *International Journal of Web Based Communities, 3*(1), 114–130. doi:10.1504/IJWBC.2007.013778

Fox, S. (2008). The engaged e-patient population. *Pew Internet & American Life Project.* Retrieved January 20, 2009, from http://www.pewinternet.org/Reports/2008/The-Engaged-Epatient-Population.aspx

Greenfield, G., & Campbell, J. (2006). Communicative practices in online communication: A case of agreeing to disagree. *Journal of Organizational Computing and Electronic Commerce, 16*(3-4), 267–277. doi:10.1207/s15327744joce1603&4_6

Hanisch, J., & Churchman, D. (2008). Virtual communities of practice: The communication of knowledge across cultural boundaries. *International Journal of Web-based Communities*, *4*(4), 418–433. doi:10.1504/IJWBC.2008.019548

Herring, S. C., Kouper, I., Paolillo, J. C., Scheidt, L. A., Tyworth, M., Welsch, P., et al. (2005). Conversations in the blogosphere: An analysis " from the bottom up". *Proceedings of the 38th Hawaii International Conference on System Sciences*. Los Alamitos: IEEE Press.

Jaschik, S. (2005, January 5). Historians reject proposed boycott. *Inside Higher Ed*. Retrieved from http://www.insidehighered.com/news/2009/01/05/boycott

Josefsson, U. (2005). Coping with illness online: The case of patients' online communities. *The Information Society*, *21*, 141–153. doi:10.1080/01972240590925357

Karlsson, L. (2007). Desperately seeking sameness. *Feminist Media Studies*, *7*(2), 137–153. doi:10.1080/14680770701287019

Kling, R., & Courtright, C. (2003). Group behavior and learning in electronic forums: A sociotechnical approach. *The Information Society*, *19*, 221–235. doi:10.1080/01972240309465

Kosonen, M. (2009). Knowledge sharing in virtual communities: A review of the empirical research. *International Journal of Web Based Communities*, *5*(2), 144–163. doi:10.1504/IJWBC.2009.023962

Kumar, R., Novak, J., Raghavan, P., & Tomkins, A. (2004). Structure and evolution of blogspace. *Communications of the ACM*, *47*(12), 35–39. doi:10.1145/1035134.1035162

Lee, D. H., Im, S., & Taylor, C. R. (2008). Voluntary self-disclosure of information on the internet: A multimethod study of the motivations and consequences of disclosing information on blogs. *Psychology and Marketing*, *25*(7), 692–710. doi:10.1002/mar.20232

Lenhart, A., & Fox, S. (2006). Bloggers: A portrait of the internet's new storytellers. *Pew Internet & American Life Projects*. Retrieved October 20, 2009, from http://www.pewinternet.org/Reports/2006/Bloggers.aspx

Papacharissi, Z. (2006). Audiences as media producers: Content analysis of 260 blogs . In Tremayne, P. (Ed.), *Blogging, citizenship, and the future of media* (pp. 21–38). New York: Routledge.

Qian, H., & Scott, C. R. (2007). Anonymity and self-disclosure on Weblogs. *Journal of Computer-Mediated Communication*, *12*(4). Retrieved from http://jcmc.indiana.edu/vol12/issue4/qian.html. doi:10.1111/j.1083-6101.2007.00380.x

Reigeluth, C. M., & Frick, T. W. (1999). Formative research: A methodology for creating and improving design theories . In Reigeluth, C. M. (Ed.), *Instructional design theories and models: A new paradigm of instructional theory* (pp. 633–652). Mahwah, NJ: Lawrence Erlbaum.

Rheingold, H. R. (2000). *The virtual community: Homesteading on the electronic fronteir*. Cambridge, MA: MIT Press.

Ridings, C. M., & Gefen, D. (2004). Virtual community attraction: Why people hang out online. *Journal of Computer-Mediated Communication*, *10*(1).

Smith, A., Schlozman, K. L., Verba, S., & Brady, H. (2009). The Internet and civic engagement. *Pew Internet & American Life Project*. Retrieved November 17, 2009, from http://www.pewinternet.org/Reports/2009/15--The-Internet-and-Civic-Engagement.aspx

Stake, R. E. (1978). The case study method in social inquiry. *Educational Researcher*, *7*(2), 5–8.

Stake, R. E. (1995). *The art of case study research*. Thousand Oaks, CA: Sage.

Takhteyev, Y., & Hall, J. (2005). *Blogging together: Digital expression in a real-life community*. Paper presented at the Social Software in the Academy Workshop, Los Angeles, CA.

Vygotsky, L. S. (1978). *Mind in society: The development of higher psychological processes*. Cambridge, MA: Harvard University Press.

Wenger, E. (1998). *Communities of practice: Learning, meaning, and identity*. Cambridge, UK: Cambridge University Press.

Wenger, E., White, N., & Smith, J. D. (2009). *Digital habitats: Stewarding technology for communities*. Portland, OR: CPSquare.

Wertsch, J. V. (1985). *Vygotsky and the social formation of mind*. Cambridge, MA: Harvard University Press.

Chapter 20
Establishing Reasoning Communities of Security Experts for Internet Commerce Security

Andrei V. Kelarev
University of Ballarat, Australia

Simon Brown
University of Ballarat, Australia

Paul Watters
University of Ballarat, Australia

Xin-Wen Wu
University of Ballarat, Australia

Richard Dazeley
University of Ballarat, Australia

ABSTRACT

The highly sophisticated and rapidly evolving area of internet commerce security presents many novel challenges for the organization of discourse in reasoning communities. This chapter suggests appropriate reasoning methods and demonstrates how establishing reasoning communities of security experts and enabling productive group discourse among them can play a crucial role in successful resolution of problems concerning the implementation, integration, deployment and maintenance of flexible local security systems for defense against malware threats in internet security. Local security systems of this sort may combine several ready open source or commercial software packages behind a common front-end and may enhance and supplement their facilities with additional plug-ins. To illustrate the diverse character of challenges the reasoning communities in internet security are likely to be faced with, this chapter concentrates on defense against phishing attacks. This example was selected as it is one of the newest and most rapidly changing application domains for the principles of organizing reasoning communities. The major group discourse methods suggested for the reasoning communities of security

DOI: 10.4018/978-1-60960-091-4.ch020

experts in this chapter include the Delphi Method, the Wideband Delphi Process, the Generic/Actual Argument Model of Structured Reasoning, Brainstorming, Reverse Brainstorming, Consensus Decision Making, Voting, Open Delphi and Open Brainstorming Methods. The Delphi Method and Wideband Delphi Process are suggested as tools for organizing a cohesive reasoning architecture, for coordinating other methods, and for preparing and allocating other methods to particular issues.

INTRODUCTION

This chapter is devoted to the development of methods for group discourse in the reasoning communities of security experts responsible for all stages of the implementation, integration, deployment and maintenance of flexible local security systems for defense against malware threats. In particular, we are looking at creating a fairly small and very flexible system to defend against phishing attacks, which may be applicable to other internet commerce security tasks, and which is adjustable enough to take into account local requirements in order to address the local information security needs of various organizations and make it easier to use uniform anti-phishing packages available on the market. For successful use of a highly flexible security system of this sort it is essential that it be supervised and guided by a small specialized group of local system administrators and security experts. The main aim of this article is to set up a framework for discourse in the reasoning communities of security experts in order to enable robust supervision of adjustments to all parameters of the local security system, as well as suggest directions for future research on associated reasoning.

There are serious differences in the operation of various organizations, their email correspondence and patterns of email usage by individual staff. Substantial differences among numerous organizations lead to the diverse character of security threats they are faced with. This provides motivation for creating a flexible system capable of complementing large uniform anti-phishing tools available on the market.

It is best to include one or more ready open source or commercial programs in the system, because these products have already been thoroughly tested and have been developed and improved for many years. Such systems can be installed centrally as well as for each individual user to take care of personal preferences. These security packages can apply a number of methods well-known in the defense against phishing. The major steps and algorithms involved in the systems include pre-processing of email messages, feature selection, classification and clustering of messages, and preparation of evidence for evaluation.

On the other hand, the installation, integration and management of several already existing large uniform programs could benefit from the development of a combined convenient front-end simplifying the configuration tasks and providing additional facilities. The effectiveness of operation of the combined local system can be further fine-tuned by developing appropriate plug-ins. This would involve various mechanisms and parameters which can be monitored and adjusted as the operating environment and current threats evolve.

The implementation, integration, deployment and maintenance of a flexible local system of this sort is impossible without successful resolution of a variety of problems by a specialized reasoning community of security experts managing these tasks. There are numerous aspects of the system which need to be monitored and adjusted. The group discourse methods for the reasoning community considered in this chapter aim to enable the reasoning community to handle all of these tasks effectively and resolve various issues required for the adaptation of a security system to local circumstances.

This chapter demonstrates that it is necessary to use a broad variety of reasoning methods in order to organize a productive discourse framework in a reasoning community of internet security experts. The methods suggested for the reasoning communities of security experts in this chapter are the Delphi Method, the Wideband Delphi Process, the Generic/Actual Argument Model of the Structured Reasoning, Brainstorming, Reverse Brainstorming, Consensus Decision Making, Voting, Open Delphi and Open Brainstorming Methods. The Delphi Method and Wideband Delphi Process are suggested as tools for organizing a cohesive reasoning architecture coordinating all other methods, preparing and allocating them to particular issues. In particular, we recommend applying the Delphi Method and Wideband Delphi Process for strategic long-term decisions concerning the architecture of the group reasoning in the community as well as for coordinating and preparing other reasoning methods, including all preparations necessary for the Generic/Actual Argument Model of Structured Reasoning.

The practical significance of various choices the security experts can make in their discourse processes will not be addressed in this chapter. A few open questions indicating possible directions for further research are included in the conclusion to our chapter.

EXAMPLES OF ISSUES FOR DISCOURSE IN REASONING COMMUNITIES OF INTERNET SECURITY EXPERTS

To illustrate the complex and dynamic character of the internet security domain this chapter concentrates on defending against phishing attacks. This domain is one of the most novel and rapidly changing areas for discourse in reasoning communities of internet security experts. The overview given here prepares motivation for recommendations in the following sections and demonstrates that a variety of reasoning methods including high level reasoning processes are necessary for reasoning communities of internet security experts.

Firstly, this chapter will overview the background information on the defense against phishing. These preliminaries are also required to explain why dealing with phishing is such a concern and why a variety of reasoning methods are required for group decisions in the area.

Phishing is an act of social engineering attempting to elicit confidential information from users by sending emails with false explanations urging users to provide private information that will be used for identity theft. The users may be requested to reply to the email, or visit a bogus web site, where they are asked to enter personal details, such as passwords, credit card numbers, tax file number or bank account numbers. More data on current trends and statistics regarding phishing activities are presented, for example, by the Anti-Phishing Working Group (2009), OECD Task Force on Spam (2009) and Wolfe, Scott and Erwin (2004).

Well known machine learning filtering programs include SpamAssassin, Bogofilter, DPHISHING, SpamBayes and CRM114. Hybrid filtering combining various methods has already been implemented in several open source programs including SpamAssassin. We refer to McDonald (2004), Schwartz (2004) and SpamAssassin (2009) for detailed information on configuration, customization, and integration of SpamAssasin. Another interesting example is the project CRM114 – the Controllable Regex Mutilator (2009). Criteria for categorization of data in CRM114 can be via a multitude of methods, including Hidden Markov Models, KNN/Hyperspace, regular expressions, approximate regular expressions, WINNOW, Correlation, Orthogonal Sparse Bigrams, or Bit Entropy. CRM114 is compatible and can be combined in one system with SpamAssassin and other email filtering software.

User education and training is probably the very first and most essential topic for group discussion. It involves many items to be implemented and maintained by the system administration group. The Anti-Phishing Working Group (2009) has compiled a list of recommendations for user education and training. For example, email users in the organization may be advised to

- avoid responding to phishing emails or clicking on any links in suspicious messages;
- avoid making their email addresses availability online whenever possible;
- modify the addresses that have to be displayed online in order to hide them from automated email harvesters by displaying all or part of the address as an image, a text logo modified using in-line CSS, or as re-ordered text restored to the original order using in-line CSS;
- disable JavaScript and HTML functionality such as display of images to prevent exploiting vulnerabilities in the HTML renderers which can be used to install spyware.

Tools for reporting of phishing need to be implemented in the local security system. Messages identified as phishing emails can be stored locally in a user's database, or they can be transferred to the central administration to be used in the development of global email filters. They can also be reported on online sites collecting information on current phishing activities.

Enforcement of technical requirements of the Simple Mail Transfer Protocol (SMTP) and strict compliance with the RFC standards can be used to reject all emails which do not adhere to the standards. Zombie computers sending phishing emails cannot always comply with all standards. Indeed they do not possess complete access to and legitimate control of all functions of the computer sending phishing emails. These options include a callback verification, which can detect if an invalid sender addresses has been indicated in the phishing email. This can be checked if the local security system makes an SMTP connection to the mail exchanger for the address in the phishing email.

Backscatter limitation may need to be managed. The senders address can also often be forged in a phishing email. Sending replies to this address will lead to backscatter. One option is not to send any replies or rejection notifications to phishing emails. However, if a challenge-response system is included and activated, then it has to send emails to the senders. In this case, the backscatter problem can be reduced by placing the challenge in a bounce message instead of creating a new reply email. The receiving mail transfer agent can give an SMTP rejection code and force the sending system to create the bounce message. This causes the bounce message to be emailed to the original sender of the email, not the one indicated in the forged sender field.

SMTP greeting delay is a deliberate pause in the operation of an SMTP server before sending the so-called SMTP greeting banner to the client. A number of phishing applications do not usually wait to receive this banner, and instead are trying to start sending data as soon as the TCP connection is opened. This can be detected to reject emails from a phishing sender.

SMTP grey listing can be used to temporarily reject emails from senders with unknown mail servers. A temporary rejection is denoted by a 4xx error code. This code is recognized by all normal servers, which then try to deliver the email again later. Senders of phishing emails usually do not store their emails and cannot therefore retry sending them again later unless they increase their operating costs storing emails for repeated sending of exactly the same copy of the email. If a message is received again, then the sender is removed from the grey list.

DNS-based anti-phishing systems can use known black lists of bad IPs and URLs available online, and local white lists of good IP addresses.

DNS black lists can also be used in the local security system to reject all correspondence from known senders of phishing emails identified and included on the lists. The DNS black lists are lists of IP addresses of known senders of phishing emails, open relays and zombie computers.

Country-based filtering can be enabled by some users to block all correspondence from certain countries if the users receive a lot of phishing emails from these countries and do not have business correspondence with legitimate users there. In this case the country of origin of each email is determined by the sender's IP address.

Header filtering can be used to inspect the headers of received emails. Every header contains information about the message and email servers transmitting it. Senders of phishing emails often spoof various fields which occur in the header in order to make it look more legitimate and hide the origin of the email. The resulting headers in phishing emails often violate the RFC 5322 standard specifying correct format of the header.

Filtering algorithms can be also used. They involve automatic analysis of the content of emails and deleting those messages which are classified as phishing. These techniques can be integrated into the central mail transfer agent. At the same time filtering can also be implemented locally on each email program of each individual user to take into account the patterns of email correspondence received by the user.

Anti-phishing programs can be incorporated in the system and used to evaluate the number and size of images in an email. Some of these programs use optical character recognition to analyze words in the images.

A central system can also be used to detect many of the disguising methods practiced by the senders of the phishing emails, and then can reject these emails without passing them on to the individual mail agents.

Limiting the allowed rate of sending emails and slowing down such transmissions can prevent a compromised local computer from sending out numerous phishing emails until a manual investigation can be undertaken.

Statistical filtering can also be used in the central system. Usually statistical filters require the creation of very large databases of emails already classified as phishing emails by the users. Such databases can be handled on a central email server, if techniques to speed up the operation of such databases by creating condensed summaries or index files are investigated and implemented.

Machine learning can be employed by requesting that users to mark undesirable messages as phishing emails. The filtering program can use machine learning methods to learn from these evaluations. A button needs to be implemented on an email client so that the users can identify phishing emails. Thereby a statistical filter takes into account individual usage and preferences with regard to content, message headers and peculiarities of the transmission protocols involved. A statistical filter can very quickly react to variations that might occur in the contents of phishing emails being received without requiring administrator's assistance or intervention.

Keywords and statistical parameters can be used in the evaluation and comparison of messages in order to detect phishing emails. Identifying keywords in the content of messages and evaluating statistical parameters of the messages have to take into account the types of legitimate emails that each individual receives and normal individual email usage patterns.

Hybrid filters use various tests for phishing and assign numerical scores to them. Each message is scanned for a number of patterns identified in the lists of rules of the program, and the resulting scores are collected in a table. Hybrid filtering has two advantages. First, it can greatly reduce the false positive rate, because it ensures that one phishing test by itself cannot label a message as a phishing message. Second, hybrid filtering makes it very difficult for the senders of phishing emails to avoid detection by specifically adjusting their emails in order to defeat any particular scanning method.

Rule-based filtering techniques can be used. They rely on the specification of lists of exact words or more sophisticated regular expressions disallowed in mail messages. It can also be implemented to extract the list of URLs which occur in an email message. The URLs in legitimate emails must resolve to valid domain names. Comparing the list of links present in an email with published lists of phishing domains is a simple way to reject many phishing emails.

Checksum-based filters can also be employed for each individual user. They maintain a database of checksums of all phishing emails previously received and identified by the user. Everything that varies between several broadcasts of the same phishing email is deleted to reduce what remains to a checksum. A local database is set up to collect the checksums of messages that each email recipient considers to be phishing messages. When a new email is received, its checksum is compared with those already stored in the database. If the checksum is already recorded in the database, the message is classified as a phishing message. These filters protect the user from repeatedly receiving many small variations of the phishing emails sent in bulk.

The reasoning community of internet security experts may consider offering commercial IP white listing as well as licensed white listing tags. In this case filtering software in the organization is set up to white list emails bearing the licensed tag. Users who purchase these tags and place them in their messages are guaranteed that their emails are delivered to the recipients.

Additional subject passwords, also called "ham" passwords, can be enabled for some users who have to handle a lot of email correspondence. When this option is activated and a new email is received from a sender who is not recognized and is not in the address book of the user, then the sender is requested to include in the subject of their email a new special password demonstrating that the email message is not a phishing email. The ham password can be provided on a web page together with the email address with an advice to include it in the subject of an email message. Ham passwords can also be used in the plus-addressing method by appending a ham password to the username in the email address.

Challenge-response filtering can be used to require new senders to pass additional tests in order to be added to the address book of the recepient. The system can hold the delivery of the first email, and respond with a request to send another email with an additional ham password in the subject line. On receipt of the second confirmation email, the senders address is then added in the address book and the first email is delivered. Filtering systems can also take into account ham passwords and make sure that emails with these passwords are never filtered out as phishing emails.

A repository of resources and information on anti-phishing and anti-spam solutions is maintained by the OECD Task Force on Spam (2009). In particular, this website provides a link to the constantly evolving OECD Anti-Spam Toolkit and its Annexes available in PDF and HTML format.

Active research has been carried out throughout the world and there are many examples of recent research results (including Kelarev, Yearwood & Watters, 2009, 2010; Kelarev et al., forthcoming, n.d.; Kelarev, Yearwood, Watters, Webb & Wu, n.d.; Layton & Watters, 2009; Layton, Brown & Watters, 2009; Ma, Yearwood & Watters, 2009; Ma et al., 2009; Watters, 2009).

REASONING METHODS FOR COMMUNITIES OF INTERNET SECURITY EXPERTS

The overview given in the preceding section demonstrates that a broad variety of reasoning processes including methods as intensive as Delphi or voting are required for reasoning communities of internet security experts. Phishing is a particularly rapidly changing area with novel challenges for effective organization of reasoning communities. The methods used by attackers and the security systems to be implemented and monitored for defense against phishing are so complex, that each participant can only be an expert in part of the system. At the same time it is necessary to monitor the rapidly developing nature of threats and adjust all reasoning processes accordingly. Thus we see that phishing involves rapid response decision making in order to respond to every particular attack.

This section is devoted to reasoning methods suggested for creating cohesive discourse architecture for a reasoning community of internet security experts adjusting and maintaining a local security system and communicating with broad communities of users. Here we consider protocols and technologies facilitating the discussion and decision making process in the reasoning community of internet security experts as well as communication with the broader communities of users.

DELPHI METHOD AND WIDEBAND DELPHI PROCESS FOR REASONING COMMUNITIES OF INTERNET SECURITY EXPERTS

The Delphi Method was developed at the RAND Corporation, which now has many downloadable publications devoted to this method on its website at RAND Corporation (2009). The method has become very well known. Nevertheless, it is often overlooked in administrative practices and is used less often than other more traditional discussion techniques. It has been adapted across many industries for various tasks including the interpretation of statistical results and marketing forecasts. A detailed exposition of the Delphi Method and its applications is given, for example, by Rowe and Wright (1999, 2001). Wideband Delphi Process increases the interaction of the experts via their discussion during meetings in addition to anonymous expression of opinions (as explained, e.g., by Stellman & Greene, 2005). A comprehensive book on the Delphi Method has been made available online (Linstone & Turoff, 2002).

We suggest the Delphi Method and Wideband Delphi Process as the center for organizing architecture for group discussions in the reasoning community of internet security experts adjusting and managing local security systems. We recommend using the Delphi Method to resolve all problems in choosing criteria for selecting an approach for reasoning on particular issues as they arise. It can also be used for all other questions concerning the reasoning architecture in community of security experts, and for resolving all other strategic issues including the development of structure for the Generic Argument approach presented below.

The Delphi Method is based on the general principle that appropriately structured reasoning in a community of experts is more likely to lead to accurate recommendations than unstructured reasoning. The experts answer questionnaires in several rounds. A facilitator provides an anonymous summary of the experts' opinions in each round. This summary includes brief explanations of the recommendations given by the experts. After that the experts are advised to revise their earlier input in view of the responses from other members of the group. During this process the recommendations of the experts gradually converge to a common opinion, which is more likely to be correct than any individual judgment. During the Delphi process it is essential to structure the information flow by providing regular feedback to the experts and guarantee the anonymity of recommendations and responses to the members of the reasoning community.

The Delphi Method can bring significant benefits to the reasoning community of internet security experts since it was designed with the particular aim of achieving consensus in the group. It can be combined with several other methods also treated in this chapter. They can be used to complement the Delphi Method and strengthen it by preparing its steps and enhancing the efficiency of implementation.

Following are the steps of the Delphi Method, designed to establish consensus in the reasoning community of internet security experts with regard to the local security system implementation, deployment, integration, and maintenance.

First, the current problems need to be formulated. The members of the reasoning community need to keep track of the current problems and future challenges they are faced with. The problems have to be written down clearly and concisely in order to make it easy for everyone to understand them. A questionnaire can be prepared in order to seek the experts' opinion on all of the problems and options to work out. Brainstorming can be used to collect new items for the questionnaire. On the other hand, the issues can be prioritized before addressing them, so that only the most urgent items are included in the questionnaire.

Second, the problems should be given to all of the internet security experts participating in the Delphi Method. A small group of additional experts can be invited to contribute their ideas. There is no need to convene a special meeting, since it is usually enough to send the description of the current problems to everyone in the group and ask them to express their opinion and write short replies, which are then regarded as anonymous messages. A procedure ensuring the anonymity in collection of responses from all members of the group has to be followed.

Third, the replies have to be processed. The replies from the members of the reasoning community of internet security experts are combined, categorized and compressed in a single anonymous file, without trying to skew the opinions or creating a bias. Headings can be used to group the responses related to similar directions.

Fourth, the combined collection of responses is sent to the same experts. This time the members of the reasoning community are asked to score each item on a suggested scale, for example, from 1 to 10, and to add possible new items if new solutions come to mind. To focus the exchange of opinions it makes sense to include scoring instructions and the original formulation of the problem at the beginning of the file. Collating all items in a table with empty columns for scores may expedite scoring process.

After that, the last two steps can be repeated iteratively in order to approach a consensus. In the cases where there is a broad range of opinions, several iterations may be required. Subsequent iterations involve aggregated analysis of the scores received in the preceding round. For, each item the scores received in the previous round can be combined using various formulas. The following formula can be recommended, because it strengthens the weight of the average and at the same time allows sufficient influence to the outliers making sure that nobody feels that their opinion is ignored by the process:

aggregated score = (lowest score + 4 x average score + highest score) / 6

Since all the responses are kept anonymous, the Delphi Method makes it easy for the administrators to express their opinion sincerely. At the same time it keeps everyone open to perception of new ideas and arguments, as well as enabling the experts to adjust their opinion without the urge to defend their original position and any feeling of being pressured.

Generic/Actual Argument Model of the Structured Reasoning for Reasoning Communities of Internet Security Experts

This section deals with the Generic/Actual Argument Model (GAAM) of the structured reasoning introduced and considered by Yearwood and Stranieri (2006). The GAAM has found numerous applications and has been actively investigated in the literature. Recently, GAAM has been used to predict judicial decisions regarding a property split following divorce (Stranieri et al., 1999), model reasoning in copyright law (Stranieri & Zeleznikow, 2001a), support interactive e-commerce (Yearwood et al., 2001), refugee status decision makers (Yearwood & Stranieri, 1999), multi-agent negotiation (Avery et al., 2001), determining eligibility for government funded legal aid (Stranieri & Zeleznikow, 2001b) and modeling reasoning that critical care nurses use (Stranieri et al., 2004). Two shell programs that implement GAAM ideas are described by Stranieri and Zeleznikow (2001b) and Yearwood and Stranieri (2002). The computational model is presented by Yearwood and Stranieri (2006), who concentrate on the benefits the structured reasoning can bring to group decision making.

Structured reasoning has been developed as an approach where group consensus on reasoning structures at various levels can be used to facilitate the comprehension of complex reasoning especially in the case where there are multiple perspectives. It leads to a formal process which allows the group members to assert their views, compare and contrast them. This makes it possible to utilize computational representation for computer modeling of reasoning and for hybrid combinations of human and computer reasoning as a coalescent framework for decision making.

The structured reasoning uses consensus and formal representations of arguments to analyze reasoning and identify issues of agreement and issues of disagreement in the reasoning community of internet security experts. The structure of reasoning on all aspects of internet security needs significant locally needs to be developed. Experts are usually able to express the structure of reasoning used in the domain separately from any particular arguments or issues. Once described the structure operates as a template to organize claims made by participants and compare them with claims made by other members and users of the security system. We recommend using the Delphi Method in developing appropriate structure for the local community of internet security experts.

Structured reasoning is defined as a layout of the reasoning possibilities in a tree structure. The root of the tree is the top level range of claims. Each node in the tree admits a defined range of claims. The goal that the reasoning is required to resolve is represented by the root. The child nodes of a node are the relevant evidence for reasoning about the parent. Structured reasoning relies on a clearly defined discursive community and purpose. Evaluation of the reasoning is based on acceptance of the agreed template or version for discourse within the reasoning community and agreement within the reasoning community space on acceptable procedures for performing inferences within the structure, sharing of the understanding of the use of concepts of the domain and the flexibility to combine human and machine inferences.

As a process, structured reasoning involves constructing a tree structure for the internet security domain. Argument maps and concept maps are examples of knowledge representations for structured reasoning that can assist in making clear the distinctions in reasoning between various participants of the community. Yearwood and Stranieri (2006) proposed a notion of structured reasoning suggesting that reasons are captured as a triple of (factor, position in the structure, context). The context reflects a whole range of background information. Firstly, the resulting notion of structured reasoning entails an articulation of generalized propositions (or proposition slots) and their relation to each other. Secondly, it handles the assignment of values (for each generalized proposition) and agreement on inference procedures that act within this structure. Thirdly, it takes care of the construction of individual actual reasoning to a conclusion (value) on the issue. The factor can be regarded as a label reflecting the concept or reason, the position captures the nexus to other elements (conclusions and evidence) in the structure. The main benefit of this approach is the understanding of the sources of disagreement or of agreement or similarity that are critical to deciding on whether consensus can be reached and on providing insight into consistency in individual reasoning within the group.

The GAAM can be used as a structured reasoning framework for group decision making. It requires the participants to describe the generalized reasoning slots, select the slots based on group consensus of the relevant factors involved, determine their inferential relationship to each other, and investigate possible ways of inferring particular propositions within these slots.

The GAAM provides abstract arguments as an effective notion of generic argument that acts as a fundamental reusable element of reasoning structure. The structure of GAAM is based on relevance and represents the structure of reasoning as a tree which is controlled by the notion of relevance as determined by the group. GAAM separates the structure and inference: In the bipartite graph the methods of inference within a piece of reasoning are maintained separate from the data/claims and together delineate the structural layout of the reasoning. It ensures flexible dialectical support for multiple actual arguments that exist within a group. These can be developed as purely computational reasoning elements, as computational elements which can be dialectically elaborated or as pure dialectical elements.

The structured reasoning approach of GAAM develops the task of generating explicit, justifiable reasoning within a domain in four stages:

1. Structure development. This stage maps out the generic argument tree as a layout of the shared understanding by the discursive community of the way in which generic arguments in the domain fit together based on relevance. This moves the problem that might be classed as ill-structured to one that is well structured. The generality of the structure permits reuse.
2. Detailing the claim slot nodes. It captures the full detail of the claims, values, and reasons for relevance and backings for each generic argument.
3. Detailing the inference slots. This stage determines the inference procedures for each generic argument and constructing them.
4. Actually reasoning. This stage instantiates generic arguments with values and possibly claim value reasons to present an actual piece of reasoning. The reasoning is then well structured as it fits into the agreed structural template or map.

GAAM allows the participants to identify the basis of their differences and similarities, thereby supporting group decision making. It helps to determine whether the group members differ on contextual information, claims or inferences, and clarify the exact nature of their difference. As a result the points of the negotiation to be undertaken become very clear. From the point of view of human reasoning GAAM enhances cognitive cooperation between members of the reasoning community of internet security experts and acts as a norm or reference or in the same way that a map does. The focus is on reasoning as a cooperative undertaking where disagreements and agreements in complex decisions can be identified, understood, negotiated, and resolved for mutual benefit of the participants.

Brainstorming for Reasoning Communities of Internet Security Experts

Brainstorming was originally invented by Osborn (1948, 1953). Here it is suggested as a method, which can be used to facilitate the collection of items and generation of new ideas for group discussion based on the Delphi Method. This method is very well known and is widely used in discussion, even though the rules of brainstorming are not always followed. Brainstorming on each particular topic can be undertaken during a meeting of the reasoning community of internet security experts.

One of the members of the reasoning community or an independent person has to be chosen as an organizer or coordinator of each brainstorming session of the reasoning community to monitor the rules. The Chair of a group meeting can often play this role. In this case it may be essential to make the selection of a Chair for each group meeting subject to a vote depending on consensus of group members. Before brainstorming begins the organizer has to remind the rules to all group members, because forgetting these rules during a discussion is quite common and usually reduces the effectiveness of the procedure. The organizer should also interfere if any of these rules are broken and restore the process.

First, criticism and debate are not allowed. People are free to express any ideas, including wrong ones. Even if an idea is impossible to implement or outright stupid, it brings the benefit of broadening the range of options, stimulating imagination, and can help to generate better solutions during discussion. On the other hand, debate and criticism can inhibit people from suggesting any new ideas.

Second, quality of ideas should not be assessed. Only quantity of ideas matters. Assessing the quality also inhibits the expression and generation of new options to look at in the future. The best ideas often turn up stimulated by previous suggestions at unpredictable time.

Third, freewheel should be used. This means that every idea already expressed can be used as a stimulus for the next one. This way of thinking helps to explore novel unexpected directions and dimensions of the problem. It is beneficial for the group discussion to encourage people to think of the ideas of other participants and use them.

Fourth, mutating and combining of ideas should be used. Mutating means deliberate modifications and grotesque distortions of ideas previously expressed. Combining uses several previous ideas in order to create a new one.

The members of the reasoning community of internet security experts attending the meeting must be prepared for brainstorming by reminding them about these rules and the problem that is the focus of the group discussion. It is a good practice to display a concise formulation of the problem so that all members could see it during the group discussion. The organizer or a secretary will have to register or write down the ideas, mediate and monitor the expression of opinions if required. It is best to organize the process so that everyone could view all ideas previously expressed to utilize them. An additional meeting can be organized on another day, since it often takes time to think of some really new and productive ideas. Only when the whole brainstorming process is complete, the suggestions can be analyzed to choose those options to bring forward towards implementation.

Note that brainstorming is one of the methods used in the well-known search conference approach due to Emery and Purser (1996). Search conferences are usually convened for establishing strategic long-term goals and directions. In contrast, the questions on internet security usually deal with searching for best ways of organizing defense in a situation where threats are rapidly changing and where it is essential to take into account the local patterns of internet usage by broad communities of employees, who cannot be invited to a conference. Membership of a small internet security group of experts often changes and depends on particular issues being implemented. This is why we have not included a recommendation to organize search conferences in the present chapter.

Reverse Brainstorming for Reasoning Communities of Internet Security Experts

Reverse brainstorming attempts to generate ideas of how the system can fail and how the problem in focus can be created. This method can be used after a brainstorming session at a separate meeting, or as an intermission, in order to bring a fresh perspective and help people to come up with novel approaches.

It should be particularly beneficial for members of reasoning communities of internet security experts in view of their strong analytical skills. Reverse brainstorming can introduce a new perspective and help to come up with more ideas. Reverse brainstorming brings the opposite viewpoint into the discussion process and the generation of ideas. In contrast with ordinary brainstorming the reverse one allows people to criticize the weaknesses in the existing arrangements, previously suggested measures and defense mechanisms. Many participants can work productively when they are encouraged to find weaknesses in the existing arrangements, criticize and judge measures introduced previously.

Instead of looking for solutions to the problem, reverse brainstorming focuses on the opposite task of making a list of ideas of how this problem can be created and how the existing defenses can be foiled. A list of ways of causing the problem needs to be created as a result of the session.

The following example strategies of the senders of phishing emails can be used as a starting point for reverse brainstorming. Senders of phishing emails can often change phrases in their emails and use incorrect spelling of words to defeat keyword detection. Senders of phishing emails can diversify their emails using unique invisible code inserted into the middle of the message, which changes its hash value and the checksum. Likewise, they can insert an invisible HTML comment in the middle of a word to foil keyword detection. They can replace words or even large parts of the email with embedded images. They can include a lot of additional legitimate text in the end of their email in order to change the number of occurrences of various words in the email and defeat statistical filters.

When reverse brainstorming is complete, the list created during the session can help to find ways of preventing the problem being caused using all methods on the reverse brainstorming list. Looking at ways of causing the problem can also enable the reasoning community of internet security experts to identify the weakest link in the systems and to verify the quality of the new measures they are planning to implement.

Consensus Decision-Making for Reasoning Communities of Internet Security Experts

Consensus decision-making can be used to expedite the resolutions of easier issues known to all members of the reasoning community, where an agreement can be achieved quickly. Consensus increases the commitment of each individual member of the internet security community to the decision. This results in stronger collaboration and more willingness to act upon the decision. A consensus decision-making process is less formal than other approaches, and it is up to the reasoning community of security experts to organize all practical details of its implementation. Usually it involves the following stages or procedures:

1. the reasoning community may set up an agenda for discussion and, optionally, the ground rules for reasoning.
2. each item of the agenda is addressed in turn. The item is discussed to identify opinions, relevant information and potential proposals for action.
3. a formal decision proposal on the issue is formed and presented to the group.
4. the facilitator of the reasoning community calls for consensus on the proposal. Each member of the group then must have the opportunity to express their agreement or veto the proposal, to avoid the community interpreting silence as agreement.
5. if consensus has not been achieved, concerns are presented and discussed, starting another round of discussion.

6. the proposal is modified, amended or rephrased to eliminate grounds for the concerns. The process then returns to the call for consensus and the cycle is repeated again.

Hand signals can be used by the members of the reasoning community for nonverbal indication of their opinions or positions during discussion of many options. Members of the reasoning community may use sets of coloured cards to speed up the expression of opinions.

Sometimes the aim of the meeting is to achieve a unanimous decision. Alternatively, the unanimity minus one, called U-1, requires that all delegates with at most one exception vote for the decision. In this case one dissenter does not have veto power, and it takes two dissenters to block a decision. Likewise, unanimity minus two or U-2, unanimity minus three or U-3, and other such systems like 80% rule or two-thirds rule for agreement can be used.

Voting for Reasoning Communities of Internet Security Experts

Voting is probably the best known and one of the most widely used decision making processes. It can supplement the Delphi Method and brainstorming at the final stages when it comes to selecting the optimal option from a list of generated ideas. There are numerous modifications and developments of voting discussed in the literature and implemented in practice. We assume that the most simple and least formal methods should be sufficient for a reasoning community of internet security experts. Let us refer to Saari (2001) for descriptions of voting systems using mathematical models. Saari (2001) advocates using the Borda count in voting. Further more recent research results, including the Quota Borda system, are presented by Emerson (2007).

The Borda count is a single-winner voting method in which the members of the reasoning community rank all options in order of preference. The Borda count determines the best option by giving each item a certain number of points which correspond to their ranks assigned by each individual member of the reasoning community of internet security experts.

This method is very quick and relatively easy to organize. It is important to precede it with other methods to make sure that the voting experts have a good understanding of all the circumstances and that the options for generating novel and unexpected alternatives have been exhausted.

If there are many ideas left on the list to choose from, then several voting rounds may be organized so that the weakest ideas are gradually eliminated and the list is reduced from round to round. In the collection of votes, crosses, numbers or ticks may be written against chosen ideas to express the voter preferences. Voting can be confidential to prevent people following the voting after the expression of the opinion of their supervisors. A completely anonymous vote can also be organized, although this takes more time and effort to carry out.

Open Delphi and Brainstorming Methods for Reasoning Communities of Internet Security Experts

The group of system administrators may decide to seek the expressions of opinions from the users to their reasoning process by opening the Delphi Method and Brainstorming sessions and inviting broader communities of users to send written submissions, or anonymous comments. It may also help to provide feedback from the users in the form of collated responses to predistributed questionnaires on materials prepared during preceding stages of the Delphi Method or Brainstorming. Voting by all users is not broadly recommended, since it would require prohibitive expenditure on education of all users with regard to all circumstances of new security options not implemented in the system yet and only considered for implementation.

FUTURE RESEARCH DIRECTIONS

We recommend applying the Delphi method and Wideband Delphi Process for strategic long-term decisions concerning the architecture of the group reasoning in the community and for deciding when and how to apply other reasoning methods. Various choices the security experts can make in organizing their discourse processes may also benefit from future experimental and theoretical work reflected in subsequent publications quantifying the practical significance and impact of the reasoning techniques. This is one of the open questions or possible directions for further research.

Future research may be devoted to the theoretical development of novel models to address rapidly evolving challenges arising in the reasoning for internet security, as well as to the experimental investigation and comparison of the effectiveness of using various existing reasoning methods in their ability to resolve particular questions in internet security.

The Delphi method and Wideband Delphi Process are also recommended for resolving all problems in choosing criteria for selecting an approach for reasoning on particular issues as they arise, as well as all other questions concerning the reasoning architecture in community of security experts, and for resolving all other strategic issues including the development of structure for the Generic Argument approach.

More research will need to be undertaken in the future to provide additional recommendations on particular steps of this process and experimental comparison of the effectiveness of various modifications of discourse in the reasoning community of internet security experts. It is also worth investigating the applications, for the uses of reasoning communities of internet security experts, of the two shell programs already implementing GAAM ideas and described by Stranieri and Zeleznikow (2001b) and Yearwood and Stranieri (2002).

CONCLUSION

This chapter has outlined an integrated discourse framework for reasoning communities of internet security experts managing all stages of the implementation, integration, deployment and maintenance of a flexible local security system. This paper has discussed relevant prerequisites on related issues of internet security. Defense against phishing attacks has been referred to as one of the most novel and most dynamically changing examples where advanced reasoning processes are required for the reasoning communities of internet security experts.

The Delphi Method and Wideband Delphi Process have been suggested as the central methods for a reasoning community of internet security experts and the main tools for preparing and organizing all other methods including Generic/Actual Argument Model of the Structured Reasoning, Brainstorming, Reverse Brainstorming, Consensus Decision Making and Voting. Possible directions for future research have been outlined.

REFERENCES

Anti-Phishing Working Group. (2009). Retrieved December 2009, from http://apwg.org/

Avery, J., Yearwood, J., & Stranieri, A. (2001). An argumentation based multiagent system for eTourism dialogue. In *Proceedings First International Workshop on Hybrid Intelligent Systems HIS 2001*, Adelaide, Australia (pp. 497–512).

CRM114 - the Controllable Regex Mutilator. (2009). Retrieved September 2009, from http://crm114.sourceforge.net/

Emerson, P. (2007). *Designing an all-inclusive democracy: Consensual voting procedures for use in parliaments, councils and committees*. Berlin: Springer.

Emery, M., & Purser, R. (1996). *The search conference*. San Francisco, CA: Joseey Bass.

Kelarev, A. V., Yearwood, J. L. Watters, P., Soukhorukova, N. V., Abawajy, J. H., & Pan, L. (n.d.). Internet security applications of the Munn rings.

Kelarev, A. V., Yearwood, J. L., & Watters, P. (2009). Rees matrix constructions for clustering of data. *Journal of the Australian Mathematics Society*, *87*, 377–393. doi:10.1017/S1446788709000299

Kelarev, A. V., Yearwood, J. L., & Watters, P. (forthcoming). Internet security applications of Gröbner-Shirshov bases. *Asian-European Journal of Mathemathics*.

Kelarev, A. V., Yearwood, J. L., Watters, P., Ma, L., Abawajy, J. H., & Pan, L. (forthcoming). A Grobner-Shirshov algorithm for applications in internet security. *Southeast Asian Bulletin of Mathematics*.

Kelarev, A. V., Yearwood, J. L., Watters, P., Webb, D., & Wu, X. W. (n.d.). Optimization of classifiers for internet security based on combinatorial syntactic semigroups.

Layton, R., Brown, S., & Watters, P. (2009, July). *Using differencing to increase distinctiveness for phishing website clustering*. Paper presented at the workshop Cybercrime and Trustworthy Computing CTC-2009 in conjunction with the 6th International Conference on Autonomic and Trusted Computing, Brisbane, Australia.

Layton, R., & Watters, P. (2009, October). *Determining provenance in phishing websites using automated conceptual analysis*. Paper presented at the 4th annual APWG eCrime Researchers Summit, Tacoma, WA.

Linstone, H. A., & Turoff, M. (2002). *The Delphi method: Techniques and applications*. Retrieved December 2009, from http://www.is.njit.edu/pubs/delphibook/

Ma, L., Ofoghi, B., Watters, P., & Brown, S. (2009). *Detecting phishing emails using hybrid features*. Paper presented at the workshop Cybercrime and Trustworthy Computing CTC-2009 in conjunction with the 6th International Conference on Autonomic and Trusted Computing, Brisbane, Australia.

Ma, L., Yearwood, J., & Watters, P. (2009, October). *Establishing phishing provenance using orthographic features*. Paper presented at the 4th annual APWG eCrime Researchers Summit, Tacoma, WA.

McDonald, A. (2004). *SpamAssassin: A practical guide to configuration, customization, and integration*. Birmingham, UK: Packt Publishing.

OECD Task Force on Spam. (2009). OECD anti-spam toolkit and its annexes. Retrieved September 2009, from http://www.oecd-antispam.org/

Osborn, A. (1948). *Your creative power*. New York: C. Scribner's Sons.

Osborn, A. (1953). *Applied imagination: Principles and procedures of creative problem solving*. New York: C. Scribner's Sons.

RAND Corporation. (2009). Retrieved December 2009, from http://www.rand.org/

Rowe, G., & Wright, G. (1999). The Delphi technique as a forecasting tool: Issues and analysis. *International Journal of Forecasting*, *15*(4). doi:10.1016/S0169-2070(99)00018-7

Rowe, G., & Wright, G. (2001). Expert opinions in forecasting: Role of the Delphi technique . In Armstrong, J. S. (Ed.), *Principles of forecasting: A handbook of researchers and practitioners* (pp. 1–32). Boston: Kluwer Academic Publishers.

Saari, D. G. (2001). *Chaotic elections!: A mathematician looks at voting*. Providence, RI: American Mathematical Society.

Schwartz, A. (2004). *SpamAssassin*. Cambridge, MA: O'Reilly & Associates.

SpamAssasin. (2009). Retrieved September 2009, from http://spamassassin.apache.org/

Stellman, A., & Greene, J. (2005). *Applied software project management*. Sebastopol, CA: O'Reilly Media.

Stranieri, A., Yearwood, J., Gervasoni, S., Garner, S., Deans, C., & Johnstone, A. (2004). Web-based decision support for structured reasoning in health. In W. Karin & B. Cesnik (Eds.), *Proceedings of the Combined Conferences of the Twelfth National Health Informatics Conference* (pp. 180–185).

Stranieri, A., & Zeleznikow, J. (2001a). Copyright regulation with argumentation agents. *Information & Communications Technology Law*, *10*(1), 123–137. doi:10.1080/13600830124950

Stranieri, A., & Zeleznikow, J. (2001b). WebShell: The development of web based expert system shells. In *Research and Development in Expert Systems XVIII. Proceedings of ES2001- The Twenty-first SGES International Conference on Knowledge Based Systems and Applied Artificial Intelligence* (pp. 245–258). London: Springer Verlag.

Stranieri, A., Zeleznikow, J., Gawler, M., & Lewis, B. (1999). A hybrid rule-neural approach for the automation of legal reasoning in the discretionary domain of family law in Australia. *Artificial Intelligence and Law*, *7*(2-3), 153–183. doi:10.1023/A:1008325826599

Stranieri, A., Zeleznikow, J., & Yearwood, J. (2001). Argumentation structures that integrate dialectical and non-dialectical reasoning. *The Knowledge Engineering Review*, *16*(4), 331–358. doi:10.1017/S0269888901000248

Watters, P. A. (2009, October) *Why do users trust the wrong messages? A behavioural model of phishing*. Paper presented at the 4th annual APWG eCrime Researchers Summit, Tacoma, WA.

Wolfe, P., Scott, C., & Erwin, M. (2004). *The anti-spam tool kit*. Emeryville, CA: McGraw-Hill Osborne Media.

Yearwood, J., & Stranieri, A. (1999). Integration of retrieval, reasoning and drafting for refugee law: A third generation legal knowledge based system. In *Proceedings of Seventh International Conference on Artificial Intelligence and Law, ICAIL'99* (pp. 117–125).

Yearwood, J., & Stranieri, A. (2002). An argumentation shell for supporting the development and drafting of legal documents. *Information & Communications Technology Law*, *11*(1). doi:10.1080/13600830220133576

Yearwood, J., & Stranieri, A. (2010). Group structured reasoning for coalescing group decisions. *Group Decision and Negotiation*, *19*(1), 77–105. doi:10.1007/s10726-009-9162-1

Yearwood, J., Stranieri, A., & Avery, J. (2001). Negotiation and argumentation based agents to facilitate eCommerce. In *Proceedings of the International Conference on Advances in Infrastructure for Electronic Business, Science and Education on the Internet, SSGRR2001* (pp. 100–109).

Yearwood, J. L., & Mammadov, M. A. (2007). *Classification technologies: Optimization approaches to short text categorization*. Hershey, PA: IGI Global.

Yearwood, J. L., & Stranieri, A. (2006). The generic/actual argument model of practical reasoning. *Decision Support Systems*, *41*(2), 358–379. doi:10.1016/j.dss.2004.07.004

KEY TERMS AND DEFINITIONS

Brainstorming: A method for generating ideals and options to consider.

Consensus Decision Making: A method for achieving a decision acceptable for all participants.

Delphi Method: A method for establishing consensus in the expert group.

GAAM: Generic/Actual Argument Model of the Structured Reasoning.

Reverse Brainstorming: A modification of the Brainstorming Method.

Voting: A method for choosing a decision from a list of options.

Wideband Delphi Process: A modification of the Delphi Method.

Chapter 21
Web Technologies and Reasoning Communities

Charlynn Miller
University of Ballarat, Australia

Philip Smith
University of Ballarat, Australia

ABSTRACT

The popularity of the Internet, coupled with a reduction in traditional community participation has resulted in maturation in the way that people use Web 2.0 technologies to support communities. Reasoning communities come together to make decisions or form courses of action on particular topics. This chapter investigates how traditional Web (1.0), Web 2.0 and Web 3.0 technologies can be used to support the four phases of the process that reasoning communities typically undergo to achieve their results.

INTRODUCTION

Communities have traditionally congregated to serve the shared interests of their members. A "reasoning community" (Yearwood & Stranieri, 2009) is one in which the members collude to make a decision or form a course of action on a particular topic. Before the availability of high speed communications most communities were, by necessity, location-centric. They were defined more by their face to face meetings than by their intent.

The massive uptake of the Internet by the mainstream community, coupled with a reduction in traditional community participation (Putnam, 1995) has resulted in maturation in the way that people use the Internet. The inception of Web 2.0 has begun to change the location-centric nature of relationships and thus communities (Mason & Rennie, 2007). The Web 2.0 community is location-less, not concerned with cultural, age and language barriers, more tolerant and are able to form and disperse at a speed never before seen in community efforts. The ability for a community to develop from individuals who were previously unknown to each other and the opportunity for this community to have an impact on the area or topic that they are addressing is now possible through the nature of Web 2.0 and its accompanying applications.

DOI: 10.4018/978-1-60960-091-4.ch021

The use of technology to assist reasoning communities is not a new concept. Many applications, including decision support systems, data mining, knowledge warehouses and artificial intelligence systems are employed to assist reasoning communities in meeting their objective of a group decision. In this chapter, the way in which traditional web, Web 2.0 and Web 3.0 applications can support reasoning communities is discussed together with the impact of these applications on the formation of reasoning communities. This discussion will involve:

- Traditional web communities.
- Reasoning community formation through Web 2.0 applications and the impact of these applications.
- Particular Web 2.0 applications that support the individual reasoning stage.
- The efficacy of some Web 2.0 applications in supporting the process of group coalescence.
- The support of group decision-making through the use of Web 2.0 technologies.

The chapter will then look to the future, exploring Web 2.0 applications from the point of view of individual needs. Finally, the potential impact of the Semantic Web (Web 3.0) on the various phases of reasoning communities will be discussed.

BACKGROUND

Communities are groups that are organised around common values. A reasoning community is one in which the participant's reason individually before coalescing as a group and using this reasoning collectively to arrive at a decision or course of action (Yearwood & Stranieri, 2009). Yearwood and Stranieri identify four stages or phases in the process a reasoning community undergoes to achieve its results:

- Engagement - in which individuals come together to form a group.
- Individual Reasoning - in which the members reason individually about the issue at hand.
- Group coalescence - in which the members come together to reason about the issue at hand.
- Decision making - in which a final decision or course of action is determined.

With the changes in the Internet to support more participation and communication between and with users, new possibilities have opened up for the use of the Internet, and more specifically, Web 2.0 applications, for reasoning communities. The ways in which Web 2.0 applications, and later Web 3.0, might be used to support each of the phases of a reasoning community will be a major focus of this chapter.

Traditional Community Structures

Traditional community development and thus the development of traditional reasoning communities were by necessity geographically based. Midgely, Hall, Hardiman and Narine (1986, p. 31) discussed community participation priorities such as "raising the level of social and political consciousness of local people" and that "mass meetings are an essential element in the promotion of community participation". Prior to the inception of the communication capabilities such as video-conferencing and the Internet, it was difficult to sustain communities whose members resided in disparate places. The cost of bringing groups together had to be considered. Reasoning communities were limited to a location-centric format in many cases. Blanchard and Horan (1998) advocated for a time when "virtual communities develop around physically based communities and when these virtual communities foster additional communities of interest".

Traditional communities have been augmented by the inception of the Internet (Web 1.0). Web 1.0, or the first phase of the Internet that was publicly available, was defined by its "one way" technologies of electronic mail, personal html-based web pages and content management (Flew, 2008). Community activity in Web 1.0 is limited to communication capabilities such as email and online databases. Some communities have developed through the use of Web 1.0 technologies. The United Nations Online Volunteering web site is an excellent example of this type of community. These communities are able to make their focus the intent of the community rather than the location through the use of online databases that form a virtual communities of volunteers and organisations seeking assistance (United Nations, 2009).

Web 2.0 Community Structures

Since it was defined in 2005, the term Web 2.0 has gained prominence (O'Reilly, 2005). The premise behind Web 2.0 is that it expands on the original concept of the Internet (Web 1.0) by incorporating aspects of collaboration: user sharing, information sharing and user-generated content. Seven core phases form Web 2.0: Service software, rich data sources, users as co-developers, collective intelligence, customer self-service, multi-device software, and "light" user interfaces (O'Reilly, 2005).

Web 2.0 bears all the markings of a phenomenon that is significantly changing the face of community development. Miller (2005) reports that, unlike Web 1.0, Web 2.0 applications are participative, work for the user, are modular, enable the location and assembly of content that meet users needs, encourages sharing, eases communication and facilitates community. The main premise of Web 2.0 is sharing: Information sharing through Wikis and Blogs, development sharing through Open Source applications, media sharing through sharing websites. Web 2.0 and its associated applications have already impacted the ways that communities develop. Mason and Rennie (2007) describe using a range of Web 2.0 applications to support community development in Scotland. Blanchard and Horan (1998) concluded that civic engagement would increase through the fostering by virtual communities of communities of interest. These authors also speculate about whether emerging virtual communities can counteract the decline of community participation. The necessity for uptake of Web 2.0 technologies is critical for success in using these technologies as community tools. White (2001) concluded that connection technologies enable people to communicate, provide feedback, exchange information, ask questions, and overall, build relationships.

Consumer Uptake of Web 2.0 Technologies

An additional area of consideration in the exploration of Web 2.0 technologies take-up for reasoning communities is the changing habits and preferences of the people who form these communities. As traditional communities formed, so those living in this era supported this form of community from a social and ideological perspective. So too the individuals forming reasoning communities today will support a very different type of community for social and ideological reasons, as online new media becomes more the mainstream option for consumers (Nielson, 2008). The demographic from 14-25 is the most prevalent group using online social media, but there is significant uptake by the 25-44 demographic as well (Nielson). In future, it is expected that the trend will be towards online social media as the mainstream option for consumers under 45 (Nielson). Attempting to employ traditional techniques rather than newer techniques for community formation is likely to result in lower take-up of community participation. Nielson reported that in the period from 2007-2008, reading Wikis increased from 37% to 61% and reading blogs increased from 48% to

60%. These trends will result in radical changes in the ways that individuals' reason and work within communities. Coleman and Gøtze (2002, p. 35) supported the potential of Web 2.0 tools coupled with the need to find ways to develop 'virtual sociability' to encourage engagement through these technology tools.

A New Way Forward for Reasoning Communities

As seen in the research presented above, Web 2.0 applications cannot only provide, through their characteristics, a platform for change in the way reasoning communities operate, but can also provide many of the collaborative tools to support the reasoning community at each stage in its deliberations. In this section a discussion of the characteristics and applications of Web 2.0 in the context of each of the four phases of reasoning communities is presented, together with a sampling of Web 2.0 applications that currently exist to support reasoning communities at each phase. Technologies will be explored in detail with regard to why they support community formation and reasoning and examples of each technology will be discussed.

Phase 1: Engagement

This is a critical introductory phase in the community reasoning process, as it determines how the reasoning community moves forward and who its participants will be. Increasing numbers of reasoning communities will define and engage participants and develop community protocols using Web 2.0 applications and technologies.

Applications

There are a great number of Web 2.0 applications available on the Internet to encourage individuals to participate in communities. Many of these applications are based partly in traditional engagement models. Others are unique within the concept of Web 2.0. This section will focus on those applications that are unique to Web 2.0 and tend to operate outside of the traditional sphere of community creation. The primary Web 2.0 applications that support and encourage engagement fall into the category of Social Networking tools. This is the area that has had the most to do with defining Web 2.0 as different from the first phase of the Internet (Web 1.0). These tools are known for their collaborative and contributory nature. They are designed to connect people together, and thus make a compelling platform for the creation of any community. They operate with a speed and global reach as yet unmatched by traditional communication applications. Their ability to connect like-minded people lie in the components of their network. Prominent examples of these tools are described below:

- **LinkedIn**: LinkedIn is a professional network of "connections". Currently used by 47 million people in 200 countries, the focus of this network is to connect professionals to make them as productive as possible through discussion of business issues, provision of professional recommendations and career advancement. The credibility of LinkedIn is much higher than that of Facebook because of the professional nature of LinkedIn. The potential to engage potential community members through LinkedIn is high, particularly if the community is business-related in some sense. (LinkedIn Inc., 2009).
- **Facebook**: Facebook is the most popular Social Networking tool with over 300 million active users. It is currently used by Reasoning Communities, such as Amnesty International, to allow cutting edge communication, collaboration, and engagement in both existing members as well as those who may be interested. Information about communities on Facebook can be re-

leased in several ways, through advertisements, status updates of participants, wall postings, group creation, event listings, and profile linking. (Facebook, 2009).

- **Twitter**: Twitter is a real time short messaging service. It allows participants to post a short message about what they are currently doing or working on, or an opinion of something happening worldwide. The power of Twitter for community engagement lies in it "trending topics" area. On the side of the main Twitter page that the user sees, there is a list of the most discussed or referenced topics on Twitter in real-time. The user may access a page that reveals all of the "tweets" on a particular topic, encouraging engagement and potential "Rapid Community Development". (Twitter, 2009).

The descriptions above illustrate that Web 2.0 applications have the power to attract and engage potential community participants. The next section will discuss the Web 2.0 applications that support the individual reasoning phase of a reasoning community.

Phase 2: Individual Reasoning

There are some very relevant, easy to use, and inexpensive applications that exist in Web 2.0 that can support both information collection and dissemination for the individual community member.

Applications

A number of Web 2.0 applications exist to support the individual reasoning stage of the reasoning community process. This section will highlight a few of the key application areas and provide an overview of some of the most prominent of these applications in existence today. The applications that are currently available will be broken down into two areas, those that apply to the collection of individual information gathering, and those that enable the communication of individual reasoning back to the group.

Social bookmarking applications are powerful tools for individual reasoning due to their three main functions: the ability to tag items of information found on the web with meaningful labels, the ability to share this information with others, and the ability to find shared bookmarks of others to enhance research into topical areas (Mathes, 2004). Social bookmarking applications are distinct from traditional "favourites" or bookmarks through their ability to be stored online, accessed from any computer connected to the Internet and shared with others (Delicious, 2009). This gives an individual the opportunity to access their own tags and those of others from anywhere, adding to the mobility and depth of information for the individual. Social bookmarking applications can also contribute to the communication of individual reasoning activities through the sharing capability of these applications.

Wikis are excellent information gathering tools. Sources of information in many topic areas can be accessed through Wikis, enabling individuals to access information about specific topics quickly and in one place. Wikis are collaborative and rely on user contributions. They are often linked to other relevant sources of information. Wikis serve a dual function for reasoning communities in that they can act as sources of information as well as a way for individuals and groups to disseminate information.

Really Simple Syndication Tools (RSS) use enabled websites to generate a feed that summarises website content. This content can be in the form of discussion boards, Wikis, news headlines and other data, and once subscribed to by the individual, is delivered automatically to the desktop or website using an aggregator program. The individual then reads the feeds through an RSS reader or other tool. An individual has the ability to add and categorise feeds to suit their needs and interests. (Lee, Miller & Newnham, 2008).

Communication applications conceived by Web 2.0 are valuable information gathering and sharing applications. These applications fall into a variety of areas and functionalities. Communication applications come in the form of Web logs (blogs), online document sharing, online storage, online presentations, media sharing and in some cases, social networking tools. These applications can all add vitality to the communication of information gathering by the individual community members in a reasoning exercise. The power of these tools lie in their anytime/anywhere access as well as their ability to share and collaborate between group members. They can reduce redundancy whilst enhancing the gathering and sharing of relevant information.

Some examples of the Web 2.0 applications discussed above are presented here:

- **Delicious**: A social bookmarking tool where a user "tags" information found with a specific name that allows them to find and share this information at a later time. It also supports searching of shared content from others. (Delicious, 2009).
- **Google Reader**: A RSS reading tool that allows the user to subscribe to RSS enabled websites, categorise and prioritise feeds to suit their reading interests. (Google, 2009a).
- **TypePad**: A free popular online blogging application that enables users to generate web logs, lists and image content without web development knowledge. TypePad also has features such as sharing, adding friends and followers and providing status updates. (Six Apart Ltd, 2009).
- **Google Documents, Spreadsheets and Presentations**: Google Docs enables individuals and groups to create, share, and edit documents, spreadsheets or presentations online with up to 200 other individuals. It also allows for publication of information that allows access to anyone by making it public. (Google, 2009a).
- **Slideshare**: A media sharing site for presentation, documents, and portable document formats. It enables uploading, downloading and sharing of media and can be used for individuals or organisations to share ideas, connect and generate business leads. Slideshows can be embedded in other sites and participants are invited to comment on the content that they view. (Slideshare Inc., 2009).
- **YouTube**: YouTube is a free video sharing application that allows users to upload their own non-copywritten videos, watch videos uploaded by others, share videos, create favourites and playlists and create and subscribe to subscription services. Videos can be shared publicly or privately and can be embedded into many other web applications such as Facebook and blogs. (YouTube, 2009).

Phase 3: Group Coalescence

This step serves three main purposes:

- clarifying the individual reasoning of participants,
- making this reasoning transparent to all,
- enabling a permanent record of the community's reasoning.

There are several applications within Web 2.0 that can support this coming together as a group, specifically within the three purposes detailed above.

Applications

Social networking sites, covered previously in the community formation and individual reasoning sections, are also useful for dissemination of information gathering to assist group coalescence. Social networking sites such as Ning allow for creation of a social network around a specific topic. This is an area used by subscribers (group members) to share and discuss information. Social networks such as LinkedIn also support this functionality with the use of groups. These applications can be open or closed depending on the community and its function. The utilisation of social networks can assist in making clear the individual participant's reasoning, in making this reasoning transparent and in maintaining a permanent record of the community's reasoning.

Concept mapping enables individuals and groups to visually represent their discussions, deliberations and decisions. The use of shared concept mapping tools can provide a transparent way for individual reasoning to be illustrated to the group. It can additionally provide a permanent visual record of this reasoning.

Cloud computing is a dynamic, online, service oriented computing environment that uses a web browser or web application to access all products (Plummer, Cearley & Smith, 2008). Cloud computing enables the use of shared documents and online storage that can assist in developing a permanent record of the community's reasoning. This information can be shared or kept private within the group through the use of in-built sharing and privacy rules in the application. Access to historical or archive information can also be assisted by the use of these online tools.

- **Ning:** Is a social platform that allows for the creating of social networks. More than 1.5 million social networks have been created by over 33 million users. Once a social network is formed, the owner invites others to join. Unlike Facebook that is itself the Network, Ning supports the creation of independent networks within their network. (Ning Inc., 2009).
- **Gliffy**: Is a collaborative tool for the creation, sharing and collaboration of diagrams including concept maps. The use of diagrams is thought to move people into the same space more quickly through tangible visual representations of thoughts. (Gliffy Inc., 2009).
- **Google Apps**: Is a business-oriented suite of products that includes email, calendars, documents, websites and archiving applications. Organisations can use Google Apps as their main infrastructure instead of onsite servers. (Google, 2009b).

Phase 4: Decision-Making

There are currently few Web 2.0 applications that support the decision-making phase of reasoning communities. This is a relatively undeveloped area of Web 2.0 for purposes of this discussion. There are a large number collective voting sites on the Internet, including those embedded within other applications, such as Facebook polling applications. However, the careful construction and consideration of these decision-making tools for reasoning communities preclude the use of many of the currently available applications.

There is a future for this area, however, as research into Government 2.0 and e-Government begins to identify and build some of the applications, such as e-voting and e-democracy (Chadwick, 2008). This emerging area of e-Government promises to produce structures and applications that will most likely support the idea of relevant Web 2.0 based decision-making tools. E-voting and e-democracy tools based on the concepts of and supported by Web 2.0 are just beginning to emerge (Chadwick, 2008). As these systems become mainstream, it is expected that some iteration of these tools will be made available for community use outside of Government entities. These new applications will enhance the ability of reasoning communities to efficiently and effectively make decisions as a group.

Applications

There are Web 2.0 applications that can support some of the needs of the decision-making phase of reasoning communities. Electronic survey and voting tools currently exist on the web, often within other Web applications. These applications can be used by reasoning communities to come to some decision, and would be particularly effective in situations where the community's members are separated by distance. Two such applications are discussed below:

- **Limesurvey**: Is a free, publicly available, open source software suite developed for creating and administering online surveys. Volunteers mainly develop it. (Limesurvey, 2009).
- **Zohopolls**: Is part of a suite of on-line business applications created by Zoho Corp. Zohopolls allow for the development of ad hoc polls for users to cast a vote. Zoho products are free for personal use. (Zoho Corporation, 2009).

Reasoning Communities & Web 2.0 Application Impacts

This section will discuss the advantages, challenges and implementation of these applications within reasoning communities. Examples and suggestions on the types of implementations that may work best with reasoning communities will be provided. The section concludes with a discussion of the future of Web 2.0 technologies, the inception of Web 3.0 and the impacts these future changes may have on reasoning communities.

Advantages

There are many benefits of using Web 2.0 applications such as Facebook and Twitter for reasoning communities. The ability to spread the word on an issue is unmatched by any previous communication means. Where it used to take days or hours to communicate, it now takes seconds, as all of these applications are available on a variety of media types, such as computers and mobile phones. Additionally, there is power in the "friend" concept of Social Networking. A potential participants discovers the issue or community opportunity through a "friend" or "friend of a friend" adding to the credibility of the invitation and the likelihood of it being taken up. The fact that many of the applications that would assist in the engagement of potential reasoning community members are free to users is an additional advantage for both community developers and potential participants.

Challenges

The dynamic nature of the Internet is one of its most prominent features. Applications that are here and popular today may be gone tomorrow. The waning popularity of the social networking application MySpace is an example of this. Placing too much emphasis on a particular application can be a disadvantage for community developers. The unique revenue models of the Internet and Web 2.0 applications, such as advertising, are the very models that make reliance on many applications in Web 2.0 risky. Harnessing these applications with any faith that they will continue to operate in the current form is risky. Add to that the possibility that free applications, such as Facebook, may incur fees at some future point. These factors are not as critical in the engagement phase of reasoning communities, but will become more important in the later phases.

Because the idea behind Web 2.0 technologies is to contribute to knowledge as well as to gather and interact with it, there is often the concern that the knowledge presented by Web 2.0 social sites is not verifiable or accurate. This can lead to a situation in which reasoning communities must be more aware of the possibility of misinformation than previously.

Implementation

Implementation of the Web 2.0 applications discussed in the four sections of this chapter would enhance the process of reasoning communities. There are two possible ways to implement these applications, either through an ad hoc method or a consolidated method. The ad hoc method of implementation would entail specific reasoning communities, or entities that initiate reasoning communities selecting Web 2.0 applications for each of the phase areas that would support the community. For example, in one community, a social network may be used as a platform for communication and engagement, with Wikis, online documents, and an e-voting site selected as add-ons to address other phases in the process. There would be no relationship between the tools selected. This ad hoc method has the advantage of using the most relevant tools at the time as well as retaining flexibility. The disadvantage of this ad hoc type of process is the lack of structure and connection between the tools, the need to potentially learn a new set of tools each time a community is formed and the risk of applications being discontinued.

The alternate method of implementation would involve either a constructed tool, or the use of Mashup technologies. Use of pre-constructed tools, such as those profiled below, would enable community-members to have one collaborative tool that performs all functions for the reasoning community. This would enable community members to learn only one set of applications for re-use in each community. This would particularly advantage reasoning communities in areas such as government and public service, which form many reasoning communities. The need for reusable tools for these organisations is clear. The disadvantage of these pre-constructed Web 2.0 tools is their static nature, which inhibits their ability to take advantage of new applications that may be developed in the ever-changing world of Web 2.0.

Tomoye Ecco is an application structure specifically designed to enable group processes within organisations. It provides a suite of Web 2.0 applications with functionality, transparency, and collaboration. The application contains tools such as document rating, question and answer sections, collaborative discussion features, blogs, social networking tools, tagging, page ranking, and syndication (RSS). (Tomoye, 2009) Tomoye Ecco focuses on a peer-driven structure, which is seen to allow more of a social networking feel to deliberation than traditional knowledge-based systems in organisations. This structure encourages community-based collaboration and decision-making. Several of the current clients of the product are using it for the development of "communities of practice" that have requirements outside of normal knowledge-management systems. One example is the U.S. Army Company Command website, which offers U.S. Army leaders a forum through which to share questions and answers with one another much more quickly than traditional structures would allow. Expansion of the use of products such as this into actual reasoning, as opposed to strictly knowledge sharing, would enable more of the phases of community reasoning to operate within these applications. As communities embrace the use of web-based tools for deliberation and reasoning, it is anticipated that products such as this will change and flourish.

The e-Democracy System (Petrik, 2009) is an application within Web 2.0 theorised to provide assistance to both government and non-government entities to aid in policy-making. The system has the following elements:

- 'Suggestion system' to suggest new policy issue.
- 'Lobby group' to enable collaboration on common interests/issues.
- 'Policy Wiki' to detail policy issue for participants.
- 'E-discussion forum' to allow participation in political debates.

- 'Evaluation system' to evaluate suggested policy strategies.
- 'Decision-making system' to allow for voting on policy strategies.
- 'Proxy representation' to determine representatives.

Whilst this system is designed for policy-making, the consolidation of Web 2.0 technologies in this system could be used within a reasoning community system. Each of the phases of a reasoning community is represented within the structure above. Further tests on this system are necessary to gauge success of the concepts presented. Further consideration of the types of structured Web 2.0 systems discussed above will be necessary to determine they are successful for reasoning communities.

The final implementation option for reasoning communities wishing to take up Web 2.0 applications is the use of Mashup technologies for semi-structured tool sets. Mashups are web applications created through the merging, transformation or aggregation of content, in which various Web 2.0 applications are melded together to form a new "mashed up" application with greater functionality (Merrill, 2006). Thus, they can be a powerful tool for reasoning communities. The ability to create Mashup applications that specifically meet the needs of the various stages of reasoning community activity would create a dynamic, powerful, semi-structured environment to encourage these communities to exist and function. Several examples of Mashups are detailed below:

- WeatherMole - is a mashup that combines weather forecasts from NOAA's Experimental National Digital Forecast Database XML Web Service and Google Maps. (Chastang, 2009).
- GuruLib - helps catalog books, movies, music, games and software online, in addition to keeping track of borrowed items. It uses a bookshelf type structure, and provides cataloguing information such as author, front cover, genre, synopsis that can be accessed automatically from Amazon or about 530 public and university libraries around the world. (Gurulib, 2009)

Web 3.0

Nova Spivack coined the term Web 3.0 to describe the third stage of the web. His belief is that the terms Web 1.0, Web 2.0 and Web 3.0 correspond to three decades of Web development (Spivack, 2007).

- **Web 1.0 between 1990 and 2000**. Web 1.0 corresponds to the static first stage of the Web which might be likened to a "giant library". This stage of the Web is characterized by web pages which present information but allow little or no user interaction.
- **Web 2.0 between 2000 and 2010**. Web 2.0 corresponds to a more dynamic stage of the Web which is characterized by user interaction with the web. In Web 2.0 the user not only reads the web page but is integral in contributing to it. Web 2.0 might be likened to "a big group of friends and acquaintances" (Strickland, 2009).
- **Web 3.0 between 2010 and 2020**. Web 2.0 was concerned with creating web pages which were primarily designed to be read by people. Computer programs can scan for keywords but cannot understand what is on web pages. The aim of Web 3.0 is to create web pages which can be read and understood by computer programs.

The term is often used synonymously with the term "Semantic Web". Web 3.0 is still in its infancy and yet to become mainstream. Some observers (Shadbolt, Hall & Berners-Lee, 2006) believe that this is an inevitability and that the appearance of Web 3.0 is imminent once "carefully developed and agreed standards" are developed.

To address the ambiguity of natural language, the concept of ontology is also an integral part of the Semantic web. An ontology creates unambiguous descriptions of concepts and of how they relate to other concepts. The standard language for creating ontologies is the Web Ontology Language (OWL). Ontologies are difficult to create and maintain and critics believe that this will greatly inhibit the growth of the Semantic Web (Strickland, 2009). However some large ontologies such as the Gene Ontology have been developed.

Web 3.0 browsers will be able to process complex requests for users through the use of intelligent software agents which crawl through the Web and returning the most relevant information. To this end Web 3.0 is sometimes described as a "giant database" or as the "Data Web" (Strickland 2009). Furthermore, Web 3.0 will build profiles of users and use them to drive their search for relevant information. Thus it is sometimes described as a "personal assistant" (Jevalins, 2009).

Web 3.0 will use artificial intelligence techniques, such as machine learning and natural language processing, to help focus searches for the most relevant material (Cho, 2008). Cho refers to this as the "Intelligent Web". Fawzi (2006) believes that the search capabilities of the Semantic Web will be such an improvement on current search engines that it will signal the end of Google.

Strickland (2009) made several predictions of how Web 3.0 might look when it finally arrives:

- That it will be a digital landscape traversed either in first person or by avatars.
- That it will be the electronic brain that computers have (erroneously) been called for decades.
- That it will extend to connect to "everything from wrist watches to television sets".
- That distinctions between entertainment forms will be lost and that the web will be the delivery platform for television, radio and other entertainment media.

As far as the effect of Web 3.0 on reasoning communities we might find the following:

- That the individual reasoning and information gathering stage is both simplified and enhanced by the Semantic Web. Furthermore, there will not be any need to store the data gathered – only the question that retrieved the data (unless, of course, a snapshot of the data at a particular point in time is required).
- This should mean that the data is always up to date and is always the most relevant. Web 3.0 has been described as a "personal assistant" which will build a profile of the user and use this profile to drive its searches.
- That the non-location centric nature of the communities will become even more pronounced through what Spivack refers to as "Ubiquitous Connectivity" (Spivack, 2007).
- That the decision-making processed is enhanced. Osman, Yearwood and Vamplew (2007) have described work in which documents (blogs) are scanned not for facts but for opinions. Perhaps, instead of creating a community specifically to provide opinions on a topic, the Semantic Web could be used to retrieve those opinions already "out there". If expert opinion is required, then perhaps a list of accepted experts in the particular domain might be included in the original request.
- It is not clear that Web 3.0 can be used to produce the final decision or course of action. However, Web 3.0 will be based on artificial intelligence techniques (Cho, 2008), so it would not appear to be outside the realms of possibility. In this case, the need for human reasoning communities would disappear altogether and decisions could be left to the web. Some might find this a frightening prospect reminiscent

of films such as The Terminator and The Matrix.

Ankoleka et al. (2008) believe in usable Web applications retaining the focus of Web 2.0 on usability and community development and drawing on the infrastructure of Web 3.0 to create information sharing through Mashup applications.

FUTURE WORK

Future work with Web 2.0 applications and their use to enhance reasoning communities will involve primarily two areas: the construction of applications or development of Mashups to enable these communities and the consideration of Web 3.0 applications in the future and their use in reasoning communities.

Future testing in the use of ad hoc, structured and Mashup Web 2.0 implementation streams suggested above will be an important addition to the discussion of the effectiveness of Web 2.0 tools for communities. Decisions about how each type of implementation, ad hoc, structured or Mashup will best serve various communities will be necessary. The greatest challenge will be how to most effectively harness Web 2.0 (and/or Web 3.0) applications for optimal gain within reasoning communities.

Young, et al (2008) predict that spending on enterprise Web 2.0 technologies across the globe will grow to $4.6 billion by 2013. This article goes on to say that the largest share of this will most likely be in the area of social networking. These figures indicate that Web 2.0 and 3.0 applications will become a core component of reasoning community structures in the future. The impact of future development in structures for support and of the web and the agents in Web 3.0 cannot be underestimated, as the nature of these future improvements will clearly advantage reasoning communities.

CONCLUSION

The aim of this chapter was to examine traditional, Web 2.0 and Web 3.0 technologies and how they could support, in whole or in part, the four phases of reasoning communities. The background presented in this chapter discussed traditional community structures and the changing nature of consumers, including the generational impacts that will drive the adoption of Web 2.0 application use for reasoning communities. A definition of Web 2.0 and how researchers see the impact of Web 2.0 on society was presented. A consideration of the future implementation of Web 3.0 applications further informed the discussion.

The four phases of reasoning communities were addressed in the context of Web 2.0 technologies. Each of the four phases was addressed with regard to the Web 2.0 applications that most suited each phase. Discussion centred around the type of application currently in existence and the way that it can advantage reasoning communities. Many examples of prominent technologies were presented to illustrate the strength of Web 2.0 technologies and potential in community use.

Consideration was given to the structure of the Web 2.0 tools that may be used for reasoning communities. More work is needed to explore these structures and the best implementation possibilities for both Web 2.0 and Web 3.0. The chapter went on to address the advantages, challenges and implementation issues surrounding the overall adoption of these technologies in reasoning communities.

Finally, this chapter discussed the future and how future iterations of the Internet may impact reasoning communities. Discussion focused on the growth of future technologies and how they may enhance possibilities with regard to reasoning communities and their adoption of current and future technologies.

The task of using and building upon the Web 2.0, and later Web 3.0 applications discussed in this chapter for use in reasoning communities is just beginning. Web 2.0 and Web 3.0 applications have great potential for enhancement of the phases that make for successful reasoning communities. This chapter has begun the process by identifying the applicable applications, some implementation options, and offering a look into the future of web technologies and their impact on reasoning communities.

REFERENCES

W3C. (2009). *W3C Semantic web activity*. Retrieved October 5, 2009, from http://www.w3.org/2001/sw/

Ankoleka, A., Krötzsch, M., Tran, T., & Vrandečić, D. (2008). The two cultures: Mashing up Web 2.0 and the Semantic Web. *Web Semantics: Science . Services and Agents on the World Wide Web*, *6*(1), 70–75.

Blanchard, A., & Horan, T. (1998). Virtual communities and social capital. *Social Science Computer Review*, *16*(3), 293–307. doi:10.1177/089443939801600306

Chadwick, A. (2008). Web 2.0: New challenges for the study of e-democracy in an Era of Informational Exuberance.

Chastang, J. (2009). *WeatherMole*. Retrieved October 2, 2009, from http://weathermole.com/WeatherMole/index.html

Cho, A. (2008). What is Web 3.0? The next generation Web: Search context for online information. Retrieved February 23, 2010, from http://internet.suite101.com/article.cfm/what_is_web_30

Coleman, S., & Gøtze, J. (2002). *Bowling together: Online public engagement in policy deliberation*. London: Hansard Society. Retrieved October 2, 2009, from http://www.bowlingtogether.net/

Delicious. (2009). *About*. Retrieved October 3, 2009, from http://delicious.com/about

Facebook. (2009). *Facebook factsheet*. Retrieved October 3, 2009, from http://www.facebook.com/cmiller61?ref=name#/press/info.php?factsheet

Fawzi, M. (2006). *Wikipedia 3.0: The end of Google?* Retrieved February 23, from http://evolvingtrends.wordpress.com/2006/06/26/wikipedia-30-the-end-of-google/

Flew, T. (2008). *New media: An introduction* (3rd ed.). Melbourne, Australia: Oxford University Press.

Gliffy, Inc. (2009). *About Gliffy, Inc*. Retrieved October 2, 2009, from http://www.gliffy.com/about.shtml

Google. (2009a). *Google docs help: Getting started guide*. Retrieved October 2, 2009, from http://docs.google.com/support/bin

Google. (2009b). *Google apps for business*. Retrieved October 1, 2009, from http://www.google.com.au/apps/intl/en/business/index.html

Gurulib. (2009). *Gurulib: About us*. Retrieved October 2, 2009, from http://www.gurulib.com/_about_us.html

Jevalins, C. (2009). What are Web 2.0 and Web 3.0 technologies? Retrieved February 23, 2010, from http://ezinearticles.com/?What-Are-Web-2.0-and-Web-3.0-Technologies?&id=2580537

Lee, M., Miller, C., & Newnham, L. (2008). RSS and content syndication in higher education: Subscribing to a new model of teaching and learning. *Educational Media International*, *45*(4), 311–322. doi:10.1080/09523980802573255

Limesurvey. (2009). *Features*. Retrieved October 4, 2009, from http://www.limesurvey.org/en/about-limesurvey/features

LinkedIn, Inc. (2009). *About us*. Retrieved October 1, 2009, from http://press.linkedin.com/about

Mason, R., & Rennie, F. (2007). Using Web 2.0 for learning in the community. *The Internet and Higher Education, 10*(3), 196–203. doi:10.1016/j.iheduc.2007.06.003

Mathes, A. (2004, December). Folksonomies: Cooperative classification and communication through shared metadata. *Computer Mediated Communication – LIS590CMC*, Graduate School of Library and Information Science, University of Illinois Urbana-Champaign.

Merrill, D. (2006). *Mashups: The new breed of Web app. An introduction to mashups.* Armonk, NY: IBM developerWorks. Retrieved February 15, 2010, from http://www.ibm.com/developerworks/xml/library/x-mashups.html

Midgely, J., Hall, A., Hardiman, M., & Narine, D. (1986). *Community participation, social development, and the state.* London: Methuen & Co. Ltd.

Miller, P. (2005). Web 2.0: Building the new library. *Ariadne, 25.* Retrieved September 26, 2009, from http://www.ariadne.ac.uk/issue45/miller/#author1

Nielsen Online. (2008, January). *Consumer generated media: Separating hype from reality. A new market study from Nielsen Online* (1st ed., p. 89). New York: Nielsen.

Ning, Inc. (2009). *About Ning.* Retrieved October 2, 2009, from http://about.ning.com/

O'Reilly, T. (2005). *What is Web 2.0.* Retrieved September 24, 2009, from http://oreilly.com/web2/archive/what-is-web-20.html

Osman, D., Yearwood, J., & Vamplew, P. (2007). Using corpus analysis to inform research into opinion detection in blogs. *Proceedings of the 6th Australasian Data Mining Conference (AusDM'07)* (pp. 65-75).

Petrik, K. (2009). Participation and e-democracy: How to utilize Web 2.0 for policy decision-making. *Proceedings of the 10th Annual International Conference on Digital Government Research: Social Networks: Making Connections between Citizens, Data and Government. ACM International Conference Proceeding Series, 390*, 254-263.

Plummer, D., Cearley, D., & Smith, D. (2008). *Cloud computing confusion leads to opportunity.* Retrieved October 3, 2009, from http://www.gartner.com/it/page.jsp?id=707508

Putnam, R. (1995). Bowling alone: America's declining social capital. *Journal of Democracy, 6*(1), 65–78. doi:10.1353/jod.1995.0002

Shadbolt, N., Hall, W., & Berners-Lee, T. (2006). The semantic Web revisited. *IEEE Intelligent Systems*, 96–101. doi:10.1109/MIS.2006.62

Six Apart Ltd. (2009). *Getting started with TypePad.* Retrieved October 3, 2009, from http://help.sixapart.com/tp/us/getting_started.html

Slideshare, Inc. (2009). *What is Slideshare?* Retrieved October 1, 2009, from http://www.slideshare.net/about

Spivack, N. (2007). *The third-generation Web -- Web 3.0.* Retrieved February 28, 2010, from http://www.intentblog.com/archives/2007/02/nova_spivack_th.html

Strickland, J. (2009). *How Web 3.0 will work.* Retrieved February 23, 2010, from http://www.howstuffworks.com/web-30.htm

Tomoye Corporation. (2009). *About Tomoye.* Retrieved October 4, 2009, from http://www.tomoye.com/about.html

Twitter. (2009). *About us.* Retrieved Oct 1, 2009, from http://twitter.com/about#about

United Nations. (2009). *Online volunteering.* Retrieved October 3, 2009, from http://www.onlinevolunteering.org/en/index.html

White, N. (2001). Facilitating and hosting a virtual community. Retrieved October 1, 2009, from http://www.fullcirc.com/community/communityfacilitation.htm

Yearwood, J., & Stranieri, A. (2009). Deliberative discourse and reasoning from generic argument structures. *AI & Society*, *23*, 353–377. doi:10.1007/s00146-006-0069-y

Young, G., Brown, E., Keitt, T., Owyang, J., Koplowitz, R., & Lo, H. (2008). *Global enterprise Web 2.0 market forecast: 2007 to 2013*. Retrieved October 2, 2009, from http://www.forrester.com/Research/Document/Excerpt/0,7211,43850,00.html

YouTube. (2009). *YouTube Fact Sheet*. Retrieved October 1, 2009, from http://www.youtube.com/t/fact_sheet

Zoho Corporation. (2009). *FAQ (Frequently Asked Questions)*. Retrieved October 4, 2009, from http://www.zoho.com/zoho_faq.html

Compilation of References

Abdel-Kader, M. G., & Dugdale, D. (2001). Evaluating investments in advanced manufacturing technology: A fuzzy set theory approach. *The British Accounting Review*, *33*, 455–489. doi:10.1006/bare.2001.0177

Aboutalib, S., & Veloso, M. (2007). Towards using multiple cues for robust object recognition. In *Proceedings of AAMAS'07*. Honolulu, Hawaii. May 14–18.

Abrahams, B., Bellucci, E., & Zeleznikow, J. (2010). Incorporating fairness into development of an integrated multi-agent online dispute resolution environment. *Group Decision and Negotiation*. Retrieved from http://www.springerlink.com/content/2jl7774266830554/

Aczel, J., & Alsina, C. (1986). On synthesis of judgments. *Socio-Economic Planning Sciences*, *20*, 333–339. doi:10.1016/0038-0121(86)90044-3

Aczel, J., & Saaty, T. L. (1983). Procedures for synthesizing ratio judgments. *Journal of Mathematical Psychology*, *27*, 93–102. doi:10.1016/0022-2496(83)90028-7

Adams, S. J., Roch, S. G., & Ayman, R. (2005). Communication medium and member familiarity: The effects on decision time, accuracy, and satisfaction. *Small Group Research*, *36*(3), 321–353. doi:10.1177/1046496405275232

Afshar, F., Yearwood, J., & Stranieri, A. (2006). A tool for assisting group decision-making for consensus outcomes in organizations . In Voges, K. E., & Pope, N. K. L. (Eds.), *Business applications and computational intelligence* (pp. 316–343). Hershey, PA: Idea Group Publishing.

Albayrakoglu, M. (1996). Justification of new manufacturing technology: A strategic approach using the analytic hierarchy process. *Production and Inventory Management Journal*, *37*(1), 71–76.

Alberts, D., Gartstka, J. J., & Stein, F. P. (2000). Network-centric warfare: Developing and leveraging information superiority (2nd ed.). Washington, DC: DoD C4ISR Co-operative Research Program. Retrieved March 1, 2007, from http://www.dodccrp.org/html3/pubs_download.html

Alexy, R. (1991). *Theorie der juristischen Argumentation: Die Theorie des rationalen Diskurses als Theorie der juristischen Begründung* (2nd ed.). Frankfurt, Germany: Suhrkamp.

Al-Ghoneim, K., & Kumar, B. V. K. V. (1998). Unified decision combination framework. *Pattern Recognition*, *31*(12), 2077–2089. doi:10.1016/S0031-3203(98)00030-2

Allen, L. E. (1957). Symbolic logic: A razor-edged tool for drafting and interpreting legal documents. *The Yale Law Journal*, *66*, 833–879. doi:10.2307/794073

Althaus, S. L. (1997). Computer-mediated communication in the university classroom: An experiment with on-line discussions. *Communication Education*, *46*(3), 158–174. doi:10.1080/03634529709379088

Aly, S., & Vrana, I. (2006). Integrating Multiple Fuzzy Expert System under Restricting Requirements. *Agricultural Economics*, *52*(4), 187–196.

Aly, S., & Vrana, I. (2008). Evaluating the knowledge, relevance and experience of expert decision-makers utilizing the Fuzzy-AHP. *Agricultural Economics*, *54*(11), 529–535.

Ambigulous. (2008, July 21). [Comment 1 to blog post: The great pretender]. *Larvatus Prodeo* [Blog], Federal election 2007. Retrieved October 1, 2009, from http://larvatusprodeo.net/2008/07/21/the-great-pretender/

Ambrosini, C., & Routhier, J. L. (2004). Objectives, methods and results of surveys carried out in the field of urban freight transport: An international comparison. *Transport Reviews*, *24*(1), 57–77. doi:10.1080/0144164032000122343

Amedeo, D., Golledge, R. G., & Stimson, R. J. (2008). *Person-environment-behavior research: Investigating activities and experiences in spaces and environments*. New York: Guilford Press.

Ankoleka, A., Krötzsch, M., Tran, T., & Vrandečić, D. (2008). The two cultures: Mashing up Web 2.0 and the Semantic Web. *Web Semantics: Science . Services and Agents on the World Wide Web*, *6*(1), 70–75.

Anti-Phishing Working Group. (2009). Retrieved December 2009, from http://apwg.org/

Aravindan, P., & Punniyamoorthy, M. (2002). Justification of advanced manufacturing technologies (AMT). *International Journal of Advanced Manufacturing Technology*, *19*, 151–156. doi:10.1007/s001700200008

Arendt, H. (1961). *Between past and future: Six exercises in political thought*. London: Faber and Faber.

Arentze, T., & Timmermans, H. (2006). Multi-agent models of spatial cognition: Learning and complex choice behavior in urban environments . In Portugali, J. (Ed.), *Complex artificial environments*. Berlin: Springer. doi:10.1007/3-540-29710-3_12

Argyris, C. (1982). *Reasoning, learning, and action*. San Francisco, CA: Jossey-Bass.

Aristotle, . (1984). *The complete works of Aristotle: The revised Oxford translation (Chapter on logic)*. Princeton, NJ: Princeton University Press.

Arrow, K. J. (1963). *Social choice and individual values*. New York: Wiley.

Arvanitis, S., & Hollenstein, H. (2001). The determinants of the adoption of advanced manufacturing technology. *Economics of Innovation and New Technology*, *10*(5), 377–414. doi:10.1080/10438590100000015

Ashley, K. D. (1990). *Modeling legal argument: Reasoning with cases and hypotheticals*. Cambridge, MA: MIT Press.

Ashley, K. D. (2006). Case-based reasoning . In Lodder, A. R., & Oskamp, A. (Eds.), *Information technology and lawyers: Advanced technology in the legal domain, from challenges to daily routine* (pp. 23–60). Dordrecht, The Netherlands: Springer.

Ashley, K. (2009). Ontological requirements for analogical, teleological, and hypothetical legal reasoning. In *Proceedings of the 12th ICAIL International Conference on Artificial Intelligence and Law* (pp. 1-10). New York: ACM Press.

Astor, H., & Chinkin, C. M. (2002). *Dispute resolution in Australia* (2nd ed.). Sydney, Australia: LexisNexis Butterworths.

Åström, J. (2004). Digital democracy: Ideas, intentions and initiatives in Swedish local governments . In Gibson, R., Römelle, A., & Ward, S. (Eds.), *Electronic democracy: Mobilisation, participation and organization via new ICTs*. London: Routledge.

Atkinson, K., Bench-Capon, T. J. M., & McBurney, P. (2005). A dialogue game protocol for multi-agent argument over proposals for action. *Autonomous Agents and Multi-Agent Systems*, *11*(2), 153–171. doi:10.1007/s10458-005-1166-x

Atkinson, K., Bench-Capon, T. J. M., & McBurney, P. (2006). Computational representation of practical argument. *Synthese*, *2006*, 157–206. doi:10.1007/s11229-005-3488-2

Auephanwiriyakul, S., Keller, J. M., & Gader, P. D. (2002). Generalized Choquet fuzzy integral fusion. *Information Fusion*, *3*, 69–85. doi:10.1016/S1566-2535(01)00054-9

Avery, J., Yearwood, J., & Stranieri, A. (2001). An argumentation based multiagent system for eTourism dialogue. In *Proceedings First International Workshop on Hybrid Intelligent Systems HIS 2001,* Adelaide, Australia (pp. 497–512).

Axelrod, R. (1976). *Structure of decision: The cognitive maps of political elites*. Princeton, NJ: Princeton University Press.

Bächtiger, A., Niemeyer, S., Neblo, M., Steenbergen, M. R., & Steiner, J. (2010). Disentangling diversity in deliberative democracy: Competing theories, their empirical blind-spots, and complementarities. *Journal of Political Philosophy*, *18*(1), 32–63. doi:10.1111/j.1467-9760.2009.00342.x

Baker, J. R., & Moore, S. M. (2008). Blogging as a social tool: A psychosocial examination of the effects of blogging. *Cyberpsychology & Behavior*, *11*(6), 747–749. doi:10.1089/cpb.2008.0053

Baltes, B. B., Dickson, M. W., Sherman, M. P., Bauer, C. C., & LaGanke, J. S. (2002). Computer-mediated communication and group decision making: A meta-analysis. *Organizational Behavior and Human Decision Processes*, *87*, 156–179. doi:10.1006/obhd.2001.2961

Bannister, D. (1960). Conceptual structure in thought disordered schizophrenics. *The Journal of Mental Science*, *106*, 1230–1249.

Bannister, D. (1965). The rationale and relevance of repertory grid technique. *The British Journal of Psychiatry*, *111*, 977–982. doi:10.1192/bjp.111.479.977

Barbanente, A., Camarda, D., Grassini, L., & Khakee, A. (2007). Visioning the regional future: Globalisation and regional transformation of Rabat/Casablanca. *Technological Forecasting and Social Change*, *74*, 763–778. doi:10.1016/j.techfore.2006.05.019

Bardossy, A., Duckstein, L., & Bogardi, J. (1993). Combination of fuzzy numbers representing expert opinions. *Fuzzy Sets and Systems*, *57*, 173–181. doi:10.1016/0165-0114(93)90157-D

Barkhi, R. (2005). Information exchange and induced cooperation in group decision support systems. *Communication Research*, *32*(5), 646–678. doi:10.1177/0093650205279352

Barkhi, R., Jacob, V. S., & Pirkul, H. (1999). An experimental analysis of face to face versus computer mediated communication channels. *Group Decision and Negotiation*, *8*(4), 325–347. doi:10.1023/A:1008621423120

Barkowsky, T., Knauff, M., Ligozat, G., & Montello, D. R. (Eds.). (2007). *Spatial cognition: Reasoning, action, interaction*. Berlin: Springer.

Barr, P. S., Stimpert, J. L., & Huff, A. S. (1992). Cognitive change, strategic action, and organizational renewal. *Strategic Management Journal*, *13*, 15–36. doi:10.1002/smj.4250131004

Barrat, M. (2004). Understanding the meaning of collaboration in the supply chain. *Supply Chain Management: An International Journal*, *9*(1), 30–42. doi:10.1108/13598540410517566

Barry, B., & Friedman, R. A. (1998). Bargainer characteristics in distributive and integrative negotiation. *Journal of Personality and Social Psychology*, *74*(2), 345–359. doi:10.1037/0022-3514.74.2.345

Basu, S., Chouhdury, T., Clarkson, B., & Pentland, A. (2001). Towards measuring human interactions in conversational settings. *Proceedings of the IEEE International Workshop on Cues in Communication*, IEEE.

Baym, N. (1998). The emergence of online community . In Jones, S. G. (Ed.), *Cybersociety 2.0* (pp. 35–68). Thousand Oaks, CA: Sage.

Beaumont, P. (2004). Multi-platform coordination and resource management in command and control. Unpublished masters thesis. Laval University, Quebec, Canada. Retrieved November 29, 2006, from http://damas.ift.ulaval.ca/projets/TeamWork/publications.en.html

Becker-Beck, U., Wintermantel, M., & Borg, A. (2005). Principles of regulating interaction in teams practicing face-to-face communication versus teams practicing computer-mediated communication. *Small Group Research, 36*(4), 499–536. doi:10.1177/1046496405277182

Behavior and Organization Design (no date). *Research in Organizational Behavior, 6*, 191-233.

Bellucci, E., & Zeleznikow, J. (2006). Developing negotiation decision support systems that support mediators: A case study of the Family_Winner system. *Journal of Artificial Intelligence and Law, 13*(2), 233–271. doi:10.1007/s10506-006-9013-1

Bellucci, E. (2008). Development of a mediation tool in family law: AssetDivider . In Francescori, E., Sartor, G., & Tiscornia, D. (Eds.), *Legal Knowledge and Information Systems* (pp. 88–97). Amsterdam: IOS Press.

Bellucci, E. (2004). *Developing compensation strategies for the construction of negotiation decision support systems*. Unpublished doctoral dissertation. La Trobe University, Victoria, Australia.

Bellucci, E., & Zeleznikow, J. (1997). *Family-negotiator: An intelligent decision support system for negotiation in Australian Family Law*. Proceedings of the Fourth Conference of the International Society for Decision Support Systems, Lausanne. *International Society for Decision Support Systems*, 359-373.

Bellucci, E., & Zeleznikow, J. (1998). A comparative study of negotiation decision support systems. In *Thirty-First Hawaii International Conference on System Sciences IEEE Computer Society* (pp. 254 – 262).

Ben-Arieh, D. (2005). Sensitivity of multi-criteria decision-making to linguistic quantifiers and aggregation means. *Computers & Industrial Engineering, 48*, 289–309. doi:10.1016/j.cie.2005.01.014

Bench-Capon, T. J. M. (2003). Persuasion in practical argument using value-based argumentation frameworks. *Journal of Logic and Computation, 13*, 429–448. doi:10.1093/logcom/13.3.429

Bench-Capon, T. J. M., & Dunne, P. E. (2007). Argumentation in artificial intelligence. *Artificial Intelligence, 171*, 619–641. doi:10.1016/j.artint.2007.05.001

Bench-Capon, T., & Prakken, H. (2006). Argumentation . In Lodder, A. R., & Oskamp, A. (Eds.), *Information technology and lawyers: Advanced technology in the legal domain, from challenges to daily routine* (pp. 61–80). Dordrecht, The Netherlands: Springer.

Bench-Capon, T., & Gordon, T. F. (2009). Isomorphism and argumentation. In *Proceedings of the 12th ICAIL International Conference on Artificial Intelligence and Law* (pp. 11-20). New York: ACM Press.

Bench-Capon, T., & Visser, P. R. S. (1997). Ontologies in legal information systems: The need for explicit specifications of domain conceptualisations. In *Proceedings of the International Conference on Artificial Intelligence and Law* (pp. 132-141). New York: ACM Press.

Benhabib, S. (1996). Toward a deliberative model of democratic legitimacy . In Benhabib, S. (Ed.), *Democracy and difference: Contesting the boundaries of the political*. Princeton, NJ: Princeton University Press.

Berger, A. (Ed.). (2009). *Access to legislation in Europe: Guide to the legal gazettes and other official information sources of the European Union and the European Free Trade Association*. Luxembourg: Publications Office of the European Union.

Bergstrom, T., & Karahalios, K. (2007). Conversation clock: Visualizing audio patterns in co-located groups. *Proceedings of the Hawaii international Conference on System Sciences*, IEEE, 78.

Berners-Lee, T., Hendler, J., & Lassila, O. (2001). The semantic web. *Scientific American, 284*(5), 34–53. doi:10.1038/scientificamerican0501-34

Beroggi, G. E. G. (2005). *Designing and evaluating e-management decision tools*. New York: Springer.

Berrah, L., Mauris, G., Montmain, J., & Cliville, V. (2008a). Efficacy and efficiency indexes for a multi-criteria industrial performance synthesized by Choquet integral aggregation. *International Journal of Computer Integrated Manufacturing*, *21*(4), 415–425. doi:10.1080/09511920701574255

Berrah, L., Mauris, G., Montmain, J., & Cliville, V. (2008b). Monitoring the improvement of an overall industrial performance based on a Choquet integral aggregation. *Omega-International Journal of Management Science*, *36*(3), 340–351. doi:10.1016/j.omega.2006.02.009

Bessette, J. (1994). *The mild voice of reason: Deliberative democracy & American national government*. Chicago: University of Chicago Press.

Bicchieri, C., & Lev-On, A. (2007). Computer-mediated communication and cooperation in social dilemmas: An experimental analysis. *Politics, Philosophy & Economics*, *6*(2), 139–168. doi:10.1177/1470594X07077267

Bichler, M., Kesten, G., & Weinhardt, C. (2003). Electronic negotiations: Foundations, systems, and experiments [Introduction to the Special Issue]. *Group Decision and Negotiation*, *12*, 85–88. doi:10.1023/A:1023064720309

Bieri, J., Atkins, A. L., Briar, S., Leaman, R. L., Miller, H., & Tripodi, T. (1966). *Clinical and social judgment: The discrimination of behavioural information*. New York: Wiley.

Bing 1998 Hypertext – the deep structure. [Unpublished invited talk, DEXA 1998. University of Vienna].

Bing, J. (1987). Designing Text Retrieval Systems for Conceptual Searching. In *Proceedings of the 1st ICAIL* (pp. 43-51). New York: ACM Press.

Bing, J. (1995). Legal Text Retrieval and Information Services. In: J. Bing & O. Torvund, 25 years Anniversary Anthology in Computers and Law (pp. 525-586). Oslo: Tano.

Bingo Bango Boingo. (2008, October 02). [Comment 22 to blog post: The state of Rudd nation]. *Larvatus Prodeo* [Blog], Federal election 2007. Retrieved October 1, 2009, from http://larvatusprodeo.net/2008/10/02/the-state-of-rudd-nation/

Bird, L. (2001, October 02). Single mind could be a liability [Letter to the editor]. *The Australian*, p. 12.

Bismarck. (2008, June 24). [Comment 2 to blog post: Marketised celebrity politics, Australian style]. *Larvatus Prodeo* [Blog], Federal election 2007. Retrieved 01 October, 2009, from http://larvatusprodeo.net/2008/06/24/marketised-celebrity-politics-australian-style/

Black, E., & Hunter, A. (2007). A generative inquiry dialogue system. In *Proceedings of the Sixth International Joint Conference on Autonomous Agents and Multi-Agent Systems* (AAMAS '07), (pp. 1010-1017). New York: ACM Press

Blackler, F. (1995). Knowledge, knowledge work and organisations: An overview and interpretation. *Organization Studies*, *16*(6). doi:10.1177/017084069501600605

Blanchard, A., & Horan, T. (1998). Virtual communities and social capital. *Social Science Computer Review*, *16*(3), 293–307. doi:10.1177/089443939801600306

Blaug, R. (1997). Between fear and disappointment: Critical, empirical and political uses of Habermas. *Political Studies*, *45*, 100–117. doi:10.1111/1467-9248.00074

Boender, C. G. E., de Graan, J. G., & Lootsma, F. A. (1989). Multi-criteria decision analysis with fuzzy pairwise comparisons. *Fuzzy Sets and Systems*, *29*, 133–143. doi:10.1016/0165-0114(89)90187-5

Bohman, J. F. (1998). The coming of age of deliberative democracy. *Journal of Political Philosophy*, *6*(4), 399–423. doi:10.1111/1467-9760.00061

Bohman, J., & Rehg, W. (Eds.). (1997). *Deliberative democracy: Essays on reason and politics*. Cambridge, MA: MIT Press.

Böker, H., Hell, D., Budischewski, K., Eppelb, A., Härtlingb, F., & Rinnertb, H. (2000). Personality and object relations in patients with affective disorders: idiographic research by means of the repertory grid technique. *Journal of Affective Disorders*, *60*, 53–59. doi:10.1016/S0165-0327(99)00161-5

Bokhorst, J. A. C., Slomp, J., & Suresh, N. C. (2002). An integrated model for part operation allocation and investments in CNC technology. *International Journal of Production Economics*, *75*(3), 267–285. doi:10.1016/S0925-5273(01)00118-9

Bonham, G. M., & Shapiro, M. J. (1984). Mapping structures of thought . In Gallhofer, I., Saris, W. E., & Melman, M. (Eds.), *Different text analysis procedures for the study of decision making* (pp. 29–51). Amsterdam: Sociometric Research Foundation.

Borri, D., & Camarda, D. (2006). Visualizing space-based interactions among distributed agents: Environmental planning at the inner-city scale. *Lecture Notes in Computer Science*, *4101*, 182–191. doi:10.1007/11863649_23

Borri, D., & Camarda, D. (2009). The cooperative conceptualization of urban spaces in AI assisted environmental planning. *Lecture Notes in Computer Science*, *5738*, 197–207. doi:10.1007/978-3-642-04265-2_29

Bos, N., Olson, J., Gergle, D., Olson, G., & Wright, Z. (2002) Effects of four computer-mediated communications Channels on trust development. In *Proceedings of Conference on Human Factors in Computing (CHI)*, 135-140.

Bouas, K., & Arrow, H. (1995). The development of group identity in computer and face-to-face groups with membership change. [CSCW]. *Computer Supported Cooperative Work*, 153–178.

Bougon, M. G., Weick, K. E., & Binkhorst, D. (1977). Cognition in organizations: An analysis of the Utrecht Jazz Orchestra. *Administrative Science Quarterly*, *22*, 606–639. doi:10.2307/2392403

Boyd, R. (1979). Metaphor and theory change: What is metaphor for? In Ortony, A. (Ed.), *Metaphor and thought* (pp. 481–532). Cambridge: Cambridge University Press.

Bozdag, C. E., Kahraman, C., & Ruan, D. (2003). Fuzzy group decision-making for selection among computer integrated manufacturing systems. *Computers in Industry*, *51*(1), 13–29. doi:10.1016/S0166-3615(03)00029-0

Bradner, E., & Mark, G. (2002). Why distance matters: Effects on cooperation, persuasion and deception. In *Proceedings of ACM Conference on Computer Supported Cooperative Work (CSCW).*

Bradshaw, J. M., Chapman, C. R., Sullivan, K. M., Zarley, D., Nims, J., Gavrin, J., et al. (1993). KS-3000: An application of DDUCKS to bone-marrow transplant patient support. *Proceedings of the Sixth Annual Florida AI Res. Symposium (FLAIRS93)*, 18–21.

Brams, S. J., & Taylor, A. D. (1996). *Fair division, from cake cutting to dispute resolution.* Cambridge, UK: Cambridge University Press.

Bratman, M. E., Israel, D. J., & Pollack, M. E. (1988). Plans and resource-bounded practical reasoning. *Computational Intelligence*, *4*, 349–355. doi:10.1111/j.1467-8640.1988.tb00284.x

Bretz, L. (2001, September 27). Court's role is to see that we act legally [Letter to the editor]. *The Australian*, p. 12.

Bromley, M. (1998). Watching the watchdogs?: The role of reader's letters in calling the press to account . In Bromley, M., & Stephenson, H. (Eds.), *Sex, lies and democracy* (pp. 147–162). London: Longman.

Brown, S. R. (1980). *Political subjectivity: Applications of Q methodology in political science.* New Haven, CT: Yale University Press.

Brown, J. S., & Duguid, P. (1991). Organizational learning and communities of practice: Toward a unified view of working, learning and innovation. *Organization Science*, 2.

Brown, S. M. (1992). Cognitive mapping and repertory grids for qualitative survey research: Some comparative observations. *Journal of Management Studies*, *29*(3), 287–307. doi:10.1111/j.1467-6486.1992.tb00666.x

Brucher, H., & Baumberger, P. (2003). Using mobile technology to support e-democracy. In *Proceedings of the 36th Annual Hawaii International Conference on System Sciences*. Washington, DC: IEEE Computer Society.

Bruhns, C. (2001, October 10). Minds, pockets agents for peace [Letter to the editor]. *The Australian*, p. 14.

Brüninghaus, S., & Ashley, K. D. (2001). Improving the representation of legal case texts with information extraction methods. In *Proceedings of the 8th ICAIL* (pp. 42-51). New York: ACM Press.

Bruns, A., Wilson, J. A., Saunders, B. J., Kirchhoff, L., & Nicolai, T. (2008, October 15-18). Australia's political blogosphere in the aftermath of 2007 federal election. Paper presented at the Association of Internet Researchers 2008: Internet Research 9.0: Rethinking Community, Rethinking Place. Copenhagen, Denmark. Retrieved September, 10, 2009, from http://eprints.qut.edu.au/15116/1/15116.pdf

Bryant, J. (1984). Modelling alternative realities in conflict and negotiation. *The Journal of the Operational Research Society*, *35*(11), 985–993.

Bryson, N. (1996). Group decision-making and the analytic hierarchy process: Exploring the consensus-relevant information content. *Computers & Operations Research*, *23*(1), 27–35. doi:10.1016/0305-0548(96)00002-H

Buckingham, D. (1997). News media, political socialization and popular citizenship: Towards a new agenda. *Critical Studies in Mass Communication*, *14*, 344–366. doi:10.1080/15295039709367023

Buder, M., Rehfeld, W., Seeger, T., & Strauch, D. (Eds.). (1997). *Grundlagen der praktischen Information und Dokumentation*. Munich, Germany: K.G. Saur.

Buede, D., & Maxwell, D. C. (1995). Rank disagreement: A comparison of multi-criteria methodologies. *Journal of Multi-Criteria Decision Analysis*, *4*, 1–21. doi:10.1002/mcda.4020040102

Bui, T., & Shakun, M. F. (1997). *Introduction to the negotiation support systems minitrack*. In Thirtieth Annual Hawaii International Conference on Systems Science–Mini track on Negotiation Support Systems, IEEE, Hawaii.

Burkhalter, S., Gastil, J., & Kelshaw, T. (2002). A conceptual definition and theoretical model of public deliberation in small face-to-face groups. *Communication Theory*, *12*(4), 398–422.

Burns, P. (2008, July 21). [Comment 2 to blog post: The great pretender]. *Larvatus Prodeo* [Blog], Federal election 2007. Retrieved October 1, 2009, from http://larvatusprodeo.net/2008/07/21/the-great-pretender/

Cabri, G., Ferrari, L., & Leonardi, L. (2005). A role-based mobile-agent approach to support e-democracy. *Applied Soft Computing*, *6*(1), 85–99. doi:10.1016/j.asoc.2004.12.001

Calhoun, C. (1992). Introduction: Habermas and the public sphere . In Calhoun, C. (Ed.), *Habermas and the public sphere* (pp. 1–50). Cambridge, MA: MIT Press.

Calvo, T., Mesiar, R., & Yager, R. R. (2004). Quantitative weights and aggregation. *IEEE Transactions on Fuzzy Systems*, *12*(1), 62–69. doi:10.1109/TFUZZ.2003.822679

Camarda, D. (2010). Beyond citizen participation in planning: Multi-agent systems for complex decision making . In Nunes Silva, C. (Ed.), *Handbook of Research on E-planning*. Hershey: IGI Global. doi:10.4018/978-1-61520-929-3.ch010

Cameron, R. (2002, January 26-27). Caring, tolerant Aussie is a thing of the past [Letter to the editor]. *The Australian*, p. 18.

Campbell, J., & Stasser, G. (2006). The influence of time and task demonstrability on decision-making in computer-mediated and face-to-face groups. *Small Group Research*, *37*, 271–294. doi:10.1177/1046496406288976

Carberry, S. (1990). *Plan recognition in natural language dialogue*. Cambridge, MA: MIT Press.

Carbogim, D., Robertson, D., & Lee, J. (2000). Argument-based applications to knowledge engineering. *The Knowledge Engineering Review*, *15*(2), 119–149. doi:10.1017/S0269888900002058

Cartwright, D., & Atkinson, K. (2008). Political engagement through tools for argumentation. In P. Besnard, S. Doutre & A. Hunter (Eds.), *Computational Models of Argument, Proceedings of COMMA-2008,* (Frontiers in Artificial Intelligence and Applications, Vol. 172, pp. 116-127). Amsterdam: Amsterdam.

Carvalho, A., Rocha, R., & Oliveira, M. A. (2004). SINUP: Using GIS to support E-democracy. In *Electronic government* ([). Berlin/Heidelberg: Springer. Retrieved from *http://www.inescporto.pt*]. *Lecture Notes in Computer Science*, *2739*, 341–344. doi:10.1007/10929179_61

Casanovas, P., Biasiotti, M. A., Francesconi, E., & Sagri, M. T. (Eds.). (2007). *Proceedings of LOAIT 07, II: Workshop on Legal Ontologies and Artificial Intelligence Techniques*. Retrieved from http://www.ittig.cnr.it/loait/LOAIT07-Proceedings.pdf

Casellas, N., Francesconi, E., Hoekstra, R., & Montemagni, S. (Eds.). (2009). *Proceedings of LOAIT 2009: 3rd Workshop on Legal Ontologies and Artificial Intelligence Techniques joint with 2nd Workshop on Semantic Processing of Legal Text*. Barcelona, Spain: IOT Series.

Castells, M. (1996). *The rise of network society* (The information age: Economy, society and culture, vol. 1). Oxford: Blackwell Publishers.

Cebrowski, A., & Gartska, J. J. (1998). Network-centric warfare: Its origins and future. *Proceedings of the Naval Institute*, *124*(1).

Chadwick, A. (2008). Web 2.0: New challenges for the study of e-democracy in an Era of Informational Exuberance.

Chaiken, S., & Stangor, C. (1987). Attitudes and attitude change. *Annual Review of Psychology*, *38*, 575–630. doi:10.1146/annurev.ps.38.020187.003043

Chakraborty, C., & Chakraborty, D. (2007). A fuzzy clustering methodology for linguistic opinions in group decision-making. *Applied Soft Computing*, *7*, 858–869. doi:10.1016/j.asoc.2006.02.006

Chambers, S. (2003). Deliberative democratic theory. *Annual Review of Political Science*, *6*(1), 307–326. doi:10.1146/annurev.polisci.6.121901.085538

Chan, F. T. S., Chan, H. K., Chan, M. H., & Humphreys, P. K. (2006). An integrated fuzzy approach for the selection of manufacturing technologies. *International Journal of Advanced Manufacturing Technology*, *27*(7–8), 747–758. doi:10.1007/s00170-004-2246-9

Chan, F. T. S., Chan, M. H., Lau, H., & Ip, R. W. L. (2001). Investment appraisal techniques for advanced manufacturing technology (AMT): A literature review. *Integrated Manufacturing Systems*, *12*(1), 35–47. doi:10.1108/09576060110361528

Chan, F. T. S., Chan, M. H., & Mak, K. L. (1999). An integrated approach to investment appraisal for advanced manufacturing technology. *Human Factors and Ergonomics in Manufacturing*, *9*(1), 69–86. doi:10.1002/(SICI)1520-6564(199924)9:1<69::AID-HFM4>3.0.CO;2-1

Chandar, S., & Zeleznikow, J. (2009). Risks to consider when negotiating IT outsourcing agreements. Proceedings of GDN2009, the 10th annual meeting INFORMS section on Group Decision and Negotiation, Toronto, Ontario, June 14-17. Retrieved August 26, 2009, from http://info.wlu.ca/~wwwmath/faculty/kilgour/gdn/papers.htm

Chandrasekaran, B., Josephson, J. R., Banerjee, B., & Kurup, U. (2002). Diagrammatic reasoning in support of situation understanding and planning. Proceedings of Army Science Conference, Dec 2-5, Orlando, Florida.

Chang, T. H., & Wang, T. C. (2009). Measuring the success possibility of implementing advanced manufacturing technology by utilizing the consistent fuzzy preference relations. *Expert Systems with Applications*, *36*, 4313–4320. doi:10.1016/j.eswa.2008.03.019

Charlton, R. (2000). *Dispute resolution guidebook*. Sydney, Australia: LBC Information Service.

Chastang, J. (2009). *WeatherMole*. Retrieved October 2, 2009, from http://weathermole.com/WeatherMole/index.html

Chayko, M. (2002). *Connecting: How we form social bonds and communities in the Internet age*. Albany, NY: State University of New York Press.

Chayko, M. (2008). *Portable communities: The social dynamics of online and mobile connectedness*. Albany, NY: State University of New York Press.

Chen, C. T. (2000). Extensions of the TOPSIS for group decision-making under fuzzy environment. *Fuzzy Sets and Systems, 114*, 1–9. doi:10.1016/S0165-0114(97)00377-1

Chen, M. F., & Tzeng, G. H. (2004). Combining grey relation and TOPSIS concepts for selecting an expatriate host country. *Mathematical and Computer Modelling, 40*, 1473–1490. doi:10.1016/j.mcm.2005.01.006

Chen, I. J., & Small, M. H. (1994). Implementing manufacturing technology: An integrated planning model. *Omega, 22*(1), 91–103. doi:10.1016/0305-0483(94)90010-8

Cheng, C. H., & Lin, Y. (2002). Evaluating the best main battle tank using fuzzy decision theory with linguistic criterion evaluation. *European Journal of Operational Research, 142*, 174–186. doi:10.1016/S0377-2217(01)00280-6

Cheng, V. S. Y., Hung, P. C. K., & Chiu, D. K. W. (2007). Enabling Web services policy negotiation with privacy preserved using XACML. Proceedings of the 40th Annual Hawaii International Conference on System Science (HICSS) (p. 33a).

Chengguo, L. L. J., & Wang, Z. (2006). Aggregation of fuzzy opinions under group-decision-making based on similarity and distance. *Journal of Systems Science and Complexity, 19*, 63–71. doi:10.1007/s11424-006-0063-y

Cheung, S. C., Hung, P. C. K., & Chiu, D. K. W. (2003). On the e-negotiation of unmatched logrolling views. Proceedings of the 36th Annual Hawaii International Conference on System Science (HICSS) (p. 29).

Chiclana, F., Herrera-Viedma, E., Herrera, F., & Alonso, S. (2007). Some induced ordered weighted averaging operators and their use for solving group decision-making problems based on fuzzy preference relations. *European Journal of Operational Research, 182*, 383–399. doi:10.1016/j.ejor.2006.08.032

Chiu, D. K. W., Cheung, S. C., Hung, P. C. K., Chiu, S. Y. Y., & Chung, A. K. K. (2005). Developing e-negotiation support with a meta-modeling approach in a Web services environment. *Decision Support Systems, 40*(1), 51–69. doi:10.1016/j.dss.2004.04.004

Cho, A. (2008). What is Web 3.0? The next generation Web: Search context for online information. Retrieved February 23, 2010, from http://internet.suite101.com/article.cfm/what_is_web_30

Choudhury, T., & Basu, S. (2004). Modeling conversational dynamics as a mixed memory Markov Process . In *Proceedings of Neural Information Processing Systems (NIPS)*. IEEE.

Christensen, E. W., & Fjermestad, J. (1997). Challenging group support systems research: The case for strategic decision making. *Group Decision and Negotiation, 6*, 351–372. doi:10.1023/A:1008664812184

Christy, W. (2002, January 24). Gaps in arrival process must be queried [Letter to the editor]. *The Australian*, p. 10.

Chu, T. C. (2002). Selecting plant location via a fuzzy TOPSIS approach. *International Journal of Advanced Manufacturing Technology, 20*, 859–864. doi:10.1007/s001700200227

Church, K. W., & Mercer, R. L. (1993). Introduction to the special issue on computational linguistics using large corpora. *Computational Linguistics, 19*, 1–24.

Chutimaskul, W., & Funilkul, S. (2004). The framework of e-democracy development. In R. Traunmuller (Ed.), *Electronic government* (Lecture Notes in Computer Science, vol. 3183, pp. 27–30). Berlin/Heidelberg: Springer.

Chuu, S. J. (2009a). Selecting the advanced manufacturing technology using fuzzy multiple attributes group decision making with multiple fuzzy information. *Computers & Industrial Engineering*, *57*(3), 1033–1042. doi:10.1016/j.cie.2009.04.011

Chuu, S. J. (2009b). Group decision-making model using fuzzy multiple attributes analysis for the evaluation of advanced manufacturing technology. *Fuzzy Sets and Systems*, *160*, 586–602. doi:10.1016/j.fss.2008.07.015

Clarkson, G. P., & Hodgkinson, G. P. (2005). Introducing Cognizer TM: A comprehensive computer package for the elicitation and analysis of cause maps. *Organizational Research Methods*, *8*(3), 317–341. doi:10.1177/1094428105278022

Clatworthy, J., Buick, D., Hankins, M., Weinman, J., & Horne, R. (2005). The use and reporting of cluster analysis in health psychology: A review. *British Journal of Health Psychology*, *10*, 329–358. doi:10.1348/135910705X25697

Cohen, J. (1986). An epistemic conception of democracy. *Ethics*, *97*(1), 26–38. doi:10.1086/292815

Cohen, B. (2002, January 29). Refugees evince hope and mental health [Letter to the editor]. *The Australian*, p. 12.

Coleman, R. (1997). The intellectual antecedents of public journalism. *The Journal of Communication Inquiry*, *21*(1), 60–76. doi:10.1177/019685999702100103

Coleman, S., & Gøtze, J. (2002). *Bowling together: Online public engagement in policy deliberation*. London: Hansard Society. Retrieved October 2, 2009, from http://www.bowlingtogether.net/

Collins, H. (2003). *Enterprise knowledge portals: Next generation portal solution for dynamic information access, better decision making and maximum results*. AMACOM.

Conen, W., & Sandholm, T. (2003). Differential-revelation VCG mechanisms for combinatorial auctions. In *Proceedings of the 4th ACM Conference on Electronic Commerce, San Diego, CA* (pp. 196-197). New York: ACM Press.

Conklin, J., & Begeman, M. L. (1988). gIBIS: A hypertext tool for exploratory policy discussion. *ACM Transactions on Office Information Systems*, *6*(4), 303–331. doi:10.1145/58566.59297

Converse, P. (1964). The nature of belief systems in mass publics . In Apter, D. (Ed.), *Ideology and discontent* (pp. 206–261). New York: The Free Press.

Converse, P. E. (1970). Attitudes and non-attitudes: Continuation of a dialogue . In Tufte, E. R. (Ed.), *The quantitative analysis of social problems*. Reading, MA: Addison-Wesley.

Cooper, M., Lambert, D. M., & Pagh, J. (1997). Supply chain management: More than a new name for logistics. *The International Journal of Logistics Management*, *8*(1), 1–13. doi:10.1108/09574099710805556

Cordella, L. P., Foggia, P., Sansone, C., Tortella, F., & Vento, M. (1999). Reliability Parameters to Improve Combination Strategies in Multi-expert Systems. *Pattern Analysis & Applications*, *2*, 205–214. doi:10.1007/s100440050029

cosmicjester. (2008, September 29). [Comment 16 to blog post: John Howard: media tart?]. *Larvatus Prodeo* [Blog], Federal election 2007. Retrieved October 1, 2009, from http://larvatusprodeo.net/2008/09/28/john-howard-media-tart/

Costa, S. E. G., & Lima, E. P. (2009). Advanced manufacturing technology adoption: An integrated approach. *Journal of Manufacturing Technology Management*, *20*(1), 74–96. doi:10.1108/17410380910925415

Courrieu, P. (2005). Function approximation on non-Euclidean spaces. *Neural Networks*, *18*, 91–102. doi:10.1016/j.neunet.2004.09.003

Crainic, T. G., & Laporte, G. (1997). Planning models for freight transportation. *European Journal of Operational Research*, *97*, 409–438. doi:10.1016/S0377-2217(96)00298-6

Crichlow, F. (2002, February 02-03). Settler families sold out and browned off [Letter to the editor]. *The Weekend Australian*, p. 16.

Crikey. (n.d.). About. Retrieved December 10, 2009, from http://www.crikey.com.au/

CRM114 - the Controllable Regex Mutilator. (2009). Retrieved September 2009, from http://crm114.sourceforge.net/

Cross, G. R., & de Bessonet, C. G. (1985). Representation of legal knowledge for conceptual retrieval. *Information Processing & Management*, *21*, 35–44. doi:10.1016/0306-4573(85)90126-8

Csikszentmihalyi, M. (1990). *Flow: The psychology of optimal experience*. New York: Harper Perennial.

Cummings, J. N., Butler, B., & Kraut, R. (2002). The quality of online social relationships. *Communications of the ACM*, *45*(7), 103–108. doi:10.1145/514236.514242

D'ıaz, M. S., Machuca, J. A. D., & Álvarez-Gil, M. J. (2003). A view of developing patterns of investment in AMT through empirical taxonomies: New evidence. *Journal of Operations Management*, *21*(5), 577–606. doi:10.1016/j.jom.2003.03.002

Dablanc, L. (2008). Urban goods movement and air quality policy and regulation issues in European cities. *Journal of Environmental Law*, *20*(2), 245–266. doi:10.1093/jel/eqn005

Daft, R. L., & Lengel, R. H. (1986). Organizational information requirements, media richness and structural design. *Management Science*, *32*(5), 554–571. doi:10.1287/mnsc.32.5.554

Daft, R. L., Lengel, R. H., & Trevino, L. K. (1987). Message equivocality, media selection, and manager performance: Implications for information systems. *Management Information Systems Quarterly*, *11*(3), 355–366. doi:10.2307/248682

Daft, R. L., & Lengel, R. H. (1984). Organizational information requirements, media richness and structural design. *Management Science*, *32*, 554–571. doi:10.1287/mnsc.32.5.554

Dagdeviren, M., & Yuksel, I. (2008). Developing a fuzzy analytic hierarchy process (AHP) model for behavior-based safety management. *Information Sciences*, *178*, 1717–1733. doi:10.1016/j.ins.2007.10.016

Dahlberg, L. (2005). The Habermasian public sphere: Taking difference seriously? *Theory and Society*, *34*, 111–136. doi:10.1007/s11186-005-0155-z

Dahlberg, L. (2001). Computer-mediated communication and the public sphere: A critical analysis. *Journal of Computer-Mediated Communication*, *7*(1). Retrieved November 30, 2009, from http://jcmc.indiana.edu/vol7/issue1/dahlberg.html

Damart, S. (2008). A cognitive mapping approach to organizing the participation of multiple actors in a problem structuring process. *Group Decision and Negotiation*, *19*(5), 505–526. doi:10.1007/s10726-008-9141-y

Damasio, A. (2005). *Descartes' error: Emotion, reason, and the human brain*. London: Penguin.

Danziger, D., & Rafal, R. (2009). The effect of visual signals on spatial decision making. *Cognition*, *110*, 182–197. doi:10.1016/j.cognition.2008.11.005

Das, A., & Jayaram, J. (2003). Relative importance of contingency variables for advanced manufacturing technology. *International Journal of Production Research*, *41*(18), 4429–4452. doi:10.1080/00207540310001595819

Datta, V., Sambasivarao, K. V., Kodali, R., & Deshmukh, S. G. (1992). Multi-attribute decision model using the analytic hierarchy process for justification of manufacturing systems. *International Journal of Production Economics*, *28*(2), 227–234. doi:10.1016/0925-5273(92)90035-6

Dawes, R. M. (1988). Not me or thee but we: The importance of group identity in eliciting cooperation in dilemma situations: Experimental manipulations. *Acta Psychologica*, *68*(1-3), 83–97. doi:10.1016/0001-6918(88)90047-9

Day, S. B., & Bartels, D. M. (2008). Representation over time: The effects of temporal distance on similarity. *Cognition*, *106*, 1504–1513. doi:10.1016/j.cognition.2007.05.013

de Hevia, M. D., & Spelke, E. S. (2009). Spontaneous mapping of number and space in adults and young children. *Cognition*, *110*, 198–207. doi:10.1016/j.cognition.2008.11.003

De Leon, E. D., & Guild, P. D. (2003). Using repertory grid to identify intangibles in business plans. *Venture Capital*, *5*(2), 135–160. doi:10.1080/1369106032000097030

De Sanctis, G. L., & Gallupe, R. B. (1987). A foundation for the study of group decision support systems. *Management Science*, *33*, 589–609. doi:10.1287/mnsc.33.5.589

Dekker, A. H. (2003). Centralisation and decentralisation in network centric warfare. *Journal of Battlefield Technology*, *6*(2), 23–28.

Delicious. (2009). *About*. Retrieved October 3, 2009, from http://delicious.com/about

Demmel, J. G., & Askin, R. G. (1992). A multiple-objective decision model for the evaluation of advanced manufacturing system technologies. *Journal of Manufacturing Systems*, *11*(3), 179–194. doi:10.1016/0278-6125(92)90004-Y

Denis, M., & Loomis, J. M. (2007). Perspectives on human spatial cognition: Memory, navigation, and environmental learning. *Psychological Research*, *71*, 235–239. doi:10.1007/s00426-006-0079-x

Denis, M., Michon, P. E., & Tom, A. (2006). Assisting pedestrian wayfinding in urban settings: Why references to landmarks are crucial in direction-giving . In Allen, G. L. (Ed.), *Applied spatial cognition: From research to cognitive technology* (pp. 25–51). Mahwah, NJ: Erlbaum.

Dennen, V. P. (2009b). Constructing academic alter-egos: Identity issues in a blog-based community. *Identity in the Information Society*, *2*(1), 23–38. doi:10.1007/s12394-009-0020-8

Dennen, V. P., & Pashnyak, T. G. (2008). Finding community in the comments: The role of reader and blogger responses in a weblog community of practice. *International Journal of Web-based Communities*, *4*(3), 272–283. doi:10.1504/IJWBC.2008.019189

Dennen, V. P. (2006). Blogademe: How a group of academics formed and normed an online community of practice . In Mendez-Vilas, A., Martin, A. S., Mesa Gonzalez, J. A., & Mesa Gonzalez, J. (Eds.), *Current developments in technology-assisted education* (*Vol. 1*, pp. 306–310). Badajoz, Spain: Formatex.

Dennen, V. P. (2009a). *Because I know you: Elements of trust among pseudonymous bloggers.* Paper presented at IADIS International Conference on Web-based Communities 2009, Algarve, Portugal.

Dennett, D. C. (1971). Intentional systems. *The Journal of Philosophy*, *68*(4), 87–106. doi:10.2307/2025382

Dennis, A. R. (1996). Information exchange and use in small group decision making: You can lead a group to information, but you can't make it think. *Small Group Research*, *27*, 532–550. doi:10.1177/1046496496274003

Dennis, A. R., Hilmer, K. M., Taylor, N. J., & Polito, A. (1996). Information exchange and use in GSS and verbal group decision making: Effects of minority influence. *Management Information Systems Quarterly*, *20*, 433–457. doi:10.2307/249563

Desanctis, G. R., & Gallupe, B. (1987, May). A foundation for the study of group decision support systems. *Management Science*, *33*(5), 589–609. doi:10.1287/mnsc.33.5.589

Dibbell, J. (1998). *My tiny life: Crime and passion in a virtual world.* New York: Henry Holt and Company.

Dickson, J. (2005). Interpretation and coherence in legal reasoning. *Stanford Encyclopedia of Philosophy.* Retrieved September 9, 2009, from http://plato.stanford.edu/entries/legal-reas-interpret/

Dietz-Uhler, B., & Bishop-Clark, C. (2001). The use of computer-mediated communication to enhance subsequent face-to-face discussions. *Computers in Human Behavior*, *17*, 269–283. doi:10.1016/S0747-5632(01)00006-1

DiMicco, J. M., Pandolfo, A., & Bender, W. (2004). Influencing group participation with a shared display. In Proceedings of *ACM Conference on Computer Supported Cooperative Work* (CSCW).

Dokas, I. M., Karras, D. A., & Panagiotakopoulos, D. C. (2009). Fault tree analysis and fuzzy expert systems: Early warning and emergency response of landfill operations. *Environmental Modelling & Software*, *24*(1), 8–25. doi:10.1016/j.envsoft.2008.04.011

Donga, Y., Xua, Y., & Yu, S. (2009). Linguistic multiperson decision-making based on the use of multiple preference relations. *Fuzzy Sets and Systems*, *160*, 603–623. doi:10.1016/j.fss.2008.08.011

Down and Out of Sài Gòn. (2009, July 01). [Comment 17 to blog post: Rudd and Gillard attack News Limited, Hartigan punches back]. *Larvatus Prodeo* [Blog], Federal election 2007. Retrieved October 1, 2009, from http://larvatusprodeo.net/2009/07/01/rudd-and-gillard-attack-news-limited-hartigan-punches-back/

Dryzek, J. S., & List, C. (2003). Social choice theory and deliberative democracy: A reconciliation. *British Journal of Political Science*, *33*(1), 1–28. doi:10.1017/S0007123403000012

Dryzek, J. (1990). *Discursive democracy: Politics, policy and political science*. Cambridge, UK: Cambridge University Press.

Dryzek, J. (2002). *Deliberative democracy and beyond: Liberals, critics, contestations*. Oxford, UK: Oxford University Press. doi:10.1093/019925043X.001.0001

Dryzek, J., & Niemeyer, S. (2006). Reconciling pluralism and consensus as political ideals. *American Journal of Political Science*, *50*(3), 634–649. doi:10.1111/j.1540-5907.2006.00206.x

Dube, L., Bourhis, A., & Jacob, R. (2006). Towards a typology of virtual communities of practice. *Interdisciplinary Journal of Information, Knowledge, and Management*, *1*.

Dubrovsky, V. J., Kiesler, S., & Sethna, B. N. (1991). The equalization phenomenon: Status effects in computer-mediated and face-to-face decision-making groups. *Human-Computer Interaction*, *6*, 119–146. doi:10.1207/s15327051hci0602_2

Dullaert, W., Bräysy, O., Goetschalckx, M., & Raa, B. (2007). Supply chain (re)design: Support for managerial and policy decisions. *European Journal of Transport and Infrastructure Research*, *7*(2), 73–92.

Duran, O., & Aguilo, J. (2008). Computer-aided machine- tool selection based on a Fuzzy-AHP approach. *Expert Systems with Applications*, *34*(3), 1787–1794. doi:10.1016/j.eswa.2007.01.046

Dutoit, A., McCall, R., Mistrík, I., & Paech, B. (2006). Rationale management in software engineering: Concepts and techniques . In Dutoit, A., McCall, R., Mistrík, I., & Paech, B. (Eds.), *Rationale management in software engineering* (pp. 1–48). Berlin: Springer. doi:10.1007/978-3-540-30998-7_1

Dzur, A. W. (2002). Public journalism and deliberative democracy. *Polity*, *34*(3), 313–336.

Edelman, M. J. (1985). *The symbolic uses of politics*. Urbana, IL: University of Illinois Press.

E-democracy.org. (n.d). About. Retrieved December 10, 2009, from http://forums.e-democracy.org/

Efstathiades, A., Tassou, S., & Antoniou, A. (2002). Strategic planning, transfer and implementation of advanced manufacturing technologies (AMT): Development of an integrated process plan. *Technovation*, *22*, 201–212. doi:10.1016/S0166-4972(01)00024-4

Egenhofer, M. J., & Golledge, R. G. (1998). *Spatial and temporal reasoning in geographic information systems*. Oxford: Oxford University Press.

Eidelman, J. A. (1993). Software for negotiations. *Law Practice Management*, *19*(7), 50–55.

Eisenhardt, K. (1989). Building theories from case study research. *Academy of Management Review*, *14*(4), 532–550. doi:10.2307/258557

Elen, J. (2001, November 30). Agenda set by losers? [Letter to the editor]. *The Australian*, p. 10.

Ellonen, H. K., Kosonen, M., & Henttonen, K. (2007). The development of a sense of virtual community . *International Journal of Web Based Communities*, *3*(1), 114–130. doi:10.1504/IJWBC.2007.013778

Elster, J. (1983). *Sour grapes: Studies in the subversion of rationality*. Cambridge: Cambridge University Press.

Elster, J. (1986). The market and the forum . In Elster, J., & Hylland, A. (Eds.), *Foundations of social choice theory* (pp. 103–122). Cambridge: Cambridge University Press.

Emerson, P. (2007). *Designing an all-inclusive democracy: Consensual voting procedures for use in parliaments, councils and committees*. Berlin: Springer.

Emery, M., & Purser, R. E. (1996). *The search conference: A powerful method for planning organisational change and community action*. San Francisco, CA: Jossey-Bass Public Administration Series.

Emery, M., & Purser, R. (1996). *The search conference*. San Francisco, CA: Joseey Bass.

English Wikipedia. (2010). Dialogue systems. Retrieved March 3, 2010, from http://en.wikipedia.org/wiki/Dialogue_system

Esmahi, L. (2008). Contract negotiation in E-marketplaces: A model based on dependency relations. *Journal of Electronic Commerce in Organizations*, *6*(3), 74–91.

Estlund, D. (1997). Beyond fairness of deliberation: The epistemic dimension of democratic authority . In Bohman, J., & Rehg, W. (Eds.), *Deliberative democracy: Essays on Reason and Politics*. Cambridge, MA: MIT Press.

Evangelou, C., & Karacapilidis, N. (2006). A multidisciplinary approach for supporting knowledge-based decision making in collaborative settings. *International Journal of Artificial Intelligence Tools*, *16*(6), 1069–1092. doi:10.1142/S0218213007003631

Fabbe-Costes, N. (2007). Systèmes d'information logistique et transport. *Techniques de l'Ingénieur, AGL2*.

Facebook. (2009). *Facebook factsheet*. Retrieved October 3, 2009, from http://www.facebook.com/cmiller61?ref=name#/press/info.php?factsheet

Farran, B. & Curnow, A. (2002, January 29). Refugees evince hope and mental health [Letter to the editor]. *The Australian*, p. 12.

Fawzi, M. (2006). *Wikipedia 3.0: The end of Google?* Retrieved February 23, from http://evolvingtrends.wordpress.com/2006/06/26/wikipedia-30-the-end-of-google/

Feixas, G., Erazo-Caicedo, M. I., Harter, S. L., & Bach, L. (2008). Construction of self and others in unipolar depressive disorders: A study using repertory grid technique. *Cognitive Therapy and Research*, *32*, 386–400. doi:10.1007/s10608-007-9149-7

Feixas, G. (1991, July). The repertory grid in psychotherapy research. Poster presented at XXII Annual Conference of the Society for Psychotherapy Research, Lyon, France.

Feixas, G., & Cornejo-Alvarez, J. M. (1996). *Manual de la técnica de rejilla mediante el programma RECORD v. 2.0*. Barcelona, Spain: Ediciones Paidos.

Feldman, A. M. (1980). *Welfare economics and social choice theory*. Boston: Kluwer.

Femia, J. (1996). Complexity and deliberative democracy. *Inquiry*, *39*(3-4), 359–397. doi:10.1080/00201749608602427

Ferber, J. (1999). *Multi-agent systems: An introduction to distributed artificial intelligence*. London: Addison-Wesley.

FIPA. (2002). FIPA ACL Message Structure Specification. Retrieved on March 22, 2006 from http://www.fipa.org

FIPA. (2002). FIPA contract Net interaction protocol. Retrieved June 14, 2006, from http://www.fipa.org/assets/XC00029G.pdf

Firestone, J. M. (2003). *Enterprise information portals and knowledge management*. Elsevier Butterworth-Heinemann.

Fischer, J. (2000). *Citizens, experts, and the environment: The politics of local knowledge*. Durham, NC: Duke University Press.

Fishburn, P. C. (1970). *Utility theory for decision-making*. New York: Operations Research Society of America.

Fisher, B. A., & Ellis, D. G. (1990). *Small group decision making: Communication and the group process*. New York: McGraw-Hill.

Fisher, R., & Ury, W. (1991). *Getting to yes: Negotiating agreement without giving in*. Boston: Houghton Mifflin.

Fishkin, J. S. (1991). *Democracy and deliberation: New directions for democratic reform*. New Haven, CT: Yale University Press.

Flanagin, A. J., Park, H. S., & Seibold, D. R. (2004). Group performance and collaborative technology: A longitudinal and multilevel analysis of information quality, contribution equity, and members' satisfaction in computer-mediated groups. *Communication Monographs*, *71*(3), 352–372. doi:10.1080/0363452042000299902

Flew, T. (2008). *New media: An introduction* (3rd ed.). Melbourne, Australia: Oxford University Press.

Forester, J. (1999). *The deliberative practitioner: Encouraging participatory planning processes*. Cambridge, MA: MIT Press.

Fox, S. (2008). The engaged e-patient population. *Pew Internet & American Life Project*. Retrieved January 20, 2009, from http://www.pewinternet.org/Reports/2008/The-Engaged-Epatient-Population.aspx

Francis Xavier Holden. (2008a, July 29). [Comment 1 to blog post: Slow progress in health funding reform]. *Larvatus Prodeo* [Blog], Federal election 2007. Retrieved October 1, 2009, from http://larvatusprodeo.net/2008/07/29/slow-progress-in-health-funding-reform/

Fransella, F. (1972). *Personal change and reconstruction*. London: Academic Press.

Fransella, F., & Adams, B. (1965). An illustration of the use of repertory grid technique in a clinical setting. *The British Journal of Social and Clinical Psychology*, *5*, 51–62.

Fransella, F., & Bannister, D. (1977). *A manual for repertory grid technique*. London: Academic Press.

Fraser, N. (1992). Rethinking the public sphere: A contribution to the critique of actually existing democracy . In Calhoun, C. (Ed.), *Habermas and the public sphere* (pp. 109–142). Cambridge, MA: MIT Press.

Freeman, J. B. (1991). *Dialectics and the macrostructure of arguments*. Berlin: Floris Publications.

French, S. (2007). Web-enabled strategic GDSS, e-Democracy and Arrow's theorem: A Bayesian perspective. *Decision Support Systems*, *43*(4), 1476–1484. doi:10.1016/j.dss.2006.06.003

Friedmann, J. (1987). *Planning in the public domain: From knowledge to action*. Princeton, NJ: Princeton University Press.

Fukuyama, F. (1995). *Trust*. New York: Free Press.

Fulk, J., Schmitz, J., & Steinfield, C. W. (1990). A social influence model of technology use . In Fulk, J., & Steinfield, C. W. (Eds.), *Organizations and communication technology* (pp. 117–140). Newbury Park, CA: Sage Publications.

Fulk, J., & Collins-Jarvis, L. (2001). Wired meeting: Technological mediation of organizational gatherings . In Jablin, F. M., & Putnam, L. L. (Eds.), *The new handbook of organizational communication: Advances in theory, research, and methods* (pp. 624–663). Thousand Oaks, CA: Sage Publications.

Gambetta, D. (1988). *Trust: Making and Breaking Cooperative Relations*. Cambridge, UK: Blackwell.

Garling, T., & Evans, G. W. (Eds.). (1991). *Environment, cognition, and action: An integrated approach*. New York: Oxford University Press.

Garnham, N. (2007). Habermas and the public sphere. *Global Media and Communication*, *3*(2), 201–214. doi:10.1177/1742766507078417

GCI & Capgemini. (2008). *Future supply chain 2016: Serving consumers in a sustainable way*. Germany: Global Commerce Initiative.

Georgiev, A., & Allen, P. K. (2004). Localization methods for a mobile robot in urban environments. [TRO]. *IEEE Transactions on Robotics and Automation*, *20*, 851–864.

Gero, J. S., & Tversky, B. (Eds.). (1999). *Visual and spatial reasoning in design*. Sydney, Australia: University of Sydney, Key Centre of Design Computing and Cognition.

Gibson, R., Römelle, A., & Ward, S. (Eds.). (2004). *Electronic Democracy: Mobilisation, participation and organization via new ICTs*. London: Routledge.

Gigone, D., & Hastie, R. (1993). The common knowledge effect: Information sharing and group judgment. *Journal of Personality and Social Psychology*, *65*, 959–974. doi:10.1037/0022-3514.65.5.959

Gillner, S., Weiss, A. M., & Mallot, H. A. (2008). Visual homing in the absence of feature-based landmark information. *Cognition*, *109*, 89–104. doi:10.1016/j.cognition.2008.07.018

Gliffy, Inc. (2009). *About Gliffy, Inc*. Retrieved October 2, 2009, from http://www.gliffy.com/about.shtml

Golden, B. L., Raghavan, S., & Wasil, E. A. (2008). *Vehicle routing: Latest advances and challenges*. Boston: Kluwer. doi:10.1007/978-0-387-77778-8

Golledge, R. G. (Ed.). (1998). *Wayfinding: Cognitive mapping and other spatial processes*. Baltimore, MD: Johns Hopkins.

Gonçalves, T., & Quaresma, P. (2005). Is linguistic information relevant for the classification of legal texts? In *Proceedings of the 10th ICAIL* (pp. 168-176). New York: ACM Press.

Goodin, R. E. (2000). Democratic deliberation within. *Philosophy & Public Affairs*, *29*(1), 81–109. doi:10.1111/j.1088-4963.2000.00081.x

Goodman, N. (1951). *The structure of appearance*. Cambridge, MA: Harvard.

Google. (2009a). *Google docs help: Getting started guide*. Retrieved October 2, 2009, from http://docs.google.com/support/bin

Google. (2009b). *Google apps for business*. Retrieved October 1, 2009, from http://www.google.com.au/apps/intl/en/business/index.html

Gordon, T. F., Prakken, H., & Walton, D. (2007). The Carneades model of argument and burden of proof. *Artificial Intelligence*, *171*, 875–896. doi:10.1016/j.artint.2007.04.010

Gordon, T. F. (1995). *The pleadings game: An artificial intelligence model of procedural justice*. Dordrecht, The Netherlands: Kluwer Academic Publishers.

Gordon, T. F., & Walton, D. (2009). Proof burdens and standards . In Rahwan, I., & Simari, G. (Eds.), *Argumentation and Artificial Intelligence* (pp. 239–260). Berlin: Springer. doi:10.1007/978-0-387-98197-0_12

Gordon, T. F., & Karacapilidis, N. I. (1997). The Zeno argumentation framework. In J. Zeleznikow, D. Hunter & K. L. Branting, (Eds.), *Proceedings of the 6th International Conference on AI and Law* (pp. 10–18). New York: ACM Press.

Gordon, T., & Walton, D. (2009). Legal reasoning with argumentation schemes. In *Proceedings of the 12th ICAIL International Conference on Artificial Intelligence and Law* (pp. 137-146). New York: ACM Press.

Gouran, D. S., & Hirokawa, R. Y. (1996). Functional theory and communication in decision-making and problem-solving groups: An expanded view . In Hirokawa, R. Y., & Poole, M. S. (Eds.), *Communication and group decision making* (2nd ed., pp. 55–80). Thousand Oaks, CA: SAGE Publications.

Gouran, D. S., & Hirokawa, R. Y. (2003). Effective decision making and problem solving in groups: A functional perspective . In Hirokawa, R. Y., Cathcart, R. S., Samovar, L. A., & Henman, L. D. (Eds.), *Small group communication* (8th ed., pp. 27–38). Los Angeles, CA: Roxbury.

Gower, J. (1975). Generalised Procrustes analysis. *Psychometrika*, *51*, 33–51. doi:10.1007/BF02291478

Grabisch, M., Kojadinovic, I., & Meyer, P. (2007). A review of methods for capacity identification in Choquet integral based multi-attribute utility theory applications of the Kappalab R package. *European Journal of Operational Research, 186*(2), 766–785. doi:10.1016/j.ejor.2007.02.025

Graetz, K. A., Boyle, E. S., Kimble, C. E., Thompson, P., & Garloch, J. L. (1998). Information sharing in face-to-face, teleconferencing, and electronic chat groups. *Small Group Research, 29*, 714–743. doi:10.1177/1046496498296003

Gray, P. (1987). Group decision support systems. *Decision Support Systems, 3*(3), 233–242. doi:10.1016/0167-9236(87)90178-3

Gray, P., Gray, X., & Zeleznikow, J. (2007). Decision negotiating logic: For richer or poorer. In R. Winkels (Ed.), *Proceedings of Eleventh International Conference on Artificial Intelligence and Law, Palo Alto, CA* (pp. 247-251). New York: ACM Press.

Gray, R. (2001, September 25). Standing force a UN imperative [Letter to the editor]. *The Australian*, p. 12.

Greenfield, G., & Campbell, J. (2006). Communicative practices in online communication: A case of agreeing to disagree. *Journal of Organizational Computing and Electronic Commerce, 16*(3-4), 267–277. doi:10.1207/s15327744joce1603&4_6

Grewendorf, G., & Rathert, M. (Eds.). (2009). *Formal linguistics and law*. Berlin: Mouton de Gruyter.

Griffen-Foley, B. (2009b). *Changing stations: The story of Australian commercial radio*. Sydney: UNSW Press.

Griffen-Foley, B. (2009a, December 2). Dial-in democracy? Forty years of talkback, from Andrea to Kyle [Audio podcast]. Sydney, Australia: The Sydney Institute. Retrieved December 10, 2009, from http://www.thesydneyinstitutepodcast.com/2009/12/02/BridgetGriffenFoleyDialinDemocracyFortyYearsOfTalkbackFromAndreaToKyle.aspx

Groessler, R. (2002, January 29). Refugees evince hope and mental health [Letter to the editor]. *The Australian*, p. 12.

Gross, T. (2000). Technological support for e-democracy: history and perspectives. In *Proceedings of 11th International Workshop on Database and Expert Systems Applications, Sept 4 – Sept 8, London, UK,* (pp. 391–395).

Gruber, T. R. (2003). A translation approach to portable ontology specifications. *Knowledge Acquisition, 5*(2), 199–220. doi:10.1006/knac.1993.1008

Guetzkow, H. (1950). Unitizing and categorizing problems in coding qualitative data. *Journal of Clinical Psychology, 6*(1), 47–58. doi:10.1002/1097-4679(195001)6:1<47::AID-JCLP2270060111>3.0.CO;2-I

Gupta, A., & Whitehouse, F. R. (2001). Firms using advanced manufacturing technology management: An empirical analysis based on size. *Integrated Manufacturing Systems, 12*(5), 346–350. doi:10.1108/09576060110398500

Gurulib. (2009). *Gurulib: About us*. Retrieved October 2, 2009, from http://www.gurulib.com/_about_us.html

Gutmann, A., & Thompson, D. (1996). *Democracy and disagreement*. Cambridge, MA: The Belknap Press.

Haas, T. (2004). The public sphere as a sphere of publics: Rethinking Habermas's theory of the public sphere. *The Journal of Communication, 54*(1), 178–184. doi:10.1111/j.1460-2466.2004.tb02621.x

Habermas, J. (1984). The theory of communicative action: Reason and the rationalization of society.: *Vol. 1. T. McCarthy, Trans.* Boston, MA: Beacon Press.

Habermas, J. (1989). *The structural transformation of the public sphere* (Burger, T., Trans.). Cambridge, UK: Polity Press.

Habermas, J. (1998). *Between facts and norms: Contributions to a discourse theory of law and democracy* (Rehg, W., Trans.). Cambridge, MA: MIT Press.

Habermas, J. (1987). *Knowledge and human interests* (Shapiro, J. J., Trans.). Cambridge, UK: Polity Press.

Habermas, J. (2006). *Time of transitions*. (C. Cronin, Trans., & M. Pensky, Ed.). Cambridge, UK: Polity Press.

Hackman, R., & Helmreich, R. L. (1987). Assessing the behavior and performance of teams in organizations: The case of air transport crews. In Peterson, D. R., & Fishman, D. B. (Eds.), *Assessment for decision*. New Brunswick, NJ: Rutgers University Press.

Haft, F. (2009). *Juristische Schreibschule: Anleitung zum strukturierten Schreiben*. Munich, Germany: Normfall.

Haft, F., Werner, M., & Gerbliner, M. (1992). *Untersuchung der Möglichkeiten des Einsatzes von Expertensystemen zur Unterstützung von Richtern*. Staatsanwälten und Rechtspflegern. [Unpublished research report, University of Tübingen]

Haft, F., Jones, R. P., & Wetter, T. (1987). A natural language based legal expert system for consultation and tutoring: The LEX project. In *Proceedings of the 1st International Conference on Artificial Intelligence & Law* (pp. 75-83). New York: ACM Press.

Hair, N., Rose, S., & Clark, M. (2009, March). Using qualitative repertory grid techniques to explore perceptions of business-to-business online customer experience. *Journal of Customer Behaviour*, *8*(1), 51–65. doi:10.1362/147539209X414380

Hájek, P. (1998). Mathematics of fuzzy logic. In *Trends in Logic* (Studia Logica library, vol. 4). Dordrecht: Kluwer Academic Publishers.

Haley (2009). Oracle and Haley. Retrieved September 17, 2009, from http://www.oracle.com/haley/index.html

Hall, M. J., Calabro, D., Sourdin, T., Stranieri, A., & Zeleznikow, J. (2005). Supporting discretionary decision-making with information technology: A case study in the criminal sentencing jurisdiction. *University of Ottawa Law & Technology Journal*, *2*(1), 1–36.

Hall, M. J. J., & Zeleznikow, J. (2003). A method for evaluating legal knowledge-based systems based upon the Context Criteria Contingency guidelines framework. In *Proceedings of the Ninth International Conference on Artificial Intelligence and Law, Edinburgh, Scotland* (pp. 274-283). New York: ACM Press.

Hall, R., & Partyka, J. (2008). On the road to mobility. *OR/MS Today*, *35*(1).

Hallandal, G. (2002, January 26-27). Caring, tolerant Aussie is a thing of the past [Letter to the editor]. *The Australian*, p. 18.

Hancock, J. T., & Dunham, P. J. (2001). Impression formation in computer-mediated communication revisited: An analysis of the breadth and intensity of impressions. *Communication Research*, *28*(3), 325–347. doi:10.1177/009365001028003004

Hanisch, J., & Churchman, D. (2008). Virtual communities of practice: The communication of knowledge across cultural boundaries. *International Journal of Web-based Communities*, *4*(4), 418–433. doi:10.1504/IJWBC.2008.019548

hannah's dad. (2008, June 23). [Comment 3 to blog post: Issues and the 2007 election]. *Larvatus Prodeo* [Blog], Federal election 2007... Retrieved October 1, 2009, from http://larvatusprodeo.net/2008/06/23/issues-and-the-2007-election/

Harker, T. H., & Ungar, L. H. (1996, May 8-10). A market-based approach to workflow automation. Paper presented at NSF Workshop in Information Systems, Athens, GA. Retrieved October 1, 2006, from http://www.cis.upenn.edu/~ungar/NSF.html

Hart, H. L. A. (1994). *The concept of law* (2nd ed.). Oxford: Clarendon Press.

Hart, R. (2001). Citizen discourse and political participation: A survey. In Bennett, L., & Entman, R. (Eds.), *Mediated politics and the future of democracy* (pp. 407–432). New York: Cambridge University Press.

Hartley, T., Trinkler, I., & Burgess, N. (2004). Geometric determinants of human spatial memory. *Cognition*, *94*, 39–75.

Hayes, A. F., & Krippendorff, K. (2007). Answering the call for a standard reliability measure for coding data. *Communication Methods and Measures*, *1*(1), 77–89.

Herrera, F., Herrera-Viedma, E., & Chiclana, F. (2001). Multi-person decision-making based on multiplicative preference Relations. *European Journal of Operational Research*, *129*, 372–385. doi:10.1016/S0377-2217(99)00197-6

Herrera, F., Herrera-Viedma, E., & Verdegay, J. L. (1996). A model of consensus in group decision-making under linguistic assessments. *Fuzzy Sets and Systems*, *78*, 73–87. doi:10.1016/0165-0114(95)00107-7

Herrera-Viedma, E., Chiclana, F., Herrera, F., & Alonso, S. (2007). Group decision-making model with incomplete fuzzy preference relations based on additive consistency. *IEEE Transactions on Systems, Man, and Cybernetics. Part B, Cybernetics*, *37*(1), 176–189. doi:10.1109/TSMCB.2006.875872

Herrestad, H. (1996). *Formal theories of rights*. Oslo: Juristforbundets Forlag.

Herring, S. C., Kouper, I., Paolillo, J. C., Scheidt, L. A., Tyworth, M., Welsch, P., et al. (2005). Conversations in the blogosphere: An analysis " from the bottom up". *Proceedings of the 38th Hawaii International Conference on System Sciences*. Los Alamitos: IEEE Press.

Hessing, M. (2003). Green mail: The social construction of environmental issues through letters to the editor. *Canadian Journal of Communication*, *28*, 25–42.

Hinds, P. J., & Bailey, D. E. (2001). Out of sight, out of sync: Understanding conflict in distributed teams. *Organization Science*, *14*(6), 615–632. doi:10.1287/orsc.14.6.615.24872

Hinkle, S., Taylor, L. A., Fox-Cardamone, D. L., & Crook, K. F. (1998). Intragroup identification and intergroup differentiation: A multicomponent approach. *The British Journal of Social Psychology*, *28*, 305–317.

Hirokawa, R. Y. (1990). The role of communication in group decision-making efficacy. *Small Group Research*, *21*(2), 190–204. doi:10.1177/1046496490212003

Hirokawa, R. Y. (1992). Communication and group decision-making efficacy . In Cathcart, R. S., & Samovar, L. A. (Eds.), *Small group communication* (6th ed., pp. 165–177). Dubuque, IA: William C. Brown.

Hirokawa, R. Y., & Salazar, A. J. (1999). Task-group communication and decision-making performance . In Frey, L. R., Gouran, D. S., & Poole, M. S. (Eds.), *The handbook of group communication theory and research* (pp. 167–191). Thousand Oaks, CA: Sage Publications, Inc.

Hirokawa, R. Y. (1987). Group communication research: Considerations for the use of interaction analysis . In Tardy, D. H. (Ed.), *A handbook for the study of human communication* (pp. 229–245). Norwood, NJ: Ablex.

Hirtle, S. C. (2003). Neighborhoods and landmarks . In Duckham, M., Goodchild, M. F., & Worboys, M. F. (Eds.), *Foundations of geographic information science* (pp. 191–230). London: Taylor & Francis. doi:10.1201/9780203009543.ch10

Hockerts, K., O'Rourke, A., & Zingales, F. (2002). Balanced scorecard and sustainability: State of art review. *CMER* [working paper series], *2002/65*. Fontainebleau, France: INSEAD.

Hodgkinson, G. P. (2004). Causal cognitive mapping in the organizational strategy field: A comparison of alternative elicitation procedures. *Organizational Research Methods*, *7*(1), 3–26. doi:10.1177/1094428103259556

Hodgkinson, G. P., Bown, N. J., Maule, A. J., Glaister, K. W., & Pearman, A. D. (1999, October). Breaking the frame: An analysis of strategic cognition and decision making under uncertainty. *Strategic Management Journal*, *20*(10), 977–985. doi:10.1002/(SICI)1097-0266(199910)20:10<977::AID-SMJ58>3.0.CO;2-X

Hodgkinson, G. P., & Maule, A. J. (2002). The individual in the strategy process: Insights from behavioural decision research and cognitive mapping . In Huff, A. S., & Jenkins, M. (Eds.), *Mapping strategic knowledge* (pp. 196–219). London: Sage.

Hoekstra, R., Breuker, J., De Bello, M., & Boer, A. (2007). The LKIF core ontology of basic legal concepts. In P. Casanovas, M. A. Biasiotti, E. Francesconi, M. T. Sagri (Eds.), *Proceedings of LOAIT 07, II: Workshop on Legal Ontologies and Artificial Intelligence Techniques* (pp. 43-64). Retrieved from http://www.ittig.cnr.it/loait/LOAIT07-Proceedings.pdf

Hollingshead, A. B., Mcgrath, J. E., & O'Connor, K. (1993). Group task performance and communication technology: A longitudinal study of computer-mediated versus face-to-face work groups. *Small Group Research*, *24*, 307–333. doi:10.1177/1046496493243003

Hopkins, J. (2009). *Creative ecologies: Where thinking is a proper job*. UQP.

Hoy, N. (2001, October 18). Public debate: voters want more [Letter to the editor]. *The Australian*, p. 12.

Hsu, H. M., & Chen, C. T. (1996). Aggregation of fuzzy opinions under group decision-making. *Fuzzy Sets and Systems*, *79*, 279–285. doi:10.1016/0165-0114(95)00185-9

Hsu, Y. L., Lee, C. H., & Kreng, V. B. (2010). The application of Fuzzy Delphi Method and Fuzzy AHP in lubricant regenerative technology selection. *Expert Systems with Applications*, *37*(1), 419–425. doi:10.1016/j.eswa.2009.05.068

Huanga, C. C., Chub, P. Y., & Chiang, Y. H. (2008). A fuzzy AHP application in government-sponsored R&D project selection. *Omega*, *36*, 1038–1052. doi:10.1016/j.omega.2006.05.003

Hudson, R. (1974). Images of the retailing environment: An example for the use of the repertory grid methodology. *Environment and Behavior*, *12*, 470–494. doi:10.1177/001391657400600403

Huff, A. S., Narapareddy, N. V., & Fletcher, K. E. (1990). Coding the causal association of concepts . In Huff, A. S. (Ed.), *Mapping strategic thought*. Chichester: Wiley.

Insua, D., Holgado, J., & Moreno, R. (2003). Multicriteria e-negotiation systems for e-democracy. *Journal of Multi-Criteria Decision Analysis*, *12*(2-3), 213–218. doi:10.1002/mcda.358

Insua, D., Fernandez, E., Rios, J., & Rivero, A. (2005). Participatory budget formation through the Web. In Michael H. Böhlen (Ed.), *E-government: Towards electronic democracy* [Proceedings of TCGOV conference, March 2-4, Bolzano, Italy]. Berlin/Heidelberg: Springer.

Ishii, K., & Sugeno, M. (1985). A model of human evaluation process using fuzzy integral. *International Journal of Man-Machine Studies*, *22*(1), 19–38. doi:10.1016/S0020-7373(85)80075-4

Ishii, H., & Ullmer, B. (1997). Tangible bits: Toward seemless interfaces between people, bits, and atoms. In *Proceedings of Conference of Human Factors in Computing*, 234-241.

Ishikawa, A., Amagasa, M., Shiga, T., Tomizawa, G., Tatsuta, R., & Mieno, H. (1993). The max–min Delphi method and fuzzy Delphi method via fuzzy integration . *Fuzzy Sets and Systems*, *55*, 241–253. doi:10.1016/0165-0114(93)90251-C

Jack. (2009a, July 01). [Comment 10 to blog post: Rudd and Gillard attack News Limited, Hartigan punches back]. *Larvatus Prodeo* [Blog], Federal election 2007. Retrieved October 1, 2009, from http://larvatusprodeo.net/2009/07/01/rudd-and-gillard-attack-news-limited-hartigan-punches-back/

Jacka, E. (2000). Public service TV: An endangered species? In Turner, G., & Cunningham, S. (Eds.), *The Australian TV book* (pp. 52–68). St. Leonards, Australia: Allen & Unwin.

Jaeger, P. (2004). The social impact of an accessible e-democracy: Disability rights laws in the development of the federal e-government. *Journal of Disability Policy Studies*, *15*(1), 19–26. doi:10.1177/10442073040150010401

Jahanshahloo, G. R., Lotfi, F. H., & Izadikhah, M. (2006). Extension of the TOPSIS method for decision-making problems with fuzzy data. *Applied Mathematics and Computation*, *181*, 1544–1551. doi:10.1016/j.amc.2006.02.057

Jain, S. P., & Maheswaran, D. (2000, March). Motivated reasoning: A depth-of-processing perspective. *The Journal of Consumer Research*, *26*, 358–371. doi:10.1086/209568

James, D. (2001, November 21). Careful vote, not racist. [Letter to the editor]. *The Australian*, p. 12.

Janis, I. (1972). *Victims of groupthink: A psychological study of foreign-policy decisions and fiascos* (2nd ed.). Boston: Houghton Mifflin.

Jaschik, S. (2005, January 5). Historians reject proposed boycott. *Inside Higher Ed*. Retrieved from http://www.insidehighered.com/news/2009/01/05/boycott

Jennings, N. R., Faratin, P., Lomuscio, A. R., Parsons, S., Wooldridge, M. J., & Sierra, C. (2001). Automated negotiations: Prospects, methods and challenges. *Group Decision and Negotiation*, *10*(2), 199–215. doi:10.1023/A:1008746126376

Jevalins, C. (2009). What are Web 2.0 and Web 3.0 technologies? Retrieved February 23, 2010, from http://ezinearticles.com/?What-Are-Web-2.0-and-Web-3.0-Technologies?&id=2580537

Jinmaro. (2008, October 02). [Comment 4 to blog post: The state of Rudd nation]. *Larvatus Prodeo* [Blog], Federal election 2007. Retrieved October 1, 2009, from http://larvatusprodeo.net/2008/10/02/the-state-of-rudd-nation/

Johannessen, J. A., Olaisen, J., & Olsen, B. (2001). Mismanagement of tacit knowledge: The importance of tacit knowledge, the danger of information technology, and what to do about it. *International Journal of Information Management*, *21*(1). doi:10.1016/S0268-4012(00)00047-5

Johnson, T., Kaye, B., Bichard, S., & Wong, W. (2008). Every blog has its day: Politically-interested Internet users' perceptions of blog credibility. *Journal of Computer-Mediated Communication*, *13*, 100–122. doi:10.1111/j.1083-6101.2007.00388.x

Josefsson, U. (2005). Coping with illness online: The case of patients' online communities. *The Information Society*, *21*, 141–153. doi:10.1080/01972240590925357

Jungk, R., & Mullert, N. (1996). *Future workshop: How to create desirable futures*. London: Institute for Social Inventions.

Kacprzyk, J., & Fedrizzi, M. (1988). A soft measure of consensus in the setting of partial (fuzzy) preferences. *European Journal of Operational Research*, *34*, 315–325. doi:10.1016/0377-2217(88)90152-X

Kacprzyk, J., Fedrizzi, M., & Nurmi, H. (1992). Group decision-making and consensus under fuzzy preferences and fuzzy majority. *Fuzzy Sets and Systems*, *49*, 21–31. doi:10.1016/0165-0114(92)90107-F

Kalai, E., & Stanford, W. (1988). Finite rationality and interpersonal complexity in repeated games. *Econometrica*, *56*(2), 397–410. doi:10.2307/1911078

Kampen, J. K., & Snijkers, K. (2003). E-democracy. *Social Science Computer Review*, *21*(4), 491–496. doi:10.1177/0894439303256095

Karlsson, L. (2007). Desperately seeking sameness. *Feminist Media Studies*, *7*(2), 137–153. doi:10.1080/14680770701287019

Karsak, E., & Tolga, E. (2001). Fuzzy multi-criteria decision-making procedure for evaluating advanced manufacturing system investments. *International Journal of Production Economics*, *69*, 49–64. doi:10.1016/S0925-5273(00)00081-5

Karsak, E. E. (2005). Choquet integral-based decision making approach for robot selection. *Knowledge-Based Intelligent Information and Engineering Systems*, 635-641.

Katsh, E., & Rifkin, J. (2001). *Online dispute resolution: Resolving conflicts in cyberspace*. San Francisco, CA: Jossey-Bass.

Kauffman, A., & Gupta, M. M. (1985). *Introduction to fuzzy arithmetic: Theory and applications*. New York: Van Nostrand Reinhold.

Kaya, İ., & Çınar, D. (2008). Facility location selection using a fuzzy outranking method. *Journal of Multiple-Valued Logic and Soft Computing*, *14*, 251–263.

Kelarev, A. V., Yearwood, J. L., & Watters, P. (2009). Rees matrix constructions for clustering of data. *Journal of the Australian Mathematics Society*, *87*, 377–393. doi:10.1017/S1446788709000299

Kelarev, A. V., Yearwood, J. L., & Watters, P. (forthcoming). Internet security applications of Gröbner-Shirshov bases. *Asian-European Journal of Mathemathics*.

Kelarev, A. V., Yearwood, J. L., Watters, P., Ma, L., Abawajy, J. H., & Pan, L. (forthcoming). A Grobner-Shirshov algorithm for applications in internet security. *Southeast Asian Bulletin of Mathematics*.

Kelarev, A. V., Yearwood, J. L. Watters, P., Soukhorukova, N. V., Abawajy, J. H., & Pan, L. (n.d.). Internet security applications of the Munn rings.

Kelarev, A. V., Yearwood, J. L., Watters, P., Webb, D., & Wu, X. W. (n.d.). Optimization of classifiers for internet security based on combinatorial syntactic semigroups.

Kelly, K. (1998). *New rules for the new economy: 10 radical strategies for a connected world*. Viking.

Kelly, G. A. (1955). *The psychology of personal constructs*. New York: Norton.

Kelly, D. M., & Bischof, W. F. (2008). Orienting in virtual environments: How are surface features and environmental geometry weighted in an orientation task? *Cognition*, *109*, 89–104. doi:10.1016/j.cognition.2008.07.012

Kelly, G. A. (1969). The role of classification in personality theory . In Maher, B. (Ed.), *Clinical psychology and personality: The selected papers of George Kelly* (pp. 289–300). New York: Wiley.

Kelsen, H. (1960). *Reine Rechtslehre* (2nd ed.). Vienna: Verlag Deuticke.

Kengpol, A., & O'Brien, C. (2001). The development of a decision support tool for the selection of advanced technology to achieve rapid product development. *International Journal of Production Economics*, *69*, 177–191. doi:10.1016/S0925-5273(00)00016-5

Kenix, L. J. (2009). Blogs as alternative. *Journal of Computer-Mediated Communication*, *14*, 790–822. doi:10.1111/j.1083-6101.2009.01471.x

Kerr, N., & Kaufman-Gililand, C. (1994). Communication, commitment, and cooperation in social dilemmas. *Journal of Personality and Social Psychology*, *66*, 688–698. doi:10.1037/0022-3514.66.3.513

Kerschner, F. (2007). *Wissenschaftliche Arbeitstechnik und Methodenlehre für Juristen* (5th ed.). Vienna: WUV Universitätsverlag.

Kersten, G. E. (1997). Support for group decisions and negotiations: An overview . In Climaco, J. (Ed.), *Multiple criteria decision making and support*. Heidelberg, Germany: Springer Verlag.

Kersten, G. E., & Lai, H. (2005). Satisfiability and completeness of protocols for electronic negotiations [INR01/05]. Montreal, Canada: InterNeg Research Center. Retrieved January 4, 2007, from http://interneg.concordia.ca/interneg/research/papers/2005/01.pdf

Khakee, A., Barbanente, A., Camarda, D., & Puglisi, M. (2002b). With or without?: Comparative study of preparing participatory scenarios using computer-aided and traditional brainstorming. *Journal of Future Research*, *6*, 45–64.

Khakee, A., Barbanente, A., & Puglisi, M. (2002a). Scenario building for Metropolitan Tunis. *Futures*, *34*, 583–596. doi:10.1016/S0016-3287(02)00002-2

Khakee, A., Hull, A., Miller, D., & Woltjer, J. (Eds.). (2008). *New principles in planning evaluation: Urban planning and environment*. Aldershot, UK: Ashgate.

Kiesler, S., Siegel, J., & McGuire, T. (1984). Social psychological aspects of computer-mediated communication. *The American Psychologist*, *39*, 1123–1134. doi:10.1037/0003-066X.39.10.1123

Kiesler, S., & Sproull, L. (1992). Group decision making and communication technology. *Organizational Behavior and Human Decision Processes*, *52*, 96–123. doi:10.1016/0749-5978(92)90047-B

Kiesler, S., Sproull, L., & Waters, K. (1996). A prisoner's dilemma experiment on cooperation with people and human-like computers. *Journal of Personality and Social Psychology.*

Kikiras, P., Tsetsos, V., & Hadjiefthymiades, S. (2006). Ontology-based user modeling for pedestrian navigation systems. *ECAI 2006 Workshop on Ubiquitous User Modeling (UbiqUM)*, Riva del Garda, Italy.

Kim, M. S., Aune, K. S., Hunter, J. E., Kim, H. J., & Kim, J. S. (2001). The effect of culture and self-construals on predispositions toward verbal communication. *Human Communication Research*, *27*(3), 382–408. doi:10.1111/j.1468-2958.2001.tb00786.x

Kim, T., Chang, A., & Pentland, A. (2008). Meeting mediator: Enhancing group collaboration with sociometric feedback. In *Proceedings of Conference on Computer Supported Collaborative Work*, 457-466.

Klemperer, P. (2002). What really matters in auction design. *The Journal of Economic Perspectives*, *16*(1), 169–189. doi:10.1257/0895330027166

Kling, R., & Courtright, C. (2003). Group behavior and learning in electronic forums: A sociotechnical approach. *The Information Society*, *19*, 221–235. doi:10.1080/01972240309465

Klion, R., & Pfenninger, D. T. (1997). Personal construct psychotherapy of addictions. *Journal of Substance Abuse Treatment*, *14*(1), 37–43. doi:10.1016/S0740-5472(96)00130-4

Klir, G. J., & Yuan, B. (1995). *Fuzzy sets and fuzzy logic: Theory and Application.* Upper Saddle River, NJ: Prentice-Hall PTG.

Kluwe, R. H., Lüer, G., & Rösler, F. (Eds.). (2003). *Principles of learning and memory*. Basel, Switzerland: Birkhäuser.

Knops, A. (2007). Debate: Agonism as deliberation – On Mouffe's theory of democracy. *Journal of Political Philosophy*, *15*(1), 115–126. doi:10.1111/j.1467-9760.2007.00267.x

Koedinger, K. R., & Anderson, J. R. (1995). Abstract planning and perceptual chunks . In Chandrasekaran, B., Glasgow, J., & Narayan, N. H. (Eds.), *Diagrammatic reasoning: Cognitive and computational perspectives* (pp. 577–625). Cambridge, MA: The MIT Press.

Kofman, F., & Senge, P. M. (1993). Communities of commitment: The heart of learning organizations. *Organizational Dynamics*, *22*(2), 5. doi:10.1016/0090-2616(93)90050-B

Koit, M., Gerassimenko, O., Kasterpalu, R., Rääbis, A., & Strandson, K. (2008). Developing a dialogue system. In P. Sojka, A. Horak, I. Kopecek, & P. Karel (Eds.), *Text, Speech and Dialogue, TSD 2008* (Lecture Notes in Artificial Intelligence 5246, pp. 593–600). Berlin: Springer.

Kojadinovic, I. (2004). Unsupervised aggregation by Choquet integral based on entropy functionals: Application to the evaluation of students. *Modeling Decisions for Artificial Intelligence*, *3131*, 163–174.

Kolodner, J. L., & Simpson, R. L. (1989). The mediator: Analysis of an early case-based problem solver. *Cognitive Science*, *13*, 507–549. doi:10.1207/s15516709cog1304_2

Komorita, S. (1995). Interpersonal relations: Mixed-motive interaction. *Annual Review of Psychology*, *46*, 183–207. doi:10.1146/annurev.ps.46.020195.001151

Kosonen, M. (2009). Knowledge sharing in virtual communities: A review of the empirical research. *International Journal of Web Based Communities*, *5*(2), 144–163. doi:10.1504/IJWBC.2009.023962

Kotha, S., & Swamidass, P. M. (1998). Advanced manufacturing technology use: Exploring the effect of the nationality variable. *International Journal of Production Research*, *36*(11), 3135–3146. doi:10.1080/002075498192337

Kowalski, R., & Sergot, M. (1991). The use of logical models in legal problem solving. *Ratio Juris*, *3*, 201–218. doi:10.1111/j.1467-9337.1990.tb00058.x

Kramer, R., & Brewer, M. (2006). Effects of group identity on resource use in a simulated commons dilemma. *Rationality and Society*, *18*(4), 443–470.

Krantz, D. H., Luce, D. R., Suppers, P., & Tversky, A. (1971). Foundation of measurement: *Vol. 1. Additive and polynomial representation*. San Diego, CA: Academic Press.

Krippendorff, K. (1980). *Content analysis: An introduction to its methodology*. Beverly Hills, CA: Sage Publications.

Kruger, J., Epley, N., Parker, J., & Ng, Z. (2005). Egocentrism over e-mail: Can we communicate as well as we think? *Journal of Personality and Social Psychology*, *89*, 925–936. doi:10.1037/0022-3514.89.6.925

Kruglanski, A. W. (1990). Lay epistemic theory in social-cognitive psychology. *Psychological Inquiry*, *1*, 181–197. doi:10.1207/s15327965pli0103_1

Kuhn, T., & Poole, M. S. (2000). Do conflict management styles affect group decision making?: Evidence from a longitudinal field study. *Human Communication Research*, *26*(4), 558–590. doi:10.1111/j.1468-2958.2000.tb00769.x

Kumar, R., Novak, J., Raghavan, P., & Tomkins, A. (2004). Structure and evolution of blogspace. *Communications of the ACM*, *47*(12), 35–39. doi:10.1145/1035134.1035162

Kutanoglu, E., & Wu, S. D. (1999). On combinatorial auction and LaGrangean relaxation for distributed resource scheduling. *IIE Transactions*, *3*(1), 813–826. doi:10.1080/07408179908969883

Kyulk, O., Wang, J., & Terken, J. (2005). Real-time feedback on nonverbal behavior to enhance social dynamics in small group meetings. In *Proceedings of the Conference of Machine Learning for Multi-modal Interaction (MLMI),* 150-161.

Laarhoven, P. J. M., & Pedrycz, W. (1983). A fuzzy extension of Saaty's priority theory. *Fuzzy Sets and Systems*, *11*, 229–241.

Labreuche, C., & Grabisch, M. (2003). Choquet integral for the aggregation of interval scales in multicriteria decision making. *Fuzzy Sets and Systems*, *137*(1), 11–26. doi:10.1016/S0165-0114(02)00429-3

Lacy, L. W. (2005). *OWL: Representing information using the Web ontology language*. Victoria, Canada: Trafford Publishing.

Lambert, D. M. (Ed.). (2008). *Supply chain management: Processes, partnerships, performance* (3rd ed.). Sarasota, FL: Supply Chain Management Institute.

Lambert, D. M., Emmelhainz, M. A., & Gardner, J. T. (1996). Developing and implementing supply chain partnerships. *The International Journal of Logistics Management*, *7*(2), 1–17. doi:10.1108/09574099610805485

Langfield-Smith, K., & Wirth, A. (1992). Measuring differences between cognitive maps. *The Journal of the Operational Research Society*, *43*(12), 1135–1150.

Larson, J. R. Jr, & Christensen, C. (1993). Groups as problem-solving units: Toward a meaning of social cognition. *The British Journal of Social Psychology*, *32*, 5–30.

Larvatus Prodeo. (n.d.). About LP Bloggers. Retrieved September 15, 2009, from http://larvatusprodeo.net/

Laudon, J. P., & Laudon, K. C. (2007). *Management information systems*. London: Pearson Prentice Hall.

Laughlin, P. R., & Ellis, A. L. (1986). Demonstrability and social combination processes on mathematical intellective tasks. *Journal of Experimental Social Psychology*, *22*, 177–189. doi:10.1016/0022-1031(86)90022-3

Laukkanen, M. (1994, August). Comparative cause mapping of organizational cognitions. *Organization Science*, *5*(3), 322–343. doi:10.1287/orsc.5.3.322

Lauritson, M., & Gordon, T. F. (2009). Towards a General Theory of Document Modeling. In *Proceedings of the 12th ICAIL* (pp. 202-211). New York: ACM Press.

Law, K. S., Wong, Chi-Sun, Wang, D., & Wang, L. (2000). Effect of supervisor-subordinate guanxi on supervisory decisions in China: An empirical investigation. *International Journal of Human Resource Management*, *11*(4), 751–765. doi:10.1080/09585190050075105

Lawrence, E., Sides, J., & Farrell, H. (2009, March). *Self-segregation or deliberation? Blog readership, participation, and polarization in American politics*. Unpublished paper. Retrieved September 10, 2009, from http://home.gwu.edu/~jsides/blogs.pdf

Layton, R., & Watters, P. (2009, October). *Determining provenance in phishing websites using automated conceptual analysis.* Paper presented at the 4th annual APWG eCrime Researchers Summit, Tacoma, WA.

Layton, R., Brown, S., & Watters, P. (2009, July). *Using differencing to increase distinctiveness for phishing website clustering.* Paper presented at the workshop Cybercrime and Trustworthy Computing CTC-2009 in conjunction with the 6th International Conference on Autonomic and Trusted Computing, Brisbane, Australia.

Lee, C. C. (1990). Fuzzy logic in control systems: Fuzzy logic control parts I & II. *IEEE Transactions on Systems, Man, and Cybernetics*, *20*, 404–435. doi:10.1109/21.52551

Lee, H. S. (2002). Optimal consensus of fuzzy opinions under group decision-making environment. *Fuzzy Sets and Systems*, *132*, 303–315. doi:10.1016/S0165-0114(02)00056-8

Lee, T., & Ghosh, S. (1996). Simulating asynchronous, decentralized military command and control. *IEEE Computational Science & Engineering*, *3*(4), 69–79. doi:10.1109/99.556514

Lee, D. H., Im, S., & Taylor, C. R. (2008). Voluntary self-disclosure of information on the internet: A multimethod study of the motivations and consequences of disclosing information on blogs. *Psychology and Marketing*, *25*(7), 692–710. doi:10.1002/mar.20232

Lee, M., Miller, C., & Newnham, L. (2008). RSS and content syndication in higher education: Subscribing to a new model of teaching and learning. *Educational Media International*, *45*(4), 311–322. doi:10.1080/09523980802573255

Leith, P. (1986). Fundamental errors in legal logic programming. *The Computer Journal*, *29*(6), 545–552. doi:10.1093/comjnl/29.6.545

Lenhart, A., & Fox, S. (2006). Bloggers: A portrait of the internet's new storytellers. *Pew Internet & American Life Projects.* Retrieved October 20, 2009, from http://www.pewinternet.org/Reports/2006/Bloggers.aspx

Leshed, G., Perez, D., Hancock, J. H., Cosley, D., Birnholtz, J., Lee, S., et al. (2009). Visualizing real-time language-based feedback on teamwork behavior in computer-mediated groups. *Proceedings of Conference of Human Factors in Computing* (CHI).

Lesser, E. L., & Storck, J. (2001). Communities of practice and organizational performance. *IBM Systems Journal*, 4.

Lew, A. R., Foster, K., Bremner, J. G., Green, M., & Slavin, S. (2005). Detection of geometric but not topological transformations in 6-12-month-old infants in a visual exploration paradigm. *Developmental Psychobiology*, *47*, 31–42. doi:10.1002/dev.20075

Li, S. S. (2007). Computer-mediated communication and group decision making: A functional perspective. *Small Group Research*, *38*(5), 593–614. doi:10.1177/1046496407304335

Limesurvey. (2009). *Features*. Retrieved October 4, 2009, from http://www.limesurvey.org/en/about-limesurvey/features

LinkedIn, Inc. (2009). *About us*. Retrieved October 1, 2009, from http://press.linkedin.com/about

Linstone, H. A., & Turoff, M. (2002). *The Delphi method: Techniques and applications*. Retrieved December 2009, from http://www.is.njit.edu/pubs/delphibook/

List, C. (2002). Two conceptions of agreement. *The Good Society*, *11*(1), 72–79. doi:10.1353/gso.2002.0011

List, C., & Pettit, P. (2002). Aggregating sets of judgments: Two impossibility results compared. *Economics and Philosophy*, *18*, 89–110.

Littleboy, A., Boughen, N., Niemeyer, S., & Fisher, K. (2006). *Societal uptake of alternative energy futures: Final report* [Technical report]. Newcastle, Australia: CSIRO Energy Transformed Research Flagship.

Lloyd, R. (2009). *Spatial cognition: Geographic environments*. Amsterdam: Springer-Verlag.

Lodder, A. R., & Oskamp, A. (Eds.). (2006). *Information technology and lawyers: Advanced technology in the legal domain, from challenges to daily routine*. Dordrecht, The Netherlands: Springer.

Lodder, A., & Zeleznikow, J. (2005). Developing an online dispute resolution environment: Dialogue tools and negotiation systems in a three step model. *The Harvard Negotiation Law Review*, *10*, 287–338.

Lourenco, S. F., & Huttenlocher, J. (2007). Using geometry to specify location: Implications for spatial coding in children and nonhuman animals. *Psychological Research*, *71*, 252–264. doi:10.1007/s00426-006-0081-3

Lovelock, J. (2001, October 19). Choice of war should be subject to vote [Letter to the editor]. *The Australian*, p. 12.

Lu, J., Zhang, G., Ruan, D., & Wu, F. (2007). *Multi-objective group decision making methods: Software and applications with fuzzy set techniques*. Singapore: Imperial Collage Press.

Luhmann, N. (1968). *Vertrauen: Ein Mechanismus der Reduktion sozialer Komplexität* [Republished 2000]. Stuttgart, Germany: Lucius & Lucius Verlagsgesellschaft.

Luhmann, N., & Baecker, D. (Eds.). (2002). *Einführung in die Systemtheorie*. Heidelberg, Germany: Carl-Auer-Systeme-Verlag.

Lynch, K. (1960). *The image of the city*. Cambridge, MA: The MIT Press.

Lynn, N., & Lea, S. (2003). A phantom menace and the new Apartheid: The social construction of asylum-seekers in the United Kingdom. *Discourse & Society*, *14*(4), 425–452. doi:10.1177/09579265030140040002

Ma, R., & Chuang, R. (2001). Persuasion strategies of Chinese college students in interpersonal contexts. *The Southern Communication Journal*, *66*(4), 267–278.

Ma, L., Ofoghi, B., Watters, P., & Brown, S. (2009). *Detecting phishing emails using hybrid features*. Paper presented at the workshop Cybercrime and Trustworthy Computing CTC-2009 in conjunction with the 6th International Conference on Autonomic and Trusted Computing, Brisbane, Australia.

Ma, L., Yearwood, J., & Watters, P. (2009, October). *Establishing phishing provenance using orthographic features*. Paper presented at the 4th annual APWG eCrime Researchers Summit, Tacoma, WA.

MacDougall, S. L., & Pike, R. H. (2003). Consider your options: Changes to strategic value during implementation of advanced manufacturing technology. *Omega*, *31*(1), 1–15. doi:10.1016/S0305-0483(02)00061-0

Macintosh, A., Robson, E., Smith, E., & Whyte, A. (2002). Electronic democracy and young people. *Social Science Computer Review*, *21*(1), 43–54. doi:10.1177/0894439302238970

MacKenzie, S. (2001, November 28). Closed culture shuts out complex world [Letter to the editor]. *The Australian*, p. 14.

Mahdavi, I., Mahdavi-Amiri, N., Heidarzade, A., & Nourifar, R. (2008). Designing a model of fuzzy TOPSIS in multiple criteria decision-making. *Applied Mathematics and Computation*, *206*, 607–617. doi:10.1016/j amc.2008.05.047

Mallgrave, H. F. (2009). *Architect's brain: Neuroscience, creativity, and architecture*. London: Wiley-Blackwell.

Manning, C. D., Raghavan, P., & Schütze, H. (Eds.). (2008). *Introduction to information retrieval*. Cambridge: Cambridge University Press.

Manville, B., & Ober, J. (2003). *A company of citizens: What the world's first democracy teaches leaders about creating great organizations*. Boston: Harvard Business School Press.

Marais, M., & Reynaud, E. (2007). Comparaison entre les entreprises françaises publiques et privées face aux exigences du développement durable. *Centre d'Etudes et de Recherche en Gestion d'Aix-Marseille (CERGAM) Working Paper Series, 800.*

Mark, G., Grudin, J., & Poltrock, S. (1999). Meeting at the desktop: An empirical study of virtually co-located teams. *Proceedings of the European Conference on Computer Supported Cooperative Work*, 159-178.

Mark. (2008b, June 24). [Comment 3 to blog post: Marketised celebrity politics, Australian style]. *Larvatus Prodeo* [Blog], Federal election 2007. Retrieved October 1, 2009, from http://larvatusprodeo.net/2008/06/24/marketised-celebrity-politics-australian-style/

Mark. (2008d, October 02). [Comment 19 to blog post: The state of Rudd nation]. *Larvatus Prodeo* [Blog], Federal election 2007. Retrieved October 1, 2009, from http://larvatusprodeo.net/2008/10/02/the-state-of-rudd-nation/

Mason, R., & Rennie, F. (2007). Using Web 2.0 for learning in the community. *The Internet and Higher Education, 10*(3), 196–203. doi:10.1016/j.iheduc.2007.06.003

Mathes, A. (2004, December). Folksonomies: Cooperative classification and communication through shared metadata. *Computer Mediated Communication – LIS590CMC*, Graduate School of Library and Information Science, University of Illinois Urbana-Champaign.

Matt, C. (2008, October 02). [Comment 5 to blog post: The state of Rudd nation]. *Larvatus Prodeo* [Blog], Federal election 2007. Retrieved October 1, 2009, from http://larvatusprodeo.net/2008/10/02/the-state-of-rudd-nation/

Matwin, S., Szpakowicz, S., Koperczak, Z., Kersten, G. E., & Michalowski, G. (1989). NEGOPLAN: An expert system shell for negotiation support. *IEEE Expert, 4*, 50–62. doi:10.1109/64.43285

Maynor, J. W. (2007, April). *Blogging for democracy: Deliberation, autonomy and reasonableness in the blogosphere*. Paper presented at the Midwest Political Science Association, Chicago, Illinois.

McBurney, P., Hitchcock, D., & Parsons, S. (2007). The eightfold way of deliberation dialogue. *International Journal of Intelligent Systems, 22*, 95–132. doi:10.1002/int.20191

McBurney, P., & Parsons, S. (2001). Chance discovery using dialectical argumentation. In T. Terano, T. Nishida, A. Namatame, S. Tsumoto, Y. Ohsawa, & T. Washio (Eds.), *New frontiers in artificial intelligence (*LNAI vol. 2253, pp. 414-424). Berlin: Springer Verlag.

McCarthy, T. (1994). Kantian constructivism and reconstructivism: Rawls and Habermas in dialogue. *Ethics, 105*(1), 44–63. doi:10.1086/293678

McCarty, L. T. (1977). Reflections on TAXMAN: An experiment in artificial intelligence and legal reasoning. *Harvard Law Review, 90*, 89–116. doi:10.2307/1340132

McDonald, A. (2004). *SpamAssassin: A practical guide to configuration, customization, and integration*. Birmingham, UK: Packt Publishing.

McGrath, J. E., & Tschan, F. (2004). *Temporal matters in social psychology: Examining the role of time in the lives of groups and individuals*. Washington, DC: APA Publications. doi:10.1037/10659-000

McGrath, J. E. (1984). Groups: Interaction and performance. Englewood Cliffs, NJ: Prentice-Hall, INC.

McGuire, T. W., Kiesler, S., & Siegel, J. (1987). Group and computer-mediated discussion effects in risk decision-making. *Journal of Personality and Social Psychology, 52*, 917–930. doi:10.1037/0022-3514.52.5.917

McKenzie, J., & van Winkelen, C. (2004). *Understanding the knowledgeable organization: Nurturing knowledge competence*. Thomson Learning.

McLean, I., & List, C. (1999). Can deliberation induce greater preference structuration?: Evidence from deliberative opinion polls . In *Deliberating about deliberative democracy*. Austin, TX: University of Texas.

McMahon, C. (2001). *Collective rationality and collective reasoning*. Cambridge, UK: Cambridge University Press.

Meikle, G. (2002). *Future active: Media activism and the internet*. New York: Routledge.

Mennecke, B. E. (1997). Using group support systems to discover hidden profiles: An examination of the influence of group size and meeting structures on information sharing and decision quality. *International Journal of Human-Computer Studies*, *47*, 387–405. doi:10.1006/ijhc.1997.0136

Mennecke, B. E., & Valacich, J. S. (1998). Information is what you make of it: The influence of group history and computer support on information sharing, decision quality, and member perceptions. *Journal of Management Information Systems*, *15*, 173–197.

Meraz, S. (2007). Analyzing political conversation on the Howard Dean candidate blog . In Tremayne, M. (Ed.), *Blogging, citizenship and the future of media* (pp. 59–81). New York, London: Routledge.

Merkel, R. (2008, July 29). [Comment 2 in blog post: Slow progress in health funding reform]. *Larvatus Prodeo* [Blog], Federal election 2007. Retrieved 01 October, 2009, from http://larvatusprodeo.net/2008/07/29/slow-progress-in-health-funding-reform/

Merkl, D., & Schweighofer, E. (1997). The exploration of legal text corpora with hierarchical neural networks: A guided tour in public international law. In *Proceedings of the 6th ICAIL* (pp. 98-105). New York: ACM Press.

Merrill, D. (2006). *Mashups: The new breed of Web app. An introduction to mashups*. Armonk, NY: IBM developerWorks. Retrieved February 15, 2010, from http://www.ibm.com/developerworks/xml/library/x-mashups.html

Mickler, S. (2004). Talkback radio, anti-elitism and moral decline: A fatal paradox? In Nile, R. (Ed.), *Us and them: Anti-elitism in Australia* (pp. 97–116). Perth, Australia: API Network, Curtin University of Technology.

Midgely, J., Hall, A., Hardiman, M., & Narine, D. (1986). *Community participation, social development, and the state*. London: Methuen & Co. Ltd.

Miller, G. A. (1956). The magical number seven plus or minus two: Some limits on our capacity for processing information. *Psychological Review*, *63*, 81–97. doi:10.1037/h0043158

Miller, P. (2005). Web 2.0: Building the new library. *Ariadne, 25*. Retrieved September 26, 2009, from http://www.ariadne.ac.uk/issue45/miller/#author1

Min, S., Roath, A. S., Daugherty, P. J., Genchev, S. E., Chen, H., Arndt, A. D., & Richey, R. G. (2005). Supply chain collaboration: What's happening. *International Journal of Logistics Management*, *16*(2), 237–256. doi:10.1108/09574090510634539

Minsky, M. (1975). A framework for representing knowledge (Reprint). In P. Winston (Ed.), *The psychology of computer vision*. New York: McGraw Hill. Retrieved from http://web.media.mit.edu/~minsky/papers/Frames/frames.html

Mital, V., & Johnson, L. (1992). *Advanced information systems for lawyers*. London: Chapman & Hall.

Mnookin, R., Peppet, S. R., & Tulumello, A. S. (2000). *Beyond winning: Negotiating to create value in deals and disputes*. Cambridge, MA: The Belnap Press of Harvard University Press.

Mockler, R. J. (1989). *Knowledge-based systems for management decisions*. Upper Saddle River, NJ: Prentice-Hall.

Modave, F., & Grabisch, M. (1998). Preference representation by Choquet integral: The commensurability hypothesis. *Proc. 7th Int. Conf. on Information Processing and Management of Uncertainty in Knowledge-Based Systems (IPMU)*, 1-8.

Moens, M. F., Uyttendaele, C., & Dumortier, J. (1997). Abstracting of legal cases: The SALOMON experience. In *Proceedings of the 6th ICAIL* (pp. 114-122). New York: ACM Press.

Mohanty, R. P., & Deshmukh, S. G. (1998). Advanced manufacturing technology selection: A strategic model for learning and evaluation. *International Journal of Production Economics*, *55*(3), 295–307. doi:10.1016/S0925-5273(98)00075-9

Monteleone, E., Raats, M. M., & Mela, D. J. (1997). Perceptions of starchy food dishes: Application of the repertory grid method. *Appetite*, *28*, 255–265. doi:10.1006/appe.1996.0081

Moon, H., Conlon, D. E., Humphrey, S. E., Quigley, N., Devers, C. E., & Nowakowski, J. M. (2003). Group decision process and incrementalism in organizational decision making. *Organizational Behavior and Human Decision Processes*, *92*(1-2), 67–79. doi:10.1016/S0749-5978(03)00079-7

Morana, J., & Gonzalez-Feliu, J. (2009). *Les décisions tactiques et opérationnelles d'une mutualisation d'un système de distribution: Le cas du groupe NMPP*. Paris: Research Workshop in Sustainable Transportation and Logistics.

Moreland, R. L., Hogg, M. A., & Hains, S. C. (1994). Back to the future: Social psychological research on groups. *Journal of Experimental Social Psychology*, *30*, 527–555. doi:10.1006/jesp.1994.1025

Morozow, O. (2006). *The heat is on: The future of energy in Australia* [Technical report]. Melbourne, Australia: Commonwealth Scientific and Industrial Research Organisation.

Morrison, A., & Love, A. (1996). A discourse of disillusionment: Letters to the editor in two Zimbabwean magazines 10 years after independence. *Discourse & Society*, *7*(1), 39–75. doi:10.1177/0957926596007001003

Muecke, N., & Stranieri, A. (2007). An argument structure abstraction for bayesian belief networks: Just outcomes in on-line dispute resolution. In J. F. Roddick & A. Hinze (Eds.), Fourth Asia-Pacific Conference on Conceptual Modelling (APCCM 2007), Ballarat, Victoria, Australia, January 2007. *Conferences in Research and Practice in Information Technology, 67*, 35-40.

Mummery, J., & Rodan, D. (2003). Discourses of democracy in the aftermath of 9/11 and other events: Protectivism versus humanitarianism. *Continuum: Journal of Media & Cultural Studies*, *17*(4), 433–443. doi:10.1080/1030431032000152014

Munzer, S., & Holmer, T. (2009). Bridging the gap between media synchronicity and task performance. *Communication Research*, *36*, 76–103. doi:10.1177/0093650208326464

Napier, N. P., Keil, M., & Tan, F. B. (2009). IT project managers' construction of successful project management practice: A repertory grid investigation. *Information Systems Journal*, *19*, 255–282. doi:10.1111/j.1365-2575.2007.00264.x

Narukawa, Y., & Murofushi, T. (2008). Choquet-Stieltjes integral as a tool for decision modeling. *International Journal of Intelligent Systems*, *23*(2), 115–127. doi:10.1002/int.20260

Neumann, D., & Weinhardt, C. (2002). Domain-independent eNegotiation design: Prospects, methods, and challenges. Proceedings of the 13th International Workshop on Database and Expert Systems Applications, DEXA (pp. 680-686).

Neumann, D., Schnizler, B., Weber, I., & Weinhardt, C. (2007). Second-best combinatorial auctions: The case of the pricing-per-column mechanism. Proceedings of the 40th Annual Hawaii International Conference on System Science, HICSS (p. 33b).

Newbourne, P. T. (1997). The role of partnerships in strategic account management. *The International Journal of Logistics Management*, *8*(1), 67–74. doi:10.1108/09574099710805592

Newell, S., Robertson, M., Scarbrough, H., & Swan, J. (2002). *Managing knowledge work*. Palgrave.

Ngwenyama, O. K., Bryson, N., & Mobolurin, A. (1996). Supporting facilitation in group support systems: Techniques for analyzing consensus relevant data. *Decision Support Systems*, *16*, 155–168. doi:10.1016/0167-9236(95)00004-6

Nielsen Online. (2008, January). *Consumer generated media: Separating hype from reality. A new market study from Nielsen Online* (1st ed., p. 89). New York: Nielsen.

Niemeyer, S. (2004). Deliberation in the wilderness: Displacing symbolic politics. *Environmental Politics*, *13*(2), 347–372. doi:10.1080/0964401042000209612

Niemeyer, S. J. (2007). *Intersubjective rationality: Measuring deliberative quality*. Canberra, Australia: Deliberative Democracy Research Group, Research School of Social Sciences, The Australian National University.

Niemeyer, S. J. (2007). *Report of the analysis of the UBC Biobank Deliberative Process*. Canberra, Australia: Deliberative Democracy Research Group, Research School of Social Sciences, The Australian National University.

Niemeyer, S. J. (2008). *New Mexico ForestERA: Analysis of a citizen and stakeholder deliberations on forest restoration issues in northern New Mexico*. Canberra, Australia: Deliberative Democracy Research Group, Research School of Social Sciences, The Australian National University.

Niemeyer, S. J., & Ayirtman, S. (2008). Achieving success in large scale deliberation: Analysis of the Fremantle Bridge community engagement process . In *Micropolitics in Deliberation Reports*. Canberra, Australia: Deliberative Democracy Research Group, Research School of Social Sciences, The Australian National University.

Niemeyer, S., & Blamey, R. (2003). *The far-north Queensland citizens' jury* [Technical Report]. Canberra, Australia: Land and Water Australia.

Niemeyer, S. J., & Dryzek, J. S. (2007). The ends of deliberation: Metaconsensus and intersubjective rationality as deliberative ideals. *Swiss Political Science Review*, *13*(4), 497–526.

Niemeyer, S. J. (2007). Intersubjective rationality: Measuring deliberative quality. Paper presented to Political Science Seminar. Canberra, Australia: The Australian National University Research School of Social Sciences.

Niemeyer, S. J. (2008). *Deliberation and the public sphere: Macro lessons from micro politics*. Paper presented at the Research School of Social Science theme seminar. Canberra, Australia: The Australian National University.

Niemeyer, S. J., & Dryzek, J. S. (2007). *Intersubjective rationality: Using interpersonal consistency as a measure of deliberative quality*. Paper presented at ECPSRR Workshop 23: Advanced Empirical Study of Deliberation, Helsinki, Finland.

Ning, Inc. (2009). *About Ning*. Retrieved October 2, 2009, from http://about.ning.com/

Nonaka, I., & Konno, N. (1998). The concept of "ba": Building a foundation for knowledge creation. *California Management Review*, 3.

Nonaka, I., Toyama, R., & Konno, N. (2000). SECI, ba and leadership: A unified model of dynamic knowledge creation. *Long Range Planning*, 33.

Nonaka, I. (1991). The knowledge creating company. *Harvard Business Review*, *69*(6), 96–104.

O'Kane, J. F., Spenceley, J. R., & Taylor, R. (2000). Simulation as an essential tool for advanced manufacturing technology problems. *Journal of Materials Processing Technology*, *107*, 412–424. doi:10.1016/S0924-0136(00)00689-0

O'Keefee, R. M. (1989). The implications of cognitive-style findings for operational research. *The Journal of the Operational Research Society*, *40*, 415–422.

O'Reilly, T. (2005). *What is Web 2.0*. Retrieved September 24, 2009, from http://oreilly.com/web2/archive/what-is-web-20.html

Ocker, R. J., & Yaverbaum, G. J. (1999). Asynchronous computer-mediated communication versus face-to-face collaboration: Results on student learning, quality and satisfaction. *Group Decision and Negotiation*, *8*(5), 427–440. doi:10.1023/A:1008621827601

OECD Task Force on Spam. (2009). OECD anti-spam toolkit and its annexes. Retrieved September 2009, from http://www.oecd-antispam.org/

Oeltjenbruns, H., Kolarik, W. J., & Schnadt–Kirschner, R. (1995). Strategic planning in manufacturing systems: AHP application to an equipment replacement decision. *International Journal of Production Economics*, *38*, 189–197. doi:10.1016/0925-5273(94)00092-O

OldSkeptic. (2008, October 06). [Comment 29 to blog post: The state of Rudd nation]. *Larvatus Prodeo* [Blog], Federal election 2007. Retrieved 01 October, 2009, from http://larvatusprodeo.net/2008/10/02/the-state-of-rudd-nation/

Olguin Olguin, D., Waber, B., Kim, T., Mohan, A., Ara, K., & Pentland, A. (2009). Sensible organizations: Technology and methodology for automatically measuring organizational behavior. *IEEE Transactions on Systems, Man, and Cybernetics. Part B: Cybernetics.*

Olszak, C. M., & Ziemba, E. (2008). A conceptual model of ICT for intellectual capital management . In Korczak, J., Dudycz, H., & Dyczkowski, M. (Eds.), *Management: Advanced information technologies for management. Wroclaw, Poland: Publishing House of the University of Economics*. Wroclaw.

Olszak, C. M., & Ziemba, E. (2006). Business intelligence systems in the holistic infrastructure development: supporting decision making in organisations. *Interdisciplinary Journal of Information, Knowledge, and Management, 1.*

Onisawa, T., Sugeno, M., Nishiwaki, M. Y., Kawai, H., & Harima, Y. (1986). Fuzzy measure analysis of public attitude towards the use of nuclear energy. *Fuzzy Sets and Systems*, *20*, 259–289. doi:10.1016/S0165-0114(86)90040-0

Orbell, J., & Kragt, A. (1988). Explaining discussion-induced cooperation. *Journal of Personality and Social Psychology*, *54*(5), 811–819. doi:10.1037/0022-3514.54.5.811

Ordoobadi, S. M., & Mulvaney, N. J. (2001). Development of a justification tool for advanced manufacturing technologies: System-wide benefits value analysis. *Journal of Engineering and Technology Management*, *18*, 157–184. doi:10.1016/S0923-4748(01)00033-9

Orley, J. H., & Loff, J. P. (1972). The effect of psychiatric education on attitudes to illness among the Ganda. *The British Journal of Psychiatry*, *121*, 137–141. doi:10.1192/bjp.121.2.137

Orlitzky, M., & Hirokawa, R. Y. (2001). To err is human, to correct for it divine: A meta-analysis of research testing the functional theory of group decision-making effectiveness. *Small Group Research*, *32*(3), 313–341. doi:10.1177/104649640103200303

Orr, S. (2002). A comparison of AMT strategies in the USA, South Africa and Germany. *International Journal of Manufacturing Technology Management*, *4*(6), 441–454. doi:10.1504/IJMTM.2002.002517

Osborn, A. (1948). *Your creative power*. New York: C. Scribner's Sons.

Osborn, A. (1953). *Applied imagination: Principles and procedures of creative problem solving*. New York: C. Scribner's Sons.

Oskamp, A., & Lodder, A. R. (2006). Introduction: Law, information technology and artificial intelligence. A. R. Lodder & A. Oskamp (Eds.), *Information technology and lawyers: Advanced technology in the legal domain, from challenges to daily routine* (pp. 1-22). Dordrecht, The Netherlands: Springer.

Osman, D., Yearwood, J., & Vamplew, P. (2007). Using corpus analysis to inform research into opinion detection in blogs. *Proceedings of the 6th Australasian Data Mining Conference (AusDM'07)* (pp. 65-75).

Palfrey, T. R., & Poole, K. T. (1987). The relationship between information, ideology, and voting behavior. *American Journal of Political Science*, *31*(3), 511–530. doi:10.2307/2111281

Papacharissi, Z. (2006). Audiences as media producers: Content analysis of 260 blogs . In Tremayne, P. (Ed.), *Blogging, citizenship, and the future of media* (pp. 21–38). New York: Routledge.

Paramà, J. R., Penabada, M. R., Brisaboaa, N. R., & Betanzosa, A. A. (1998). EECB: A knowledge elicitation tool based on repertory grid and city block metric. *Expert Systems with Applications*, *14*, 249–258. doi:10.1016/S0957-4174(97)00078-X

Park, B. C. (2001). An optimal dwell point policy for automated storage/retrieval systems with uniformly distributed, rectangular racks. *International Journal of Production Research*, *39*(7), 1469–1480. doi:10.1080/00207540010023583

Parker, J. R. (1995). Voting methods for multiple autonomous agents. Proceedings of the Third Australian and New Zealand Conference on Intelligent Information Systems, ANZIIS (pp. 128-133).

Pateman, C. (1970). *Participation and democratic theory*. Cambridge, UK: University Press.

Patier, D. (2004). *La place du transport de marchandises en compte propre*. La Direction de La Recherche et des Affaires Scientifiques et Techniques & Centre pour la Communication Scientifique, Hyper Article en Ligne.

Patier, D., Dufour, J. G., & Routhier, J. L. (2007). *Du transport de marchandises en ville à la logistique urbaine*. Techniques de l'Ingénieur.

Paul Burns. (2008, July 21). [Comment 2 to blog post: The great pretender]. *Larvatus Prodeo* [Blog], Federal election 2007. Retrieved 01 October, 2009, from http://larvatusprodeo.net/2008/07/21/the-great-pretender/

Paul Walter. (2008, October 02). [Comment 21 to blog post: The state of Rudd nation]. *Larvatus Prodeo* [Blog], Federal election 2007. Retrieved 01 October, 2009, from http://larvatusprodeo.net/2008/10/02/the-state-of-rudd-nation/

Paulus, P. B., & Brown, V. R. (2007). Toward more creative and innovative group idea generation: A cognitive-social-motivational perspective of brainstorming. *Social and Personality Psychology Compass*, *1*, 248–265. doi:10.1111/j.1751-9004.2007.00006.x

Paulus. (2008a, October 02). [Comment 18 to blog post: The state of Rudd nation]. *Larvatus Prodeo* [Blog], Federal election 2007. Retrieved 01 October, 2009, from http://larvatusprodeo.net/2008/10/02/the-state-of-rudd-nation/

Pedersen, E., McCall, K., Moran, T., & Halasz, F. (1993). Tivoli: An electronic whiteboard for informal workgroup meetings. *Proceedings conference on Human factors in computing systems*, ACM Press, 391-398.

Peters, J. (1993). Distrust of representation: Habermas on the public sphere. *Media Culture & Society*, *15*, 541–571. doi:10.1177/016344393015004003

Peterson, M., & Waterman, D. (1985). Evaluating civil claims: An expert systems approach to evaluating product liability cases . In Walter, C. (Ed.), *Computer power and legal reasoning* (pp. 627–659). St. Paul, MN: West Publishing Company.

Petit, P. (2001). Deliberative democracy and the discursive dilemma. *Philosophical Issues*, *11*, 268–299. doi:10.1111/j.1758-2237.2001.tb00047.x

Petrik, K. (2009). Participation and e-democracy: How to utilize Web 2.0 for policy decision-making. *Proceedings of the 10th Annual International Conference on Digital Government Research: Social Networks: Making Connections between Citizens, Data and Government. ACM International Conference Proceeding Series*, *390*, 254-263.

Phillips, K. (1996). The spaces of public dissension: Reconsidering the public sphere. *Communication Monographs*, *63*, 231–248. doi:10.1080/03637759609376391

PinkyOz. (2008, October 02). [Comment 23 to blog post: The state of Rudd nation]. *Larvatus Prodeo* [Blog], Federal election 2007. Retrieved 01 October, 2009, from http://larvatusprodeo.net/2008/10/02/the-state-of-rudd-nation/

Plummer, D., Cearley, D., & Smith, D. (2008). *Cloud computing confusion leads to opportunity.* Retrieved October 3, 2009, from http://www.gartner.com/it/page.jsp?id=707508

Polletta, F. (2006). *It was like a fever: Storytelling in protest and politics*. Chicago: University of Chicago Press.

Polletta, F., & Stark, D. (2005). NSF project highlights: Policy made public: Technologies of Deliberation and Representation in Rebuilding Lower Manhattan. In Proceedings of the 2005 National Conference on Digital Government Research. *ACM International Conference Proceedings, 89*, 153-154.

Possum. (n.d.). About. Retrieved December 10, 2009, from http://possumcomitatus.wordpress.com/about-possum/

Postmes, T., Spears, R., & Cihangir, S. (2001). Quality of decision making and group norms. *Journal of Personality and Social Psychology, 80*, 918–930. doi:10.1037/0022-3514.80.6.918

Pouget, A., Ducom, J. C., Torri, J., & Bavelier, D. (2002). Multisensory spatial representations in eye-centered coordinates for reaching. *Cognition, 83*, B1–B11. doi:10.1016/S0010-0277(01)00163-9

Prakken, H. (2006). Formal systems for persuasion dialogue. *The Knowledge Engineering Review, 21*, 163–188. doi:10.1017/S0269888906000865

Prakken, H. (1997). *Logical tools for modelling legal argument: A study in defeasible reasoning in law*. Dordrecht, The Netherlands: Kluwer Academic Publishers.

Procter, H. G. (1985). Repertory grid techniques in family therapy and research . In Beail, N. (Ed.), *Repertory grid technique and personal construct application in clinical and educational setting*. London: Croom Helm.

Professor rat. (2008, October 02). [Comment 7 to blog post: The state of Rudd nation]. *Larvatus Prodeo* [Blog], Federal election 2007. Retrieved 01 October, 2009, from http://larvatusprodeo.net/2008/10/02/the-state-of-rudd-nation/

Prosser, A., & Krimmer, R. (2004). Electronic voting in Europe . In Prosser, A., & Krimmer, R. (Eds.), *Electronic Voting in Europe – Technology, Law, Politics and Society Workshop of the ESF TED Programme together with GI and OCG July, 7th–9th, 2004 in Schloß Hofen/Bregenz, Lake of Constance, Austria onstance*. Austria.

Pruitt, D. G. (1981). *Negotiation behaviour*. New York: Academic Press.

Punniyamoorthy, M., & Ragavan, P. V. (2003). A strategic decision model for the justification of technology selection. *International Journal of Advanced Manufacturing Technology, 21*(1), 72–78. doi:10.1007/s001700300008

Putnam, R. (1995). Bowling alone: America's declining social capital. *Journal of Democracy, 6*(1), 65–78. doi:10.1353/jod.1995.0002

Putnam, L. L., & Fairhurst, G. T. (2001). Discourse analysis in organizations: Issues and concerns . In Jablin, F. M., & Putnam, L. L. (Eds.), *The new handbook of organizational communication: Advances in theory, research, and methods* (pp. 78–136). Thousand Oaks, CA: Sage Publications.

Qian, H., & Scott, C. R. (2007). Anonymity and self-disclosure on Weblogs. *Journal of Computer-Mediated Communication, 12*(4). Retrieved from http://jcmc.indiana.edu/vol12/issue4/qian.html. doi:10.1111/j.1083-6101.2007.00380.x

Raboy, M., Abramson, B. D., Proulx, S., & Welters, R. (2001). Media policy, audiences and social demand. *Television & New Media, 2*(2), 95–115. doi:10.1177/152747640100200202

Raeymaeckers, K. (2005). Letters to the editor: A feedback opportunity turned into a marketing tool: An account of selection and editing practices in the Flemish daily press. *European Journal of Communication, 20*(2), 199-221.

Raghu, T. S., Ramesh, R., Chang, A. M., & Whinston, A. B. (2001). Collaborative decision making: A connectionist paradigm for dialectical support. *Information Systems Research, 12*(4), 363–383. doi:10.1287/isre.12.4.363.9705

Raiffa, H., Richardson, J., & Metcalfe, D. (2002). *Negotiation analysis: The science and art of collaborative decision making*. Reading, MA: Addison Wesley.

Ramanathan, R., & Ganesh, L. S. (1994). Group preference aggregation methods employed in AHP: An evaluation and an intrinsic process for deriving members' weight ages. *European Journal of Operational Research, 79*, 249–265. doi:10.1016/0377-2217(94)90356-5

RAND Corporation. (2009). Retrieved December 2009, from http://www.rand.org/

Rasmusen, E. (2001). Auctions . In *Games and information: An introduction to game theory* (3rd ed., pp. 404–454). Malden, MA: Blackwell Publishers.

Reed, C., & Rowe, G. (2004). Araucaria: Software for argument analysis, diagramming and representation. *International Journal of Artificial Intelligence Tools*, *14*(3–4), 961–980. doi:10.1142/S0218213004001922

Reger, R. (1990). Managerial thought structures and competitive positioning . In Huff, A. S. (Ed.), *Mapping strategic thought* (pp. 71–88). Chichester: Wiley.

Reicher, S. D., Spears, R., & Postmes, T. (1995). A social identity model of deindividuation phenomena. *European Review of Social Psychology*, *6*, 161–197. doi:10.1080/14792779443000049

Reigeluth, C. M., & Frick, T. W. (1999). Formative research: A methodology for creating and improving design theories . In Reigeluth, C. M. (Ed.), *Instructional design theories and models: A new paradigm of instructional theory* (pp. 633–652). Mahwah, NJ: Lawrence Erlbaum.

Reisinger, L. (1987). *Strukturwissenschaftliche Grundlagen der Rechtsinformatik*. Graz, Austria: Leykam-Verlag.

Rescher, N. (1993). *Pluralism: Against the demand for consensus*. Oxford: Clarendon Press.

Reynolds, M. & McDonald-Toone, T. (2001, October 16). Prized for peace [Letter to the editor]. *The Australian*, p. 12.

Rheingold, H. R. (2000). *The virtual community: Homesteading on the electronic fronteir*. Cambridge, MA: MIT Press.

Rice, T. J. (1978, April). Cognitive mapping: Small group technique in teaching. *Teaching Sociology*, *5*(3), 259–274. doi:10.2307/1317202

Richards, T., & King, B. (2000). An alternative to the fighting frame in news reporting. *Canadian Journal of Communication*, *25*, 479–496.

Ridings, C. M., & Gefen, D. (2004). Virtual community attraction: Why people hang out online. *Journal of Computer-Mediated Communication*, *10*(1).

Rinner, C. (2001). Argumentation maps: Gis-based discussion support for on-line planning. *Environment and Planning. B, Planning & Design*, *28*, 847–863. doi:10.1068/b2748t

Rittel, H., & Webber, M. (1973). Dilemmas in a general theory of planning. *Policy Sciences*, *4*, 155–169. doi:10.1007/BF01405730

Rittel, H. J., & Webber, M. M. (1984). Planning problems are wicked problems . In Cross, N. (Ed.), *Developments in design methodology* (pp. 135–144). Hoboken, NJ: John Wiley and Sons.

Robert, H. M. I. (2000). *Robert's rules of order newly revised*. Cambridge, MA: Perseus Publishing.

Roberts, F. S. (1979). *Measurement theory with applications to decisionmaking, utility and social sciences*. Reading, MA: Addison-Wesley.

Robertson, S., Zachary, W., & Black, J. (1990). *Cognition, computing and cooperation*. Norwood, NJ: Ablex Publishing Cooperation.

Rocco, E. (1998). Trust breaks down in electronic contexts but can be repaired by some initial face-to-face contact. *Proceedings of Conference of Human Factors in Computing*, 496-502.

Rodan, D., & Balnaves, M. (2009). Democracy to come: Active forums as indicator suites for e-participation and e-governance. In A. Macintosh & E. Tambouris (Eds.), *Proceedings of First International Conference, ePart 2009*, (pp. 175–185). Berlin: Springer-Verlag

Rodwin, L., & Hollister, R. M. (1984). *Cities of the mind: Images and themes of the city in the social sciences*. Berlin: Springer.

Rokeach, M. (1979). *Understanding human values: Individual and societal*. New York: Free Press.

Romano, N. C., Jr., & Nunamaker, J. F. (2002). Meeting analysis: Findings from research and practice. *Proceedings of System Sciences*, IEEE Press.

Roschelle, J., & Teasley, S. D. (1994). *The construction of shared knowledge in collaborative problem solving*. NATO ASI Series F Computer and Systems Sciences.

Rose, D. E. (1994). *A Symbolic and Connectionist Approach to Legal Information Retrieval*. Hillsdale, NJ: Lawrence Erlbaum Associates Publishers.

Rosenberg, S. (2005). The empirical study of deliberative democracy: Setting a research agenda. *Acta Politica, 40*, 212–224. doi:10.1057/palgrave.ap.5500105

Ross, L., Lepper, M. R., & Hubbard, M. (1975). Perseverance in self-perception and social perception: Biased attributional processes in the debriefing paradigm. *Journal of Personality and Social Psychology*, *32*, 880–892. doi:10.1037/0022-3514.32.5.880

Roth, E. (2000). Game theory as a tool for market design . In Patrone, F., Garcia-Jurado, I., & Tijs, S. (Eds.), *Game practice: Contributions from applied game theory* (pp. 7–18). Dordrecht: Kluwer.

Rousseau, J. J. (1968). *The Social Contract*. Harmondsworth: Penguin.

Rowe, G., & Wright, G. (1999). The Delphi technique as a forecasting tool: Issues and analysis. *International Journal of Forecasting*, *15*(4). doi:10.1016/S0169-2070(99)00018-7

Rowe, G., & Wright, G. (2001). Expert opinions in forecasting: Role of the Delphi technique . In Armstrong, J. S. (Ed.), *Principles of forecasting: A handbook of researchers andpractitioners* (pp. 1–32). Boston: Kluwer Academic Publishers.

Russel, C. G., & Cox, D. N. (2003). A computerised adaptation of the repertory grid methodology as a useful tool to elicit older consumers' perceptions of foods. *Food Quality and Preference*, *14*, 681–691. doi:10.1016/S0950-3293(03)00060-0

Russel, C. G., & Cox, D. N. (2004). Understanding middle-aged consumers' perceptions of meat using repertory grid methodology. *Food Quality and Preference*, *15*, 317–329. doi:10.1016/S0950-3293(03)00073-9

Saad, I., & Hammadi, S. (2008). Choquet integral for criteria aggregation in the flexible job-shop scheduling problems. *Mathematics and Computers in Simulation*, *76*(5-6), 447–462. doi:10.1016/j.matcom.2007.04.010

Saari, D. G. (2001). *Chaotic elections!: A mathematician looks at voting*. Providence, RI: American Mathematical Society.

Saaty, T. L. (2000). *Fundamentals of decision making with the analytic hierarchy process*. Pittsburgh, PA: RWS Publications.

Saaty, T. L. (2005). *Theory and applications of the analytic network process*. Pittsburgh, PA: RWS Publications.

Saaty, T. L. (2010). *Principia Mathematica Decernendi*. Pittsburgh, PA: RWS Publications.

Saaty, T. L., & Peniwati, K. (2008). *Group decision making: Drawing out and reconciling differences*. Pittsburgh, PA: RWS Publications.

Saaty, T. L. (1980). *The analytic hierarchy process*. New York: McGraw-Hill.

Sadler, R., Hayllar, T., & Powell, C. (1985). *Senior language*. South Melbourne, Australia: The MacMillan Company.

Salmon, P., Arnold, J. M., & Collyer, Y. M. (1972). What do the determinants determine?: The internal validity of the Rorschach. *Journal of Personality*, *36*, 33–38.

Salton, G. (1989). *Automatic Text Processing: The Transformation, Analysis, and Retrieval of Information by Computer*. Reading, MA: Addison-Wesley.

Sambasivarao, K. V., & Deshmukh, S. G. (1997). A decision support system for selection and justification of advanced manufacturing technologies. *Production Planning and Control*, *8*(3), 270–284. doi:10.1080/095372897235325

Sampson, P. (1972, February). Using the repertory grid test. *JMR, Journal of Marketing Research*, *9*, 78–81. doi:10.2307/3149614

Sanchez, E. (Ed.). (2006). *Fuzzy logic and the semantic Web*. Amsterdam: Elsevier.

Sanders, L. (1997). Against deliberation. *Political Theory*, *25*, 347–376. doi:10.1177/0090591797025003002

Sandholm, T. W. (2000). Agents in electronic commerce: Component technologies for automated negotiation and coalition formation. *Autonomous Agents and Multi-Agent Systems*, *3*(1), 73–96. doi:10.1023/A:1010038012192

Sandholm, T. W. (1993). An implementation of the Contract Net protocol based on marginal cost calculations. In proceedings 11th national conference on artificial intelligence (pp. 256-262).

Sandholm, T. W. (1996). Negotiation among self-interested computationally limited agents. Unpublished doctoral dissertation. Department of Computer Science, University of Massachusetts, USA. Retrieved December 9, 2009, from http://www.cs.cmu.edu/~sandholm/dissertation.ps

Sator, G., Rudnianski, M., Rotolo, A., Riveret, R., & Mayor, E. (2009). Why lawyers are nice (or nasty): A game-theoretical argumentation exercise. In *Proceeding of the 12th ICAIL International Conference on Artificial Intelligence and Law* (pp. 108-117). New York: ACM Press.

Saunders, H. (2001, October 11). PM acted well in affairs not of his making [Letter to the editor]. *The Australian*, p. 12.

Scerri, P., Vincent, R., & Mailler, R. (2004). Comparing three approaches to large scale coordination. In *Proceedings of the First Workshop on the Challenges in the Coordination of Large Scale Multi-agent Systems* (pp. 1-21). Retrieved on August 28, 2009, from http://citeseerx.ist.psu.edu/viewdoc/download?doi=10.1.1.61.1753&rep=rep1&type=pdf

Schank, R. C., & Abelson, R. P. (1977). *Scripts, plans, goals and understanding*. Hillsdale, NJ: Erlbaum.

Schmandt-Besserat, D. (1996). *How writing came about*. Austin, TX: University of Texas Press.

Schmitt, E., Bombardier, V., & Wendling, L. (2008). Improving fuzzy rule classifier by extracting suitable features from capacities with respect to Choquet integral. *IEEE Transactions on Systems, Man, and Cybernetics. Part B, Cybernetics*, *38*(5), 1195–1206. doi:10.1109/TSMCB.2008.925750

Schmulow, A. (2001, October 04). Stand and fight for free society values [Letter to the editor]. *The Australian*, p. 10.

Schneider, K. (2006). Rationale as a by-product . In Dutoit, A., McCall, R., Mistrík, I., & Paech, B. (Eds.), *Rationale management in software engineering* (pp. 91–109). Berlin: Springer. doi:10.1007/978-3-540-30998-7_4

Schön, D. A. (1983). *The reflexive practitioner*. New York: Basic Books.

Schudson, M. (1997). Why conversation is not the soul of democracy. *Critical Studies in Mass Communication*, *14*(4), 1–13. doi:10.1080/15295039709367020

Schultz, T., Kaufmann-Kohler, G., Langer, D., & Bonnet, V. (2001). *Online dispute resolution: State of the art, issues, and perspectives* [E-Com research project]. Geneva, Switzerland: University of Geneva Law School.

Schultz, J. (1994). Media convergence and the fourth estate . In Schultz, J. (Ed.), *Not just another business: Journalists, citizens and the media* (pp. 15–33). Leichhardt, Australia: Pluto Press.

Schutz, A. (1967). *The Phenomenology of the Social World*. Evanston, IL: Northwestern University Press.

Schwartz, A. (2004). *SpamAssassin*. Cambridge, MA: O'Reilly & Associates.

Schweighofer, E. (1999). *Legal knowledge representation, automatic text analysis in public international and European law*. The Hague: Kluwer Law International.

Schweighofer, E. (2008). Reduktion von Komplexität: Durch Recht und IKT . In Schweighofer, E., Geist, A., Heindl, G., & Szücs, C. (Eds.), *Komplexitätsgrenzen der Rechtsinformatik: Tagungsband des 11. Internationalen Rechtsinformatik Symposions IRIS 2008* (pp. 42–48). Stuttgart, Germany: Boorberg.

Schweighofer, E., Haneder, G., Rauber, A., & Dittenbach, M. (2002). Improvement of vector representation of legal documents with legal ontologies. In *Proceedings of the 5th BIS.* [CD-ROM]. Poznan: Poznan University of Economics Press.

Schweighofer, E., Liebwald, D., & Drachsler, M. (2003). Electronic notification of block exempted state aid. In Proceedings of the 16th Annual Conference on Legal Knowledge and Information Systems (pp. 147-148). Amsterdam: IOS Press

Schweighofer, E., Rauber, A., & Dittenbach, M. (2001). Automatic text representation, classification and labeling in European law. In *Proceeding of the 8th ICAIL International Conference on Artificial Intelligence and Law* (pp. 78-87). New York: ACM Press.

Searle, J. R. (1990). Ist der menschliche Geist ein Computerprogramm? *Spektrum der Wissenschaft*, *3*, 40–47.

Seiersen, N. (2006). Systèmes d'information logistique. *Techniques de l'ingénieur*.

Semet, F., Routhier, J. L., Meunier, F., & Stumm, M. (Eds.). (2009). *Logistique Urbaine Mutualisée Durable: Etat de l'art scientifique*. Report LUMD – Advancity.

Sena, J. A., & Shani, A. B. (1999). Intellectual capital and knowledge creation: Towards an alternative framework . In Liebowitz, J. (Ed.), *Knowledge management handbook*. CRC Press.

Sener, Z., & Karsak, E. E. (2007). A decision model for advanced manufacturing technology selection using fuzzy regression and fuzzy optimization. *IEEE International Conference on Systems, Man and Cybernetics*, 565-569.

Sener, Z., & Karsak, E. E. (2008). A decision making approach based on fuzzy regression and fuzzy multiple objective programming for advanced manufacturing technology selection. *IEEE International Conference Industrial Engineering and Engineering Management*, 964-968.

Senge, P. M. (1990). *The fifth discipline: The art and practice of the learning organization*. New York: Doubleday.

Shadbolt, N., Hall, W., & Berners-Lee, T. (2006). The semantic Web revisited. *IEEE Intelligent Systems*, 96–101. doi:10.1109/MIS.2006.62

Shakun, M. F. (1988). *Evolutionary systems design: Policy making under complexity and group decision support systems*. Oakland, CA: Holden-Day.

Shang, J., & Sueyoshi, T. (1995). A unified framework for the selection of a flexible manufacturing system. *European Journal of Operational Research*, *85*, 297–315. doi:10.1016/0377-2217(94)00041-A

Shaw, M. E. (1932). A comparison of individuals and small groups in the rational solution of complex problems. *The American Journal of Psychology*, *44*, 491–504. doi:10.2307/1415351

Shettleworth, S. J. (2009). *Cognition, evolution, and behavior*. New York: Oxford University Press.

Shieh, J. I., Wu, H. H., & Liu, H. C. (2009). Applying a complexity-based Choquet integral to evaluate students' performance. *Expert Systems with Applications*, *36*(3), 5100–5106. doi:10.1016/j.eswa.2008.06.003

Shirani, A. I. (2006). Sampling and pooling of decision-relevant information: Comparing the efficiency of face-to-face and GSS supported groups. *Information & Management*, *43*, 521–529. doi:10.1016/j.im.2006.01.001

Shubsachs, A. P. W. (1975). To repeat or not to repeat? Are frequently used constructs more important to the subject? A study of the effect of allowing repetition of constructs in a modified Kelly repertory test. *The British Journal of Medical Psychology*, *48*, 31–37.

Shum, S., Selvin, A., Sierhuis, M., Conklin, J., Haley, C., & Nuseibeh, B. (2006). Hypermedia support for argumentation-based rationale . In Dutoit, A., McCall, R., Mistrík, I., & Paech, B. (Eds.), *Rationale management in software engineering* (pp. 111–132). Berlin: Springer. doi:10.1007/978-3-540-30998-7_5

Siau, K., & Tan, X. (2005). Improving the quality of conceptual modelling using cognitive mapping techniques. *Data & Knowledge Engineering, 55*, 343–365. doi:10.1016/j.datak.2004.12.006

Siau, K., & Sheng, H. X. (2007, November). Important characteristics of software development team members: An empirical investigation using repertory grid. *Information Systems Journal.*

Simatupang, T. M., & Sridharan, R. (2005). An integrative framework for supply chain collaboration. *International Journal of Logistics Management, 16*(2), 257–274. doi:10.1108/09574090510634548

Simitis, S. (1970). *Informationskrise des Rechts und Datenverarbeitung*. Karlsruhe, Germany: C. F. Müller.

Simon, H. (1996). *The sciences of the artificial*. Cambridge, MA: The MIT Press.

Simonot, P. Y., & Roure, J. (2007). *Logistique collaborative. Une question d'avenir. Paris: Economica. TL&Associés & LET (2009). Flexibilité et Impacts de la Demande de transport des différents secteurs Economiques, et simulation de Scénarios d'évolution.* Paris: PREDIT.

Six Apart Ltd. (2009). *Getting started with TypePad*. Retrieved October 3, 2009, from http://help.sixapart.com/tp/us/getting_started.html

Skyrme, D. (1999). *Knowledge networking: Creating the collaborative enterprise*. Butterworth-Heinemann.

Slideshare, Inc. (2009). *What is Slideshare?* Retrieved October 1, 2009, from http://www.slideshare.net/about

Small, M. H., & Chen, I. J. (1997). Economic and strategic justification of AMT inferences from industrial practices. *International Journal of Production Economics, 49*, 65–75. doi:10.1016/S0925-5273(96)00120-X

Smith, C. A. P., & Hayne, S. C. (1997). Decision making under time pressure: An investigation of decision speed and decision quality of computer-supported groups. *Management Communication Quarterly, 11*(1), 97–126. doi:10.1177/0893318997111005

Smith, J. C., Gelbart, D., Maccrimmon, K., Atherton, B., Mcclean, J., Shinehoft, M., & Quintana, L. (1995). Artificial intelligence and legal discourse: The Flexlaw legal text management system. *Artificial Intelligence and Law, 3*(1-2), 55–95. doi:10.1007/BF00877695

Smith, R. G. (1980). The contract Net protocol: High level communication and control in a distributed problem solver. *IEEE Transactions on Computers, 29*(12), 1104–1113. doi:10.1109/TC.1980.1675516

Smith, R. G., & Davis, R. (1981). Frameworks for cooperation in distributed problem solving. *IEEE Transactions on Systems, Man, and Cybernetics, 11*(1). doi:10.1109/TSMC.1981.4308579

Smith, A., Schlozman, K. L., Verba, S., & Brady, H. (2009). The Internet and civic engagement. *Pew Internet & American Life Project.* Retrieved November 17, 2009, from http://www.pewinternet.org/Reports/2009/15--The-Internet-and-Civic-Engagement.aspx

Smolíková, R., & Wachowiak, M. P. (2002). Aggregation operators for selection problems. *Fuzzy Sets and Systems, 131*, 23–34. doi:10.1016/S0165-0114(01)00252-4

SofaMan. (2008, October 02). [Comment 11 to blog post: The state of Rudd nation]. *Larvatus Prodeo* [Blog], Federal election 2007. Retrieved 01 October, 2009, from http://larvatusprodeo.net/2008/10/02/the-state-of-rudd-nation/

Sotillo, S. M., & Starace-Nastasi, D. (1999). Political discourse of a working-class town. *Discourse & Society, 10*(2), 249–276. doi:10.1177/0957926599010002006

SpamAssasin. (2009). Retrieved September 2009, from http://spamassassin.apache.org/

Sparks, M. K., & Perez, E. (1991). Women not heard in letters columns. *The Masthead*, *43*(2), 14–15.

Spivack, N. (2007). *The third-generation Web -- Web 3.0.* Retrieved February 28, 2010, from http://www.intentblog.com/archives/2007/02/nova_spivack_th.html

Stake, R. E. (1978). The case study method in social inquiry. *Educational Researcher*, *7*(2), 5–8.

Stake, R. E. (1995). *The art of case study research.* Thousand Oaks, CA: Sage.

Start an Issues Forum. (n.d.). Retrieved December 10, 2009, from http://forums.e-democracy.org/

Stasser, G., & Stewart, D. (1992). Discovery of hidden profiles by decision-making groups: Solving a problem versus making a judgment. *Journal of Personality and Social Psychology*, *63*, 426–434. doi:10.1037/0022-3514.63.3.426

Stasser, G., Stewart, D., & Wittenbaum, G. M. (1995). Expert roles and information exchange during discussion: The importance of knowing who knows what. *Journal of Experimental Social Psychology*, *31*, 244–265. doi:10.1006/jesp.1995.1012

Stasser, G., Taylor, L. A., & Hanna, C. (1989). Information sampling in structured and unstructured discussion of three- and six-person groups. *Journal of Personality and Social Psychology*, *57*, 67–78. doi:10.1037/0022-3514.57.1.67

Stasser, G., & Titus, W. (1985). Pooling of unshared information in group decision making: Biased information sampling during discussion. *Journal of Personality and Social Psychology*, *48*, 1467–1478. doi:10.1037/0022-3514.48.6.1467

Stasser, G., & Titus, W. (2003). Hidden profiles: A brief history. *Psychological Inquiry*, *14*, 304–313. doi:10.1207/S15327965PLI1403&4_21

Stasser, G., & Birchmeier, Z. (2003). Group creativity and collective choice . In Paulus, P., & Nijstad, B. (Eds.), *Group Creativity* (pp. 85–109). New York: Oxford University Press. doi:10.1093/acprof:oso/9780195147308.003.0005

Steen, L. A., & Seebach, J. A. (1995). *Counterexamples in topology*. New York: Dover Publications.

Steenbergen, M., Bachtinger, A., Sporndliand, M., & Steiner, J. (2003). Measuring political deliberation: A discourse quality index. *Comparative European Politics*, *1*(1), 21–48. doi:10.1057/palgrave.cep.6110002

Steiner, J. (2008). Concept stretching: The case of deliberation. *European Political Science*, *7*(2), 186–190. doi:10.1057/palgrave.eps.2210186

Steiner, I. D. (1972). *Group process and productivity*. San Diego, CA: Academic Press.

Steiner, J., Bachtiger, A., Sporndli, M., & Steenbergen, M. (2004). *Deliberative politics in action: Analysing parliamentary discourse*. Cambridge, UK: Cambridge University Press.

Steinmüller, W. (1993). *Informationstechnologie und Gesellschaft: Einführung in die Angewandte Informatik*. Darmstadt, Germany: Wissenschaftliche Buchgesellschaft.

Stellman, A., & Greene, J. (2005). *Applied software project management*. Sebastopol, CA: O'Reilly Media.

Stewart, T. (1996). Good maintainers and poor maintainers: a personal construct approach to an old problem. *Journal of Fluency Disorders*, *21*, 33–48. doi:10.1016/0094-730X(95)00043-7

Stranieri, A., Zeleznikow, J., Gawler, M., & Lewis, B. (1999). A hybrid—neural approach to the automation of legal reasoning in the discretionary domain of family law in Australia. *Artificial Intelligence and Law*, *7*(2-3), 153–183. doi:10.1023/A:1008325826599

Stranieri, A., & Zeleznikow, J. (2001a). Copyright regulation with argumentation agents. *Information & Communications Technology Law*, *10*(1), 123–137. doi:10.1080/13600830124950

Stranieri, A., Zeleznikow, J., & Yearwood, J. (2001). Argumentation structures that integrate dialectical and non-dialectical reasoning. *The Knowledge Engineering Review*, *16*(4), 331–358. doi:10.1017/S0269888901000248

Stranieri, A., & Zeleznikow, J. (2006). Knowledge discovery from legal databases: Using neural networks and data mining to build legal decision support systems . In Lodder, A. R., & Oskamp, A. (Eds.), *Information technology and lawyers: Advanced technology in the legal domain, from challenges to daily routine* (pp. 81–118). Dordrecht, The Netherlands: Springer.

Stranieri, A., & Zeleznikow, J. (2001b). WebShell: The development of web based expert system shells. In *Research and Development in Expert Systems XVIII. Proceedings of ES2001 - The Twenty-first SGES International Conference on Knowledge Based Systems and Applied Artificial Intelligence* (pp. 245–258). London: Springer Verlag.

Stranieri, A., Yearwood, J., & Mays, H. (2008). Explicit representations of reasoning to support deliberation within groups. In Burstein, F., Johanson, G & Thompson, H (Eds), Harnessing Knowledge Management to Build Communities. Proceedings of the 11th Australian Conference on Knowledge Management and Intelligent Decision Support, 2008. Monash University, Australia, pp 1-14.

Stranieri, A., Yearwood, J., Gervasoni, S., Garner, S., Deans, C., & Johnstone, A. (2004). Web-based decision support for structured reasoning in health. In W. Karin & B. Cesnik (Eds.), *Proceedings of the Combined Conferences of the Twelfth National Health Informatics Conference* (pp. 180–185).

Straus, S. G. (1996). Getting a clue: The effects of communication media and information distribution on participation and performance in computer-mediated and face-to-face groups. *Small Group Research*, *27*(1), 115–142. doi:10.1177/1046496496271006

Straus, S. G., & McGrath, J. E. (1994). Does the medium matter?: The interaction of task type and technology on group performance and member reactions. *The Journal of Applied Psychology*, *79*, 87–97. doi:10.1037/0021-9010.79.1.87

Streitz, A., Geissler, J., Haake, J., & Hol, J. (1994). DOLPHIN: Integrated meeting support across liveBoards, local and remote desktop environments. *Proceedings of the ACM conference on Computer supported Cooperative Work*, 345-358.

Stretton, H. (2002, February 09-10). Detainees must have legal refuge [Letter to the editor]. *The Weekend Australian*, p. 18.

Strickland, J. (2009). *How Web 3.0 will work*. Retrieved February 23, 2010, from http://www.howstuffworks.com/web-30.htm

Stromer-Galley, J. (2007). Measuring deliberations content: A coding scheme. *Journal of Public Deliberation*, *3*(1), 1–37.

Sturgis, P., & Roberts, C. (2005). A different take on the deliberative poll: Information, deliberation, and attitude constraint. *Public Opinion Quarterly*, *69*(1), 30–65. doi:10.1093/poq/nfi005

Sugeno, M. (1974). *Theory of fuzzy integrals and its applications*. Unpublished doctoral dissertation. Tokyo Institute of Technology, Japan.

Sullivan, D. (2003). *Proven portals: Best practices for planning, designing, and developing enterprise portals*. Reading, MA: Addison-Wesley.

Sunstein, C. R. (1995). Incompletely theorized agreements. *Harvard Law Review*, *108*, 1733–1772. doi:10.2307/1341816

Supple, A. & Supple C. (2002, February 11). Refugee stories reach happy ending [Letter to the editor]. *The Australian*, p. 12.

Susskind, R. (1987). *Expert systems in law*. Oxford: Oxford University Press.

Suthers, D., Lesgold, A., Katz, S., Weiner, A., Toth, E., & Harrigal, K. ...DeLeon, Y (1997). An interface for collaborative and coached approaches to learning critical inquiry. IUI '97, Proceedings of the 1997 International Conference on Intelligent User Interfaces, January 6-9, 1997, Orlando, Florida, USA. *ACM*, 217-220.

Swann, W. B. Jr, & Read, S. J. (1981). Acquiring self-knowledge: The search for feedback that fits. *Journal of Personality and Social Psychology*, *85*, 1095–1106. doi:10.1037/0022-3514.85.6.1095

Sycara, K. (1993). Machine learning for intelligent support of conflict resolution. *Decision Support Systems*, *10*, 121–136. doi:10.1016/0167-9236(93)90034-Z

Tajfel, H., & Turner, J. C. (1986). The social identity theory of inter-group behavior . In Worchel, S., & Austin, L. W. (Eds.), *Psychology of Intergroup Relations*. Chicago: Nelson-Ha.

Takhteyev, Y., & Hall, J. (2005). *Blogging together: Digital expression in a real-life community*. Paper presented at the Social Software in the Academy Workshop, Los Angeles, CA.

Talluri, S., Whiteside, M. M., & Seipel, S. J. (2000). A nonparametric stochastic procedure for FMS evaluation. *European Journal of Operational Research*, *124*(3), 529–538. doi:10.1016/S0377-2217(99)00188-5

Talluri, S., & Yoon, K. P. (2002). A cone-ratio DEA approach for AMT justification. *International Journal of Production Economics*, *66*(2), 119–129. doi:10.1016/S0925-5273(99)00123-1

Tan, F. B., & Tung, L. L. (2003). Exploring website evaluation criteria using the repertory grid technique: A web designers' perspective. Conference Proceedings of the Second Annual Workshop on HCI Research in MIS, Seattle, WA, 65-9, AUT University.

Tang, J., Yankelovich, N., Begole, J., Kleek, M., Li, F., & Bhalodia, J. (2001). ConNexus to Awarenex: Extending awareness to mobile users. *Proceedings of Conference on Human Factors in Computer Systems*, 221-228.

Tang, Y., & Parsons, S. (2006). Argumentation-based multi-agent dialogues for deliberation. In S. Parsons, N. Maudet, P. Moraitis & I. Rahwan (Eds.), *Proceedings of the Second International Workshop on Argumentation in Multi-Agent Systems (ArgMAS '05)* (LNAI vol. 4049, pp. 229-244). Berlin: Springer.

Tasa, K., & Whyte, G. (2005). Collective efficacy and vigilant problem solving in group decision making: A non-linear model. *Organizational Behavior and Human Decision Processes*, *96*(2), 119–129. doi:10.1016/j.obhdp.2005.01.002

Tastle, W. J., & Wierman, M. J. (2007). Consensus and dissention. A measure of ordinal dispersion. *International Journal of Approximating Reasoning*, *45*, 531–545. doi:10.1016/j.ijar.2006.06.024

Terra, J. C., & Gordon, C. (2002). *Realizing the promise of corporate portals: Leveraging Knowledge for Business Success*. Butterworth-Heinemann.

Tesh, S. N. (2000). *Uncertain hazards: Environmental activists and scientific proof*. Ithaca, NY: Cornell University Press.

Thakur, L. S., & Jain, V. K. (2008). Advanced manufacturing techniques and information technology adoption in India: A current perspective and some comparisons. *International Journal of Advanced Manufacturing Technology*, *36*, 618–631. doi:10.1007/s00170-006-0852-4

The 12th International Conference on Artificial Intelligence and Law. (2009). [Proceedings]. New York: ACM.

The Media Report [Television show]. (2001, July 12). Radio National. Sydney, Australia: ABC.

Thiessen, E. M., & McMahon, J. P. (2000). Beyond win-win in cyberspace. *Ohio State Journal on Dispute Resolution*, *15*, 643.

Thomson, M. (2009). *Annual report of the telephone dispute resolution service to the attorney general*. Queensland: Relationships Australia.

Tolchinsky, P., Modgil, S., Cortes, U., & Sanchez-Marre, M. (2006). CBR and argument schemes for collaborative decision making. In P. E. Dunne & T. J. M. Bench-Capon (Eds.), *Computational Models of Argument: Proceedings of COMMA 2006,* (pp. 71-82). Amsterdam: IOS Press.

Tomoye Corporation. (2009). *About Tomoye*. Retrieved October 4, 2009, from http://www.tomoye.com/about.html

Torra, V. (2004). OWA operators in data modeling and re-identification. *IEEE Transactions on Fuzzy Systems*, *12*(5), 652–660. doi:10.1109/TFUZZ.2004.834814

Toulmin, S. (1958). *The uses of argument*. Cambridge: Cambridge University Press.

Travers Scott, D. (2007). Pundits in muckrakers' clothing: Political blogs and the 2004 US presidential election . In Tremayne, M. (Ed.), *Blogging, citizenship and the future of media* (pp. 39–57). New York, London: Routledge.

Triantaphyllou, E., & Lin, C. T. (1996). Development and evaluation of five fuzzy multi attribute decision-making methods. *International Journal of Approximate Reasoning*, *14*, 281–310. doi:10.1016/0888-613X(95)00119-2

Tsabadze, T. (2006). A method for fuzzy aggregation based on group expert evaluations. *Fuzzy Sets and Systems*, *157*, 1346–1361. doi:10.1016/j.fss.2005.11.015

Tsai, H. H., & Lu, I. Y. (2006). The evaluation of service quality using generalized Choquet integral. *Information Sciences*, *176*(6), 640–663. doi:10.1016/j.ins.2005.01.015

Tsui, A. S., & Farh, J. L. L. (1997). Where guanxi matters: Relational demography and guanxi in the Chinese context. *Work and Occupations*, *24*(1), 56–79. doi:10.1177/0730888497024001005

Tucker, D. M., Hartry-Speiser, A., McDougal, L., Phan, L., & Degrandpre, D. (1999). Mood and spatial memory: Emotion and right hemisphere contribution to spatial cognition. *Biological Psychology*, *50*, 103–125. doi:10.1016/S0301-0511(99)00005-8

Turner, G., Tomlinson, E. & Pearce, S. (2006). Talkback Radio: Some notes on format, politics and influence. *Media International Australia incorporating Culture & Policy, 118*, 107-119.

Turtle, H. (1995). Text retrieval in the legal world. *Artificial Intelligence and Law*, 3(1-2), 5–54. doi:10.1007/BF00877694

Tversky, A., & Kahneman, D. (1981). The framing of decisions and the psychology of choice. *Science*, *21*, 453–458. doi:10.1126/science.7455683

Tversky, B., & Hard, B. M. (2009). Embodied and disembodied cognition: Spatial perspective-taking. *Cognition*, *110*, 124–129. doi:10.1016/j.cognition.2008.10.008

Twitter. (2009). *About us*. Retrieved Oct 1, 2009, from http://twitter.com/about#about

United Nations. (2009). *Online volunteering*. Retrieved October 3, 2009, from http://www.onlinevolunteering.org/en/index.html

Unnikrishnan Nair, K. (2001). Cognitive maps of managers and complex problem solving . In Lant, T. K., & Shapira, Z. (Eds.), *Organizational cognition: Computation and interpretation* (pp. 211–240). London: Lawrence Erlbaum Associates Publishers.

Valacich, J. S., Dennis, A. R., & Connolly, T. (1994). Group versus individual brainstorming: A new ending to an old story. *Organizational Behavior and Human Decision Processes*, *57*, 448–467. doi:10.1006/obhd.1994.1024

Van den Braak, S. W., Van Oostendorp, H., Prakken, H., & Vreeswijk, G. (2006, August 28-29). A critical review of argument visualization tools: Do users become better reasoners? Workshop Notes of the ECAI-06 Workshop on Computational Models of Natural Argument Riva del Garda, Italy.

van Gelder, T. (2003). Enhancing deliberation through computer supported argument mapping . In Kirschner, P. A., Shum, S. J. B., & Carr, C. S. (Eds.), *Visualizing argumentation: Software tools for collaborative and educational sense-making*. London: Springer Verlag.

Van Gelder, T. (2009, August 10). *Deliberative decisions,* (pp. 5-6). Retrieved August 18, 2009, from http://timvangelder.com

Van Mill, D. (1996). The possibility of rational outcomes from democratic discourse. *The Journal of Politics*, *58*, 734–752. doi:10.2307/2960440

Vaníček, J., Vrana, I., & Aly, S. (2009). Fuzzy aggregation and averaging for group decision-making: A generalization and survey. *Knowledge-Based Systems*, *22*, 79–84. doi:10.1016/j.knosys.2008.07.002

Velagapudi, P., Prokopyev, O., Scerri, P., & Sycara, K. (2009). A token-based approach to sharing beliefs in a large multiagent team. In M. J. Hirsch, C. W. Commander, P. M. Pardalos, & R. Murphey (Eds.), Optimization and cooperative control strategies. *Lecture Notes in Control and Information Systems, 381*, 417-429.

Verter, V., & Dasci, A. (2002). The plant location and flexible technology acquisition problem. *European Journal of Operational Research, 136*(2), 366–382. doi:10.1016/S0377-2217(01)00023-6

Vygotsky, L. S. (1978). *Mind in society: The development of higher psychological processes*. Cambridge, MA: Harvard University Press.

W3C. (2009). *W3C Semantic web activity*. Retrieved October 5, 2009, from http://www.w3.org/2001/sw/

Wahl Jorgensen, K. (2001). Letters to the editor as a forum for public deliberation: Modes of publicity and democratic debate. *Critical Studies in Media Communication, 18*(3), 303–320. doi:10.1080/07393180128085

Wahl-Jorgensen, K. (2002a). The normative-economic justification for public discourse: Letters to the editor as a "wide open" forum. *Journalism & Mass Communication Quarterly, 79*(1), 121–133.

Wahl-Jorgensen, K. (2002b). Understanding the conditions for public discourse: Four rules for selecting letters to the editor. *Journalism Studies, 3*(1), 69–81. doi:10.1080/14616700120107347

Wahl-Jorgensen, K. (2004). A "legitimate beef" or "raw meat"? Civility, multiculturalism, and letters to the editor. *Communication Review, 7*, 89–105.

Waissmann, F. (1955). Verifiability . In Flew, A. (Ed.), *Logic and language*. Oxford: Blackwell.

Walther, J. B., Anderson, J. F., & Park, D. W. (1994). Interpersonal effects in computer-mediated interaction: A meta-analysis of social and antisocial communication. *Communication Research, 21*(4), 460–487. doi:10.1177/009365094021004002

Walther, J. B., Slovacek, C., & Tidwell, L. C. (2001). Is a picture worth a thousand words?: Photographic images in long term and short term virtual teams. *Communication Research, 28*, 105–134. doi:10.1177/009365001028001004

Walther, J. B. (1993). Impression development in computer-mediated interaction. *Western Journal of Communication, 57*, 381–398.

Walther, J. B. (1996). Computer-mediated communication: Impersonal, interpersonal, and hyperpersonal interaction. *Communication Research, 23*, 3–43. doi:10.1177/009365096023001001

Walther, J. B., Anderson, J. F., & Park, D. W. (1994). Interpersonal effects in computer-mediated interaction. *Communication Research, 21*, 460–487. doi:10.1177/009365094021004002

Walther, J. B., & Burgoon, J. K. (1992). Relational communication in computer-mediated interaction. *Human Communication Research, 19*, 50–88. doi:10.1111/j.1468-2958.1992.tb00295.x

Walther, J. (1996). Computer mediated communication: Impersonal, interpersonal, and hyperpersonal interaction. *Journal of Communication Research, 23*(1), 3–43. doi:10.1177/009365096023001001

Walton, D. (2006). How to make and defend a proposal in deliberation dialogue. *Artificial Intelligence and Law, 14*, 177–239. doi:10.1007/s10506-006-9025-x

Walton, D. (2007). *Dialogue theory for critical argumentation*. Amsterdam: John Benjamins.

Walton, D., Reed, C., & Macagno, F. (2008). *Argumentation schemes*. Cambridge: Cambridge University Press.

Walton, D., & Krabbe, E. (1995). *Commitment in dialogue: Basic concepts of interpersonal reasoning*. Albany, NY: State University of New York Press.

Walton, D., Atkinson, K., Bench-Capon, T. J. M., Wyner, A., & Cartwright, D. (2009). Argumentation in the framework of deliberation dialogue . In Bjola, C., & Kornprobst, M. (Eds.), *Arguing Global Governance*. London: Routledge.

Walton, D. N. (1990). *Practical reasoning: Goal-driven, knowledge-based, action-guiding argumentation* (Studies in Epistemology and Cognitive Theory, vol. 2). Lanham, MD: Rowman & Littlefield Publishers.

Wang, T. C., & Lee, H. D. (2009). Developing a fuzzy TOPSIS approach based on subjective weights and objective weights. *Expert Systems with Applications*, *36*, 8980–8985. doi:10.1016/j.eswa.2008.11.035

Wang, W. P. (2009). Evaluating new product development performance by fuzzy linguistic computing. *Expert Systems with Applications*, *36*(6), 9759–9766. doi:10.1016/j.eswa.2009.02.034

Wang, Y. M., Luo, Y., & Hua, Z. (2007). Aggregating preference rankings using OWA operator weights. *Information Sciences*, *177*, 3356–3363. doi:10.1016/j.ins.2007.01.008

Ward, I. (1995). *Politics of the media*. South Melbourne, Australia: Macmillan.

Warnke, G. (1995). Communicative rationality and cultural values . In White, S. K. (Ed.), *The Cambridge companion to Habermas* (pp. 120–142). Cambridge, UK: Cambridge University Press.

Warren, K. (1995). Exploring competitive futures using cognitive mapping. *Long Range Planning*, *28*(5), 10–21. doi:10.1016/0024-6301(95)00034-G

Warren, T., & Gibson, E. (2002). The influence of referential processing on sentence complexity. *Cognition*, *85*, 79–112. doi:10.1016/S0010-0277(02)00087-2

Waterman, D. A., Paul, J., & Peterson, M. (1986). Expert systems for legal decision making. *Expert Systems: International Journal of Knowledge Engineering and Neural Networks*, *3*(4), 212–226. doi:10.1111/j.1468-0394.1986.tb00203.x

Watters, P. A. (2009, October) *Why do users trust the wrong messages? A behavioural model of phishing*. Paper presented at the 4th annual APWG eCrime Researchers Summit, Tacoma, WA.

Weber, S. F. (1993). A modified analytic hierarchy process for automated manufacturing decisions. *Interfaces*, *23*(4), 75–84. doi:10.1287/inte.23.4.75

Weisband, S. P. (1992). Group discussion and first advocacy effects in computer-mediated and face-to-face decision making groups. *Organizational Behavior and Human Decision Processes*, *53*, 352–380. doi:10.1016/0749-5978(92)90070-N

Wellman, M., Walsh, W., Wurman, P., & Mackie-Mason, J. (2001). Auction protocols for decentralized scheduling. *Games and Economic Behavior*, *35*, 271–303. doi:10.1006/game.2000.0822

Wenger, E. C., McDermott, R., & Snyder, W. M. (2002). *Cultivating communities of practice: A guide to managing knowledge*. Boston: Harvard Business School Press.

Wenger, E. C., & Snyder, W. M. (2000). *Communities of practice: The organizational frontier. Harvard Business Review*. January-February.

Wenger, E. (1998). *Communities of practice: Learning, meaning, and identity*. Cambridge, UK: Cambridge University Press.

Wenger, E., White, N., & Smith, J. D. (2009). *Digital habitats: Stewarding technology for communities*. Portland, OR: CPSquare.

Wensley, R. (2003). Strategy as intention and anticipation . In Cummings, S., & Wilson, D. (Eds.), *Images of strategy* (pp. 105–135). Malden, MA: Blackwell.

Wermus, H. (1997, October). Representation, beliefs, and hindrance in understanding. *Cognitive Systems*, *5*(1), 1–16.

Wertheim, E., Love, A., Littlefield, L., & Peck, C. (1992). *I win: You win* (pp. 130–135). Ringwood, Australia: Penguin Books.

Wertsch, J. V. (1985). *Vygotsky and the social formation of mind*. Cambridge, MA: Harvard University Press.

Weyns, D., Omicini, A., & Odell, J. (2007a). Environment as a first class abstraction in multiagent systems. *Autonomous Agents and Multi-Agent Systems*, *14*, 5–30. doi:10.1007/s10458-006-0012-0

Weyns, D., Parunak, H. V. D., & Michel, F. (Eds.). (2007b). *Environments for multi-agent systems*. Berlin: Springer. doi:10.1007/978-3-540-71103-2

White, N. (2001). Facilitating and hosting a virtual community. Retrieved October 1, 2009, from http://www.fullcirc.com/community/communityfacilitation.htm

Whitehead, S., & Ballard, D. H. (2004). Learning to perceive and act by trial and error. *Machine Learning*, *7*, 45–83. doi:10.1007/BF00058926

Wieberneit, N. (2008). Service network design for freight transportation: A review. *OR-Spektrum*, *30*, 77–112. doi:10.1007/s00291-007-0079-2

Wilhelm, A. G. (2000). *Democracy in the digital age: Challenges to political life in the digital age*. New York: Routledge.

Wilkinson, J. H. (1964). *The algebraic eigenvalue problem*. Oxford: Clarendon Press.

Wilson, D. S., Timmel, J. J., & Miller, R. M. (2004). Cognitive cooperation: When the going gets tough, think as a group. *Human Nature (Hawthorne, N.Y.)*, *15*(3), 225–250. doi:10.1007/s12110-004-1007-7

Wimmer, R. D., & Dominick, J. R. (2003). *Mass media research: An introduction* (7th ed.). Belmont, CA: Wadsworth Publishing Company.

Winter, D. A. (1985). Repertory grid technique in the evaluation of therapeutic outcome . In Beail, N. (Ed.), *Repertory grid technique and personal constructs* (pp. 154–170). London: Croom Helm.

Wittenbaum, G. M., & Bowman, J. M. (2004). A social validation explanation for mutual enhancement. *Journal of Experimental Social Psychology*, *40*, 169–184. doi:10.1016/S0022-1031(03)00091-X

Wittenbaum, G. M., Hubbell, A. P., & Zuckerman, C. (1999). Mutual enhancement: Toward an understanding of the collective preference for shared information. *Journal of Personality and Social Psychology*, *77*, 967–978. doi:10.1037/0022-3514.77.5.967

Wittenbaum, G. M., & Park, E. S. (2001). The collective preference for shared information. *Current Directions in Psychological Science*, *10*, 70–73. doi:10.1111/1467-8721.00118

Wittenbaum, G. M., & Stasser, G. (2000). Management of information in small groups . In Nye, J. L., & Brower, A. M. (Eds.), *What's social about social cognition?* (pp. 1–28). London: Sage.

Wittgenstein, L. (1984). *Tractatus logico-philosophicus, Werkausgabe* (*Vol. 1*). Frankfurt am Main, DE: Suhrkamp.

Wolfe, P., Scott, C., & Erwin, M. (2004). *The anti-spam tool kit*. Emeryville, CA: McGraw-Hill Osborne Media.

Wombat. (2009, July 01). [Comment 13 to blog post: Rudd and Gillard attack News Limited, Hartigan punches back]. *Larvatus Prodeo* [Blog], Federal election 2007. Retrieved October 1, 2009, from http://larvatusprodeo.net/2009/07/01/rudd-and-gillard-attack-news-limited-hartigan-punches-back/

Woods, E. (2005). Collaboration software: Evolution and revolution . In Rao, M. (Ed.), *Knowledge management tools and techniques: Practitioners and experts evaluate knowledge management solutions*. Butterworth-Heinemann.

Wooldridge, M. (2002). *An introduction to multi-agent systems*. London: Wiley.

Wooldridge, M. J., Jennings, N. R., & Kinny, K. (2000). The Gaia methodology for agent-oriented analysis and design. *Autonomous Agents and Multi-Agent Systems*, *3*(3), 285–312. doi:10.1023/A:1010071910869

World Bank. (2009). Knowledge economy index, 2009. Retrieved November 10, 2009, from http://www.worldbank.org/kam

World Legal Information Institute. (2007). Retrieved from http://www.worldlii.org/

Wright, S. (2006). Electrifying democracy?: 10 years of policy and practice. *Parliamentary Affairs*, *59*(2), 236–249. doi:10.1093/pa/gsl002

Wu, D. (2009). Supplier selection in a fuzzy group setting: A method using grey related analysis and Dempster–Shafer theory. *Expert Systems with Applications*, *36*, 8892–8899. doi:10.1016/j.eswa.2008.11.010

Wurth, B. (2002, January 26-27). Caring, tolerant Aussie is a thing of the past [Letter to the editor]. *The Australian*, p. 18.

Xenos, M. (2008). New mediated deliberation: Blog and press coverage of the Alito nomination. *Journal of Computer-Mediated Communication*, *13*, 485–503. doi:10.1111/j.1083-6101.2008.00406.x

Xu, Y., Scerri, P., Lewis, M., & Sycara, K. (2008). Token-based approach for scalable team coordination . In Pardalos, P. M., Grundel, D., & Murphey, R. A. (Eds.), *Cooperative networks: Control and optimization* (pp. 5–29). Cheltenham, UK: Edward Elgar Publishing.

Xu, Y., Scerri, P., Yu, B., Okamoto, S., Lewis, M., & Sycara, K. (2005). An integrated token-based algorithm for scalable coordination. Proceedings of the fourth international joint conference on Autonomous agents and multiagent systems (pp. 407 – 414).

Yager, R. R. (1988). An ordered weighted averaging aggregation operators in multi-criteria decision-making . *IEEE Transactions on Systems, Man, and Cybernetics*, *18*, 183–190. doi:10.1109/21.87068

Yankelovich, N., Kaplan, J., Provino, J., Wessler, M., & DiMicco, J. M. (2006). Improving audio conferencing: Are two ears better than one? *Proceedings of the 2006 Conference on Computer Supported Cooperative Work*, 333-342.

Yearwood, J. L., & Stranieri, A. (2006). The generic/actual argument model of practical reasoning. *Decision Support Systems*, *41*, 358–379. doi:10.1016/j.dss.2004.07.004

Yearwood, J. L., & Stranieri, A. (2009). Deliberative discourse and reasoning from generic argument structures. *AI & Society*, *23*(3), 353–377. doi:10.1007/s00146-006-0069-y

Yearwood, J., & Stranieri, A. (2002). An argumentation shell for supporting the development and drafting of legal documents. *Information & Communications Technology Law*, *11*(1). doi:10.1080/13600830220133576

Yearwood, J., & Stranieri, A. (2010). Group structured reasoning for coalescing group decisions. *Group Decision and Negotiation*, *19*(1), 77–105. doi:10.1007/s10726-009-9162-1

Yearwood, J. L., & Mammadov, M. A. (2007). *Classification technologies: Optimization approaches to short text categorization*. Hershey, PA: IGI Global.

Yearwood, J., & Stranieri, A. (2008) A study of frameworks for structured reasoning on reasoning quality and knowledge, Learning & Teaching: an international journal in classroom pedagogy, Vol.1, No.1, pp.71-90.

Yearwood, J., & Stranieri, A. (1999). Integration of retrieval, reasoning and drafting for refugee law: A third generation legal knowledge based system. In *Proceedings of Seventh International Conference on Artificial Intelligence and Law, ICAIL'99* (pp. 117–125).

Yearwood, J., Avery, J., & Stranieri, A. (2006). Interactive narrative by thematic connection of dramatic situations. In *Proceedings of The Third Narrative and Interactive Learning Environments Conference* University of Edinburgh, (pp. 98 – 110).

Yearwood, J., Stranieri, A., & Avery, J. (2001). Negotiation and argumentation based agents to facilitate eCommerce. In *Proceedings of the International Conference on Advances in Infrastructure for Electronic Business, Science and Education on the Internet, SSGRR2001* (pp. 100–109).

Yong, J. S. (2005). *e-Government in ASIA: Enabling public service innovation in the 21st century*. Singapore: Times Media.

Yorke, D. M. (1985). Indexes of stability in repertory grids: A small-scale comparison. *British Educational Research Journal*, *11*(3), 221–225. doi:10.1080/0141192850110303

Young, I. M. (2000). *Inclusion and democracy*. New York: Oxford University Press.

Young, G., Brown, E., Keitt, T., Owyang, J., Koplowitz, R., & Lo, H. (2008). *Global enterprise Web 2.0 market forecast: 2007 to 2013*. Retrieved October 2, 2009, from http://www.forrester.com/Research/Document/Excerpt/0,7211,43850,00.html

YouTube. (2009). *YouTube Fact Sheet*. Retrieved October 1, 2009, from http://www.youtube.com/t/fact_sheet

Yuan, Y., Rose, B. J., Archer, N., & Suarga, H. (1998). A Web-based negotiation support system. *EM - . Electronic Markets, 8*(3), 13–17. doi:10.1080/10196789800000033

Yurdakul, M. (2004). Selection of computer-integrated manufacturing technologies using a combined analysis hierarchy process and goal programming model. *Robotics and Computer-integrated Manufacturing, 20*, 329–340. doi:10.1016/j.rcim.2003.11.002

Yusuff, R. M., Yee, K. P., & Hashmi, M. S. J. (2001). A preliminary study on the potential use of the analytical hierarchical process (AHP) to predict advanced manufacturing technology (AMT) implementation. *Robotics and Computer-integrated Manufacturing, 17*, 421–427. doi:10.1016/S0736-5845(01)00016-3

Zadeh, L. A. (1965). Fuzzy sets. *Information and Control, 8*, 338–353. doi:10.1016/S0019-9958(65)90241-X

Zambonelli, F., Jennings, N. R., & Wooldridge, M. J. (2003). Developing multi-agent systems: The Gaia methodology. *ACM Transactions on Software Engineering and Methodology, 12*(3), 317–369. doi:10.1145/958961.958963

Zeleznikow, J., & Hunter, D. (1994). *Building intelligent legal information systems: Representation and reasoning in law. Computer Law Series, 13*. Deventer, The Netherlands: Kluwer.

Zeleznikow, J., & Stranieri, A. (2005). *Knowledge discovery from legal databases*. Dordrecht, The Netherlands: Springer.

Zeleznikow, J., Meersman, R., Hunter, D., & van Helvoort, E. (1995). *Computer tools for aiding legal negotiation. ACIS95* Sixth Australasian Conference on Information Systems, Curtin University of Technology, Perth, Australia (pp. 231-251).

Zhang, Z., & Chu, X. (2009). Fuzzy group decision-making for multi-format and multi-granularity linguistic judgments in quality function deployment. *Expert Systems with Applications, 36*, 9150–9158. doi:10.1016/j.eswa.2008.12.027

Zhao, S. (2004). Consociated contemporaries as an emergent realm of the lifeworld: Extending Schutz's phenomenological analysis to cyberspace. *Human Studies, 27*, 91–105. doi:10.1023/B:HUMA.0000012246.33089.68

Zimmerman, H. J. (1987). *Fuzzy sets, decision-making, and expert systems*. Boston, MA: Kluwer Academic Publications.

Zimmermann, H. J. (1991). *Fuzzy set theory and its applications*. Dordrecht, Germany: Kluwer Academic.

Zoho Corporation. (2009). *FAQ (Frequently Asked Questions)*. Retrieved October 4, 2009, from http://www.zoho.com/zoho_faq.html

About the Contributors

John Yearwood is Professor of Informatics in the Graduate School of Information Technology and Mathematical Sciences and Director of the Centre for Informatics and Applied Optimization at the University of Ballarat in Australia. His research spans areas of data mining, argumentation, reasoning and decision support and its applications in health and law. In 2001 he received a Queen Elizabeth II Fellowship from the Australian Research Council to work on argumentation models and narrative models. This research has been exploited in the development of computational interactive learning environments that deploy narrative automatically generated from domain knowledge to enhance learning outcomes. Application of these models in health will appear in the Journal of Interactive Learning Environments. Yearwood with Andrew Stranieri developed the Generic/Actual Argument Model (GAAM) for the representation of practical reasoning (Decision Support Systems). This model has been used in modelling decision making in discretionary domains such as law and has been developed and applied in a number of situations in family law, refugee law, copyright law and also in sentencing. It has been used in the development of several decision support and advisory systems. It was used to develop GetAid, an electronic decision support system used by Victoria Legal Aid to assist lawyers in determining clients eligibility for legal aid. Also based on this model is the web-based Office of Training & Tertiary Education (OTTE) Career Exploration Decision Support System, which is online and part of the Victorian State Governments strategy to assist people in exploring their career options. The reasoning shell JustReason, based on the model, allows the construction of software systems using the model. It has been used by others; for example, the Advokate system in the development of systems in Scottish family law and in eyewitness selection. His work in data mining and computational intelligence has led to the development of new learning and hybrid learning algorithms for artificial neural networks, which have been published in the International Journal of Neural Learning. These learning algorithms, based on the Discrete Gradient method and hybridisations with global methods such as genetic algorithms, have demonstrated improved performance over the more standard back propagation learning algorithms in multilayer perceptron networks. Contributions have also been made in terms of the most effective combination strategies for these hybridisations. He has also been Chief Investigator on a number of Australian Research Council Grants in these areas. Yearwood was a CI on the ARC funded Linkage Project Enhanced prescribing decision support: data mining the Australian adverse drug reactions database and providing on-line access to information. This led to the development of prototype software that is currently being commercialised through Datascreen Pty Limited, a University of Ballarat start-up company. He has worked with a wide range of industry partners including Telstra, Transfield Services, Ballarat Health Services, St John of God Hospital, Department of Sustainability and Environment, the Australian Defence Force, Victorian State Revenue Office, IBM, Westpac Banking Corporation, Woodside Petroleum and Shell Australia. He has been a reviewer for the ARC, the

NHMRC grant program, and for the Dutch Government in the assessment of their NWO/ToKeN2000 program (Netherlands Organisation of Scientific Research/Access to Knowledge and its Enhancement. Professor Yearwood is Editor for the Journal of Research and Practice in Information Technology. He has over 160 refereed journal and conference publications.

Andrew Stranieri has published three books and over eighty journal and conference articles and book chapters, over the last eighteen years in the fields of artificial intelligence and data mining applied to law, health and group reasoning. The argumentation based formulation of legal reasoning he developed led to the use of machine learning methods to model judicial decision making. With John Yearwood, he extended and formalised an argumentation based approach for modelling structure and inference called the Generic Actual Argument Model. That model led to the formation of a spin-out company he managed that commercially developed intelligent systems in tax law, intensive care nursing, legal aid, career counselling, witness selection, patient's needs assessment and assessing the suitability of volunteers. The argumentation model ideas evolved into approaches that groups of people can use to enhance collective decision making. Currently, Stranieri is a leader within the University of Ballarat's Centre for Informatics and Applied Optimisation where he conducts research in health informatics and reasoning communities.

****

Shady Aly received his MSc in Production Engineering from Helwan University in Cairo and his PhD from the Czech University of Life Sciences in Prague. His research is oriented into artificial intelligence, intelligent decision support systems, fuzzy systems and their application in expert systems with a fuzzy group decision making.

Elarbi Badidi is an Assistant Professor of Computer Science at the College of Information Technology (CIT) of United Arab Emirates University. Before joining the CIT, he held the position of Bioinformatics Group Leader at the Biochemistry Department, University of Montreal from 2001 to July 2004. He received a PhD in Computer Science from University of Montreal, Quèbec (Canada). His research interests include Web services and Service Oriented Computing, Middleware and Bioinformatics data and tools integration.

Emilia Bellucci has been a researcher in Negotiation Support Systems for 15 years. She is currently a lecturer at Victoria University, Melbourne, Australia. She completed her PhD under the supervision of Professor Zeleznikow in 2004; her research was the focus of much media attention in 2005. The Family_Winner software system attracted the attention of over a dozen radio programs in all Australian states and on BBC Radio 5 and separately the BBC World Service, and print journalists from the Sydney Morning Herald, the Times of London, the Australian Financial Review and the Economist. The inventors were asked to compete on ABC (Australian Broadcasting Commission) *New Inventors* television show on November 16 2005. Bellucci has published 3 refereed journal articles, 24 referred conference papers and 3 refereed workshop papers. She has attended and presented her research at numerous international conferences and workshops.

Dino Borri has been Full Professor in Town Planning at the Polytechnic University of Bari since 1994. From 1996 to 2009 he was Head of the Department of Architecture and Town Planning at the Poly, where he teaches Town and Country Planning and Environmental Engineering. His main research and scientific interests are in planning theory, interactive planning, futures research, application of information technology to spatial planning, intelligent planning agents. He has been scientific coordinator of many EU funded research projects among European and Euro-Asiatic Institutions on planning issues. He has written several books and more than a hundred scientific papers for international journals and conferences worldwide.

Domenico Camarda has been assistant professor at the Polytechnic University of Bari, Italy since 2004. At the Poly he teaches Town and Country Planning, as well as Planning Techniquess. He has a degree in Civil Engineering, a PhD in Spatial Planning and an MSc in Economic Policy and Planning. His research interests are environmental planning, multi-agent planning systems, decision-support systems. He has been research fellow in some EU funded research projects among European and Euro-Asiatic Institutions on planning issues. He has written about 40 national and international papers and co-edited five international books. Also, he has participated to several international conferences worldwide, delivering research papers.

Jamonn Campbell is an Associate Professor in the Department of Psychology at Shippensburg University. He earned his PhD in Social Psychology at Miami University. He teaches courses in social psychology, social influence, group dynamics, and the psychology of computers and the internet. His research interests include decision-making, social influence and impression formation during computer-mediated communication, as well as sport fan identification and behavior. Recent publications have appeared in *Small Group Research, Social Influence, Computers in Human Behavior,* and *Journal of Sport Behavior.*

Marco Castellani is Lecturer of Sociology in the University of Brescia (Italy), from which he also received his doctorate degree in Economic Sociology. He is a member of GECS (Research Group on Experimental and Computational Sociology), based in the Department of Social Sciences of the University of Brescia. His research is focused on individual and collective decision making, bounded rationality, and cognitive processes in organizational choices. He is the author of a book on bounded rationality and several journal articles.

Selçuk Cebi received his PhD in industrial engineering from Istanbul Technical University. He is currently a Research Assistant in Karadeniz Technical University Department of Industrial Engineering. His main research areas are decision making, axiomatic design and fuzzy set theory.

Richard Dazeley received his PhD in 2007 from the University of Tasmania in the area of knowledge acquisition with Ripple-Down Rules applied to prudence analysis. He has published numerous papers in the areas of philosophy of knowledge, knowledge acquisition with dynamic contexts, prudence analysis and knowledge acquisition for small to medium sized enterprises. Currently he is the Deputy Leader of the Data Mining and Informatics Research Group in the Centre for Informatics and Applied optimisation at the University of Ballarat and is project leader in research projects in the areas of data stream mining, virtual reality visualisation and knowledge based systems.

Vanessa Paz Dennen is an Associate Professor of Instructional Systems in the Department of Educational Psychology and Learning Systems at Florida State University. Prior to joining the faculty at Florida State University she was an Assistant Professor of Educational Technology at San Diego State University and earned a PhD in Instructional Systems Technology from Indiana University. Her research, broadly described, investigates the nexus of cognitive, motivational, and social elements in computer-mediated communication. Within this larger area, her work has concentrated on two major problems: the study of learner participation in computer-supported collaborative learning activities; and, more recently, the study of interactions, norm development, and support networks within online communities of practice.

Larbi Esmahi is an Associate Professor of the School of Computing and Information Systems at Athabasca University. He was the Graduate Program Coordinator at the same School during 2002-2005. He holds a PhD in Electrical Engineering from Ecole Polytechnique, University of Montreal. His current research interests are in e-services, e-commerce, multiagent systems, and intelligent systems. He is member of the Editorial Board of many journals in the field of e-services and intelligent systems (IJISSC, IJISSS, IJAIS, JETWI, IJESMA, JDET, IJWLTT, and TKJSE). He is also member of the Editorial Advisory Board of the Advances in Web-Based Learning Book Series, IGI Global. His articles appeared in the Journal of Electronic Commerce in Organizations, International Journal of Web-Based Learning and Teaching Technologies and the International Journal on Web Services Practices. He also published several book chapters and conference papers.

Jesus Gonzalez-Feliu is a Post-Doctoral Researcher at the French National Centre of Scientific Research (CNRS), and member of the Laboratoire d'Economie des Transports (LET). He recieved his PhD in 2008 in operations research for researches in urban freight distribution solutions and two-echelon vehicle routing problems. His research interests include city logistics planning and policy, decision support systems, transportation systems optimisation and performance, sustainable supply chain management, and collaborative transportation.

Lin-Mei Huang (PhD, University of Iowa, 1993) is Professor of Speech Communication Department at Shih Hsin University, Taiwan. Her major research area is conflict management and resolution behavior in East Asia. Her research has appeared in *Intercultural Communication Studies*, *Chinese Journal of Communication Research* and *Mass Communication Research*, and she has co-authored one book entitled *Bargaining and Negotiation.*

Cengiz Kahraman received his MSc and PhD degrees on industrial engineering from Istanbul Technical University. He is in the editorial board of some journals such as *Journal of Human and Ecological Risk Assessment*, *International Journal of Computational Intelligence Systems*, *New Mathematics and Natural Computation*, *Journal of Enterprise Information Management*, *Technological and Economic Development of Economy*, *Baltic Journal on Sustainability*, *International Journal of Strategic Property Management*, and *Journal of Multiple Valued Logic and Soft Computing* and served as guest editor for special issues of *Information Sciences*, *Journal Of Enterprise Information Management*, and *International Journal Of Approximate Reasoning.* He is the editor of the Springer books such as *Fuzzy Applications in Industrial Engineering*, *Fuzzy Multicriteria Decision Making: Theory and Applications with Recent Developments*, *Fuzzy Engineering Economics with Applications* and *Production Engineering and Management under Fuzziness.*

İhsan Kaya received his PhD in industrial engineering from Istanbul Technical University. He is currently a Research Assistant at Selcuk University Department of Industrial Engineering. His main research areas are fuzzy set theory, process capability analyses, quality control and fuzzy statistics.

Andrei Kelarev received his PhD in 1989 and has 15 years of full-time teaching experience in several universities in Australia and USA. He was the Chief Investigator of one Discovery grant of Australian Research Council. He has published two books, a volume of refereed conference proceedings, and over 160 refereed articles, mainly in international journals. Kelarev has been working on several research grants at the Graduate School of Information Technology and Mathematical Sciences in the University of Ballarat. His research is devoted to the optimization of various constructions for classification and clustering of data and applications in internet security.

Taemie Kim is a PhD candidate at the MIT Media Lab interested in using automatic measurement of social signals to help distributed collaborations. Her PhD research focuses on applying wearable sensing technologies and human-computer-interaction design techniques to provide real-time mobile feedback. The effectiveness of her system has been verified in both social psychological lab experiments as well as real-world organizations. She received a BS in Electrical Engineering and Computer Science from KAIST (Korea) with a focus on hardware circuit design and an MS in Electrical Engineering from Stanford University with a focus in human-computer interaction in 2003 and 2006 respectively.

Shu-Chu Sarrina Li is the Director and a professor at the Institute of Communication Studies at National Chiao Tung University in Taiwan. She has received grants from U.S. Fulbright Foundation and Taiwan's National Science Council for doing research projects in media economics, media management, and digital TV in Taiwan. She has published many articles in internationally well known journals such as *Journal of Broadcasting & Electronic Media*, *Journal of Media Economics*, and *New Media & Society*. Her research interests include new technologies and media management, cable and satellite TV. Li also served as the consultant for various government agencies and has been frequently involved with media policy making for Taiwan's terrestrial TV, cable TV, and satellite TV. Li obtained her master's degree in television and radio studies at Texas Tech University in 1985 and her PhD in Communication Studies at the University of Iowa in 1993. She was a Visiting Scholar in the Fairbank Center at Harvard University in 2005.

Yi-Ching Liu (Master's, National Chiao Tung University, 2007) is a research assistant in the Institute of Communication Studies at National Chiao Tung University, Taiwan. Her research interests include media performance and group communication.

Charlynn Miller (BSc, Post-Bach HRD, MEd VCU, PhD UVA) is the Deputy Head of School in the Graduate School of Information Technology & Mathematical Sciences at the University of Ballarat Australia. Her research involves the use of emerging technologies for enhancement business and learning. She also conducts research in the area of Cyber-Safety. Miller has a number of publications regularly presents in the area of emerging technologies and their impact on learning and business. She worked in the private IT sector for a number of years. Miller is a member and certified professional with the Australian Computer Society and a member of the ACS Women's Committee for Victoria.

Joëlle Morana is a university lecturer at the Université Lumière Lyon 2 and a member of the Laboratoire d'Economie des Transports (LET). She earned her PhD in 2002 for research in balanced scorecards applied to Supply Chain Management. Her research interests include performance management in a logistical context, SCM strategy and information systems, sustainable supply chain management, and collaborative transportation. She has published several papers on these topics.

Jane Mummery is a Senior Lecturer in Philosophy at the University of Ballarat, Victoria, Australia. Her principal research interests are in the emancipatory movements sponsored by post-metaphysical philosophy and in how these movements can help in the analysis of cotemporary public discourses. She has published a number of articles in these fields in international and Australasian journals (several of which are with her co-author Debbie Rodan), the books, *The Post to Come: an Outline of Post-metaphysical Ethics* (Peter Lang, 2005) and, with Peta Bowden, *Understanding Feminism* (Acumen, 2009).

Simon Niemeyer is a Research Fellow at the Centre for Deliberative Democracy and Global Governance, located in the School of Politics and International Relations, Research School of Social Sciences, The Australian National University. His research interests span deliberative democracy (preference transformation, institutionalisation), environmental governance, and adaptation to and mitigation of climate change. He was one of the first in the field of deliberative democracy to systematically examine the processes of preference transformation of individuals participating in democratic discourse. His research findings challenge a number of assumptions regarding how deliberation works in practice, which have significantly contributed to deliberative theory. As part of his current research he is also leading a multidisciplinary team of ANU researchers investigating public response and institutional adaptation to climate change and a chief investigator in the ARC Linkage Citizens' Parliament project.

Celina M. Olszak is a Professor of Management Information Systems at the University of Economics in Katowice, Poland. She is Chair of the Department of Business Informatics and Vice-Dean in Charge of Research at the Faculty of Economics. She was scholarship holder at Swiss Federal Institute of Technology in Zurich in Switzerland and scholarship holder of Deutsche Akademische Austausch Dienst at Trier University in Germany. She visited and gave different courses at the universities and colleges in Europe, the USA and Australian. She has published numerous articles in information systems, decision support systems, and systems development strategies. Her current research interests include knowledge management, Business Intelligence Systems, e-business and information society. She is a member of the Informing Science Institute in California, USA.

Alex "Sandy" Pentland is a pioneer in organizational engineering, mobile information systems, and computational social science. Pentland's focus is the development of human-centered technology, and the creation of ventures that take this technology into the real world. He directs the Human Dynamics Lab at the MIT Media Lab, helping companies to become more productive and creative through organizational engineering, and the Media Lab Entrepreneurship Program, which helps translate cutting-edge technology into real-world impact around the world. He is among the most-cited computer scientists in the world, and in 1997, *Newsweek* magazine named him one of the 100 Americans likely to shape this century.

Debbie Rodan is a Senior Lecturer in Media & Cultural Studies at Edith Cowan University, Perth, Australia. Her principal research interest is in citizen participation as well as communicative spaces in mass media forums and she has published a number of articles (several with her co-author Jane Mummery) in Australasian and international journals. Debbie is currently researching communicative spaces on the Internet, such as media active forums and political blog sites, which aim to advance citizen participation. She is the author of *Identity and Justice: Conflicts, Contradictions and Contingencies* (Peter Lang, 2004) and a chief investigator on an Australia Research Council linkage grant, *Construction of the subject as heart patient in the context of the gift economy.*

Thomas Saaty holds the Chair of Distinguished University Professor at the University of Pittsburgh and is member of the National Academy of Engineering, USA. He is internationally recognized for his decision making process, the Analytic Hierarchy Process (AHP) and its generalization to network decisions, the Analytic Network Process (ANP). He won the Gold Medal from the International Society for Multicriteria Decision Making for his contributions to the field, the Akao Prize from Japan and the Impact prize from INFORMS. His first career included a substantial body of work in mathematics and in operations research and many of his books in these fields were translated to other languages. He was also awarded prizes for works in both mathematics and operations research. His work is in decision making, planning, conflict resolution and in neural synthesis.

Erich Schweighofer is Associate Professor at Vienna University where he teaches legal informatics, international and European law. Since 1986 at Vienna University, head of the Centre for Legal Informatics, Main organiser of the IRIS conferences (Internationales Rechtsinformatik Symposion - International Legal Informatics Symposion). Besides an academic career he has extensive practice of 9 years in administration (Ministry of Foreign Affairs, Federal Academy of Administration, European Commission). He is the author of many books and articles, in particular of Legal Knowledge Representation.

Phil Smith (BSc Hons UNE, PhD Deakin) is a senior lecturer in IT in the Graduate School of Information Technology & Mathematical Sciences at the University of Ballarat Australia. Smith has publications in IT education and Distributed Simulation. He is a member of the Australian Computer Society and the Association for Computing Machinery.

Jiří Vaníček received his MSc in Mathematics from Charles University in Prague, his PhD in Computer Science from Czech Technical University, is a member of Czechoslovak Academy of Sciences, and is a Professor of Information Management and Czech University of Life Sciences in Prague. He was an Assistant Professor for Mathematics (1960-1965), Head of the Software Research Division in the Czechoslovak Research Institute for Computers (1965–1989), Director of the Government Institute for Informatics in Education (1990–1994), Associate Professor for Computer Science at the Czech Technical University in Prague (1994–1997), Professor for Information Management, Department of Information Engineering (Deputy Head), Faculty of Economics and Management, Czech University of Life Sciences in Prague (1997–present), and has served as President of the Czech National Technical Standardization Committee for Information Technology since 1990. His research is oriented to mathematical background of computer science.

Ivan Vrana received his MSc, PhD, and DrSc degrees in radio electronics from the Czech Technical University in Prague. Since 1990 he has been a Full Professor and head of the Department of Information Engineering, Faculty of Economics and Management at Czech University of Life Sciences in Prague. His research is on systems engineering and projecting of information systems. Recently, his research priority is a group decision making in conditions of ambiguity. He is a founder and a president of the EUNIS-CZ (European University Information Systems Organization in Czech Republic).

Douglas Walton is a Canadian academic and author, well known for his many widely published books and papers on argumentation, logical fallacies and informal logic. He presently holds the Assumption Chair in Argumentation Studies and is Distinguished Research Fellow of the Centre for Research in Reasoning, Argumentation and Rhetoric (CRRAR) at the University of Windsor, Canada. His latest book is *Argumentation Schemes,* coauthored with Chris Reed and Fabrizio Macagno (Cambridge University Press, 2008). Walton's work has been used to better prepare legal arguments and to help develop artificial intelligence. His books have been translated worldwide and he attracts students from many countries to study with him.

Xin-Wen Wu received his PhD from the Chinese Academy of Sciences in 1995. He has been working with the Academy of Mathematical and System Sciences, Chinese Academy of Sciences, University of Louisiana, Lafayette, and University of California, San Diego. At the beginning of 2003, he joined the University of Melbourne, as a research fellow. In 2005, he joined the University of Ballarat as a faculty member, and he is now at Griffith University. His research interests include Information Security, Applied Cryptography, Coding Techniques for Data Communications, and related areas. He has published extensively in these areas. Many of his papers have been published in top ranked international journals and conferences, such as IEEE transactions. He is a coauthor of two books. His second will be published by the Cambridge University Press. Wu is an IEEE member.

John Zeleznikow is an internationally recognised expert with over 30 years experience in research on developing decision support systems including the last 15 years working on negotiation decision support and machine learning in law. In eighteen years as a computer science and information systems academic, Zeleznikow has received over $A7,000,000 in competitive research grants; published three books, forty-five refereed journal articles, one hundred and twenty-five refereed conference papers, twenty-four refereed workshop articles and eight book chapters; conducted research and taught in Australia, France, Israel, Netherlands, United Kingdom and United States; and given invited talks and tutorials at over twenty international conferences. Zeleznikow's research has broadly focused upon using intelligent systems to support human decision-making in legal domains. His best known example is in the use of such techniques to predict how marital property is distributed following divorce in Australia. Similar work has been conducted in Copyright Law, Eligibility for Legal Aid, Sentencing and Workers Compensation.

Ewa Ziemba is a Professor of Management Information Systems at the University of Economics in Katowice, Poland. She is an author of numerous articles published in journals, conference proceedings and as book chapters. She has written three books on e-business systems, web designing and corporate portals for knowledge management. Her current research concentrates mostly on information systems development methodologies, ebusiness and information society, information systems for knowledge management, especially designing corporate portals for knowledge-based organizations. She has participated in several European research project relating to e-business systems and e-learning. She has visited and given lectures at the universities and colleges in Europe, the USA and Australia. Her academic qualifications have been combined with practical experience – she has been working as the IT Project Manager for over ten years. Furthermore, she is a member of Informing Science Institute in California, USA.

Index

D

E

F

G

H

I

J

K

L

M

N

O

P

R

S

T

Breinigsville, PA USA
23 November 2010
249730BV00003BA/5/P